NOBEL LECTURES

PHYSIOLOGY OR MEDICINE

1996–2000

NOBEL LECTURES

INCLUDING PRESENTATION SPEECHES
AND LAUREATES' BIOGRAPHIES

PHYSICS
CHEMISTRY
PHYSIOLOGY OR MEDICINE
LITERATURE
PEACE
ECONOMIC SCIENCES

NOBEL LECTURES

INCLUDING PRESENTATION SPEECHES AND LAUREATES' BIOGRAPHIES

PHYSIOLOGY OR MEDICINE

1996–2000

EDITOR

Hans Jornvall

Department of Medical Biochemistry and Biophysics

Karolinska Institutet

World Scientific

New Jersey • London • Singapore • Hong Kong

Published for the Nobel Foundation in 2003 by

World Scientific Publishing Co. Pte. Ltd.
5 Toh Tuck Link, Singapore 596224
USA office: Suite 202, 1060 Main Street, River Edge, NJ 07661
UK office: 57 Shelton Street, Covent Garden, London WC2H 9HE

NOBEL LECTURES IN PHYSIOLOGY OR MEDICINE (1996–2000)

ISBN 981-238-005-1
ISBN 981-238-006-X (pbk)

Printed by FuIsland Offset Printing (S) Pte Ltd, Singapore

FOREWORD

Since 1901 the Nobel Foundation has published annually "Les Prix Nobel" with reports from the Nobel award ceremonies in Stockholm and Oslo as well as the biographies and Nobel lectures of the laureates. In order to make the lectures available for people with special interests in the different prize fields the Foundation gave Elsevier Publishing Company the right to publish in English the lectures for 1901–1970, which were published in 1964–1972 through the following volumes:

Physics 1901–1970	4 vols.
Chemistry 1901–1970	4 vols.
Physiology or Medicine 1901–1970	4 vols.
Literature 1901–1967	1 vol.
Peace 1901–1970	3 vols.

Thereafter, and until the year 2000, the Nobel Foundation has given World Scientific Publishing Company the right to bring the series up to date and also publish the Prize lectures in Economics from the year 1969. The Nobel Foundation is very pleased that the intellectural and spiritual message to the world laid down in the laureates' lectures, thanks to the efforts of World Scientific, will reach new readers all over the world.

Bengt Samuelsson
Chairman of the Board

Michael Sohlman
Executive Director

Stockholm, August 2001

PREFACE

This volume gives an account of great discoveries in Physiology or Medicine towards the end of the twentieth century. It details Nobel Laureates and Prizes within this field for the years 1996–2000, recognizing the progress of natural science in relation to life and mankind, or in the words of Alfred Nobel: discoveries for the "greatest benefit on mankind" within the "domain of physiology or medicine". The scientists and fields awarded by the Nobel Assembly at Karolinska Institutet cover central discoveries in these Prize areas:

* "the specificity of the cell mediated immune defence" — in 1996 to Drs Peter Doherty and Rolf Zinkernagel — marks discoveries regarding T-cells, one of our tools for survival by combating infections and tumor development.

* "prions — a new biological principle of infection" — in 1997 to Dr Stanley Prusiner — challenged the prevailing concept of infectious disease agents, and showed the molecular nature of the agent behind mad cow disease and its human counterpart.

* "nitric oxide as a signaling molecule in the cardiovascular system" — in 1998 to Drs Robert Furchgott, Louis Ignarro and Ferid Murad — showed the presence of yet another bioactive gas in the human body, nitrous oxide, NO, explaining the effectiveness of treatment of angina with nitroglycerin and other drugs.

* "intrinsic signals" in proteins that "govern their transport and localization in the cell" — in 1999 to Dr Günter Blobel — showed that cells, with their compartmentalization of functions into organelles, distribute the working machinery — the proteins — to the correct places by intrinsic address tags.

* "signal transduction in the nervous system" — in 2000 to Drs Arvid Carlsson, Paul Greengard and Eric Kandel — illustrates another breakthrough in contemporary science, now reaching also the central nervous system — the organ behind behavior, thought and memory, concepts long not accessible to molecular science.

This 5-year period, and its Awardees and Awards, gives an excellent illustration of the diversity, the progress, and the richness of natural science, leading to molecular understanding and novel treatments of human disease, with new perspectives on life and mankind.

The Awards also illustrate that even at this time of large scientific teams and global science, when popular belief sometimes suggests that it is becoming more and more difficult to identify individual achievements, creativity and separate contributions can still be recognizable. During this 5-year period, the Award was given twice to single scientists, once to two, and just twice to three. We are grateful to Alfred Nobel for his vision of promoting science by recognition of important discoveries, and to the Laureates for opening novel prospects to human well-being by molecular discoveries in health and disease.

Hans Jörnvall
Professor of Physiological Chemistry
Secretary of the Nobel Assembly at Karolinska Institutet

CONTENTS

Physiology or Medicine 1996

PETER C. DOHERTY and ROLF M. ZINKERNAGEL

"for their discoveries concerning the specificity of the cell mediated immune defence"

THE NOBEL PRIZE IN PHYSIOLOGY OR MEDICINE

Speech by Professor Lars Klareskog of the Nobel Committee at the Karolinska Institute.
Translation of the Swedish text.

Your Majesties, Your Royal Highness, Ladies and Gentlemen,

Peter Doherty and Rolf Zinkernagel are receiving their Prizes for their discoveries concerning the specificity of the cell mediated immune defense, or to be more precise–they discovered how white blood cells recognize and kill virus-infected cells. During the following few minutes I will try to describe why this is so remarkable.

Let us thus try to imagine the extremely dangerous environment in which we live. We are surrounded by a wide array of microorganisms which are universally present in our environment. Our remedy is the immune system which, however, faces a formidable task. Not only are microorganisms numerous and ubiquitous. The immune system must be able to distinguish between friendly microbes and hostile, between foreign material and self structures. And not only that. Certain invaders–for example viruses–utilize host cells for their own replication and hide inside these same cells.

When Peter Doherty and Rolf Zinkernagel began their research in the late 1960s and early 1970s, it had already been discerned how antibodies, the circulating defense molecules, recognize and kill their targets such as bacteria. It was far less well understood, however, how the cellular immune effectors, white blood cells, recognize and kill virus-infected cells without destroying the normal uninfected cells.

Another observation that gave rise to both fascination and problems was the finding of each individual's immunological uniqueness. The small, but important inter-individual differences that exist between molecules called transplantation antigens are what our white blood cells recognize as "self" or "foreign". But the reason behind nature's creation of this immunological uniqueness, this obstacle to transplantation, remained a big mystery.

Rolf Zinkernagel began in 1973 the kind of "journey for knowledge" which is open to the curious scientist. He came from Switzerland to Robert Blanden's laboratory at the John Curtin School of Medical Research in Canberra, Australia, where he met Peter Doherty. Together they studied basic aspects of the immune defense against virus infections in different strains of mice. They found that white blood cells–or more precisely killer T cells–from one mouse strain recognize and kill virus-infected cells from another mouse strain only if the two mouse strains carry the same variant of transplantation antigens. From this seemingly technical and simple observation Doherty and Zinkernagel, and subsequently a whole generation of immunologists, were able to find new approaches and new solutions to a series of fundamental immunological problems.

It became possible to understand that the true function of transplantation antigens is not to provide an obstacle to transplantation. Instead, their function is to bind and present molecules from viruses and other microorganisms to white blood cells in such a way that the white blood cells understand whether they should become aggressive or stay calm. As a consequence it became obvious how each individual, thanks to his/her unique set of transplantation antigens, also carries his/her unique immune system.

It also became possible to understand why evolution has created these large immunological differences between us as individuals within a species. Immunological diversity is advantageous for the single individual as well as for the species. Thus, there will always be some individuals that survive even severe epidemics. In return, individuals carrying a certain variant of transplantation antigens have an increased susceptibility to autoimmune diseases such as rheumatoid arthritis or multiple sclerosis, and this is possibly the price that these individuals pay for the fact that their forefathers survived a severe epidemic.

It is most encouraging, however, that this and subsequent research now makes it easier not only to understand, but also to change–if not the world, then the immune system. Our increased knowledge enables us to strengthen beneficial immune reactions, for example in cases of otherwise insufficient responses towards invading microorganisms or against cancer metastasis. Our knowledge should also enable us to diminish or change unwanted immune reactions towards the body's own tissue, such as those occurring in rheumatic diseases.

The discoveries made by this year's Nobel Laureates thus illustrate how findings within basic biological research may give rise to conclusions with wide implications for society–such as the biological advantage of diversity–and to fundamentally new therapeutic approaches in clinical medicine.

Peter Doherty and Rolf Zinkernagel,

Your discoveries concerning the specificity of the cell-mediated immune defense have deeply influenced the understanding of fundamental issues in basic and clinical immunology. On behalf of the Nobel Assembly at the Karolinska Institute I wish to convey to you my warmest congratulations and I now ask you to step forward and receive your Nobel Prizes from the hands of His Majesty the King.

PETER C. DOHERTY

My childhood was spent on the outskirts of the sub-tropical city of Brisbane. I have a younger brother, Ian, and we grew up as part of a traditional, extended family that was very much influenced by the values of our two grandmothers. The one was a devout Methodist, the other a lapsed Quaker who was born in England and embraced the informal Australian life style with great enthusiasm. My parents (Linda and Eric) were first and second generation Australians, the various elements of the family coming from County Louth (in the 1840's), Lancashire and Essex. Eric Doherty, a clever and entertaining man, trained initially as a telephone mechanic and was an administrator involved in the planning of telephone services. His mother had been left in straitened financial circumstances when my grandfather succumbed to pneumonia during the 1919 influenza epidemic. My father communicated his frustration at not having received an adequate formal education and, with his strong encouragement, the desire to learn and understand became the major focus of my life. Linda Byford was a piano teacher who, with her two brothers, spent much of her youth on the tennis court. After marriage she cared for her family, played Chopin, Debussy, and Beethoven, and grew roses. She gave me an appreciation, and emotional need for, classical music, but did not pass on the genes for tennis. The Byfords were devastated by the death of the eldest son, who was captured at the fall of Singapore and lost in a Japanese transport torpedoed by an American submarine. I remember my other Byford uncle shivering with recurrent malaria that he contracted during the fighting in New Guinea. I share Alfred Nobel's conviction that war is the greatest of all human disasters. Infectious disease runs a good second.

My Irish genetic heritage gave me a very fair skin, making me totally unsuited for life in a city that is known as the melanoma capital of the world. This limited my participation in the outdoor-oriented Australian way of life, and caused me to spend a great deal of time reading anything and everything. Even so, the Australian landscape was at our back door, there were adventures with home made canoes, I played tennis and Australian Rules football, and the extended family went to the beach for at least three weeks each year. The two things that I miss most when living out of Australia are the bush and the Pacific coast, especially fishing in the surf at night! My father had a workshop and I learned to be a carpenter, a skill that has resulted in the manufacture of some very substantial coffee tables and a fair amount of time working on houses. My most ambitious project as a teenager was the construction of a photographic enlarger and darkroom, but all the photographs that I took at that time seem to have been lost after my father's early death (in 1961) and the selling of the family house.

I went to the local public schools and Methodist church. The commute to the high school involved daily trips on a steam train. I played basketball, and was a sergeant in the army cadet corps. Physics and chemistry came easily, but my natural inclinations were towards literature and history. Growing up without much money, however, also left me with the conviction that I needed to get some sort of reasonable job. An older cousin, Ralph Doherty was a brilliant scholar who was in the process of establishing himself as a leading viral epidemiologist. What he was doing seemed fascinating, but my contacts with the local general practitioners left me with no great enthusiasm for the idea of following his path to medical school. A visit to an "open day" at the University Veterinary School was my first real contact with biology in the formal sense: the subject could only be studied by girls in the Queensland public schools of that era. Another major influence at the time of my matriculation, was that I was reading Aldous Huxley, Jean Paul Sartre and Ernest Hemingway simultaneously. I decided to be the man of action rather than the philosopher, and resolved to graduate in veterinary science and pursue a research career. At this stage I was just 17 years old, and would probably have made a very different decision if I had been more mature.

The then vice-chancellor of the University of Queensland pursued a policy of open admission, from the conviction that matriculation results bore little relationship to later academic performance. As a consequence, the veterinary school had a number of mature students who had worked in the bush, while more than 50% (mostly school leavers) did not make it past the examinations at the end of the first year. This was one of two veterinary schools in Australia, and the survivors were joined by a spectrum of students from other states, New Zealand and south-east Asia (under the Colombo plan) at the beginning of year two of the five-year program. Being exposed in this milieu, together with spending the long vacations employed on sheep and cattle properties and seeing practice with rural veterinarians, caused me to grow up quickly. I soon discovered that I had little interest in small animal medicine or surgery, but retain a sense of nostalgia for the satisfaction and physical challenge of working with large domestic animals.

The veterinary school was staffed by a fairly young group of teachers, many of whom had strong research backgrounds. Courses in the physical sciences, zoology, botany and biochemistry were taught from the science faculty, and physiology in the medical school. I was introduced to the principles of ecology in first year zoology, with the commitment of my professor to marine biology almost causing me to switch to that discipline. Infectious disease was taught by John Francis, who communicated great enthusiasm for research, and immunology by the parasitologist, J.F.A. Sprent. Another course that influenced me strongly was population genetics given by Glenorchy MacBride. I also read F.M. Burnet and R.M. Stanley's books on viruses, some of Burnet's writings on immunology and cancer and wrote a final year thesis on the UV-induced squamous cell carcinoma (cancer eye) of Hereford cattle. Burnet's teleological Darwinism, the idea that the body is a set of ecosystems and the realisation that good science involves quantitation have stayed with me from those early days.

When I graduated, I was contracted (under the terms of a "bonded" scholarship) to work for several years in the Queensland Department of Agriculture and Stock. I expressed enthusiasm for laboratory-based research, so the Department immediately sent me to the country as a rural veterinary officer. I spent some months driving large distances to postmortem cattle and pigs that had died of unknown causes, and to survey cattle for various venereal diseases. This resulted in the diagnosis of Trichomoniasis in an area where it was thought that complete eradication had been achieved. Realizing that I was a danger to their regulatory effort, the Department quickly brought me back to the state veterinary laboratory, the Animal Research Institute (ARI), Yeerongpilly. My task was to conduct a large-scale, externally-funded experiment on the epidemiology of bovine leptospirosis. This project involved injecting several cows with *Leptospira pomona,* then watching the spread of the disease throughout the herd. I became adept at dark-field microscopic analysis of urine for spirochetes, the histology of the bovine kidney and the serological test for the organism. This work was submitted for a master's thesis and published in local journals. I was also involved in the diagnostic veterinary pathology service.

The ARI was in the process of establishing a facility for diagnostic virology, and had employed a very attractive young microbiology graduate, Penny Stephens, to develop the laboratory. We married in 1965. Knowing of my interest in virology the ARI Director, Les Newton, sent me to Melbourne for six weeks to learn basic techniques. I worked with Toby St. George in the Commonwealth Scientific and Industrial Research Organization (CSIRO) laboratory of Dr. E.L. French, spent time in the virology group at the Commonwealth Serum Laboratories and (en route) visited F.J. Fenner's Department of Microbiology at the John Curtin School of Medical Research (JCSMR), Canberra. The latter was motivated by the desire to meet C.A. Mims, whose work on viral pathogenesis had considerably influenced my thinking about disease processes. On returning to Brisbane, I realised fairly quickly that I am an experimentalist at heart. A career as a diagnostic virologist was not for me!

I tried for a Ph.D. scholarship to work with Cedric Mims, but was told to apply again later because he already had a "scholar" and would take only one student at a time. At about the same time I got to know J.A. Roberts, who had done an excellent series of experiments with Mims on the ectromelia model and had recently returned from Cornell to a position as a research parasitologist in the CSIRO laboratories on the Yeerongpilly site. John Roberts told me that he had been very impressed by a visit that he had made to the Moredun Research Institute in Edinburgh, where there was a major research effort on scrapie, the then enigmatic "slow virus" disease. The Moredun also trained graduate students, who were affiliated with the University of Edinburgh. The following week a job in the Department of Experimental

Pathology at the Moredun was advertised in *Nature*. We sailed for Britain in early 1967, on a very slow and cheap ship.

The experimental pathology position at the Moredun required that I do research and help to run the diagnostic neuropathology program that the institute operated for the Scottish Veterinary Investigation Service. I learned neuropathology from the head of the department, R. M. Barlow, and from Hugh Fraser who was doing seminal studies with Alan Dickinson defining the genetics of the scrapie mouse model. Dick Barlow also taught me to write clear, concise English. My initial intention was to work on sheep scrapie, but I quickly realised that this was not a good project area for an experimentalist. My abiding interest in infectious disease caused me to focus on the tickborne flavivirus, louping-ill virus, which was then regarded as problematic because of concerns about the safety of the vaccine, first developed at the Moredun many years previously. I enroled as a part-time graduate student at the University of Edinburgh medical school and, after I had been working with the virus for some time, developed a collaboration with another young veterinary graduate, H.W. Reid. Hugh Reid did the virology and serology aspects of the ensuing sheep experiments, while I concentrated on light and ultrastructural pathology. Part of Hugh's role was to test my blood for the presence of virus and antibody when I injected myself in the finger, an inadvertent human experiment that I later wrote up for the *Lancet*.

We greatly enjoyed living in Edinburgh. Penny worked with E.C.R. Reeve at the Institute for Animal Genetics until the birth of our two sons, James and Michael. The Edinburgh Festival and the Traverse Theatre were high points and, for the first time in my life, I could spend the whole day outside without the penalty of sunburn. Our long vacations were used for camping holidays in Europe, including our first trip to Scandinavia and Stockholm with a young child in the back of a Volkswagen van. I went to veterinary research and neuropathology meetings, and we came very close to staying permanently in Britain.

Eric French visited the Moredun, and raised the possibility of a permanent appointment in the veterinary virology group at the CSIRO laboratories in Melbourne. At about this time I heard a seminar by Mel Greaves at the Metchnikoff Club, an Edinburgh group organized by Spedding Micklem and Angus Stewart, that convinced me I had no real understanding of contemporary immunology. Cedric Mims also came through and talked about the work that he and R.V. Blanden had been doing on T cell responses in virus infections. Shortly afterwards a junior academic appointment was advertised in the Department of Microbiology at the JCSMR, with a job description that seemed to fit me reasonably well. Fenner's successor as head of the department, G.L. Ada, had actually written it for Bob Blanden, but offered me the only other position that he had available, a postdoctoral fellowship to work with Cedric Mims. I left my permanent research position, and turned down the offer of another, to take this opportunity to learn basic immunology. My long-term intention was to return to veterinary research. My only formal involvement in the veterinary world since then has been to serve (1987–1992) on the board of the International Laboratory for Research In Animal Diseases,

Nairobi, Kenya. This was an enormously broadening experience, and I learned a great deal from (in particular) my African colleagues.

We moved from Edinburgh to Canberra in December 1971. Cedric Mims had by then decided to take the Chair in Microbiology at Guy's Hospital Medical School so, though we overlapped by six months or so, we did not ever formally work together. At first I studied the pathogenesis of Semliki Forest virus infection in the mouse, then switched to the lymphocytic choriomeningitis virus (LCMV) model which was a much more powerful tool for immunological analysis. I first met Rolf Zinkernagel when he arrived to work with Bob Blanden in 1973, and Gordon Ada (for space reasons) moved him into the laboratory with me. We also lived in the same university housing complex, and shared rides to the JCSMR with a physicist from Trondheim, a Japanese pharmacologist, and a biochemist (Bob Gerdes) who is now working at the Karolinska Institute. The story of our scientific interaction through that time is told in the accompanying articles, and in an account that we wrote jointly some time back that is yet to appear in "Immunology Today". We were then (and have remained) good friends, though we don't always agree on everything.

The basis of the "single T-cell receptor altered self" hypothesis was fairly much worked out by the time of the Second International Immunology Meeting in Brighton, England. I traveled through the United States and gave the same talk in about 20 institutions. Among my hosts were Alan Rosenthal at NIH, Bethesda, and David Katz at Harvard. I also met Gene Shearer, who had results comparable to ours with haptenated cells. This was probably the first time that the immunology establishment became fully aware of what we were saying. Our ideas both contradicted the accepted North American model for the role of immune response genes, and turned the perception of the transplantation system on its head. Many people have told me years later that they heard this seminar, came away with the sense that the findings were significant, but did not fully grasp the import. Evidently some were also infuriated by what we were saying. Rolf traveled more extensively through Europe, and I also visited a number of institutions in England and traveled to Stockholm to speak to Göran and Erna Möller's group at the Wallenberg Laboratory in Lilla Frescati.

Despite the fact that we had made a major breakthrough, the rigidities imposed by the excessive use of tenured contracts through the earlier years at the JCSMR had made any prospects of long-term appointments there fairly remote. Rolf accepted a faculty position at the Scripps Institute, and Hilary Koprowski called on my 34th birthday to offer me an Associate Professorship at the Wistar Institute. I had visited Hilary, who was a good friend of Cedric Mims, during my publicity tour en route to Brighton earlier that year. We moved to Philadelphia in 1975, and I quickly became involved with the outstanding Immunology Graduate Group headed by Darcy Wilson and Norman Klinman at the University of Pennsylvania. The Wistar/Penn axis was a highly interactive, and very open, intellectual environment. I collaborated extensively with Walter Gerhard on the influenza model, did some experiments with the late Tad Wiktor in Hilary Koprowski's rabies program and was part of a large, campus-wide multiple sclerosis research effort. I talked a lot with Jon Sprent, the son of my parasitology professor in Brisbane, who taught me how to do lymph duct cannulation in mice. Penny went back to school, and developed a new career in the area of drug information. I wrote grants, was a member of the immunology circuit, worked with outstanding graduate students and became an established scientist and academic.

My self confidence was such that I made the major mistake of accepting an offer to return to the JCSMR as Head of the Department of Experimental Pathology, intending to build a vital program comparable to that Gordon Ada had been able to create in the early 1970's. However, the era that this was possible had passed, and my decision was made on emotional grounds rather than on the basis of what was actually being offered. The less said about this time (1982–1988) the better, as I am still trying to get the overall experience in perspective. The most positive aspect was my interactions with some excellent colleagues, particularly Jane Allan and Rhodri Ceredig. With the passage of the years, the retirement of many of the tenured staff, the adoption of a more flexible appointment structure, and the return from Denver of Kevin Lafferty as Director, things at the JCSMR are now greatly improved. At the stage that I was there the situation looked hopeless. I decided to move back to a scientific world that I knew I could handle.

The opportunity to rebuild my research career came with the resources offered to me by J.V. Simone, then the Director of St. Jude Children's Research Hospital (SJCRH). I had first visited SJCRH during my swing through the USA in 1974. At that stage it was still a small institution, with a strong virology department headed by Alan Granoff. My contact was via Rob Webster, who had trained with Stephen Fazekas de St. Groth in Frank Fenner's program at the JCSMR and remains a close colleague of the JCSMR virologist, Graeme Laver. Alan and Rob engineered my move to Memphis, and Rob has been an outstanding friend and collaborator. This is a superb, open, research environment, that is extremely well funded. The two problems are that there is too much sunshine, and that we are too far from the Pacific Ocean. Such is life!

My characteristics as a scientist stem from a non-conformist upbringing, a sense of being something of an outsider, and looking for different perceptions in everything from novels, to art to experimental results. I like complexity, and am delighted by the unexpected. Ideas interest me. I was influenced early on by reading Arthur Koestler and Edward de Bono, and more recently by the writings of Karl Popper and Thomas Kuhn. My research career has been highly unconventional, and I have not been a full-time student in the academic sense since I was 22 years old. I have never had a powerful mentor who saw me as the product (or continuation) of his program, a situation that probably helped to determine the outcome of my two attempts to return to Australia. Intellectually, I march to the beat of my own drum and have little interest in competing in "races." There are too few people working in the area of viral pathogenesis and immunity, too little funding, too many problems and too little time.

CELL MEDIATED IMMUNITY IN VIRUS INFECTIONS

Nobel Lecture, December 8, 1996

by

PETER C. DOHERTY

St. Jude Children's Research Hospital, Memphis, Tennessee 38104, USA

INTRODUCTION

Many key concepts concerning the nature of immunity have originated from the very practical need to control virus infections. This year, 1996, has been designated the "Year of the Vaccine" commemorating the 200th anniversary of Edward Jenner's vaccination of James Phipps with cowpox virus, and subsequent challenge with smallpox virus. Insight into the nature of viruses, and how viruses interact with mammalian cells, has evolved since the turn of the century. Our concepts of immunity developed concurrently, beginning with Pasteur's treatment of Joseph Meissner with "aged" rabies virus. Antibody-mediated protection conferred by attenuated, live yellow fever virus won the Nobel Prize for Max Theiler in 1951. Perhaps the most exciting area of immunology when I graduated from the University of Queensland Veterinary School in 1962 was the nature of virus neutralization (1), a topic that is still being resolved with the monoclonal antibody (mAb) technology originated by Georges Köhler and César Milstein (Nobel Prize 1985). Crystallographic analysis of influenza virus neuraminidase-mAb complexes, and variants of neuraminidase selected by mAbs, led to the clear demonstration that Ig molecules normally bind to tertiary structure on proteins (2).

The present award is for our discovery (3–6) from experiments with lymphocytic choriomeningitis virus (LCMV) that the nature of T cell-mediated immunity (CMI) is essentially different, focusing on the recognition of cell-surface major histocompatibility complex (MHC) glycoproteins that have been modified as a consequence of infection. My intention here is to place these findings in historical context, and to develop some aspects to the present day in the context of viral immunity and pathology.

THE AUSTRALIAN SCHOOL OF VIRAL PATHOGENESIS AND IMMUNITY

The strengths in both virology and immunology at the John Curtin School of Medical Research (JCSMR) where we did the LCMV experiments were a direct consequence of themes developed in Australia by F.M. (Sir Mac) Burnet. Over a period of more than 20 years, Sir Mac built the Walter and Eliza Hall Institute (WEHI) in Melbourne into a major international center for virus (particularly influenza) research. In the late 1940's the only person working there on another virus disease was Frank Fenner, who at Burnet's suggestion studied the epidemiology and pathogenesis of ectromelia (mousepox). Following Burnet's demonstration that ectromelia could be titrated by pock assay on chick chorioallantoic membrane, Fenner did careful quantitative studies of virus distribution in a range of tissues (7). His conclusion was that the virus replicated initially at the site of inoculation in the dermis, then the regional lymph nodes, then generalized via the liver and spleen, and again to the skin, where it produced a smallpox-like rash. He also noted that delayed type "allergy" appeared earlier than serum antibody. In 1949 Fenner became head of the Department of Microbiology in the JCSMR at the Australian National University, which had just been formed largely as a consequence of initiatives taken by Sir Howard (later Lord) Florey (Nobel Prize 1945). Fenner built a very strong center for virology research.

In 1957, having just enunciated the clonal selection theory (8), Burnet abandoned virology and re-directed the work of the WEHI to basic immunology, a focus continued with great distinction by his protégé G.J.V. Nossal. Nossal recruited J.F.A.P. Miller, who established a strong program in T-cell immunology that was to have major influence both locally and on the international scene. Some of the WEHI virologists joined Fenner's department in Canberra. One of the few remaining in Melbourne was G.L. Ada, who switched to immunology. Gordon Ada then moved to Canberra in 1968, replacing Fenner who had become Director of the JCSMR. By the time that I arrived, the microbiology department in Canberra had a strong research effort in immunology, interfacing with the virologists remaining from the Fenner era.

Fenner's ectromelia model was subsequently pursued at the JCSMR in two ways that were to influence our discovery. The first was that C.A. Mims made extensive use of fluorescence microscopy to define patterns of virus growth in different organ sites (9), helping to keep the general area of whole-animal viral pathogenesis alive through the 1960's, when most virologists (following Ender's, Weller and Robbins, Noble Prize 1954) turned their attention to *in vitro* tissue culture systems. Fenner's final contribution to experimental viral immunology was to recruit R.V. Blanden back to Australia from the Trudeau Institute, where he had been working with George Mackaness (a Florey trainee and former JCSMR scientist) on CMI in bacterial infections (10). Bob Blanden applied the approaches that he had developed with Mackaness to the ectromelia model, using both adoptive transfer experiments and depletion with anti-thymocyte serum to show the crucial role of T lymphocytes in controlling the infection (11–13).

I had followed Mim's papers for many years, and returned to Australia from Edinburgh at the end of 1971 to work with him. He had warned me that he might be moving to a Chair in Microbiology in London, and left before the end of 1972. I inherited his laboratory, his technician (Gail Essery) and the LCMV model that had been brought to Canberra some years before by Fritz Lehmann-Grube. Rolf Zinkernagel arrived in 1973, to work with Blanden on CMI in Salmonellosis. Due to space requirements, Gordon Ada moved Rolf into the laboratory with me, which is how our collaboration started. Rolf was later to use our experimental data for his Ph.D. thesis but, though I helped him with the writing, the relationship was always as equal colleagues rather than supervisor and student.

EARLY EXPERIMENTS WITH LCMV AND THE DISCOVERY OF MHC RESTRICTION

My initial focus with the LCMV model was to combine contemporary T-cell immunology approaches with the capacity to quantitate inflammatory pathology (14) by counting cells in mouse cerebrospinal fluid (CSF), using a CSF tap technique that I had learned from R.I. Carp (15). This proved to be a very powerful approach (Fig. 2), providing strong support for the theme developed initially by W.P. Rowe and J.E. Hotchin (and later by D. H. Gilden, G.A. Cole and N. Nathanson) that clinical LCM is an immunopathological disease (16–18). The most visually satisfying experiment done at this time was to use Evan's blue injection to show that the virus-immune T-cells were causing a total breakdown of the blood-brain barrier to protein, an experiment that reaches back to the dye studies of Paul Ehrlich (Nobel Prize 1908). Much of this work never appeared in the primary literature, and was only published (3) in an article that G. Möller requested for *Transplantation Reviews.* This also carried the first, tentative description of the MHC-restriction finding.

The formal collaboration with Rolf Zinkernagel started when I suggested that we might look to see if the CSF cells that I was obtaining (Fig. 2) from clinically affected mice could be shown to have cytotoxic T lymphocyte (CTL) activity. Having worked in Lausanne, he was very familiar with the ^{51}Cr release assay that was being used so effectively by J.-C. Cerottini and K.T. Brunner (19) to study alloreactivity and had tried (unsuccessfully) to apply the approach to the *Salmonella* model. The CTL assay had already been used with spleen cells from LCMV-infected mice by M.B.A. Oldstone, O. Marker, M. Volkert and colleagues (20,21). Fortuitously, we were using H-2^{k}- compatible CBA/H mice and L929 fibroblasts (L cells) as a source of T lymphocytes and virus-infected targets respectively. The first experiment worked beautifully, and we were able to show that the lytic activity was mediated by Thy-1 positive cells (22).

At about this time, the very active group of cellular immunologists that made up Gordon Ada's weekly "bible class" had been discussing recent publications on immune response (Ir) gene effects from B. Benacerraf (Nobel Prize 1980) and D.H. Katz at Harvard , which used a very complex *in vivo* mouse model of T-cell/B cell collaboration (23). We later read of the experiments of E. Shevach and A. Rosenthal at NIH, Bethesda, who were looking at the Ir gene question using *in vitro* stimulation of guinea pig T-cells (24). The paradigm (25) in this East Coast USA immunology axis was that the Ir genes, which had been mapped to the I-region (now MHC class II) between the loci for the "strong" transplantation antigens (H-2K and H-2D, now MHC class I), encoded all (or part of) the "enigmatic" T-cell receptor (TCR). This interpretation would probably have quickly been revised when it became apparent that the Ir gene product (Ia antigen) was expressed on macrophages (26).

We then saw a report by H. McDevitt, G.R. Mitchell and M.B.A. Oldstone that there was an "Ir gene" effect in the LCM immunopathology model (27). This stimulated us to accumulate all the mouse strains that were available (CBA/H, BALB/c, and C57BL/6J) in Canberra to see if we could correlate the level of CTL activity with H-2 type. The big surprise was that only the H-2^k T-cells were lytic for the MHC compatible, LCMV-infected L cells (3,4). We had no appropriate H-2^b or H-2^k cell lines available, so used primary peritoneal macrophages (28) to provide a source of LCMV-infected targets to demonstrate reciprocal exclusion of T-cell recognition. The next important experiment was with the H-K^kD^d A/J mouse strain (29) , establishing that CTL recognition mapped to H-$2D^d$ and (perhaps) to H-$2K^k$. Later studies on immunodominance hierarchies (30,31) were to show us how lucky we had been to have a system available where both the H-2K and H-2D alleles were associated with a potent LCMV-specific CTL response. The basic findings with LCMV were rapidly replicated for ectromelia by Bob Blanden's student Ian Gardner (32). Bob also used his contacts as a more established immunologist to bring in a range of H-2 recombinant and mutant mouse strains over the next 12–18 months that were used (with both the LCMV and ectromelia models) to map virus-specific CTL effector function to the MHC class I alleles (33,34). The more detailed description of the actual experiments, the preceding history of the MHC and the way that the TCR repertoire develops during ontogeny are being covered in the accompanying article by Rolf Zinkernagel (35).

THE "SINGLE TCR ALTERED SELF" HYPOTHESIS

Right from the time that we made the discovery (3–6), we considered that there might be two possible explanations for our findings (Fig. 1). The first was based on the "TCR" type hypothesis of Katz and Benacerraf, which was formulated as the various dual recognition models proposed through the 1970's and early 1980's. The second was that the virus was in some way modifying the MHC molecule (4–6), either by complexing with it on the cell surface or by inducing some other (perhaps allosteric) change. This idea of an "altered self" molecule recognized by a single TCR seemed reasonable for virus-infected cells, though we were a bit perturbed when the recognition of "minor" histocompatibility antigens was later shown to obey the same rules (36,37). Evidence against the dual recognition model as originally stated (like-like interaction) came from adoptive transfer studies, where we showed that LCMV-primed T-cells from H-2 heterozygous mice ("AxB" F1) which would kill virus (v) infected A and B targeT-cells partitioned into "clonotypes" lytic for A+v alone or B+v alone following further stimulation in irradiated, virus-infected mice of type A or B, respectively (5). Later variants of "two recep-

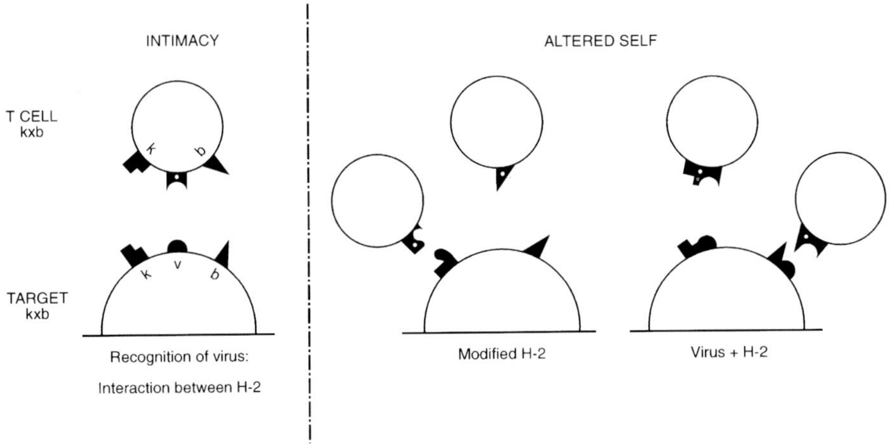

Figure 1. This formulation of the "single T-cell receptor altered self" hypothesis was published in the second article that appeared in *Nature* (4), about a year after the initial experiment showing MHC restriction of CTL activity. In this paper we discredited the "intimacy" model, which argued that there might be some form of "like-like" interaction between H-2 glycoproteins expressed on the T lymphocyte and the target cell. The experiments, and the intellectual context, are dealt with in much greater detail in the accompanying article by Rolf Zinkernagel (35).

tor" models got around this problem by arguing that the postulated recognition molecules seeing A or B molecules were themselves clonally excluded.

The "single TCR-altered self" hypothesis allowed us to assign a biological role to the strong transplantation antigens, and to explain alloreactivity, Ir gene hierarchies, and the extreme polymorphism of the MHC. We believed from the outset that we had found the mechanistic basis of immunological surveillance of self, though we used the term in a somewhat different context (5,6,38,39) from that employed by Lewis Thomas and Burnet in their discussion of susceptibility to cancer (40). Many people did not accept the "altered-self" idea, including some of our colleagues in the JCSMR group (34). It was to remain controversial, and regarded as heretical by most T-cell immunologists for the next 10 years or so (discussed in, 41,42). The "two receptor" models that were generally favoured continued, however, to have the problem that they offered no satisfactory generalization that would accommodate alloreactivity and the Ir gene mechanism.

The resolution of the dilemma came with the finding of A.R. Townsend, A.J. McMichael and colleagues that the class I MHC molecules are presenting viral peptides processed via the "endogenous" pathway (43). This, together with the publication of the 3-dimensional structure of a class I MHC molecule by P. Bjorkman, J. Strominger and D. Wiley (44), and the definition of the two chain TCR by M. Davis, S. Hedrick, T. Mak, J. Allison, P. Marrack, J. Kappler, S. Tonegawa and others (45), put the final nail in the coffin of the various "dual receptor" TCR models. The recent characterization of the tertiary structure of the TCR provides a very satisfactory conclusion to the debate that started with the "altered self" hypothesis (46,47).

THE LCMV MODEL AND T-CELL TARGETING *IN VIVO*

The MHC-restriction of CTL effector function indicated from the outset that the virus-immune T-cell must interact directly with the virus-infected target cell. This had already been obvious for alloreactivity, but no one had suggested that the strong transplantation antigens were in any way involved in the immune response to pathogens. Thinking about CMI in infectious diseases had been very much constrained by experiments with *Listeria monocytogenes,* which emphasized the need for T-cells in macrophage recruitment and activation (10). The numbers and the status (in terms of heat shock-protein mRNA expression) of monocyte/macrophages that localize to the lungs of mice infected with an influenza A virus are, for example, clearly a function of the concurrent CD4+ and/or CD8+ T-cell response (48). Macrophages activated during the T-cell-mediated elimination of LCMV or ectromelia virus can rapidly destroy *L. monocytogenes* (49).

The MHC-restriction findings with *in vitro* CTL assays were quickly translated into the *in vivo* situation. One of the reasons that the LCMV model was so powerful was that we had a very clean *in vivo* system for analysing an inflammatory process. Unlike many viruses, LCMV causes little damage and the levels of "background" cellular infiltration independent of the T-cell response (3,50) are low (Fig. 2). Combining this with the capacity to quantitate

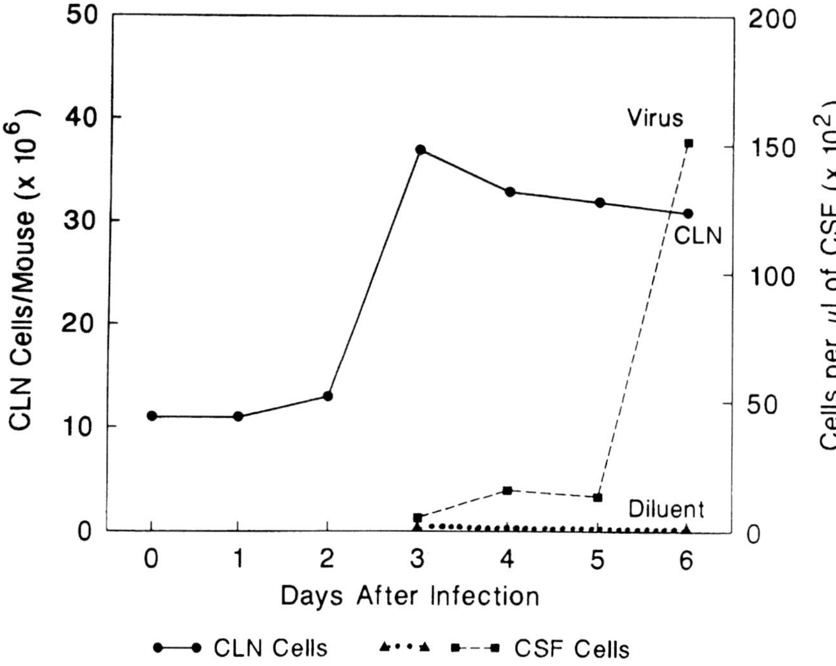

Figure 2. The finding that led to the discovery of MHC-restriction was the demonstration that the inflammatory population recovered from the CSF of mice with clinical LCM contains potent CTL effectors (22). The data shown are from a much later experiment (50), which illustrates the massive influx of cells (dotted line) into the CSF accompanying the onset of CTL activity between days 5 and 6 after infection of the central nervous system. Also shown (solid line) is the preceding increase in cellularity (4-fold) of the cervical lymph nodes (CLN).

very accurately by counting cells in the CSF allowed us to determine rapidly by adoptive transfer experiments that the same rules apply for T-cell recognition *in vitro* and *in vivo*. This was shown for differences in MHC haplotype, particular MHC class I alleles and for H-2 mutant mice (34,51,52).

The interpretation that the T-cell must bind directly *in vivo* to the virus-infected target rather than just to an appropriate antigen-presenting stimulator cell has, however, continued to be subject to challenge. The alternative idea is that cytokines released as a consequence of such interactions are the key effector molecules, an idea that seems to work for the control of a hepatitis virus transgene by interferon-mediated mechanisms (53). So far, this hepatitis model seems to be unique. Adoptively transferred LCMV-immune T-cells only caused severe meningitis in virus-infected, chimeric mice when the appropriate MHC restriction element was present on virus-infected epithelial cells in the brain. Neither secondary stimulation of the virus-specific CTL populations in lymphoid tissue, nor interaction with inflammatory monocyte/macrophages in the site of pathology, were alone sufficient to cause the massive cellular extravasation characteristic of LCM (54–56).

Perhaps the final indication that the LCMV-immune effector CTL must interact directly with virus infected CNS epithelia *in vivo* has come from experiments with perforin -/- mice, which failed to develop the classical symptoms of T-cell-mediated immunopathology (57). Cytokine-mediated mechanisms may, however, be responsible for the chronic wasting disease (58–60) that develops in LCMV-infected CD8 "knockout"(-/-) mice and $CD8^+$ T-cell-deficient mice (61) that are -/- for β2-microglobulin (β2-m), the light chain of the MHC class I glycoproteins. Both these genetically- manipulated mouse strains mice are unable to clear LCMV, leading to a persistent confrontation between virus-infected stimulators and the immune $CD4^+$ population. The studies with LCMV, and other experiments with respiratory viruses (see below), indicate that the effector mechanisms used by $CD4^+$ and $CD8^+$ T-cells to deal with viruses are fundamentally different.

THE NON-SELF COMPONENT: ANALYSIS WITH RESPIRATORY VIRUSES

The division of labor in the LCMV experiments at the JCSMR was that Rolf Zinkernagel did the *in vitro* CTL analysis, while I was responsible for the *in vivo* immunopathology experiments and for writing the manuscripts with, of course, the benefit of his immediate and constant critique. I thus used respiratory infection with the parainfluenza type 1 virus, Sendai virus, to develop a facility with the CTL assay (62), and to confirm that MHC restriction was true for more than LCMV and the poxviruses (27, 32, 63). I returned to the Sendai model (64, 65) years later at St Jude Children's Research Hospital, principally because of the molecular virology expertise available in the laboratory of A. Portner.

When I moved to the Wistar Institute in mid-1975, it quickly became apparent that the animal facility was not sufficiently secure to allow extensive experimentation with LCMV, which can easily induce clinically "silent" infections in mouse colonies. The CTL response to Sendai virus was being studied by others at the Wistar, so I took the opportunity to collaborate with W.E. Gerhard on the analysis of T-cell-mediated immunity to the influenza A viruses (Fig. 3). Walter Gerhard had learned his influenza virology at the Basel Institute for Immunology from Stephen Fazekas de St Groth (1), who had formerly been a leading figure in Fenner's virology department at the JCSMR.

The aim was to exploit "recombinant" influenza A viruses to dissect the other half of the T-cell specificity equation, for viral components rather than for MHC glycoproteins. The so-called recombinants result from growing two different influenza A viruses concurrently in the one cell (66). The eight segments of the influenza genome re-package to give novel variants, the mechanism responsible for the phenomenon of "antigenic shift". The principal focus of immunology-related influenza research, which tends to focus on vaccine studies, was on the viral hemagglutinin (H) and neuraminidase (N) glycoproteins. These virion surface proteins are subject to the antibody-mediated selection pressure that results in "antigenic shift," a further mechanism that can give rise to novel pandemic strains.

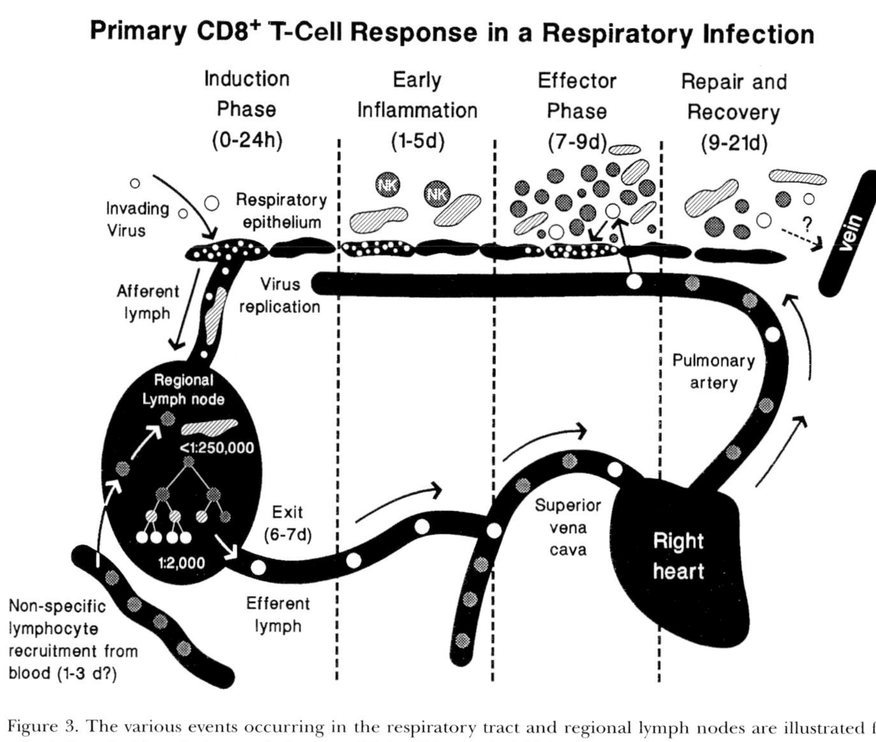

Figure 3. The various events occurring in the respiratory tract and regional lymph nodes are illustrated for mice infected with an influenza A virus, or with Sendai virus (64,98,83). Infectious virus cannot generally be recovered from the lung subsequent to day 9 after infection. The CD8+ and CD4+ T-cells are responsible for most of the monocyte/macrophage recruitment (48). The CD8+ set predominates in the inflammatory exudate, but most of these lymphocytes are memory T-cells specific for other antigens. Probably <1:100 are reactive to the inducing pathogen (99). Potent CTL effectors, and large numbers of cytokine-producing cells, are found mainly in the virus-infected lung.

We thought that, by using serologically distinct variants such as the H1N1 and H3N2 viruses which do not cross-neutralize, we would be able to map CTL activity to H or N, both of which are expressed on the plasma membrane of the virus-infected cell. The aim then was to use "drifted" variants to map fine specificity. The initial experiments were done by my first graduate student, R.B. Effros and were later pursued by my second student J.R. Bennink. Much to our surprise, Rita found that the CTL response for these two viruses was almost totally cross-reactive (67,68). Though we were unaware of it at the time, similar findings were being recorded in B.A. (Ita) Askonas' laboratory at the National Institute for Medical Research in London (69). Our findings with LCMV had induced Ita to drop her long-term studies of the B cell responses to concentrate on virus-specific T-cell mediated immunity. Her laboratory was to make an enormous contribution over the next 10 years.

Though we failed at this stage to define the nature of the antigenic entity recognized by the influenza specific CTL, these early experiments further supported the conclusion from the MHC restriction analysis that the specificity profiles of T and B lymphocytes are fundamentally different (34,70). In addition, we quickly established that exposure to any one influenza A virus will prime T-cell memory for a secondary response (Fig. 4) to any other in-

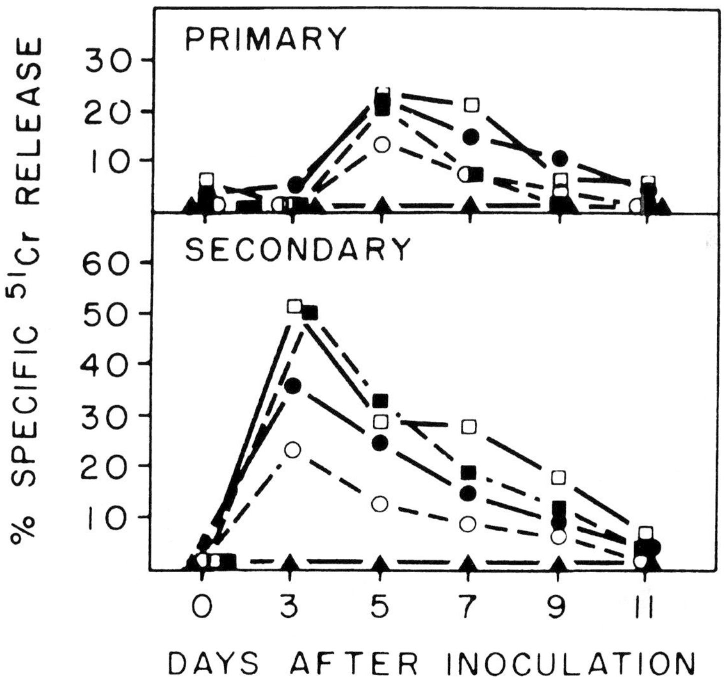

Figure 4. Immunologically naive (primary) CBA/J (H-2^k) mice, or mice that had been infected with the PR8 (H1N1) influenza A virus 52 days previously (secondary), were challenged with the HKx31 (H3N2) influenza A virus (68). Spleen cells were assayed (100:1) at the intervals shown on H-2 compatible targets infected with a range of influenza A viruses, or with an influenza B virus (triangles). Both the enhanced kinetics and potency of the recall response are very apparent. Later experiments with HKx31 infection of C57BL/6J (H-2^b) mice have shown very little CTL activity in lymphoid tissue following primary exposure (98).

fluenza A virus (68,71). This meant that infection with a "drifted" or "shifted" influenza A virus in an adult person is likely to be proceeding in the context of a secondary CTL response and a primary humoral response. The albeit imperfect protection conferred by influenza-specific memory T-cells (72–74) probably explains why young to middle-aged adults are generally much less likely to die from influenza than are small children or elderly people. These results were quickly confirmed for other viruses which had, after the influenza experiments, also been found to induce CTL specificity profiles that were not predicted by serological analysis (75).

Other significant experiments that were done with the influenza model at about this time were the extension of MHC-restriction to the rat model (76) and, most importantly, the clear demonstration by A.J. McMichael (from the Askonas laboratory) that this is also true for humans (77). Further pursuit of this unpredicted cross-reactivity for the influenza viruses led to A. Townsend's (again from the Askonas laboratory) seminal finding (43,78) that MHC class I restricted T-cells are responding to a peptide derived from the influenza nucleoprotein (NP), which finally explained the molecular mechanism underlying our "altered self" hypothesis (Fig. 1) and started the whole field of antigen processing in the "endogenous" compartment. Analysis at the Wistar Institute by Jack Bennink and J.R. Yewdell with recombinant vaccinia viruses also showed that much of the CTL recognition in the influenza model is directed at internal components of the virus (79).

Latterly, the non-lethal respiratory infections have emerged as the experimental system of choice for analysing localised, transient viral infections (80–83). These models have the advantage that it is easy to obtain both the regional lymphoid tissue, and the inflammatory cells from the pneumonic lung by bronchoalveolar lavage (BAL). There is also no obvious way that either the virus or the viral genome can persist after infection with such negative strand RNA viruses (84), which is important when we are considering the difficult area of T-cell memory (85).

Some recent experiments addressed again the question of T-cell targeting *in vivo* that we had analysed earlier with the LCMV model (Fig. 2). Using adoptive transfer protocols and bone marrow radiation chimeras made between β2-m -/- and +/+ mice, we showed that the clearance of Sendai virus depends on the $CD8^+$ effector T-cells interacting directly with virus-infected, MHC class I^+ respiratory epithelium (86). The opposite conclusion was reached concerning the capacity of $CD4^+$ T-cells (87–89) to control influenza virus infection in the absence of the $CD8^+$ subset: expression of the MHC class II glycoproteins that target the $CD4^+$ T-cells is not essential in the respiratory tract (90). More recent experiments performed by David Topham are confirming the theme developed by Walter Gerhard (91) that complete clearance of the virus by $CD4^+$ T-cells requires the concurrent presence of antibody-forming cells.

QUANTITATING INFLAMMATION AND T-CELL-MEDIATED IMMUNITY

The essence of the virology-based approach to pathogenesis has always been quantitation, a theme that was started by Mac Burnet with his various plaque assays and continued by Cedric Mims for the fluorescence microscopic localization of virus-infected cells in different organ sites (9). Measuring the inflammatory process that is the consequence of CMI has, however, traditionally been (at best) semi-quantitative, The classical delayed-type hypersensitivity foot-pad swelling assay is a very blunt instrument.

Analysing inflammation

My early experience in ultrastructural pathology and immunocytochemistry (92,93) resulted in an intense curiosity concerning the nature of the lymphocyte populations that invade into tissue sites of virus growth (Fig. 5). The

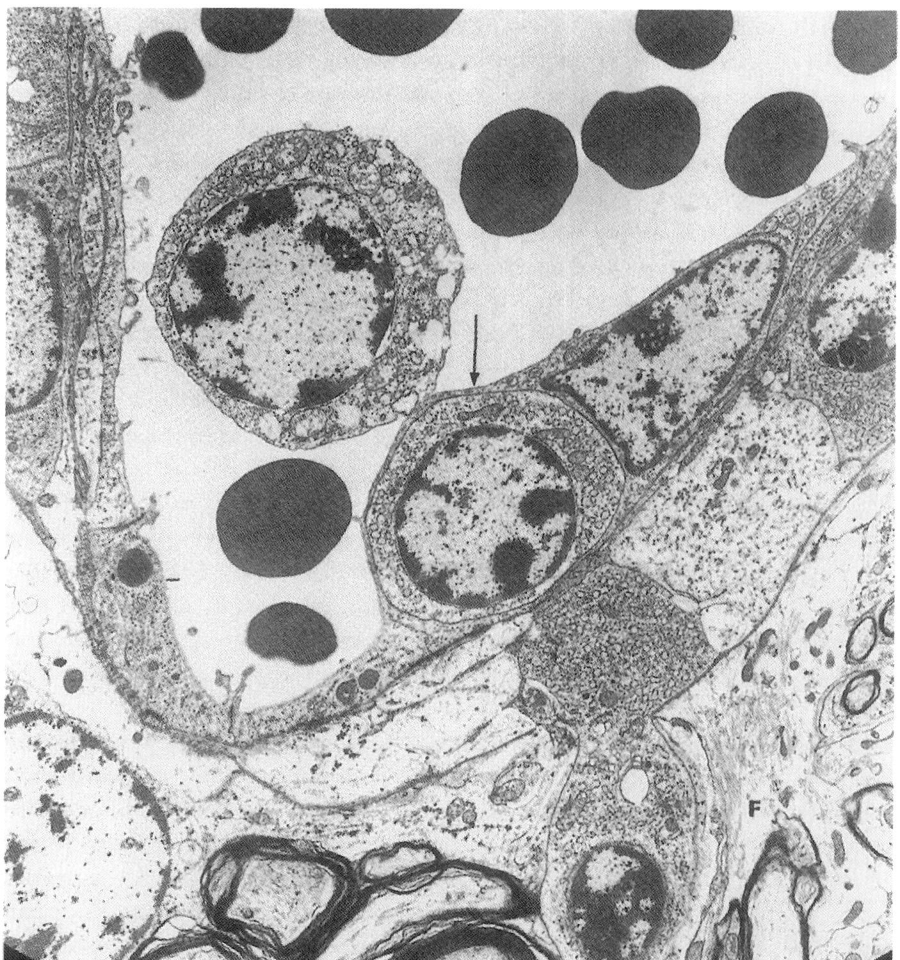

Figure 5. My interest in the nature of virus-specific CMI was stimulated by the experimental studies of louping-ill encephalomyelitis that Hugh Reid and I did with sheep at the Moredun Research Institute, Edinburgh (92–94). The electron micrograph shows cells of undetermined identity extravasating from a capillary into the brain parenchyma. Lymphocyte invasion into tissue sites of virus growth is the central feature of CMI.

experiments that H.W. Reid and I did with the sheep louping-ill (tick-borne flavivirus) model at the Moredun Research Institute in Edinburgh led to the conclusion that antibody forming cells were extravasating into the CNS and were responsible for a substantial, *in situ* virus-specific Ig response (92, 94). Much later studies by D.E. Griffin have shown that such locally-produced antibody controls Sindbis virus infection of mouse brain (95).

The dual frustrations with morphological approaches are the lack of functional analysis and the difficulty of quantitation, though my current perception is that there is a great need to focus more attention on the anatomical localisation of events in immunity that is only possible with microscopy-based protocols. My early attempts at quantitating inflammatory pathology caused me to seize on the CSF tap technique that Richard Carp (15) had developed to look for enzymatic activity in CSF from scrapie-infected mouse brain and apply it to the study of viral meningitis (Fig. 2). Experiments with this simple model induced us to develop the LCMV-specific CTL assay, and allowed the later *in vivo* dissection of MHC-restricted T-cell effector function. The capacity to measure statistically significant differences for quite small effects enabled us to operate with unique sensitivity in the *in vivo* situation (3, 34). However, because the approach was technically demanding and only really useful with the LCMV model, nobody else adopted this analytical system. The other, great advantage of being able to obtain inflammatory cells directly from the CSF (or from the lung by the BAL) is that the lymphocytes do not need to be freed from tissue by enzymatic techniques. This has been of great benefit for later experiments concentrating on flow cytometric analysis and FACS separation to define functional T-cell subsets (56, 96, 97), experiments that were first started for LCMV (with R. Ceredig in Canberra) and have been a major focus of the respiratory infection studies (98,99) in Memphis (Fig. 3).

Measuring the T-cell response

Though the *in vitro* ^{51}Cr release CTL assay provides numbers, the precursor (p) frequencies that give comparable levels of lytic activity (both during the course of *in vivo* infection or as a consequence of *in vitro* culture) can vary enormously (100,101). The potency of the CD8$^+$ CTL set in the BAL population from mice with influenza pneumonia is, for example, apparently equivalent by day 10 after infection for CD4$^+$ T-cell deficient MHC class II -/- mice and for MHC class II +/+ controls (101,102). However, the CTLp numbers in both the BAL and regional lymph nodes are much lower for the MHC class II -/- mice, and virus clearance may be slightly delayed. The much smaller CTLp pool in the MHC class II -/- mice is almost totally consumed to provide the level of CTL effector function needed to deal with the infection. A conclusion based on the CTL data alone that the absence of the CD4$^+$ set does not greatly modify the magnitude of the CD8$^+$ T-cell response would thus be misleading. Careful, kinetic analysis is essential in these *in vivo* pathogenesis studies.

We first realised this when we tried to quantitate the Ir gene hierarchy that we discovered for the MHC class I alleles H-2K^k, H-2K^b and H-2D^b with the influenza A viruses (30,31). The situation here is that the CTL response associated with H-2D^b +NP peptide is dominant in mice that express H-2K^b, but apparently absent when H-2K^k is present. The same thing happens with vaccinia virus. Jack Bennink and I tried hard to work out the underlying mechanism, using thoracic duct cannulation and negative selection *in vivo* to remove the alloreactive T-cells (103), a very tedious protocol that we learned from J.A. Sprent. The lack of any useful resolution is a good example of the fact that (though important insights can be generated) molecular mechanisms cannot ultimately be worked out by biological experiments. My current guess is that H-2K^k must greatly out-compete H-2D^b for binding some constant molecule involved in the MHC class I antigen processing pathway (104), though not to the extent that H-2D^b +NP peptide is no longer present at sufficient levels to be recognized by the appropriate effector CTL on an H-K^{k}D^b targeT-cell.

When J. A. Owen joined the laboratory in Philadelphia she took on the task of developing a quantitative limiting dilution analysis (LDA) for determining influenza virus-specific CTLp frequencies (105), an approach that has been central to my research program over the subsequent years. Judy Owen and Michelle Allouche then applied this technique to the H-2K^{k}D^b hierarchy problem and found that, for mice that had been primed a month or more previously, the CTLp numbers specific for influenza virus components expressed in the context of H-2K^k and H-2D^b may be fairly comparable (106). Recent experiments with this experimental system by R.A. Tripp in Memphis indicate that the reason that the H-2K^k-restricted CTL effectors are preferentially generated is that this component of the response emerges more quickly.

The rate problem in immunology

The preceding is a good example of one of the major difficulties that we face in developing a detailed understanding of immunity. How do we measure the true kinetics of responses in terms of lymphocyte generation and loss, and in the context of transit times through the various tissue sites that are sampled? A major insight developed from studies with superantigens was that the majority of the T-cells that proliferate following such stimulation die as a consequence (107,108). We initially thought that this was not the case for "conventional" antigens, such as influenza virus or Sendai virus epitopes, largely because sequential LDA studies indicated a remarkable constancy (Fig. 6) in CTLp frequencies (1:2–3,000 cells in a lymph node) from about day 7 after infection through to long-term memory (85,99,109). This was somewhat different from the situation that was concurrently being described for LDA studies with LCMV by Rafi Ahmed (110), but we agreed that the difference was probably due to the fact that systemic LCMV infection undoubtedly gives a much greater antigenic stimulus.

The thymidine analogue bromodeoxyuridine (BrDU) incorporates into the DNA of multiplying cells and causes the formation of toxic thymidine dimers following exposure to bright light, a protocol that was used many years ago to "suicide" *in vitro*-stimulated T-cells (111). Ralph Tripp found that load-

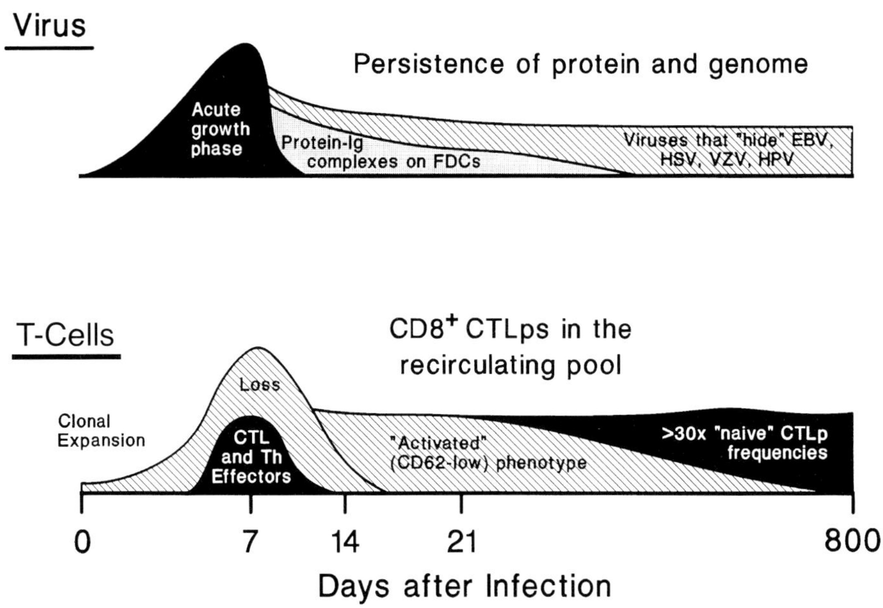

Figure 6. *Virus.* Replication of the negative strand RNA viruses is generally terminated by the immune response within 7–10 days (84,118), with the only trace of the infectious agent generally being the long-term maintenance of protein-Ig complexes on the surface of follicular dendritic cells (110). These do not seem to be necessary for the maintenance of influenza-specific $CD4^+$ T-cell memory (119), but are probably essential for B cell memory. Other pathogens, such as the herpes viruses, may persist in a latent form in neurons or B lymphocytes and periodically reactivate into lytic phase (120).

T-cells. Clonal expansion of the $CD4^+$ THp and $CD8^+$ CTLp during the acute phase of the response is greatly in excess of the numbers required to generate the cytokine-producing TH or effector CTL populations that deal with the infection. Many of the CTLp, in particular, either die or are excreted, and CTLp frequencies may remain remarkably stable thereafter for the life of a laboratory mouse. However, the activation phenotype of these T-cells may (as first shown by Sam Hou, 121) change with time.

ing proliferating T-cells with BrDU *in vivo* led to the same effect when the lymphocytes were exposed to the laser beam of the flow cytometer (112). Use of this protocol indicates that the CTLp numbers generated during the acute phase of the response to an influenza A virus are more than 10-fold in excess of what would be expected from estimates of CTLp frequencies (Fig. 6). Some of these T-cells will give rise to the CTL effectors that are found in the virus-infected respiratory tract, but it seems that many others are lost. Measuring a virus specific $CD8^+$ T-cell response by effector CTL assays takes no account of the CTLp numbers that are generated. However, the more extensive quantitation of CTLp frequencies shows only the balance between CTLp generation and loss, and does not give a true estimate of the real magnitude of an acute host response. This is less a problem with T-cell memory, where the cell populations are turning over at a much slower rate (113).

Homeostasis: on the edge of chaos?

The above experience, together with trying to understand the nature of T-cell memory, has left me with the conviction that the major challenge for cellular immunology is to develop a much clearer understanding of lymphocyte homeostasis (114). In the past, many of us would have followed Burnet's insight that the balance between responsiveness and tolerance is the key question, but recent studies (particularly with transgenic mice) have clarified the issues, blurred the distinction and made the tolerance/response mantra less compelling (115). It could be argued that tolerance/response is simply a more refined statement of the homeostasis problem, but language conditions, perceptions and defining the area in this way focuses attention solely towards antigen and away from the more physiological mechanisms and quantitative considerations that are also likely to be enormously important.

An obsession with homeostasis is dangerous territory, as it has led in the past to some of the least useful and expensive efforts in immunology. Even so, it may be time for our experimentally- based discipline to take greater cognizance of the contribution that can be made by theoreticians, particularly those who are more mathematically inclined. A good example of the way that quantitative analysis can contribute to the development of better predictive models has been provided by recent determinations of the numbers of virus infected cells (116) in people infected with the human immunodeficiency virus (HIV). Influenza virus-specific memory $CD8^+$ T-cells seem to show a remarkable constancy in frequency over a very long period (Fig. 6). Is this an example (117) of a chaotic system? Now that we are starting to generate useful numbers, we need the help of people whose business is numbers.

CONCLUSIONS

The need to deal with pathogens has driven the evolution of the vertebrate immune system, so it should not be surprising that experiments with infectious agents have often illuminated key elements of the underlying mechanisms. The discovery of MHC restriction and the development of the single TCR "altered" self hypothesis is a classical example of how interfacing different scientific disciplines and ways of thinking, an inevitable consequence of studying viral pathogenesis, can lead to a major paradigm shift. The progress that was made over the subsequent 10 years, and the intellectual directions that were followed as a consequence of this simple, operational hypothesis tell us a great deal about the power of ideas. This is not to decry the importance of technology. Our success depended totally on the availability of the ^{51}Cr release CTL assay and inbred mouse strains, both of which had been developed to study alloreactivity. The experiments were done in the context of an experimental framework that was being used to dissect T-cell responsiveness, the immunopathology of LCMV infection, and the nature of CMI to *L. monocytogenes* and ectromelia virus. The local intellectual environment was strong, and heavily focused on the then current forefront of immunology research. Being isolated in those pre-FAX and e-mail days in Australia was a great advantage, as it allowed time to discuss and to think things through. We had outspoken and informed local critics, the freedom and resources to pursue our own ideas and were given full credit for our efforts. Those of us who have senior roles in science need to do everything possible to ensure that comparable opportunities and environments remain available to young scientists.

ACKNOWLEDGMENTS

My research career has been supported by the taxpayers of Australia, the United Kingdom and the United States, the National Multiple Sclerosis Societies of the USA and Australia, and by the American Lebanese Syrian Associated Charities who provide much of the funding for St. Jude Children's Research Hospital. I would have achieved nothing without the devotion and encouragement of my wife, Penny, who is my best friend and critic. I owe a very great deal to my students, fellows, colleagues and collaborators over the years.

REFERENCES

1. Fazekas de St Groth, S. 1962. The neutralization of viruses. *Adv. Virus Res.* 9:1.
2. Colman, P. M., W. G. Laver, J. N. Varghese, a. T. Baker, P. a. Tulloch, G. M. Air, and R. G. Webster. 1987. Three-dimensional structure of a complex of antibody with influenza virus neuraminidase. *Nature* 326:358.
3. Doherty, P. C. and R. M. Zinkernagel. 1974. T-cell-mediated immunopathology in viral infections. *Transplant. Rev.* 19:89.
4. Zinkernagel, R.M., and P.C. Doherty. 1974. Restriction of in vitro T-cell-mediated cytotoxicity in lymphocytic choriomeningitis within a syngeneic or semiallogeneic system. *Nature* 248:701.
5. Zinkernagel, R. M. and P. C. Doherty. 1974. Immunological surveillance against altered self components by sensitised T lymphocytes in lymphocytic choriomeningitis. *Nature* 251:547.
6. Doherty, P. C. and R. M. Zinkernagel. 1975. A biological role for the major histocompatibility antigens. *Lancet* 1:1406.
7. Fenner, F. 1948. The clinical features and pathogenesis of mouse-pox (infectious ectromelia of mice). *J. Path. Bact.* 4:529.
8. Burnet, F. M. 1957. A modification of Jerne's theory of antibody production using the concept of clonal selection. *Aust. J. Sci.* 20:67.
9. Mims, C. A. 1966. Pathogenesis of rashes in virus diseases. *Bacteriol. Rev.* 30:739.
10. Mackaness, G. B. and R. V. Blanden. 1967. Cellular immunity. *Prog. Allergy* 11:89.
11. Blanden, R. V. 1970. Mechanisms of recovery from a generalized viral infection: mousepox. I. The effects of anti-thymocyte serum. *J. Exp. Med.* 132:1035.
12. Blanden, R. V. 1971. Mechanisms of recovery from a generalized viral infection: mousepox. II. Passive transfer of recovery mechanisms with immune lymphoid cells. *J. Exp. Med.* 133:1074.
13. Blanden, R. V. 1971. Mechanisms of recovery from a generalized viral infection: mousepox. 3. Regression infectious foci. *J. Exp. Med.* 133:1090.
14. Doherty, P. C. 1973. Quantitative studies of the inflammatory process in fatal viral meningoencephalitis. *Am. J. Pathol.* 73:607.
15. Merz, P. A., G. S. Merz, and R. I. Carp. 1973. Higher frequency of a protein band in the cerebrospinal fluid from scrapie mice. *Res. Vet. Sci.* 14:392.
16. Rowe, W.P., P. H. Black, and R. H. Levey. 1963. Protective effect of neonatal thymectomy on mouse LCM infection. *Proc. Soc. Exp. Biol. Med.* 114:248.
17. Hotchin, J. and E. Sikora. 1964. Protection against the lethal effects of lymphocytic choriomeningitis virus in mice by neonatal thymectomy. *Nature* 202:214.

18. Gilden, D. H., G. A. Cole, and N. Nathanson. 1972. Immunopathogenesis of acute central nervous system disease produced by lymphocytic choriomeningitis virus. II. Adoptive immunization of virus carriers. *J. Exp. Med.* 135:874.
19. Cerottini, J. C. and K. T. Brunner. 1974. Cell-mediated cytotoxicity, allograft rejection, and tumor immunity. *Adv. Immunol.* 18:67.
20. Oldstone, M. B. and F. J. Dixon. 1970. Tissue injury in lymphocytic choriomeningitis viral infection: virus-induced immunologically specific release of a cytotoxic factor from immune lymphoid cells. *Virology* 42:805.
21. Marker, O. and M. Volkert. 1973. Studies on cell-mediated immunity to lymphocytic choriomeningitis virus in mice. *J. Exp. Med.* 137:1511.
22. Zinkernagel, R. M. and P. C. Doherty. 1973. Cytotoxic thymus-derived lymphocytes in cerebrospinal fluid of mice with lymphocytic choriomeningitis. *J. Exp. Med.* 138:1266.
23. Katz, D. H., T. Hamaoka, M. E. Dorf, and B. Benacerraf. 1973. Cell interactions between histoincompatible T and B lymphocytes. The H-2 gene complex determines successful physiologic lymphocyte interactions. *Proc. Natl. Acad. Sci. U. S. A.* 70:2624.
24. Rosenthal, A. S. and E. M. Shevach. 1973. Function of macrophages in antigen recognition by guinea pig T lymphocytes. I. Requirement for histocompatible macrophages and lymphocytes. *J. Exp. Med.* 138:1194.
25. Benacerraf, B. and D. H. Katz. 1975. The histocompatibility-linked immune response genes. *Adv. Cancer Res.* 21:121.
26. Klein, J. 1975. *"Biology of the Mouse Histocompatibility-2 Complex"*, Springer-Verlag, Berlin and New York.
27. Oldstone, M. B., F. J. Dixon, G. F. Mitchell, and H. O. McDevitt. 1973. Histocompatibility-linked genetic control of disease susceptibility. Murine lymphocytic choriomeningitis virus infection. *J. Exp. Med.* 137:1201.
28. Zinkernagel, R. M. and P. C. Doherty. 1975. Peritoneal macrophages as targeT-cells for measuring virus-specific T-cell mediated cytotoxicity in vitro. *J. Immunol. Methods* 8:263.
29. Doherty, P. C. and R. M. Zinkernagel. 1975. H-2 compatibility is required for T-cell-mediated lysis of targeT-cells infected with lymphocytic choriomeningitis virus. *J. Exp. Med.* 141:502.
30. Doherty, P. C., W. E. Biddison, J. R. Bennink, and B. B. Knowles. 1978. Cytotoxic T-cell responses in mice infected with influenza and vaccinia viruses vary in magnitude with H-2 genotype. *J. Exp. Med.* 148:534.
31. Zinkernagel, R. M., A. Althage, S. Cooper, G. Kreeb, P. A. Klein, B. Sefton, L. Flaherty, J. Stimpfling, D. Shreffler, and J. Klein. 1978. Ir-genes in H-2 regulate generation of anti-viral cytotoxic T-cells. Mapping to K or D and dominance of unresponsiveness. *J. Exp. Med.* 148:592.
32. Gardner, I. D., N. A. Bowern, and R. V. Blanden. 1975. Cell-mediated cytotoxicity against ectromelia virus-infected targeT-cells. III. Role of the H-2 gene complex. *Eur. J. Immunol.* 5:122.
33. Blanden, R. V., P. C. Doherty, M. B. Dunlop, I. D. Gardner, R. M. Zinkernagel, and C. S. David. 1975. Genes required for cytotoxicity against virus-infected targeT-cells in K and D regions of H-2 complex. *Nature* 254:269.
34. Doherty, P. C., R. V. Blanden, and R. M. Zinkernagel. 1976. Specificity of virus-immune effector T-cells for H-2K or H-2D compatible interactions: implications for H-antigen diversity. *Transplant. Rev.* 29:89.
35. Zinkernagel, R. M.. 1996. Cellular immune recognition and the biological role of major transplantation antigens. *Nobel Lecture.*
36. Bevan, M. J. 1975. The major histocompatibility complex determines susceptibility to cytotoxic T-cells directed against minor histocompatibility antigens. *J. Exp. Med.* 142:1349.
37. Gordon, R. D., E. Simpson, and L. E. Samelson. 1975. In vitro cell-mediated immune response to the male specific (H-Y) antigen in mice. *J. Exp. Med.* 142:1108.
38. Doherty, P. C. and R. M. Zinkernagel. 1975. Enhanced immunological surveillance in

mice heterozygous at the H-2 gene complex. *Nature* 256:50.
39. Doherty, P. C., B. B. Knowles, and P. J. Wettstein. 1984. Immunological surveillance of tumors in the context of major histocompatibility complex restriction of T-cell function. *Adv. Cancer Res.* 42:1.
40. Burnet, F. M. 1971. Immunological surveillance in neoplasia. *Transplant. Rev.* 7:3.
41 Bennink, J. R. and P. C. Doherty. 1981. The response to H-2-different virus-infected cells is mediated by long-lived T lymphocytes and is diminished by prior virus priming in a syngeneic environment. *Cell. Immunol.* 61:220.
42 Doherty, P. C. 1987. Historical developments in understanding the function of class I MHC genes. p341 in, "*H-2 Antigens, Genes Molecules and Functions*", C. S. David ed, NATO ASI series, Plenum Press, New York and London.
43. Townsend, A. R., J. Rothbard, F. M. Gotch, G. Bahadur, D. Wraith, and A. J. McMichael. 1986. The epitopes of influenza nucleoprotein recognized by cytotoxic T lymphocytes can be defined with short synthetic peptides. *Cell* 44:959.
44. Bjorkman, P. J., M. A. Saper, B. Samraoui, W. S. Bennett, J. L. Strominger, and D. C. Wiley. 1987. The foreign antigen binding site and T-cell recognition regions of class I histocompatibility antigens. *Nature* 329:512.
45. Davis, M. M. 1990. T-cell receptor gene diversity and selection. *Annu. Rev. Biochem.* 59:475.
46. Garcia, K. C., M. Degano, R. L. Stanfield, A. Brunmark, M. R. Jackson, P. A. Peterson, L. Teyton, and I. A. Wilson. 1996. An $\alpha\beta$T-cell receptor structure at 2.5 A and its orientation in the TCR-MHC complex. *Science* 274:209.
47. Garboczi, D. N., P. Ghosh, Q. R. Fan, W. E. Biddison, and D. C. Wiley. 1996. Structure of the complex between human T-cell receptor, viral peptide and HLA-A2. *Nature* 384:134.
48. Allan, W., S. R. Carding, M. Eichelberger, and P. C. Doherty. 1993. hsp65 mRNA$^+$ macrophages and $\gamma\delta$T-cells in influenza virus-infected mice depleted of the CD4$^+$ and CD8$^+$ lymphocyte subsets. *Microb. Pathog.* 14:75.
49. Blanden, R. V. and C. A. Mims. 1973. Macrophage activation in mice infected with ectromelia or lymphocytic choriomeningitis viruses. *Aust. J. Exp. Biol. Med. Sci.* 51:393.
50. Lynch, F., P. C. Doherty, and R. Ceredig. 1989. Phenotypic and functional analysis of the cellular response in regional lymphoid tissue during an acute virus infection. *J. Immunol.* 142:3592.
51. Doherty, P. C. and R. M. Zinkernagel. 1975. Capacity of sensitized thymus-derived lymphocytes to induce fatal lymphocytic choriomeningitis is restricted by the H-2 gene complex. *J. Immunol.* 114:30.
52. Doherty, P. C., M. B. Dunlop, C. R. Parish, and R. M. Zinkernagel. 1976. Inflammatory process in murine lymphocytic choriomeningitis is maximal in H-2K or H-2D compatible interactions. *J. Immunol.* 117:187.
53. Guidotti, L. G., P. Borrow, M. V. Hobbs, B. Matzke, I. Gresser, M. B. A. Oldstone, and F. V. Chisari. 1996. Viral cross talk: Intracellular inactivation of the hepatitis B virus during an unrelated viral infection of the liver. *Proc. Natl. Acad. Sci. U.S.A.* 93:4589.
54. Doherty, P. C. and J. E. Allan. 1986. Role of the major histocompatibility complex in targeting effector T-cells into a site of virus infection [published erratum appears in Eur J Immunol 1986 Dec;16(12):1646]. *Eur. J. Immunol.* 16:1237.
55. Doherty, P. C., R. Ceredig, and J. E. Allan. 1988. Immunogenetic analysis of cellular interactions governing the recruitment of T lymphocytes and monocytes in lymphocytic choriomeningitis virus-induced immunopathology. *Clin. Immunol. Immunopathol.* 47:19.
56. Doherty, P. C., J. E. Allan, F. Lynch, and R. Ceredig. 1990. Dissection of an inflammatory process induced by CD8+ T-cells. *Immunol. Today* 11:55.
57. Kagi, D., B. Ledermann, K. Burki, P. Seiler, B. Odermatt, K. J. Olsen, E. R. Podack, R. M. Zinkernagel, and H. Hengartner. 1994. Cytotoxicity mediated by T-cells and natural killer cells is greatly impaired in perforin-deficient mice. *Nature* 369:31.
58. Fung-Leung, W. P., T. M. Kundig, R. M. Zinkernagel, and T. W. Mak. 1991. Immune

response against lymphocytic choriomeningitis virus infection in mice without CD8 expression. *J. Exp. Med.* 174:1425.
59. Doherty, P. C., S. Hou, and P. J. Southern. 1993. Lymphocytic choriomeningitis virus induces a chronic wasting disease in mice lacking class I major histocompatibility complex glycoproteins. *J. Neuroimmunol.* 46:11.
60. Lehmann-Grube, F., J. Lohler, O. Utermohlen, and C. Gegin. 1993. Antiviral immune responses of lymphocytic choriomeningitis virus-infected mice lacking CD8+ T lymphocytes because of disruption of the beta 2-microglobulin gene. *J. Virol.* 67:332.
61. Zijlstra, M., M. Bix, N. E. Simister, J. M. Loring, D. H. Raulet, and R. Jaenisch. 1990. Beta 2-microglobulin deficient mice lack CD4-8+ cytolytic T-cells. *Nature* 344:742.
62. Doherty, P. C. and R. M. Zinkernagel. 1976. Specific immune lysis of paramyxovirus-infected cells by H-2-compatible thymus-derived lymphocytes. *Immunology* 31:27.
63. Koszinowski, U. and H. Ertl. 1975. Lysis mediated by T-cells and restricted by H-2 antigen of targeT-cells infected with vaccinia virus. *Nature* 255:552.
64. Hou, S., P. C. Doherty, M. Zijlstra, R. Jaenisch, and J. M. Katz. 1992. Delayed clearance of Sendai virus in mice lacking class I MHC-restricted $CD8^+$ T-cells. *J. Immunol.* 149:1319.
65. Hou, S., X. Y. Mo, L. Hyland, and P. C. Doherty. 1995. Host response to Sendai virus in mice lacking class II major histocompatibility complex glycoproteins. *J. Virol.* 69:1429.
66. Webster, R. G., W. J. Bean, O. T. Gorman, T. M. Chambers, and Y. Kawaoka. 1992. Evolution and ecology of influenza A viruses. *Microbiol. Rev.* 56:152.
67. Effros, R. B., P. C. Doherty, W. Gerhard, and J. Bennink. 1977. Generation of both cross-reactive and virus-specific T-cell populations after immunization with serologically distinct influenza A viruses. *J. Exp. Med.* 145:557.
68. Doherty, P. C., R. B. Effros, and J. Bennink. 1977. Heterogeneity of the cytotoxic response of thymus-derived lymphocytes after immunization with influenza viruses. *Proc. Natl. Acad. Sci. U. S. A.* 74:1209.
69. Zweerink, H. J., S. A. Courtneidge, J. J. Skehel, M. J. Crumpton, and B. A. Askonas. 1977. Cytotoxic T-cells kill influenza virus infected cells but do not distinguish between serologically distinct type A viruses. *Nature* 267:354.
70. Zinkernagel, R. M. and P. C. Doherty. 1979. MHC-restricted cytotoxic T-cells: studies on the biological role of polymorphic major transplantation antigens determining T-cell restriction-specificity, function, and responsiveness. *Adv. Immunol.* 27:51.
71. Effros, R. B., J. Bennink, and P. C. Doherty. 1978. Characteristics of secondary cytotoxic T-cell responses in mice infected with influenza A viruses. *Cell. Immunol.* 36:345.
72. Andrew, M. E., B. E. Coupar, D. B. Boyle, and G. L. Ada. 1987. The roles of influenza virus haemagglutinin and nucleoprotein in protection: analysis using vaccinia virus recombinants. *Scand. J. Immunol.* 25:21.
73. Doherty, P. C., W. Allan, D. B. Boyle, B. E. Coupar, and M. E. Andrew. 1989. Recombinant vaccinia viruses and the development of immunization strategies using influenza virus. *J. Infect. Dis.* 159:1119.
74. Lawson, C. M., J. R. Bennink, N. P. Restifo, J. W. Yewdell, and B. R. Murphy. 1994. Primary pulmonary cytotoxic T lymphocytes induced by immunization with a vaccinia virus recombinant expressing influenza A virus nucleoprotein peptide do not protect mice against challenge. *J. Virol.* 68:3505.
75. Zinkernagel, R. M. and K. L. Rosenthal. 1981. Experiments and speculation on antiviral specificity of T and B cells. *Immunol. Rev.* 58:131.
76. Marshak, A., P. C. Doherty, and D. B. Wilson. 1977. The control of specificity of cytotoxic T lymphocytes by the major histocompatibility complex (AG-B) in rats and identification of a new alloantigen system showing no AG-B restriction. *J. Exp. Med.* 146:1773.
77. McMichael, A. J., A. Ting, H. J. Zweerink, and B. A. Askonas. 1977. HLA restriction of cell-mediated lysis of influenza virus-infected human cells. *Nature* 270:524.
78. Townsend, A. R., A. J. McMichael, N. P. Carter, J. A. Huddleston, and G. G. Brownlee.

1984. Cytotoxic T-cell recognition of the influenza nucleoprotein and hemagglutinin expressed in transfected mouse L cells. *Cell* 39:13.
79. Bennink, J. R. and J. W. Yewdell. 1990. Recombinant vaccinia viruses as vectors for studying T lymphocyte specificity and function. *Curr. Top. Microbiol. Immunol.* 163:153.
80. Lukacher, A. E., V. L. Braciale, and T. J. Braciale. 1984. In vivo effector function of influenza virus-specific cytotoxic T lymphocyte clones is highly specific. *J. Exp. Med.* 160:814.
81. Openshaw, P. J. 1989. Flow cytometric analysis of pulmonary lymphocytes from mice infected with respiratory syncytial virus. *Clin. Exp. Immunol.* 75:324.
82. Doherty, P. C., W. Allan, M. Eichelberger, and S. R. Carding. 1992. Roles of alpha beta and gamma delta T-cell subsets in viral immunity. *Annu. Rev. Immunol.* 10:123.
83. Doherty, P. C. 1996. Immune response to viruses. p 535 in, *Clinical Immunology, Principles and Practice.* R. R. Rich, T. A. Fleischer, B. D. Schwartz, W. T. Shearer, and W. Strober eds, Mosby, St Louis.
84. Eichelberger, M. C., M. Wang, W. Allan, R. G. Webster, and P. C. Doherty. 1991. Influenza virus RNA in the lung and lymphoid tissue of immunologically intact and CD4-depleted mice. *J. Gen. Virol.* 72:1695.
85. Doherty, P. C. , D. J. Topham, and R. A. Tripp. 1996. Establishment and persistence of virus-specific $CD4^+$ and $CD8^+$ T-cell memory. *Immunol. Rev.* 150:23.
86. Hou, S. and P. C. Doherty. 1995. Clearance of Sendai virus by $CD8^+$ T-cells requires direct targeting to virus-infected epithelium. *Eur. J. Immunol.* 25:111.
87. Eichelberger, M., W. Allan, M. Zijlstra, R. Jaenisch, and P. C. Doherty. 1991. Clearance of influenza virus respiratory infection in mice lacking class I major histocompatibility complex-restricted $CD8^+$ T-cells. *J. Exp. Med.* 174:875.
88. Bender, B. S., T. Croghan, L. Zhang, and P. A. Small, Jr. 1992. Transgenic mice lacking class I major histocompatibility complex-restricted T-cells have delayed viral clearance and increased mortality after influenza virus challenge. *J. Exp. Med.* 175:1143.
89. Epstein, S. L., J. A. Misplon, C. M. Lawson, E. K. Subbarao, M. Connors, and B. R. Murphy. 1993. 2-microglobulin-deficient mice can be protected against influenza A infection by vaccination with vaccinia-influenza recombinants expressing hemagglutinin and neuraminidase. *J. Immunol.* 150:5484.
90. Topham, D. J., R. A. Tripp, S. R. Sarawar, M. Y. Sangster, and P. C. Doherty. 1996. Immune $CD4^+$ T-cells promote the clearance of influenza virus from major histocompatibility complex class II -/- respiratory epithelium. *J. Virol.* 70:1288.
91. Palladino, G., K. Mozdzanowska, G. Washko, and W. Gerhard. 1995. Virus-neutralizing antibodies of immunoglobulin G (IgG) but not of IgM or IgA isotypes can cure influenza virus pneumonia in SCID mice. *J. Virol.* 69:2075.
92. Doherty, P. C., H. W. Reid, and W. Smith. 1971. Louping-ill encephalomyelitis in the sheep. IV. Nature of the perivascular inflammatory reaction. *J. Comp. Pathol.* 81:545.
93. Doherty, P. C. and H. W. Reid. 1971. Louping-ill encephalomyelitis in the sheep. II. Distribution of virus and lesions in nervous tissue. *J. Comp. Pathol.* 81:531.
94. Reid, H. W., P. C. Doherty, and A. M. Dawson. 1971. Louping-ill encephalomyelitis in the sheep. 3. Immunoglobulins in cerebrospinal fluid. *J. Comp. Pathol.* 81:537.
95. Tyor, W. R. and D. E. Griffin. 1993. Virus specificity and isotype expression of intraparenchymal antibody-secreting cells during Sindbis virus encephalitis in mice. *J. Neuroimmunol.* 48:37.
96. Ceredig, R., J. E. Allan, Z. Tabi, F. Lynch, and P. C. Doherty. 1987. Phenotypic analysis of the inflammatory exudate in murine lymphocytic choriomeningitis. *J. Exp. Med.* 165:1539.
97. Doherty, P. C., J. E. Allan, and R. Ceredig. 1988. Contributions of host and donor T-cells to the inflammatory process in murine lymphocytic choriomeningitis. *Cell. Immunol.* 116:475.
98. Allan, W., Z. Tabi, A. Cleary, and P. C. Doherty. 1990. Cellular events in the lymph node and lung of mice with influenza. Consequences of depleting CD4+ T-cells. *J. Immunol.* 144:3980.
99. Tripp, R. A., S. Hou, A. McMickle, J. Houston, and P. C. Doherty. 1995. Recruitment

and proliferation of CD8$^+$ T-cells in respiratory virus infections. *J. Immunol.* 154:6013.

100. Hou, S., L. Hyland, K. W. Ryan, A. Portner, and P. C. Doherty. 1994. Virus-specific CD8+ T-cell memory determined by clonal burst size. *Nature.* 369:652.
101. Tripp, R. A., S. R. Sarawar, and P. C. Doherty. 1995. Characteristics of the influenza virus-specific CD8$^+$ T-cell response in mice homozygous for disruption of the H-21A^b gene. *J. Immunol.* 155:2955.
102. Bodmer, H., G. Obert, S. Chan, C. Benoist, and D. Mathis. 1993. Environmental modulation of the autonomy of cytotoxic T lymphocytes. *Eur. J. Immunol.* 23:1649.
103. Bennink, J. R. and P. C. Doherty. 1979. Reciprocal stimulation of negatively selected high-responder and low- responder T-cells in virus-infected recipients. *Proc. Natl. Acad. Sci. U. S. A.* 76:3482.
104. Androwlewicz, M. J., and P. Creswell. 1996. How selective is the transporter associated with antigen processing? *Immunity* 5:1
105. Owen, J. A., M. Allouche, and P. C. Doherty. 1982. Limiting dilution analysis of the specificity of influenza-immune cytotoxic T-cells. *Cell. Immunol.* 67:49.
106. Allouche, M., J. A. Owen, and P. C. Doherty. 1982. Limit-dilution analysis of weak influenza-immune T-cell responses associated with H-2K^b and H-2D^b. *J. Immunol.* 129:689.
107. Webb, S., C. Morris, and J. Sprent. 1990. Extrathymic tolerance of mature T-cells: clonal elimination as a consequence of immunity. *Cell* 63:1249.
108. McCormack, J. E., J. E. Callahan, J. Kappler, and P. C. Marrack. 1993. Profound deletion of mature T-cells in vivo by chronic exposure to exogenous superantigen. *J. Immunol.* 150:3785.
109. Tripp, R. A., S. Hou, and P. C. Doherty. 1995. Temporal loss of the activated L-selectin-low phenotype for virus-specific CD8$^+$ memory T-cells. *J. Immunol.* 154:5870.
110. Ahmed, R. and D. Gray. 1996. Immunological memory and protective immunity: Understanding their relation. *Science* 272:54.
111. Paul, W. E., E. M. Shevach, S. Pickeral, D. W. Thomas, and A. S. Rosenthal. 1977. Independent populations of primed F1 guinea pig T lymphocytes respond to antigen-pulsed parental peritoneal exudate cells. *J. Exp. Med.* 145:618.
112. Tripp, R. A., J. M. Lahti, and P. C. Doherty. 1995. Laser light suicide of proliferating virus-specific CD8+ T-cells in an in vivo response. *J. Immunol.* 155:3719.
113. Tough, D. F. and J. Sprent. 1994. Turnover of naive- and memory-phenotype T-cells. *J. Exp. Med.* 179:1127.
114. Freitas, A. A. and B. B. Rocha. 1993. Lymphocyte lifespans: homeostasis, selection and competition. *Immunol. Today* 14:25.
115. Matzinger, P. 1994. Tolerance, danger, and the extended family. *Annu. Rev. Immunol.* 12:991.
116. Perelson, A. S., A. U. Neumann, M. Markowit, J. M. Leonard, and D. D. Ho. 1996. HIV-1 dynamics in vivo: virion clearance rate, infected cell life-span and viral generation time. *Science* 271:1582.
117. Waldrop, M. W. 1992. "*Complexity: the emerging science at the edge of chaos.*" Simon and Shuster, New York.
118. Doherty, P. C. 1995. Immunity to viruses. *The Immunologist* 3:231.
119. Topham, D. J., R. A. Tripp, A-M. Hamilton-Easton, S. R. Sarawar, and P. C. Doherty. 1996. Quantitative analysis of influenza virus-specific CD4$^+$ T-cell memory in the absence of B cells and Ig. *J. Immunol.* 157:2947.
120. Cardin, R. D., J. W. Brooks, S. R. Sarawar, and P.C. Doherty. 1996. Progressive loss of CD8$^+$ T-cell-mediated control of a -herpesvirus in the absence of CD4$^+$ T-cells. *J. Exp. Med.* 184:863.
121. Hou, S. and P. C. Doherty. 1993. Partitioning of responder CD8$^+$ T-cells in lymph node and lung of mice with Sendai virus pneumonia by LECAM-1 and CD45RB phenotype. *J. Immunol.* 150:5494.

ROLF M. ZINKERNAGEL

I was born in 1944 in Riehen, a village near Basel, and spent almost all of my ftrst twenty-five years with my family in the same house. My grandfather on my father's side had bought this house in 1918 after moving with his family from Tübingen to Basel to become Professor of German Literature at the University of Basel. My father grew up in Basel, went through the schools there, and studied biology, finishing with a thesis under the guidance of Prof. A. Portmann. Portmann was an outstanding zoologist-palaeontologist, with a very broad perspective on human development seen in an evolutionary context, not only anatomically, but also psychologically. With this training my father became the first PhD to be employed by the JR Geigy AG–one of the former four big pharmaceutical companies in Basel–not as a chemist, but as a biologist. This in a way heralded a new era of biologically oriented pharmaceutical research and development.

My mother grew up in La Chaux-de-Fonds, in the French-speaking Jura mountains, raised by parents whose family was in the watch-making business and in banking. After moving to Basel, my mother became a lab technician and met my father at work. I was the middle child of three, my brother Peter, born in 1942, became an architect and my younger sister Anne-Marie, born in 1945, became a lab technician.

I went through public school in Riehen, then in Basel, to the Mathematisch-Naturwissenschaftliches Gymnasium, the same school attended by both my father and father-in-law. Since this school didn't teach Latin as a compulsory subject, which was at the time still necessary for several disciplines, such as medicine or law, I took four years of voluntary Latin as well as the school's more mathematically and science-oriented subjects. During that time I had a great number of hobbies: I was introduced by a chemist and collaborator of my father's–who is also a gifted painter–to the prehistory of the Basel region. This was extremely interesting, because during the last glacial period this area was not covered with ice, so that many sites of the previous inter-glacial period have survived. At the same time I also attended several handicraft courses, learning cabinet-making and smithing, as well as enjoying dancing and going to the mountains with the Swiss Alpine Club. My father sent my brother and me on a holiday exchange program to England to learn English. I read a lot and was allowed to do a fair amount of travelling through England, France and the Scandinavian countries, between the ages of twelve to sixteen. When I obtained my matura in 1962, I was uncertain as to what to study. The two areas I favoured were either medicine or chemistry and, because of the greater range of choices the medical profession could offer, i.e. research, clinical activity, or private practice in the mountains, medicine was my target for the next 6 years. I first had to acquire my matura in Latin, however, and in parallel with the medical studies I also had to do my military service. I somehow managed all this by working hard during the first two or three years of medical school. During that time I met my wife, Kathrin, who was studying in the same class, also at the University of Basel. We took our final exams together, which we had prepared with a very nice group of four medical students. In November of 1968, two weeks after the final board exams, we got married. We had originally wanted to go to Africa, where I would have liked to work and learn about leprosy. We applied to the WHO in Geneva and other international organisations, but we were not accepted because of our lack of experience. On the first of January 1969 I began to work at the surgery department at one of the hospitals in Basel, and Kathrin started at the University of Basel Eye Clinic. However, within that first year I somehow became aware that surgery might not be the career I should pursue for the rest of my life and I started to look around for alternatives. After many discussions about my career with several researchers (including A. Pletscher, J. Lindenman and many others), to find another goal, I applied to the postgraduate course in Experimental Medicine at the University of Zürich. To fill in the time between surgery and this course, I did some studies on capillary growth in the epiphysis of the long bones in the Institute of Anatomy at the University of Basel, under the direction of R. Schenk and U. Riede. The course in Zürich is a unique institution that is financed by the Swiss National Science Foundation and the state of Zürich, it gave some ten medical students from all over Switzerland the opportunity to learn more about modern science, in particular molecular biology, biochemistry, genetics. neurobiology and immunology, and to catch up with what had been missed during medical school. Starting in October 1970, I spent two years in the Department of Biochemistry at the University of Lausanne, under the direction of H. Isliker, learning about immunology, immuno-chemistry and the frustrations of experimental lab work. In Lausanne, I was asked to apply to bacteria a technique that had been made popular by T. Brunner and then by J.C. Cerottini, the ^{51}Cr release assay to monitor the destruction of the immunological effector functions of host cells. This test involved the labelling of cells with a radioactive isotope, to monitor destruction of cells by the immune system. The process was measured by determining the relative amount of radioactivity released from the dying cells. This project proved very difficult and did not produce conclusive results: The chromium was not properly absorbed by the bacteria and this release assay was therefore not easily feasible. Nevertheless some work was accomplished on the role of IgA, which was obtained from hyperimmunised cows that release a significant amount of IgA into the colostral milk. I was evaluating whether such hyperimmune milk products were able to protect in an ileal loop model against the entrotoxin-releasing entropathogenic *E. coli*. This confrontation with an infectious disease and the potential of immune responses to protect against it., motivated me to look for a second postdoc position in the same field. Together, Kathrin and I sent about fifty applications to various labs throughout the world, including ones in the UK, the USA, and in Australia, but we either got no answer back or only negative ones. At that time we already had two children and my wife was also trying to find a position to pursue her own career as an eye doctor.

In 1972, while I was looking around for positions, H. Isliker discussed my plans with Professor G. Ada, Head of Department of Microbiology of the John Curtin School of Medical Research in Canberra at the Australian National University, and with Robert Blanden, a professorial fellow at the same institution. G. Ada and H. Isliker were working together at the International Union Against Cancer in Lyon, and R. Blanden came by to teach at the WHO course on immunology at our institution, which was hosting the WHO training lab in Lausanne. This juxtaposition opened up the possibility that I could join the Department of Microbiology in Canberra with the condition that I brought my own salary. The post-doctoral fellowship from the Swiss Foundation for Biomedical Fellowships granted me 32,000 SFr. per year for two years to go to Australia. Fortunately Kathrin did not object to such a drastic move with our two small children Christine, 2 ½ years, and Annelies 11 months old. We flew to Canberra in January of 1973. My plan was to work with R. Blanden on cell-mediated immunity against Salmonella and Listeria to learn more about the role of cell-mediated versus antibody-dependent immune effector mechanisms in these infectious disease models. When we arrived in Canberra we were very lucky and happy to find a generous infrastructure offered by the Australian National University, which provided us with a detached four-room family house within a group of some thirty houses lovingly called the "University ghetto", in Hughes. It provided an extremely nice and congenial environment for students, young and middle-aged post-doctoral, predoctoral and professorial visitors from all over the world. Kathrin soon found a position as a part-time physician at the emergency room of the Woden Valley Hospital, the kids found access to play-groups and kindergarten, I spent all day in the lab studying immunity to infectious diseases and we made many friends amongst the Hughes community.

Within the department, the only empty space in the small labs at the John Curtin School, was in the lab occupied by Peter Doherty. He had arrived as a post-doc from Edinburgh at the end of 1971, his interests being mostly in inflammatory processes in the brain particularly in mice infected with the Semliki Forest virus or with lymphocytic choriomeningitis virus (LCMV). We started to cooperate on immune responses against the LCMV virus; he was tapping the cerebrospinal fluid and doing the inflammatory and immunopathological analyses in the brains, while I was doing the cytotoxicity assays, since I had become familiar with them in Lausanne. This collaboration resulted not only in the discovery of the MHC restriction, as will be detailed in our lectures, but also encouraged me to eventually enrol at the age of 28 as a PhD student at ANU. I had two reasons: one, obviously, was to add a PhD to the doctorate that I had earned with a thesis on the clinical problems of neuritis of the plexus brachialis at the University of Basel. The second motivation was that the exchange rate of 5.2 Swiss Francs for 1 Australian dollar made life rather difficult at the time. The PhD scholarship added a welcome 2000 Australian dollars to our annual budget.

The two-and-a-half years in Canberra were particularly successful because the group of people that had come together in G. Ada's department (including R. Blanden, K. Lafferty, A. Cunningham, P. Pletscher, P. McCullagh and many others), was just the right mix of investigative, critical if not aggressive, intelligent if not inquisitive, humorous if not bitter, and enjoyable minds working together and making sure that one's feeling of being right was constantly questioned and challenged. Of course, the fact that all these people–or at least most–worked with biological model situations, either involving infectious diseases or the transplantation of organs, made all of us very aware that immunology really had to deal with defence *in vivo* and not with artificial antigens in an *in vitro* setting.

As soon as our first papers were published in *Nature* I had to look around for the next post-doctoral or professional situation and was contacted by F. Dixon of the Scripps clinic in La Jolla, who was looking for an assistant professor to join Scripps to work on cell-mediated immunity of auto-immune mice; he had heard from J.C. Cerottini in Lausanne about our work in Canberra. I had also approached B. Benacerraf at Harvard to find out whether there would be any chance of working in his Department of Pathology to continue studying infectious disease immunology. At the second International Congress of Immunology in Brighton in 1974 P. Doherty and I had various opportunities to report on our findings of MHC restriction. I met with F. Dixon and B. Benacerraf. After several discussions and because I could no longer continue to work with the virus in the Boston lab, but also because California, the sea and sun seemed attractive, the decision to join the lab of F. Dixon in La Jolla was easy.

Our two children were very happy in Canberra and they both spoke the most beautiful Australian English. Our second daughter, Annelies, however, went through repeated colds and middle-ear infections, one of them causing a near-lethal Haemophilus influenza meningitis, signalling a selective IgA defect (that turned out to be transient). On December 9th, 1974 Kathrin gave birth to our Australian son, Martin, at Woden Valley Hospital, while I was summarising our experiments on MHC-restricted T-cell recognition during the annual meeting of the Australian Society for Immunology assembled in Canberra.

Kathrin moved back to Switzerland in early January 1975 to spend a few months with our parents and to get another 6 month's training in ophthalmology, while I had to write up my PhD thesis and Peter Doherty was sweating to correct my offences to the English language. So for another 3 months I finished up several studies in the lab in Canberra and then travelled through Australia by train and bus. The two following months in Switzerland were spent by renovating a 16th century house in the Jura mountains with my brother Peter. In early July 1975, we all moved to La Jolla with a green card, i.e. as US-immigrants, to join the Scripps Clinic of Medical Research. The decision to apply for a green card was suggested by F. Dixon not only to avoid time restrictions during my stay in the USA but also because the prospects for finding an interesting job in Switzerland were virtually nil. Work at Scripps started well, not least because Alana Althage, one of F. Dixon's technicians, was assigned to me. She has been helping me ever since by preparing much of the experimental work and by keeping the lab running for the past 20 years. I continued studying cell-mediated immunity to LCMV. This virus had also been studied for several years at Scripps by M.B.A. Oldstone and F. Dixon. We were using experimental surgical techniques to evaluate whether or not the MHC of the thymus played a role or not in the selection and expression of the T-cell specificity for self. A series of similar data was obtained by experiments done in parallel by M. Bevan at the MIT. They resulted in the discovery that the MHC-molecules of the thymus dictated the restriction specificity of T-cells for self-MHC. These results caused major excitement in the immunological community because they fitted in nicely with what one knew about the role of thymus in T-cell maturation as originally described by J.F.A.P. Miller in England and in Australia.

My wife worked as a voluntary collaborator at the ophthalmology clinic at Scripps for about 10 to 20 % of her time; she kept her medical skills alive, particularly in 1978, when she decided to study for–and successfully passed–the US-medical board exams.

The University of Zürich had approached me in 1976 to look into the possibility of taking over a position that had been freed in 1975 by Professor G. Zbinden, Head of Toxicology of the University and the ETH. This division of Experimental Pathology within the Department of Pathology was an attractive chance to go back to Switzerland and to start a larger group. Although the Medical Faculty of the University of Zürich had voted on a finalist for this position sometime in 1977 and had put me in first position, it took another two years of tough negotiations with the government before we could move in the fall of 1979. This required a lot of patience from Kathrin, F. Dixon and all of us; the signed contract only arrived about 10 days before the planned starting date in Zürich!

For the past 17 years we have lived near Zürich, first in Zollikon, then, since 1972, in Zumikon, in a cosy old house with huge woodstoves, a beautiful flower garden and a handy vegetable plot, chickens, and an Appenzeller dog. Kathrin finished her ophthalmology specialisation in the first few years, then started up her own practice. The children adapted reasonably well to schools here and are just about to finish medical school.

Work in the lab was difficult at first because we had to organise all installations, equipment and animal quarters for infectious experiments. This was made much easier when Hans Hengartner, a molecular biologist from the ETH, joined the lab; he had spent 4 years at the Basel Institute of Immunology. In 1978, he approached me to plan a possible move together to Zürich. This was another lucky event in my life, comparable to the moves to Australia and to Scripps; the collaboration with Hans, first in the division of Experimental Pathology and subsequently in our Institute of Experimental Immunology, as associate–and eventually as full professor of both the University and the ETH–has been extremely productive, exciting, stimulating, straightforward and mutually complementary. Joining our molecular, immunological and physiological capabilities and expertise has helped to achieve far more than either of us could have done alone. Together, we continue to follow viruses in infected hosts to find out more about how the host immune system functions and how viruses and immune system have co-evolved.

CELLULAR IMMUNE RECOGNITION AND THE BIOLOGICAL ROLE OF MAJOR TRANSPLANTATION ANTIGENS

Nobel Lecture, December 8, 1996

by

ROLF M. ZINKERNAGEL

Institute of Experimental Immunology, University of Zürich, Zürich, Switzerland

THE ORIGINAL EXPERIMENT ON T-CELL SPECIFICITY

While Immunology began as an attempt to understand immunity against infectious disease, in the 1960s and early 1970s it was largely preoccupied with antibody and T-cell responses against easily available foreign protein antigens or chemically defined small molecules (haptens)[1]. The mechanisms of foreign-organ graft rejection were also intensively studied, although the biological function of the highly polymorphic major transplantation molecules coded by the major histocompatibility complex (MHC, HLA in humans, H-2 in mice)[2,3] was largely unclear. Only a few people studied immunity against infectious agents. The Department of Microbiology at the Australian National University in The John Curtin School of Medical Research in Canberra, headed by G. Ada, was one place where virologists and immunologists worked together on antibacterial and antiviral T-cell-mediated immunity, particularly on the capacity of immunized cytotoxic $CD8^+$ T-cells to destroy either virus infected or allogeneic target cells *in vitro*. Peter C. Doherty from Brisbane, who had returned to Australia from Edinburgh at the end of 1971, was working in this Department. He was interested in inflammatory cell responses in the brain against virus infections. When I arrived in Canberra in early 1973 I started working on *Listeria* with R. V. Blanden, and joined forces with Peter to work on cell-mediated immunity to the lymphocytic choriomeningitis virus (LCMV)[4–6]. I had come from an institution that was involved in establishing the ^{51}Cr release assays by T. Brunner and J. C. Cerottini in Lausanne, Switzerland[7,8], and I therefore attempted to establish the cytotoxicity assay against LCMV in the same way as R. V. Blanden and I. Gardner were using it to monitor cellular immune responses against ectromelia virus (mouse pox) in mice[9]. Because many papers on cytotoxic T-cell responses against LCMV had already been published by various groups[10,11] and since studies had already been carried out on mouse pox in Canberra by R. V. Blanden's group, members of the Department were sceptical that we should and could run with the strong competition. Although we had some initial problems because of our limited experience with LCMV we successfully established the test with some help from I. D. Gardner. We then used this test to find out whether inflammatory cells in the cerebrospinal fluid of mice infected intracerebrally with LCMV were cytolytic *in vitro* and whether there was a correlation between cytotoxic T-cell activity and severity of choriomeningitis. P. C. Doherty was very good at collecting a few microliters of cerebrospinal fluid from the cisterna magna of a mouse. I miniaturized the ^{51}Cr release assay to be able to measure the activity of such small numbers of cells in microwell plates. These experiments revealed that cytotoxic T-cells specifically destroying LCMV infected target cells could be found in the cerebral spinal fluid of normal mice, but not in that of nude mice lacking a thymus and T-cells; this implied that T-cells probably also destroyed infected meningeal and ependymal cells *in vivo* and that this was the essential pathogenic mechanism causing lethal choriomeningitis. These findings were published in the *Journal of Experimental Medicine* in March 1973[12]. In this journal a paper by Oldstone, McDevitt and collaborators had just appeared, showing that mice with different major histocompatibility gene complexes differed with respect to their susceptibility to LCM disease after intracerebral infection[13]. We therefore checked whether a correlation existed between the virus specific cytotoxic T-cell activity in mice and their susceptibility to disease. Six to eight mice of each inbred and cross-bred strain available at the school were infected intracerebrally with LCMV. Two of each were sacrificed on day 7 after infection, when the first mice became sick, to test antiviral cytotoxic T-cell activities in spleens (Table 1). The remaining mice were monitored for the development of lethal disease during the next ten days. While all the mice died of choriomeningitis by day 10, surprisingly only some strains of mice generated virus-specific cytotoxic T-cell activity that was measurable in our *in vitro* assay (Table 1). This result either signalled that cytotoxic T-cells had nothing to do with lethal choriomeningitis, or, alternatively, that our test was in some way inadequate. The latter interpretation proved to be correct: We had used mouse L-929 cells infected with LCMV as target cells to assess cytotoxic T-cell activities. This was a fibroblast cell line used by virologists at the John Curtin School of Medical Research to quantitate viruses by a plaque assay[14]. Since this was the only mouse-cell line usually used in the Department, along with a Vero-cell line derived from monkeys or the BHK-cell line derived from hamsters, we chose it to provide target cells in the virus-specific cytotoxic T-cell assays. By chance, and fortunately, the mice that were most commonly used in the Department were of the CBA strain, while the L-cells had been derived from a closely related mouse strain C3H some 50 years earlier. Also by chance, both mice possessed the same MHC-molecules (H-2^k). Our critical test now revealed that LCMV-immune spleen cells from all mice that possessed the H-2^k haplotype (as do e.g. CBA mice), including cross-breeds with H-2^k mice, lysed L929 (H-2^k) cells infected with the virus, but did not lyse uninfected targets, or those infected with a third-party virus; all spleen cells derived from immunized mice that were not of the H-2^k type failed to do so[15].

Two additional experiments carried out within the next few weeks promptly confirmed these findings. It was important to show that LCMV-immune lymphocytes from non-H-2^k strains of mice were able to lyse LCMV-infected

Table 1. Experiments demonstrating MHC-restriction specificity and positive selection in the thymus for antiviral cytotoxic T-cells.

	Stem cells (MHC, H-2)	Thymus (MHC, H-2)	Other host cells (MHC, H-2)	Virus-specific CD8+ T-cells specific for (MHC, H-2)		Lethal CD8+-T-cell-mediated choriomeningitis after i.c. infection
I	*Original experiments with normal mice*			k	d	
1.	k	k	k	+	–	100 %
2.	b	b	b	–	–	100 %
3.	d	d	d	–	+	100 %
4.	k/d	k/d	k/d	+	+	100 %
II	*Thymus transplantation to thymus-deficient recipient mice:*					
1.	k/d	–	k/d	–	–	0 %
2.	k/d	k	k/d	+	–1	100 %
3.	k/d	d	k/d	–	+	100 %
4.	d	k	d	–	+	100 %

Normal mice (I) vs. mice lacking a thymus that therefore have no T-cells and were given a thymus graft under the kidney capsule (II), were infected with LCMV intracerebrally. All mice died, except those that had no T-cells (II.1.), because of the lethal $CD8^+$ T-cell-mediated LCMV-specific immunopathology. The fact that mice without functional T-cells survived shows that the noncytopathic LCMV does not cause disease directly.

The LCMV-specific cytotoxic T-cells lysed infected target cells that share MHC-molecules (e.g. I.1.; Immune T-cells from infected CBA mice) ($H\text{-}2^k$) lysed infected L929 fibroblast cells ($H\text{-}2^k$)[15].

The MHC of the thymus determines which MHC is recognised by mature T-cells. The experimental group II.3. shows that ($H\text{-}2^k$ x $H\text{-}2^d$) mice with a thymus $H\text{-}2^d$ generated virus-specific T-cells that lysed virus infected $H\text{-}2^d$ target cells but not infected $H\text{-}2^k$ cells[75]. There are, however, experimental exceptions to this rule (II.4.). Thymus-deficient nude mice ($H\text{-}2^d$) reconstituted with an $H\text{-}2^k$ thymus generated $H\text{-}2^d$-restricted T-cells [76].

target cells of the corresponding MHC-type. This did not prove to be easy, because the other available mouse-cell lines in the Department, such as the $H\text{-}2^d$ mastocytoma P815 or the $H\text{-}2^b$ thymoma EL4, could not be infected with LCMV. Because of my work with R. Blanden on *Listeria*–which infects macrophages and is essentially controlled by cell-mediated activation of macrophages as shown by Mackaness[16]–we tried to use macrophages directly isolated from peritoneal washings of mice as target cells for these tests. Macrophages adhered well to plastic, and could be readily infected and labelled with ^{51}Cr. Proper criss-cross experiments showed that LCMV-immune T-cells from $H\text{-}2^b$ mice lyse LCMV-infected macrophages of $H\text{-}2^b$ origin but not those of other H-2 types, and vice versa. The report on these findings was sent off via J. Humphrey as a letter to *Nature* in early December[15]; it was accepted in January 1974, and was published in April 1974. The first public presentation of our data outside Australia was at a Keystone meeting, in Squaw Valley, attended by A. Cunningham in February and a Brook Lodge meeting attended by G. Ada in March 1974. A letter sent back to Canberra by A. Cunningham summarized data from G. Shearer showing that TNP-specific cytotoxic T-cells lysed syngeneic TNP-lated targets more efficiently than allogeneic TNP-lated targets; these data were submitted to the *European Journal of*

Immunology[17] at about the time our report appeared in *Nature*. Obviously, the two findings had emerged independently.

INTERPRETATIONS OF THE DATA

The biological function of MHC and of transplantation antigens was largely unknown in the early 1970s. Their function was obviously not simply to frustrate transplantation surgeons. Transplantation antigens have been defined by P. A. Gorer[18], and by G. Snell[19], based on the work of C. Little, L. Strong and others, who had developed many inbred strains of mice in order to be able to define the rules of transplantation and rejection of tissue and cell grafts (reviewed by J. Klein[2]). Haematologists, particularly J. Dausset[3] and J. van Rood[20], defined lymphocyte surface antigens in humans as being similar to red blood cell antigens, and called them human lymphocyte antigens (HLA). Once many patients had been typed for their transplantation antigens it became apparent that several disease susceptibilities were somehow linked to the transplantation antigen types. It was revealed in studies by B. Benacerraf[21], and later in great and critical detail by H. McDevitt and coworkers[22,23] and by F. Lilly[24], that inbred strains of guinea pigs and mice differed in their responses to some of the model antigens or tumor studied. Because of the availability of well-defined inbred strains of mice, this was readily mapped to the MHC and even to subregions of the MHC in mice by H. McDevitt et al.[25]. In the early 1970s, transplantation antigens were widely discussed because of these findings. MHC polymorphism was thought either to prevent mutual parasitism or transmission of tumor cells[26], or to prevent viruses or other pathogens from mimicking transplantation antigens and, therefore, from eliminating the species[27-29]; alternatively, it was proposed that transplantation antigens functioned as enzymes or as generators of antibody diversity[30]. A most fascinating proposal had been formulated by H.S. Lawrence in 1959[31]. He proposed that infectious agents complexed with transplantation antigens and formed a (*self* + *x*) complex–a fantastically prophetic view of what was to be found later!

There is no doubt that the experiments, which were to reveal the essential role of MHC and T-cell recognition in all, depended upon the foundation built by tumor and transplantation immunologists. Without inbred and MHC (H-2)–congenic or H-2 mutant mouse strains, as developed by G. Snell[19] and C. Bailey[32] and coworkers, respectively, this problem would not at the time have been accessible to analysis. There is also no doubt that MHC-restricted, T-cell recognition would have been discovered by others, using a different approach, a few years later. This would certainly have happened once cloned effector T-cells were developed by M. Schreier, H. Hengartner, H. von Boehmer and G. Fathman[33,34], when T-cell hybridomas were developed by J. Kappler and R. Marrack, or when T-cell receptors were first successfully analyzed by J. Kappler, P. Marrack and J. Allison[35-37] and then were molecularly defined by M. Davis and T. Mak[38,39].

From our very first experiments showing the double specificity of cytotoxic

T-cells for MHC and virus, we immediately knew that we had discovered something important. Our results were not the only ones that hinted at the biological role of major transplantation antigens, and they fitted several observations made during 1972/73. In addition to the suggestions from cytotoxic T-cell studies with leukemia, ectromelia and LCM viruses[9,40,41,41a], there were the experiments by B. Kindred and D. Shreffler[42], who had reported that H-2-incompatible T-helper cells transfused to T-cell-deficient nude mice were not able to help nude B cells make antibodies. Also, P. J. McCullagh and D. H. Katz, H. Hamaoka and B. Benacerraf had shown that histoincompatible B cells and T-cells were not interacting successfully to produce a good IgM to IgG switch[43,44]. In parallel experiments with inbred strains of guinea pigs, A. S. Rosenthal and E. M. Shevach analyzed antigen-specific proliferative T-cell responses, and found them only when primed T-cells and antigen-presenting cells were from guinea pigs with the same MHC type[45]. Most of these experiments were, however, complicated and difficult to interpret, because the mixing of T-cells and B-cells or antigen-presenting cells of different MHC type initiated allogeneic mixed lymphocyte reactions, resulting in non-specific signals. These findings were, therefore, only accepted with hesitation by immunologists; this changed when our data appeared. The simplicity of the *in vitro* assays, the parallel observation by G. Shearer using TNP[17], its easy reproducibility by R. V. Blanden and I. D. Gardner using ectromelia virus[46], by U. Koszinowski using vaccinia virus[47], by E. Simpson, S. Gordon and H. von Boehmer using the male H-Y-antigen[48,49] and by M. J. Bevan using various minor histocompatibility antigens[50], all helped to convince immunologists of the general character of MHC-restricted T-cell recognition. Also MHC-restriction of T-cell recognition was soon confirmed *in vivo* to be also important for antiviral protection transmissible from immune to naive recipients by immune T-cells. By 1977, A. McMichael and coworkers in I. Askonas's lab showed that human influenza virus-specific cytotoxic T-cells[51] and E. Goulmy proved that male antigen H-Y-specific cytotoxic T-cells were HLA-restricted[52].

Our results triggered heated discussions in the Department. We thought that the virus somehow altered the normal cells' MHC molecules and that this virus-specific alteration was recognized by cytotoxic T-cells in a similar way to that of foreign transplantation antigens. Everyone's imagination and intellect was stimulated to come up with a more general, simpler and more convincing explanation for the findings. Discussions were particularly lively because at the same time K. Lafferty and A. Cunningham developed their ideas on second signals necessary to induce responses against foreign transplantation antigens[53], G. Ada and R. V. Blanden where studying other virus-specific T-cell responses, L. Pilarski, P. Bretscher were thinking about B-cell responses and signal requirements along the two-signal theories of P. Bretscher and M. Cohn[54,55] and C. Parish, I. D. Gardner, I. Ramshaw, A. Happle, S. Kirov, W. Davidson, M. Dunlop and Y. Rosenberg, working on B-cell and T-cell response in various virus infections, were discussing the role of enzymes modifying carbohydrates or other self-surface structures to explain the experiments.

FURTHER ANALYSIS

We interpreted our findings to signify that virus infection somehow caused alterations of transplantation antigens on the cell surface by forming a complex of viral antigen with MHC-molecules, or undefined structural alterations or complexing of the two, and that these alterations were recognized by T-cell receptors (Fig. 1)[56]. Foreign transplantation antigens (so-called alloantigens) could then be viewed as a genetically altered form of self-transplantation antigens. This view differed from the then-favoured possibility that lymphocytes and target cells interacted mutually via transplantation antigens (see Figure), i.e. H-2^k interacted best with H-2^k, H-2^b best with H-2^b molecules in a symmetrical like-like complementarity. This intimacy model was soon excluded by the "F_1-experiment" showing that virus-specific cytotoxic T lymphocytes from heterozygote (H-2^k x H-2^b) F_1 mice consisted of at least two

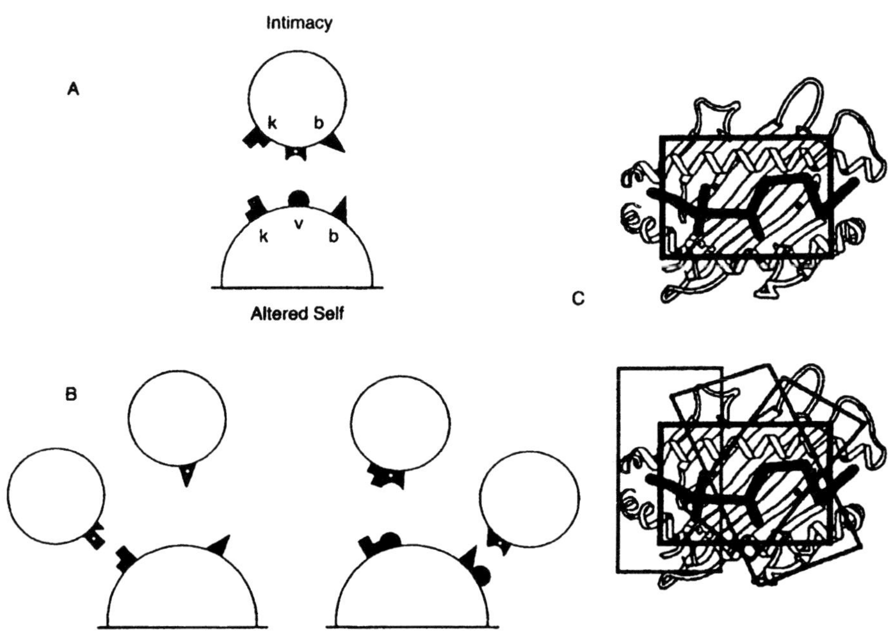

Fig. 1. *Models originally proposed to explain MHC-restricted T-cell recognition.* The idea that MHC-molecules interacted mutually to permit T-cell and target cell to come close enough together (A) was disproven by the demonstration that in an F1 between H-2^k and H-2^b (H-2^k $\times$ H-2^b) mice there were two clonally specific T-cells that were specific for virus plus H-2^k and a second population that was specific for H-2^b. Therefore T-cell receptors were either specific for the virally modified MHC molecule where neither details of either virus antigen or MHC molecules were recognized in its original form (B) or T-cells were specific for a complex formed between MHC molecule and viral antigen, that the T-cell receptor recognized parts of the viral antigen and parts of the MHC molecule. It is now clear in 1996[69, 70] that the T-cell receptor is composed of two chains Vα and Vβ that recognize a viral peptide presented by the MHC molecule[63]. What is less unclear (C), whether the T-cell receptor, here drawn as a square, interacts with the peptide plus MHC complex is always in the same directed fashion, so that certain hypervariable regions of the receptor always interact with corresponding parts of the peptide, or of the two MHC domains forming the groove (more likely), or whether alternatively the receptor interacts with the complex formed by MHC plus peptide in many possible random ways. reproduced with permission of Nature, *251*, 547, 1974, Copyright (1974) Macmillan Magazines Ltd. and JAMA, *274*, 1070, 1995.

subpopulations, one of each being specific for infected H-2^k, and the other for infected H-2^b targets. Since both MHC types were co-dominantly expressed on the lymphocyte surfaces on all host cells, this indicated that some T-cell receptors of one T-cell subpopulation were probably specific for H-2^k-plus-virus and the other subpopulation was specific for H-2^b-plus-virus.

Further experiments together with R. Blanden carried out with the help of mouse geneticists (including C. David and H. McDevitt in the U.S.A.) showed that the H-2D and H-2K regions coding for class I MHC molecules were involved in virus-specific cytotoxic T-cell recognition[46]. These findings separated the MHC-restricted recognition by virus-specific cytotoxic T-cells from the MHC class II -(H-2I)-linked immune response phenomena regulating T-cell–B-cell, or T-cell–macrophage interactions. Analyses of cytotoxic T-cell interactions *in vivo* causing lethal immunopathology[57], antiviral protection[58], and protection against *Listeria monocytogenes*[59], all confirmed that MHC restriction was also valid *in vivo.*

In our second letter to *Nature*, we therefore concluded that T-cells might function essentially by surveying the integrity of transplantation antigens. Recognition of cell surface alteration due to virus infection, chemical modification or genetic differences (i.e. alloantigens) may then be accomodated within the same model. A general hypothesis was formulated in *Lancet* in 1975[60], which stated explicitly that the function of the major histocompatibility molecules is to signal modifications of self-MHC to the immune system.

What we had discovered and tried to explain for cytotoxic T-cells, we also extended to helper T-cells, proposing that they might recognize antigen-induced modifications of Ia (as the MHC class II molecules were called at the time) on macrophages and B cells. Importantly, the results offered an explanation of the reasons for the extensive polymorphism of MHC molecules; it minimized both the possibility that some cell-destroying pathogens failed to cause immunogenic modification and the risk of there being general unresponsiveness in a population. Obviously, what was unknown at the time was that MHC molecules or transplantation antigens are the antigen-presenting molecules that are recognized as a complex with the antigenic peptide. This became known in the subsequent ten years, mostly thanks to the work of E. Unanue[61] and H. Grey and coworkers[62] for class II antigens, as well as on the particularly revealing and eye-opening work of A. Townsend[63]. He showed that class I molecules of virus-infected cells present peptides, 9–10 amino acids in length, to virus-specific cytotoxic T-cells. Similar results were obtained by J. Maryanski and coworkers[64]. These peptides became very real when they were first eluted from target cells by the group of H. G. Rammensee[65]. Peptides were also soon shown to be involved in anti-tumor CTL responses by T. Boon's group[66]. All became convincingly clear in the classical studies of P. Björkman, J. Strominger and D. Wiley in 1987, when the X-ray cristallography of a class I HLA-molecule revealed the peptide binding cleft[67,68]. It is probably not only by chance that within a few weeks of the news of this year's Nobel Prize for specificity of the cell-mediated immune response, the first studies appeared in *Science* and *Nature*, revealing the X-ray structure of the complete complex of the T-cell receptor-MHC class I plus the bound peptide by I. A. Wilson and by D. Wiley and their coworkers[69,70]. What is less unclear, in 1996 is, which part of the TCR, and whether always corresponding parts of the TCR, recognized the peptide and the MHC molecule in the same general position[71,72] (Fig. 1).

IMPLICATIONS

The role of the thymus in MHC-restricted T-cell repertoire selection: Experiments first published by M. Bevan at MIT[73] and conducted in parallel in my lab at Scripps with the help of J. Callahan, G. Dennert and J. Klein signalled a role of thymic MHC in selection of the MHC-restricted T-cell specificity[74,75]. Reconstitution of lethally radiated H-2^b recipient mice with bone-marrow stem cells of (H-2^k x H-2^b) F_1 origin resulted in bone marrow chimeras that were tolerant of H-2^k and H-2^b; when immunised, these chimeras reacted against H-2^b plus minor histocompatibility antigens, or, in our experiments, against H-2^b plus virus only (Table 1). This indicated that MHC-restricted T-cells were specifically selected during T-cell maturation according to the MHC expressed in the thymus. This was formally shown when MHC-restricted T-cell specificity was studied in mice that lacked a thymus and therefore did not have mature T-cells. When these mice of (H-2^k x H-2^b) F_1 type were given a thymus of fetal H-2^k origin, they eventually generated effector T-cells that recognized virus-infected H-2^k, but not infected H-2^b target cells. Surprisingly, thymus deficient nude mice given a histoincompatible thymus generated T-cells that were specific for the nude mice's MHC[76], but in general the MHC (apparently of the radio-resistant part of the thymus) selected the restriction specificity of virus-specific cytotoxic T-cells.

These thymus and bone-marrow-grafting experiments had an immediate impact on clinical medicine by providing rational rules for the reconstitution of immunodeficiency disease. Accordingly, it is not only necessary to deplete T-cells in order to avoid lethal graft-versus-host disease but, in addition, the host and the transplanted bone marrow cells and the host's own or the transplanted thymus grafts must share MHC molecules. Otherwise, T-cells capable of recognizing antigen-plus-MHC molecules on infected epithelial, mesenchymal cells, macrophages or the corresponding B cells would not develop and function appropriately in such reconstituted hosts. These rules for positive selection of T-cells according to the MHC of the thymus were subsequently elegantly–and even more convincingly–confirmed with T-cell receptor transgenic mice, by H. von Boehmer and collaborators[77]. In addition several other groups also analyzed transgenic T-cell receptor expressing mice; D. Loh with an alloreactive T-cell receptor[78] and H. P. Pircher, H. Hengartner, T. Mak and K. Bürki with an LCMV-specific receptor[79]; others used a variety of additional specificities.

New vaccines: The recognition that peptides derived from viruses, bacteria, or classical parasites are presented to T-cells via MHC class I or class II molecules immediately suggested that instead of live, and therefore potentially harmful, infectious agents, peptides could possibly be used as vaccines to induce T-cell responses[80]. This was first formally shown for virus-specific peptides in studies by M. Schulz, P. Aichele, and H. Hengartner[81], (Table 2, Fig. 2) and then by K. Melief and M. Kast and coworkers[82]. The main problem which slowed things down after the discovery of the key role of peptides by A. Townsend, was the fact that the half-life of such peptides is usually short, and protective T-cells could therefore only be induced with the help of adjuvants that guaranteed the relatively slow, long-term release of peptides, triggering T-cells over a prolonged period of time[83]. Peptide treatments were first shown to prevent EAE under various conditions[84,85]. More recently, the capacity of peptides to induce cytotoxic T-cells so exhaustively and completely that they are deleted has been shown by D. Kyburz, P. Aichele, H.P. Pircher and H. Hengartner, and D. Moskophidis[86-89]. T-cells were either induced or exhausted, depending on the relative amount and kinetics of the available antigen within a recipient mouse (Fig. 2). Thus, with too much peptide, specific T-cells could be deleted for as long as the peptide persisted–even permanently, in the case of a thymectomized host. This signalled the possibility of a "negative" vaccination-strategy (Fig. 2, Table 2): Instead of increasing T-cell precursor frequencies to enhance protection (positive vaccination), one could also reduce or delete T-cells by means of excess peptides (negative vaccination)[90]. The latter possibility may allow the immunopathological, disease-causing T-cells to be exhausted and deleted. Although one such example has been documented in an MHC class I-specific immunopathological T-cell-mediated transgenic diabetes model[87], attempts to achieve the same in already primed hosts, before or after initiation of disease, have only met with partial success.

T-cell epitope escape mutant viruses: It could be expected that noncytopathic vi-

Table 2. The role of antigen localisation, dose and time on "positive" and "negative" vaccination.

vaccination	Infection with				
	cytopathic viruses		noncytopathic viruses		
	localized	generalized	localized		generalized
			choriomeningitis	subcutaneous	
none	survival	death	death	disease	disease but survive (virus eliminated)
positive vaccination	protection	protection	protection (virus eliminated)	disease (shorter)	death or protection (virus eliminated)
negative vaccination by "exhaustion"	disease and death dependent upon critical role of cytotoxic T-cells		Treated mice do not develop disease and survive but do not eliminate virus (develop carrier status).		

Summarized from P. Aichele, 1994[87]; M. Schultz, 1991[81]; St. Oehen[88], P. Aichele, 1995[88a], P. Aichele, 1990[114].

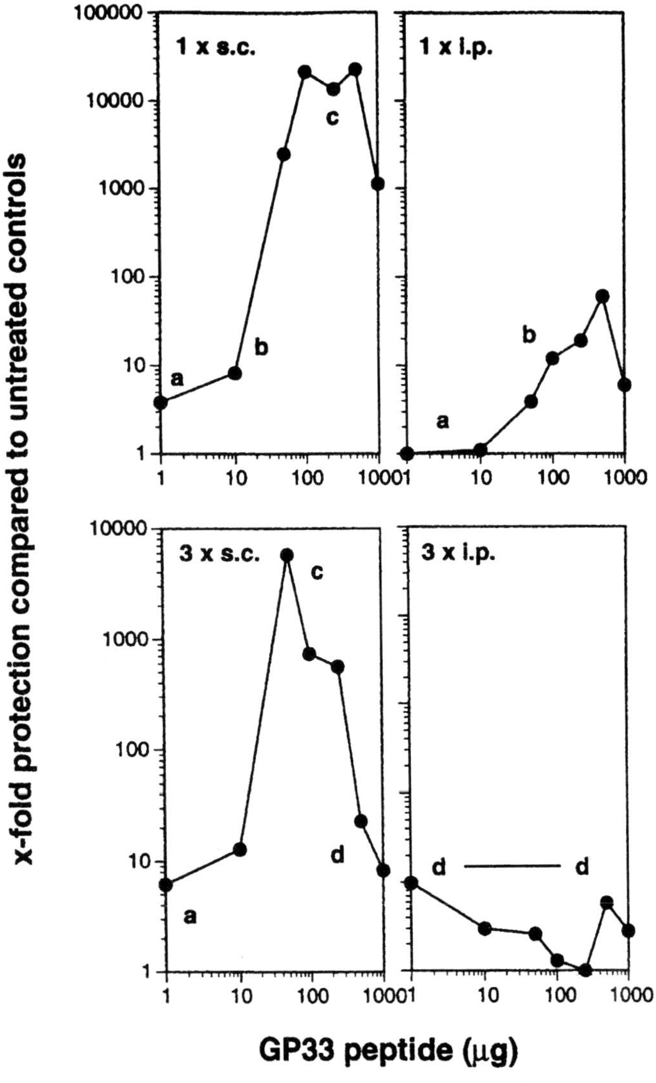

Fig. 2. *The role of antigen localisation and of dose and time on T-cell responses:* The LCMV-glycoprotein peptide GP_{33-41} presented by MHC class I D^b in C57 BL/6 $H\text{-}2^b$ mice was used as vaccine. Its half-life *in vivo* is <<12 hours; the read-out chosen in this experiment to assess T cytotoxic specific, LCMV specific cytotoxic T-cell immunity is the fold protection compared with controls; protection was titrated against amount of peptide with adjuvans given either 1 time or 3 times subcutaneously (s.c.) or 1 time or 3 times intraperitoneally (i.p.). No induction was seen if antigen is available in too small amounts (a), or for too short a period of time (b), Induction of a protective immune T-cell response occurs if sufficient antigen is available for long enough (c). In contrast, too much, excessive, induction of all inducible T-cells by too much antigen spread throughout the organism and available for too long causes exhaustion/deletion of all specific T-cells (d). Modified from Aichele et al., 1995[88a].

ruses would mutate the 9-10 amino acid peptides recognized by T-cells in the context of the crucial MHC class I antigen. Mutation of this peptide, so that either its presentation by MHC molecules or its recognition by T-cells is no longer possible, could help viruses to escape immune surveillance. A first example of this possibility was found by chance when H. P. Pircher, D. Moskophidis and H. Hengartner analyzed T-cell receptor transgenic mice that expressed a T-cell receptor specific for the LCMV glycoprotein peptide 33-41 presented by the MHC class I (D^b) molecule[91,92] (Fig. 3). When we infected such mice in the footpad we found a very early swelling reaction caused by immunopathological cytotoxic T-cell response by day 2 to 4, which, however, waned, and a second CTL-mediated response was measurable after day 8 (Fig. 3). When this unexpected double peak of the T-cell-mediated footpad swelling reaction was analyzed further, it became clear that the virus had mutated by day 6 of infection; it no longer expressed the original gp 34-41, but exhibited various mutations within this epitope, presented by the MHC class I (D^b) molecule[91]. Apparently the vehement virtually monoclonal antiviral $CD8^+$ T-cell response in the T-cell-receptor-transgenic mouse had quickly selected the T-cell-epitope-mutant virus that had escaped the transgenic T-cells. A similar mutant virus that escaped the T-cell response has subsequently also been found in patients infected with HIV by R. Phillips, A. McMichael and coworkers[93] and in HBV-infected patients by Bertoletti *et al.*[94].

MHC-disease associations: The linkage between some disease susceptibilities and certain HLA-types was one of the first findings signalling the important role of MHC-molecules in immunity (reviewed in[95,96]). These diseases are often of autoimmune or immunopathological nature and are often linked to HLA class I, rather than class II, molecules[97]. The critical role of modified MHC molecules in T-cell recognition explained why different allelic forms of

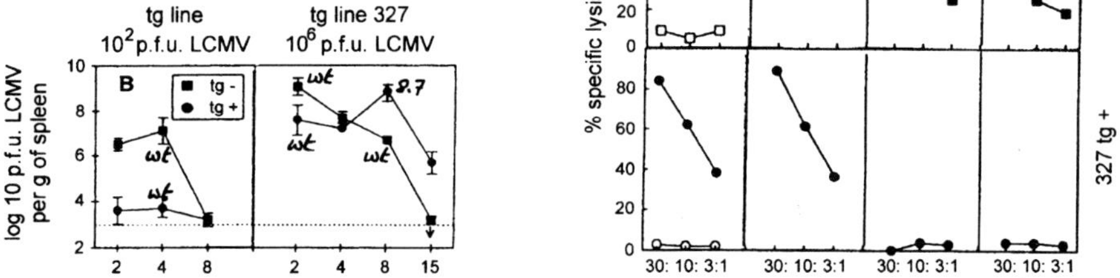

Fig. 3. *Selection of cytotoxic-T-cell-epitope escape mutants in T-cell receptors of transgenic (TCR-tg327) mice.* LCMV-WE wild-type (wt) was injected into the footpad (A). The control. C57BL/6 developed the expected $CD8^+$ T-cell-dependent footpad swelling by day 6–7, the TCR-tg327 mice exhibited an early swelling caused by the tg-TCR specific for LCMV-GP_{33-41} plus D^b. This early response selected a mutant virus 8.7 by about day 7–8, that induced a swelling reaction by the endogenous non-tg T-cells specific for other peptides. The virus titers found in the spleen of mice infected with a low (10^2 pfu) or a high (10^6 pfu) dose of LCMV-WE is shown in (B). All mice were infected with LCMV wt; a variant virus 8.7 was selected in tg line327 mice infected with 10^6 pfu of LCMV wt by day 8 (B). The cytotoxic T-cell activity found in control C57BL/6 or TCR-tg327 mice on day 8 is shown in C. The peptides of the T-cell-epitope escape variant virus are not recognized by the tg-TCR, but are still partially (Tyr→Phe) or not (Val→Leu) recognized by effector T-cells from control C57BL/6 mice. Summarized from Pircher et al., 1990[91].

MHC are randomly distributed in the population and made it likely that infectious agents and their peptides are presented by at least one of the four to ten MHC molecules expressed by an individual; this minimizes the possibility that a virus escapes immune surveillance, endangering survival of the entire population. The fact that some differences with respect to antigenicity and immunogenicity are linked to the MHC, and correlate directly with different strengths of the T-cell responses, shows that different MHC molecules directly determine and regulate resistance to diseases.

Obviously, cytopathic viruses must be controlled efficiently by the immune system, otherwise the host species dies. Therefore, hosts with non-responder or non-presenting MHC-molecules have probably been eliminated long ago by natural selection, leaving only high responders to survive. In contrast, for noncytopathic viruses, diseases are not caused by the infectious agent itself, but rather by the damaging effect of protective T-cell responses. Because such agents do not directly cause disease, they do not exert a direct selective pressure on survival. But since these viruses may induce immunopathological cytotoxic T-cell responses, differences in MHC may influence the severity of the disease. In fact, many of the diseases exhibiting some association with MHC have an aura of being autoimmune or immunopathologically mediated. This has been directly shown for known noncytopathic viruses[97,98] (Table 3).

Table 3. MHC-disease association reflecting $CD8^+$ T-cells mediated immunopathology against host cells infected with noncytopathic LCMV.

Mouse strain	A. Maximal increase of serum transaminases after i.v. infection with LCMV (pfu): T-cell mediated hepatitis				B. Mortality after intracerebral infection with LCMV	
	WE_{UBC} A		WE_{UBC} D		WE_{UBC} A	WE_{UBC} D
	10^2	10^6	10^2	10^6		
BI0.G H-$2D^qL^q$	–	–	–	++	100 %	100 %
BI0.BR H-$2D^k$	–	–	++	±	100 %	0 %

Mice were infected i.v. with the indicated doses of LCMV WE_{UBC} A or D. Maximal increases of serum transaminases were indicated with –: < 3 fold, +: 10-20 fold, ++: > 100 fold. Damage of liver cells or of choriomeningeal or ependymal cells caused by LCMV-specific $CD8^+$ T-cells. In absence of $CD8^+$ T- no immunopathology is seen.

C. Role of virus characteristics increasing likelihood to establish persistent LCMV-infection in mice

Virus parameters:	Increased replication rate, increased resistence to interferons, loss of T-cell epitopes by mutation, increased tropism for lymphohemopoietic cells.
Host parameters:	Decrease in relative interferon levels, loss of presenting MHC class I molecules (e.g BALB/c dm2 mice), loss of $CD8^+$ T-cells (e.g. DBA/2 mice), loss of T-help, lack of IL-2.

Summarized from R. M. Zinkernagel, 1985[98]; T. P. Leist, 1989[100]; D. Moskophidis, 1994[101].

As pointed out above, one of the motivations for testing various mouse strains for cytotoxic T-cell activity, resulting in the discovery of MHC-restricted T-cell recognition, was the variable and weak evidence of susceptibility differences of mice to lethal choriomeningitis that somehow correlated with MHC (Table 4). The comparison of a slowly replicating neurotropic strain of LCMV (UBC-A), with a rapidly replicating viscerotropic isolate UBC-D characterized by Ch. Pfau[99], revealed a dramatic and strict correlation with MHC when the different diseases were studied. After intracerebral infection of mice, UBC-A virus caused death in all recipients, irrespective of MHC. UBC-D caused immunopathological disease in all those possessing the MHC class I H-2D^qL^q allele but not in the others[98,100]. Furthermore, the susceptibility of mice to becoming persistently infected virus-carriers could be linked by D. Moskophidis to the absence of class I MHC molecules that were able to present the critically important dominant immunogenic viral peptide[101,102]. These studies, in addition, demonstrated that virus dose and virus strain (including T-cell epitope variants) also played an important role in the overall virus-host balance and in MHC-disease association.

Table 4. Transmission of protective immunity from mother to offspring.

Because of MHC-restriction of T-cell recognition, T-cell maturation in offspring is slow and starts after birth to prevent graft-versus host disease. During this period of physiological immuno incompetence transfer of maternal antibody-memory is essential for protection of offspring against many infections.

	preexistent antibodies	"activated" T-cells
placenta	maternal antibodies transmissible except with chorio-epithelial double layered placenta (bovines)*	non-transmissible, danger of host-versus-graft disease and rejection
milk	transmissible via gut of offspring (temporarly) passive protection in gut*	evidence poorly analyzed
function	*"altruistic"*: protection of *offspring* against infectious diseases during the period of T-cell immunodeficiency and protect mothers during pregnancy	*egoistic:* protection of *original host* against spread of non-cytopathic infectious agents (controls immunopathology and tumor cells within host)

*Summarized in Brambell[106a].

Thus MHC (in humans HLA-)–disease association may not be found readily for acute cytopathic viral or bacterial infections, because natural selection has favoured good MHC-presenter-virus combinations[90]. In contrast, for non-cytopathic viruses, the selection pressures are much weaker and various virus-host immune response balances resulting in more or less immunopathological disease are acceptable for species survival. Because cytotoxic and protective T-cell responses directly depend upon class I MHC presentation of peptides, susceptibility to consequential disease may correlate directly with the MHC class I allele, depending upon the localisation of infection (choriomeningitis vs. establishment of a virus carrier status)[98,101].

One may therefore extrapolate that at least some autoimmune diseases may eventually turn out to be caused by immunopathological T-cell responses against viruses that are poorly cytopathic or noncytopathic[95, 96] (Fig. 3, 4). Hepatitis B, C, or D virus infections or possibly immunodeficiency virus (HIV) infections in man may represent this type of chronic infection leading to a protective but immunopathological T-cell response, causing MHC-regulated immunopathologic disease. Alternatively, infectious agents that are either unknown or not yet recognized may be involved initially in triggering the disease by immunopathological or autoimmune mechanisms. For example, it is suspected that also common cytopathic viruses may be involved in the pathogenesis of some autoimmunity diseases (Fig. 4, 5). Infections of MHC class-I-positive epithelial or neuroendocrine cells should be controlled primarily via $CD8^+$ T-cells; their efficiency will, therefore, not only determine the extent and kinetics of destruction of infected host cells, but also whether an autoimmune T-cell and B cell response against such sequestered self-antigens is induced.

The reason for MHC class I linkage of autoimmune diseases is, therefore, probably in some diseases that the autoimmune effector T-cells are class I restricted (Fig. 4), and in other situations that modulations of common or unrecognized infections by the efficiency of class-I-restricted effector T-cells may indirectly regulate autoantibody responses (Fig. 5).

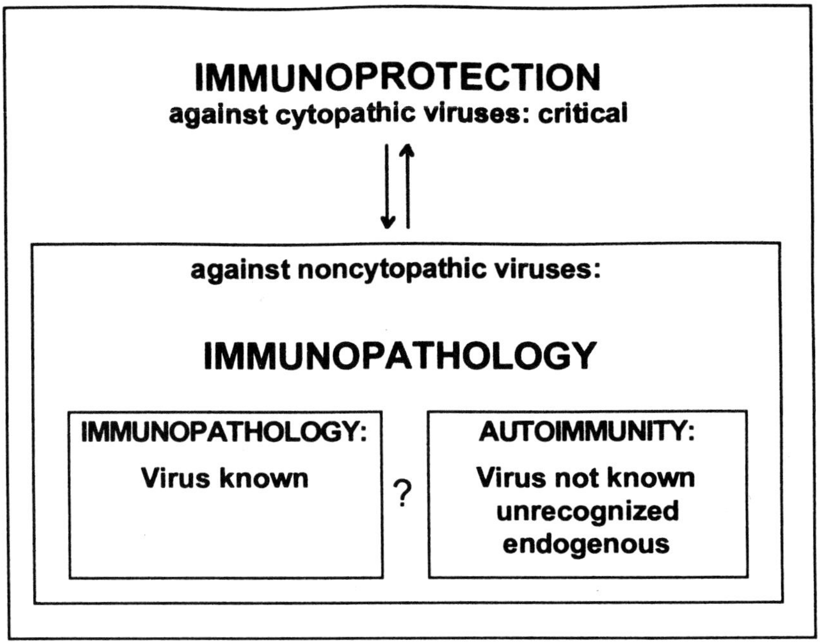

Fig. 4. *Immunoprotection and immunopathology:* Efficient immunoprotection is mandatory for survival of the host infected with a cytopathic virus. Since noncytopathic viruses do not cause cell damage directly, pathology is caused by immune mechanisms. If the virus is not recognized or is unknown, the disease may be mistaken as an autoimmune disease.

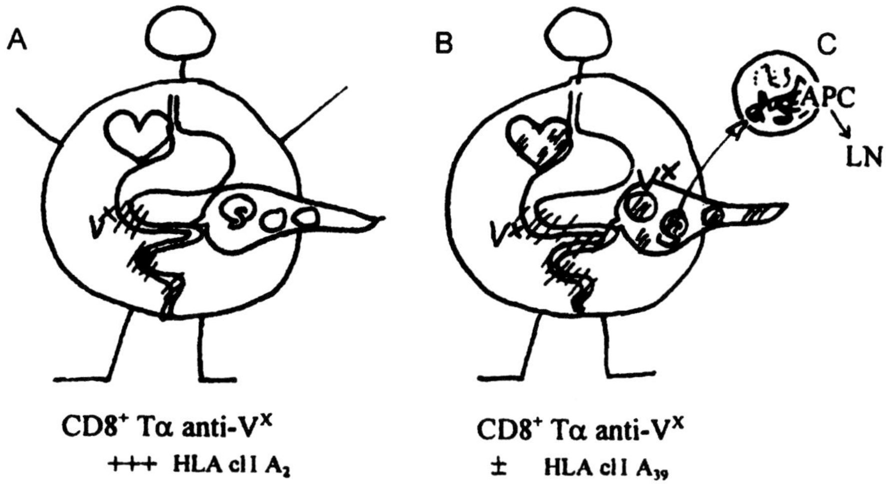

Fig. 5. *Autoimmunity induced by viruses.* If a cytopathic or noncytopathic virus is stopped early enough by MHC-class I-(HLA-cl I A_2)-restricted effector T-cells (A), sensitive target cells are not destroyed. If the protective HLA-cl I A_{39}-restricted T-cell response is slow (B), virus may reach sensitive target organs and cells that may be destroyed directly or induce disease by T-cell-mediated immunopathology, i.e. cause release of self (S)-antigen. The self-antigen may then be picked up and presented by APC (C) and in local lymph node or in the spleen induce autoimmune T-cells and/or autoantibodies. Autoantibody production needs T-help that may be provided by newly induced $CD4^+$ T-cells specific either for viral antigens that are linked to self-antigens or are specific for so far segregated immunologically ignored self-antigens.

Consequences for immunological memory: The discovery that T-cell specificity monitored MHC molecules, that it was positively selected in the thymus, and that this explained the transplantation reaction, offered a basis for understanding the role of protective immunological memory. Immunological memory by B and T-cells is an important hallmark of immunity and has been exploited successfully by vaccinations (reviewed in[103-106]). However, before vaccinations were used, the idea of immunological memory had two raisons d'être. One could argue that if a host dies during the primary infection, he does not need an immunological memory. Also, if he survives the primary infection, he does not need memory because the system has proven itself efficient. What then is the function of immunological memory? The first and most critical one is via transfer of immune antibodies of the mother to protect offspring during the phase of maturation of the immune system after birth, the second is to protect mothers from infections during pregnancy (reviewed in[106a, b]). I will only discuss the first aspect further here.

Calves are born without antibodies because, as for all vertebrates, the immune system is not yet mature enough to produce its own antibody response[106a] (Table 4). In addition, in calves, maternal antibodies cannot be transmitted because of the complete doubly layered placenta. All protective antibodies are transmitted via colostral milk from the mother within 24 hours of birth. If this does not happen, the calf dies within a few weeks as a result of common bacterial infections.

Why then, are newborn vertebrates immunoincompetent? (Table 4). My explanation is, that because of MHC-restricted T-cell recognition, T-cells of the fetus should not mature, so as not to endanger pregnancy, or cause graft-versus-host disease against maternal MHC molecules[90,106]. The passively acquired antibodies provide protection during the critical period after birth, during which T-cell and B-cell maturation slowly takes place over 3–6 months in humans, or four weeks in mice. Since maternal antibodies must cover a wide spectrum of relevant infectious agents that may not be encountered during pregnancy, immunological memory carried by antibodies is mandatory for at least two reasons. Firstly, pregnancy is relatively short compared to the time taken for the host to reach sexual maturity, and therefore antibody memory must be developed over a long period. Secondly, the relevant infections must not occur during pregnancy, because many infections cause damage to and abortion of the foetus.

Evidence has been found in various models and infections that persistence of antigens both, in antibody-antigen complexes and exposed on follicular dendritic cells, maintains antibody responses for many (Table 5) years, to guarantee levels of transmissible protection[107–109].

Table 5. "Memory B-cells" are not protective whereas preexistent neutralizing levels of antibodies protect naive recipients against cytopathic infectious agents.

Adoptive transfer to naive mice	Booster immunization 2 d before challenge infections	Result of challenge infection
B-cells, naive or normal	no	disease
	yes	disease
"primed" or "memory" B-cells	no	disease
	yes	disease
immune serum with neutralizing antibodies	no	protection
	yes	protection

Summarized from experiments performed in mice with vesicular stomatitis virus (VSV) in refs.: U. Steinhoff, 1995[108]; M. F. Bachmann, 1994[115].

What then is the role of T-cell memory?

Memory T-cells are probably necessary to maintain memory B cells but this is not yet certain. Although increased precursor frequencies of T-cells provide some protection, which should suffice to protect mothers during pregnancy, there is a second very important aspect of T-cell memory. T-cells are critical for controlling persisting noncytopathic viruses that hide in peripheral epithelial or mesenchymal cells and prevent them from reemerging to trigger T-cell mediated immunopathological disease (Table 4). There is good evidence available now that these protective T-cells are also maintained by antigens[106,110,111], although this simple notion is very controversial at the present time[102,112,113]. Thus, low levels of ongoing T-cell responses reflect antigen-driven activated effector T-cells and not special memory T-cells. These serve to protect the host itself and not progeny; these *egoistic* protective memory T-cells cannot be transmitted, due to MHC-differences between mother and offspring (Table 4). Taken together, the summarized evidence suggests that because of MHC-restricted T-cell recognition which causes the transplantation reaction, prolonged physiological immunodeficiency of the offspring is necessary; protection during the period of maturation of the immune system is provided by passively transferred *altruistic* antibodies, which necessitate a mandatory antigen-driven elevated antibody level in mothers (Table 4).

CONCLUSIONS

In the complex balance between viruses and hosts T-cells play an important role that varies with different host and viral characteristics. The unexpected finding, by chance and necessity, of MHC-restricted T-cell recognition, in a field of biological science prepared by immunologists, geneticists and virologists, has triggered a great number of subsequent important studies by many other groups, resulting in an excellent molecular understanding of T-cell recognition of virally infected target cells. These combined findings have helped to understand better both immunological specificity and immunological memory. They have furthered our understanding of disease pathogenesis and made possible applications of the acquired knowledge for improving protective immunity and for diminishing immunopathological T-cell responses.

ACKNOWLEDGEMENTS

Looking back over the past 28 years of my professional life since graduating from medical school and marrying my wife Kathrin, I must admit life has been very good to me. I was given the chance to work and spend time with many excellent scientists and good friends as well as many eager young students and postdocs. I am part of a caring family, I have the luck to discover new things every once in a while, while I am paid a salary for enjoying researching (in the true sense of the word) the secrets of nature.

I thank my families, my mentors, and my collaborators for all these years. I would also like to thank the universities of Basel, Zürich and Lausanne, and especially the John Curtin School of Medical Research, Professor G. Ada and R. V. Blanden. I am also greatly indebted to the Schweizerische Stiftung für Biologisch-medizinische Stipendien, the Australian National University in Canberra, the SCRIPPS Clinic and Research Foundation in La Jolla, CA, the NIH, the Canton of Zürich, the Swiss Federal Government, the Swiss National Science Foundation, the University of Zürich, and the Fondation Jeantet for their generous financial support.

BIBLIOGRAPHY

1. Landsteiner, K. *The specificity of serological reactions* (Harvard University Press, Boston, 1945).
2. Klein, J. *Biology of the mouse histocompatibility-2 complex* (Springer Ltd. 1975).
3. Dausset, J. *Acta Haematol. (Basel)* **20**, 156–166 (1958).
4. Hotchin, J. *Cold Spring Harbor Symp. Quant. Biol.* **27**, 479–499 (1962).
5. Lehmann-Grube, F. *Virol. Monogr.* **10**, 1–173 (1971).

6. Buchmeier, M. J., Welsh, R. M., Dutko, F. J. & Oldstone, M. B. A. *Adv. Immunol.* **30**, 275–331 (1980).
7. Brunner, K. T., Mauel, J., Cerottini, J. C. & Chapuis, B. *Immunology* **14**, 181–196 (1968).
8. Cerottini, J. C., Nordin, A. A. & Brunner, K. T. *Nature* **228**, 1308–1309 (1970).
9. Gardner, I., Bowern, N. A. & Blanden, R. V. *Eur. J. Immunol.* **4**, 63–67 (1974).
10. Marker, O. & Volkert, M. *J. Exp. Med.* **137**, 1511–1525 (1973).
11. Cole, G. A. Nathanson, N, f. Prendergast, R. A. *Nature* **238;** 335–337 (1972).
12. Zinkernagel, R. M. & Doherty, P. C. *J. Exp. Med.* **138**, 1266–1269 (1973).
13. Oldstone, M. B., Dixon, F. J., Mitchell, G. F. & McDevitt, H. O. *J Exp. Med.* **137**, 1201–1212 (1973).
14. Blanden, R. V. *J. Exp. Med.* **133**, 1074–1089 (1971).
15. Zinkernagel, R. M. & Doherty, P. C. *Nature* **248**, 701–702 (1974).
16. Mackaness, G. B. *J. Exp. Med.* **116**, 381–406 (1962).
17. Shearer, G. M. *Eur. J. Immunol.* **4**, 527–533 (1974).
18. Gorer, P. A. *Brit. J. Exp. Pathol.* **17**, 42–50 (1936).
19. Snell, G. D. *J.Genet.* **49**, 87–108 (1948).
20. Van Rood, J. J. & Van Leeuwen, A. *J. Clin. Invest.* **42**, 1382–1390 (1963).
21. Levine, B. B., Ojeda, A. P. & Benacerraf, B. *J. Exp. Med.* **118**, 953–957 (1963).
22. McDevitt, H. O. & Sela, M. *J. Exp. Med.* **122**, 517–531 (1965).
23. McDevitt, H. O. & Chinitz, A. *Science* **163**, 1207–1210 (1969).
24. Lilly, F., Boyse, E. A. & Old, L. J. *Lancet* **2**, 1207–1209 (1964).
25. McDevitt, H. O., Deak, B. D., Shreffler, D. C., Klein, J., Stimpfling, J. H. & Snell, G. D. *J. Exp. Med.* **135**, 1259–1278 (1972).
26. Burnet, F. M. *Nature* **245**, 359–361 (1972).
27. Bodmer, W. F. *Nature* **237**, 139–145 (1972).
28. Benacerraf, B. & McDevitt, H. O. *Science* **175**, 273–279 (1972).
29. Amos, D. B., Bodmer, W. F., Ceppellini, R., Condliffe, P. G., Dausset, J. & Fahey, J. L. *Fed. Proc.* **31**, 1087–1104 (1972).
30. Jerne, N. K. *Eur. J. Immunol.* **1**, 1–9 (1971).
31. Lawrence, H. W. *Physiol. Rev.* **39**, 811–859 (1959).
32. Bailey, D. W., Snell, D. G. & Cherry, M. *Proceedings of the Symposium on Immunogenetics of the H-2 System. S. Karger AG. , Basel, Switzerland* **55**, (1971).
33. Fathman, C. G. & Hengartner, H. *Nature* **272**, 617–618 (1978).
34. von Boehmer, H., Hengartner, H., Nabholz, M., Lernhardt, W., Schreier, M. H. & Haas, W. *Eur. J Immunol.* **9**, 592–597 (1979).
35. Kappler, J. W., Skidmore, B., White, J. & Marrack, P. *J Exp. Med.* **153**, 1198–1214 (1981).
36. Kappler, J., Kubo, R., Haskins, K., et al. *Cell* **35**, 295–302 (1983).
37. McIntyre, B. W. & Allison, J. P. *Cell* **34**, 739–746 (1983).
38. Hedrick, S. M., Cohen, D. I., Nielsen, E. A. & Davis, M. M. *Nature* **308**, 149–153 (1984).
39. Yanagi, Y., Yoshikai, Y., Leggett, K., Clark, S. P., Aleksander, I. & Mak, T. W. *Nature* **308**, 145–149 (1984).
40. Leclerc, J. C., Gomard, E., Plata, F. & Levy, J. P. *Int. J. Cancer* **11**, 426–432 (1973).
41. Lavrin, D. H., Herbermann, R. B., Nunn, M. & Soares, N. *J. Natl. Cancer Inst.* **51**, 1497–1508 (1973).
41.a Cole, G. A. Prendergast, R. A. & Henney, C. S. *Fed.Proc.* **32**, 964 (1973a).
42. Kindred, B. & Shreffler, D. C. *J. Immunol.* **109**, 940–943 (1972).
43. McCullagh, P. J. *J. Exp. Med.* **132**, 916–925 (1970).
44. Katz, D. H., Hamaoka, T. & Benacerraf, B. *J. Exp. Med.* **137**, 1405–1418 (1973).
45. Rosenthal, A. S. & Shevach, E. M. *J. Exp. Med.* **138**, 1194–1212 (1973).
46. Blanden, R. V., Doherty, P. C., Dunlop, M. B., Gardner, I. D., Zinkernagel, R. M. & David, C. S. *Nature* **254**, 269–270 (1975).
47. Koszinowski, U. & Ertl, H. *Nature* **255**, 552–554 (1975).

48. Gordon, R. D., Simpson, E. & Samelson, L. E. *J. Exp. Med.* **142**, 1108–1120 (1975).
49. von Boehmer, H., Haas, W. & Jerne, N. K. *Proc. Natl. Acad. Sci. U. S. A.* **75**, 2439–2442 (1978).
50. Bevan, M. J. *Nature* **256**, 419–421 (1975).
51. McMichael, A. J., Ting, A., Zweerink, H. J. & Askonas, B. A. *Nature* **270**, 524–526 (1977).
52. Goulmy, E., Termijtelen, A., Bradley, B. A. & Van Rood, J. J. *Nature* **266**, 544–545 (1977).
53. Lafferty, K. J. & Cunningham, A. J. *Aust. J. Exp. Biol. Med. Sci.* **53**, 27–42 (1975).
54. Bretscher, P. & Cohn, M. *Science* **169**, 1042–1049 (1970).
55. Cohn, M. & Langman, R. E. *Immunol. Rev.* **115**, 11–147 (1990).
56. Zinkernagel, R. M. & Doherty, P. C. *Nature* **251**, 547–548 (1974).
57. Doherty, P. C. & Zinkernagel, R. M. *J. Immunol.* **114**, 30–33 (1975).
58. Gardner, I. D., Bowern, N. A. & Blanden, R. V. *Eur. J. Immunol.* **5**, 122–127 (1975).
59. Zinkernagel, R. M. *Nature* **251**, 230–233 (1974).
60. Doherty, P. C. & Zinkernagel, R. M. *Lancet* **1**, 1406–1409 (1975).
61. Babbitt, B. P., Allen, P. M., Matsueda, G., Haber, E. & Unanue, E. R. *Nature* **317**, 359–360 (1985).
62. Buus, S., Colon, S., Smith, C., Freed, J. H., Miles, C. & Grey, H. M. *Proc. Natl. Acad. Sci. U. S. A.* **83**, 3968–3971 (1986).
63. Townsend, A. R. M., Rothbard, J., Gotch, F. M., Bahadur, G., Wraith, D. & McMichael, A. J. *Cell* **44**, 959–968 (1986).
64. Maryanski, J. L., Pala, P., Corradin, G., Jordan, B. R. & Cerottini, J. C. *Nature* **324**, 578–579 (1986).
65. Rotzschke, O., Falk, K., Deres, K., et al. *Nature* **348**, 252–254 (1990).
66. Lurquin, C., van Pel, A., Mariame, B., et al. *Cell* **58**, 293–303 (1989).
67. Bjorkman, P. J., Saper, M. A., Samraoui, B., Bennett, W. S., Strominger, J. L. & Wiley, D. C. *Nature* **329**, 506–511 (1987).
68. Bjorkman, P. L., Saper, M. A., Samraoui, B., Bennet, W. S., Strominger, J. L. & Wiley, D. C. *Nature* **329**, 512–518 (1987).
69. Garcia, K. C., Degano, M., Stanfield, R. L., et al. *Science* **274**, 209–219 (1996).
70. Garboczi, D. N., Ghosh, P., Utz, U., Fan, Q. R., Biddison, W. E. & Wiley, D. C. *Nature* **384**, 134–141 (1996).
71. Jorgensen, J. L., Esser, U., de St.Groth, B. F., Reay, P. A. & Davis, M. M. *Nature* **355**, 224–230 (1992).
72. Sant'Angelo, D. B., Waterbury, G., Preston-Hurlburt, P., et al. *Immunity* **4**, 367–376 (1996).
73. Bevan, M. J. *Nature* **269**, 417–418 (1977).
74. Zinkernagel, R. M., Callahan, G. N., Klein, J. & Dennert, G. *Nature* **271**, 251–253 (1978).
75. Zinkernagel, R. M., Callahan, G. N., Althage, A., Cooper, S., Klein, P. A. & Klein, J. *J. Exp. Med.* **147**, 882–896 (1978).
76. Zinkernagel, R. M., Althage, A., Waterfield, E., et al. *J. Exp. Med.* **151**, 376–399 (1980).
77. Kisielow, P., Teh, H. S., Bluthmann, H. & von Boehmer, H. *Nature* **335**, 730–733 (1988).
78. Sha, W. C., Nelson, C. A., Newberry, R. D., Kranz, D. M., Russell, J. H. & Loh, D. Y. *Nature* **336**, 73–76 (1988).
79. Pircher, H., Mak, T. W., Lang, R., et al. *EMBO J* **8**, 719–727 (1989).
80. Deres, K., Schild, H., Wiesmüller, K. H., Jung, G. & Rammensee, H. G. *Nature* **342**, 561–564 (1989).
81. Schultz, M., Zinkernagel, R. M. & Hengartner, H. *Proc. Natl. Acad. Sci. U. S. A.* **88**, 991–993 (1991).
82. Kast, W. M., Roux, L., Curren, J., *et al. Proc. Natl. Acad. Sci. U. S. A.* **88**, 2283–2287 (1991).

83. Widmann, C., Maryanski, J. L., Romero, P. & Corradin, G. *J. Immunol.* **147**, 3745–3751 (1991).
84. Clayton, J. P., Gammon, G. M., Ando, D. G., Kono, D. H., Hood, L. & Sercarz, E. E. *J. Exp. Med.* **169**, 1681–1691 (1989).
85. Gaur, A., Wiers, B., Liu, A., Rothbard, J. & Fathman, C. G. *Science* **258**, 1491–1494 (1992).
86. Kyburz, D., Aichele, P., Speiser, D. E., Hengartner, H., Zinkernagel, R. M. & Pircher, H. *Eur. J. Immunol.* **23**, 1956–1962 (1993).
87. Aichele, P., Kyburz, D., Ohashi, P. S., et al. *Proc. Natl. Acad. Sci. U. S. A.* **91**, 444–448 (1994).
88. Oehen, St, Hengartner, H. & Zinkernagel, R. M. *Science* **251**, 195–198 (1991).
88a. Aichele, P., Brduscha Riem, K., Zinkernagel, R. M., Hengartner, H. & Pircher, H. P. *J. Exp. Med.* **182**, 261–266 (1995).
89. Moskophidis, D., Lechner, F., Pircher, H. & Zinkernagel, R. M. *Nature* **362**, 758–761 (1993).
90. Zinkernagel, R. M. *Science* **271**, 173–178 (1996).
91. Pircher, H., Moskophidis, D., Rohrer, U., Bürki, K., Hengartner, H. & Zinkernagel, R. M. *Nature* **346**, 629–633 (1990).
92. Aebischer, T., Moskophidis, D., Rohrer, U. H., Zinkernagel, R. M. & Hengartner, H. *Proc. Natl. Acad. Sci. U. S. A.* **88**, 11047–11051 (1991).
93. Phillips, R. E., Rowland-Jones, S., Nixon, D. F., Gotch, F. M., Edwards, J. P., Ogunlesi, A. O. & McMichael, A. J. *Nature* **354**, 453–459 (1991).
94. Bertoletti, A., Sette, A., Chisari, F. V., Penna, A., Levrero, M. & De Carli, M. *Nature* **369**, 407–410 (1994).
95. Dausset, J. & Svejgaard, A. *HLA and disease* (1977). Munksgaard, Copenhagen.
96. Möller, E., Böhme, J., Valugerdi, M.A., Ridderstad, A. & Olerup, O. *Imm. Rev.* **118,** 5–19 (1990).
97. Zinkernagel, R. M. & Doherty, P. C. *Adv. Immunol.* **27**, 51–177 (1979).
98. Zinkernagel, R. M., Pfau, C. J., Hengartner, H. & Althage, A. *Nature* **316**, 814–817 (1985).
99. Pfau, C. J., Valenti, J. K., Pevear, D. C. & Hunt, K. D. *J. Exp. Med.* **156**, 79–89 (1982).
100. Leist, T. P., Althage, A., Haenseler, E., Hengartner, H. & Zinkernagel, R. M. *J. Exp. Med.* **170**, 269–277 (1989).
101. Moskophidis, D., Lechner, F., Hengartner, H. & Zinkernagel, R. M. *J. Immunol.* **152**, 4976–4983 (1994).
102. Moskophidis, D., Battegay, M., van den Broek, M. F., Laine, E., Hoffmann Rohrer, U. & Zinkernagel, R. M. *J. Gen. Virol.* **76**, 381–391 (1995).
103. Gray, D. & Leanderson, T. *Curr. Top. Microbiol. Immunol.* **159**, 1–17 (1990).
104. Jamieson, B. D. & Ahmed, R. *J. Exp. Med.* **169**, 1993–2005 (1989).
105. Beverley, P. C. *Immunol. Today* **11**, 203–205 (1990).
106. Zinkernagel, R. M., Bachmann, M. F., Kündig, T. M., Oehen, S., Pircher, H. & Hengartner, H. *Annu. Rev Immunol.* **14**, 333–367 (1996).
106a. Brambell, R.W.R. The Transmission of Immunity from Mother to Young, Amsterdam, North Holland (1970).
106b. Möller, G. Immunology of Feto-Maternal Relationship *Immunol. Rev.* **75**, 5–230 (1983).
107. Bachmann, M. F., Odermatt, B., Hengartner, H. & Zinkernagel, R. M. *J. Exp. Med.* **183**, 2259–2269 (1996).
108. Steinhoff, U., Müller, U., Schertler, A., Hengartner, H., Aguet, M. & Zinkernagel, R. M. *J Virol.* **69**, 2153–2158 (1995).
109. Gray, D. & Skarvall, H. *Nature* **336**, 70–73 (1988).
110. Bachmann, M. F., Kündig, T. M., Hengartner, H. & Zinkernagel, R. M. *Proc. Natl. Acad. Sci. U. S. A.* **94,** 640–645 (1997).
111. Kündig, T. M., Bachmann, M. F., Ohashi, P. S., Pircher, H., Hengartner, H. & Zinkernagel, R. M. *Immunol. Rev.* **150**, 63–90 (1996).

112. Hou, S., Hyland, L., Ryan, K. W., Portner, A. & Doherty, P. C. *Nature* **369**, 652–654 (1994).
113. Lau, L. L., Jamieson, B. D., Somasundaram, T. & Ahmed, R. *Nature* **369**, 648–652 (1994).
114. Aichele, P., Hengartner, H., Zinkernagel, R. M. & Schultz, M. *J. Exp. Med.* **171**, 1815–1820 (1990).
115. Bachmann, M. F., Kündig, T. M., Odermatt, B., Hengartner, H. & Zinkernagel, R. M. *J. Immunol.* **153**, 3386–3397 (1994).

Erratum

Page 4, line 23 should read as:

responses towards invading microorganisms or against cancer metastases. Our

Physiology or Medicine 1997

STANLEY B. PRUSINER

"for his discovery of prions — a new biological principle of infection"

THE NOBEL PRIZE IN PHYSIOLOGY OR MEDICINE

Speech by Professor Ralf F. Pettersson of the Nobel Committee at the Karolinska Institute.
Translation of the Swedish text.

Your Majesties, Your Royal Highnesses, Ladies and Gentlemen,

This year's Nobel Prize in Physiology or Medicine has been awarded to Stanley B. Prusiner for his discovery of prions–a new biological principle of infection. What is a prion? It is a small infectious protein capable of causing fatal dementia-like diseases in man and animals. It has been known for approximately a century that infectious diseases can be caused by bacteria, viruses, fungi and parasites. All these infectious agents possess a genome, the hereditary material that provides the basis for their replication. The ability to replicate is essential for the manifestation of the diseases they cause. The most remarkable feature of prions is that they are able to replicate themselves without possessing a genome; prions lack hereditary material. Until prions were discovered, duplication without a genome was considered impossible. This discovery was unexpected and provoked controversy.

Although the existence of prions was not known until the work of Stanley Prusiner, many prion diseases have been previously documented. On Iceland, scrapie, a disease affecting sheep was first described in the 18th century. In the 1920s, the neurologists Hans Creutzfeldt and Alfons Jakob discovered a similar disease in man. During the 1950s and 60s Carleton Gajdusek studied kuru, a disease that was spread through cannibalistic rituals practised by the Fore people in New Guinea. Presently attention is focused on mad cow disease, which has affected approximately 170,000 cows in Britain. These diseases exhibit common pathologies. They are inevitably fatal due to the destruction of the brains of infected individuals. The incubation times may last for several years, during which the affected regions of the brain become gradually spongy in appearance. Gajdusek discovered that kuru and Creutzfeldt-Jakob disease could be transmitted to monkeys demonstrating that these diseases are contagious. In 1976, when Gajdusek received his Nobel Prize, the nature of the infectious agent was completely unknown. At this time, these diseases were assumed to be caused by a new unidentified virus, termed a slow or unconventional virus. During the 1970s, no significant advances regarding the nature of the agent were made, that is, not until Stanley Prusiner took on the problem.

Prusiner set out to purify the infectious agent, and after 10 years of hard work he obtained a pure preparation. To his great surprise, he found that the agent consisted only of a protein, which he named prion, a term derived from proteinaceous infectious particle. Strangely enough, he found that the protein was present in equal amounts in the brains of both diseased and healthy individuals. This discovery was confusing and it was generally concluded that Prusiner must have arrived at the wrong conclusion. How could a protein cause disease if it was present both in diseased and healthy brains? The answer to this question came when Prusiner showed that the prion protein from diseased brains had a completely different three-dimensional conformation. This led Prusiner to propose a hypothesis for how a normal protein could become a disease-causing agent by changing its conformation. The process he proposed may be compared to the transformation of Dr Jekyll to Mr Hyde–the same entity, but in two manifestations, a kind innocuous one, and a vicious lethal one. But how can a protein replicate without a genome? Stanley Prusiner suggested that the harmful prion protein could replicate by forcing the normal protein to adopt the shape of the harmful protein in a chain reaction-like process. In other words, when a harmful protein encounters a normal protein, the normal protein is converted into the harmful form. A remarkable feature of prion diseases is that they can arise in three different ways. They can occur spontaneously, or be triggered by infection, or occur as a consequence of hereditary predisposition.

The hypothesis that prions are able to replicate without a genome and to cause disease violated all conventional conceptions and during the 1980s was severely criticised. For more than 10 years, Stanley Prusiner fought an uneven battle against overwhelming opposition. Research during the 1990s has, however, rendered strong support for the correctness of Prusiner's prion hypothesis. The mystery behind scrapie, kuru, and mad cow disease has finally been unravelled. Additionally, the discovery of prions has opened up new avenues to better understand the pathogenesis of other more common dementias, such as Alzheimer's disease.

Stanley Prusiner,

Your discovery of the prions has established a novel principle of infection and opened up a new and exciting area in medical research. On behalf of the Nobel Assembly at the Karolinska Institute I wish to convey to you my warmest congratulations and I now ask you to step forward to receive your Nobel Prize from the hands of His Majesty the King.

Stanley Prusiner

STANLEY B. PRUSINER

My history is not atypical of many Americans: born in the midwest, educated in the East, and now living in the West. My early years were shared between Des Moines, Iowa and Cincinnati, Ohio. Shortly after I was born on May 28, 1942 in Des Moines, my father, Lawrence, was drafted into the United States Navy. I was named for my father's younger brother who died of Hodgkin's disease at the age of 24. We moved to Boston briefly where my father enrolled in Naval officer training school before being sent to the south Pacific. He served as a communications officer for the remainder of World War II on an island called Eniwetok where the first hydrogen bomb was detonated a decade later.

During my father's absence, my mother, Miriam, and I lived in Cincinnati where her mother, Mollie Spigel, also lived. Prior to moving to Cincinnati, Mollie had lived in Norfolk, Virginia, where she raised three children after her husband Benjamin was killed at age 50 in a traffic accident. Besides many special memories of my maternal grandmother, I have many fond reminiscences of my paternal grandfather, Ben, who emigrated to the United States in 1896 as a young boy from Moscow. He grew up in Sioux City, Iowa, as did my father with many other Russian Jews. Shortly after the end of World War II, we returned to Des Moines where I attended primary school and my brother, Paul, was born. In 1952, we moved back to Cincinnati with the hope that my father would be able to find a much better job as an architect. In Cincinnati, he practiced architecture for the next 25 years, which enabled him to provide a very comfortable home for his family.

During my time at Walnut Hills High School, I studied Latin for five years, which was to help me immensely later in the writing of scientific papers. But I found high school rather uninteresting and was most fortunate to be accepted by the University of Pennsylvania where I majored in Chemistry.

The intellectual environment of the University of Pennsylvania was extraordinary–there were so many internationally renowned scholars who were invariably receptive to the intrusions of undergraduate students even before the days of student evaluations of the faculty. The small size of the undergraduate student body undoubtedly contributed to the accessibility of the faculty. Besides numerous science courses, I had the opportunity to study philosophy, the history of architecture, economics, and Russian history in courses taught by extraordinarily knowledgeable professors. Although I was among the smallest of the heavyweight crew team members and thus had no chance of rowing in the varsity boat, I greatly enjoyed the many hours that I spent at this wonderful sport.

During the summer of 1963 between my junior and senior years, I began a research project on hypothermia in the Department of Surgery with Sidney Wolfson. I quickly became fascinated by the project and continued working on it throughout my senior year. I decided to remain at Penn for Medical School largely because of the wonderful experience of doing research with Sidney Wolfson. During the second year of medical school, I decided to ask Britton Chance if he would allow me to study the surface fluorescence of brown adipose tissue in Syrian golden hamsters as they arose from hibernation. Chance had reported that the surface fluorescence of other organs reflected the oxidation-reduction state of those tissues. As anticipated, large changes in the fluorescence of brown fat were found during non-shivering thermogenesis.

My research on brown fat allowed me to spend much of the fourth year of medical school at the Wenner-Gren Institute in Stockholm working with Olov Lindberg on the metabolism of isolated brown adipocytes. This was an exciting time and I began to consider seriously a career in biomedical research. Early in 1968, I returned to Philadelphia to complete my medical studies and to contemplate my options. The previous spring, I had been given a position at the NIH once I completed an internship in medicine. It was the height of the Vietnam war with 500,000 young Americans trying to control the spread of Communism in southeast Asia. But I was facing an internship at the University of California San Francisco (UCSF) that would require me to work every other night for an entire year, a prospect about which I was not enthusiastic. The privilege of serving in the US Public Health Service at the NIH clearly outweighed the unpleasant prospects of an internship. Although the workload was awesome, I managed to survive because San Francisco was such a nice place to live. During that year, I met my wife, Sandy Turk, who was teaching mathematics to high school students.

At the NIH, I worked in Earl Stadtman's laboratory where I studied glutaminases in *E. coli.* My three years at the NIH were critical in my scientific education. I learned an immense amount about the research process: developing assays, purifying macromolecules, documenting a discovery by many approaches, and writing clear manuscripts describing what is known and what remains to be investigated. As the end of my time at the NIH began to near, I examined postdoctoral fellowships in neurobiology but decided a residency in Neurology was a better route to developing a rewarding career in research. The residency offered me an opportunity to learn about both the normal and abnormal nervous system.

In July 1972, I began a residency at the University of California San Francisco in the Department of Neurology. Two months later, I admitted a female patient who was exhibiting progressive loss of memory and difficulty performing some routine tasks. I was surprised to learn that she was dying of a "slow virus" infection called Creutzfeldt-Jakob disease (CJD) which evoked no response from the body's defenses. Next, I learned that scientists were unsure if a virus was really the cause of CJD since the causative infectious agent had some unusual properties. The amazing properties of the presumed causative "slow virus" captivated my imagination and I began to think that defining the molecular structure of this elusive agent might be a wonderful research project. The more that I read about CJD and the seemingly related diseases–kuru of the Fore people of New Guinea and scrapie of sheep–the more captivated I became.

Over the next two years I completed an abbreviated residency while reading every paper that I could find about slow virus diseases. In time, I developed a passion for working on these disorders. As I plotted out a course of action, the task became more and more daunting. The tedious, slow, and very expensive assays in mice for the scrapie agent had restricted progress and I had no clever idea about how to circumvent the problem. I did think that after working with the scrapie agent for some time that I might eventually be able to develop such an assay.

Since both Sandy and I liked living in San Francisco, I accepted the offer of an assistant professor position from Robert Fishman, the Chair of Neurology, and began to set up a laboratory to study scrapie in July 1974. Although many people cautioned me about the high risk of studies on scrapie due to the assay problems, such warnings did not dull my enthusiasm. To gain a base of research support from the NIH, I initially wrote grant proposals on glutamate metabolism in the choroid plexus. Such proposals were dull but were readily funded because I had worked on glutaminases earlier. Eventually, I managed to gain modest NIH support for my scrapie studies but this was not without considerable difficulty. To rebut the disapproval of my first NIH application on scrapie, I set up a collaboration with William Hadlow and Carl Eklund who were working at the Rocky Mountain Laboratory in Hamilton, Montana. They taught me an immense amount about scrapie and helped me initiate studies on the sedimentation behavior of the scrapie agent.

I had anticipated that the purified scrapie agent would turn out to be a small virus and was puzzled when the data kept telling me that our preparations contained protein but not nucleic acid. About this time, I was informed by the Howard Hughes Medical Institute (HHMI) that they would not renew their support and by UCSF that I would not be promoted to tenure. When everything seemed to be going wrong, including the conclusions of my research studies, it was the unwavering, enthusiastic support of a few of my closest colleagues that carried me through this very trying and difficult period. Fortunately, the tenure decision was reversed and I was able to continue my work. Although my work was never supported by HHMI again, I was extremely fortunate to receive much larger funding from the R. J. Reynolds Company through a program administered by Fred Seitz and Macyln McCarty and shortly thereafter from the Sherman Fairchild Foundation under the direction of Walter Burke. While the vast majority of my funding always came from the NIH, these private sources were crucial in providing funds for the infrastructure which was the thousands of mice and hamsters that were mandatory.

As the data for a protein and the absence of a nucleic acid in the scrapie agent accumulated, I grew more confident that my findings were not artifacts and decided to summarize that work in an article that was eventually published in the spring of 1982. Publication of this manuscript, in which I introduced the term "prion", set off a firestorm. Virologists were generally incredulous and some investigators working on scrapie and CJD were irate. The term prion derived from protein and infectious provided a challenge to find the nucleic acid of the putative "scrapie virus." Should such a nucleic acid be found, then the word prion would disappear! Despite the strong convictions of many, no nucleic acid was found; in fact, it is probably fair to state that Detlev Riesner and I looked more vigorously for the nucleic acid than anyone else.

While it is quite reasonable for scientists to be skeptical of new ideas that do not fit within the accepted realm of scientific knowledge, the best science often emerges from situations where results carefully obtained do not fit within the accepted paradigms. At times the press became involved since the media provided the naysayers with a means to vent their frustration at not being able to find the cherished nucleic acid that they were so sure must exist. Since the press was usually unable to understand the scientific arguments and they are usually keen to write about any controversy, the personal attacks of the naysayers at times became very vicious. While such scorn caused Sandy considerable distress, she and my two daughters, Helen and Leah, provided a loving and warm respite from the torrent of criticism that the prion hypothesis engendered. During the winter of 1983, I herniated a disc in my lumbar spine while skiing and this slowed the pace of my work for much of the year. After a laminectomy, I began swimming regularly, which brought relaxation and a much needed quiet time to my life.

Just prior to my back problem, the protein of the prion was found in my laboratory and the following year, a portion of the amino acid sequence was determined by Leroy Hood. With that knowledge, molecular biological studies of the prions ensued and an explosion of new information followed. I collaborated with Charles Weissmann on the molecular cloning of the gene encoding the prion protein (PrP) and with George Carlson and David Kingsbury on linking the PrP gene to the control of scrapie incubation time in mice. About the same time, we succeeded in producing antibodies that provided an extremely valuable tool that allowed us to discover the normal form of PrP. In a very important series of studies, the antibodies were used by Stephen DeArmond to study the pathogenesis of prion disease in transgenic mice. Steve brought the much needed talents of an outstanding neuropathologist to these studies. As more data accumulated, an expanding edifice in support of the prion concept was constructed. Ruth Gabizon dispersed prions into liposomes and purified scrapie infectivity on columns with PrP antibodies. Karen Hsiao discovered a mutation in the PrP gene that caused familial disease and reproduced the disease in transgenic mice while Michael Scott produced transgenic mice abrogating the prion species barrier and later artificial prions from chimeric PrP transgenes. Indeed, no experimental findings that might overturn the prion concept were reported from any laboratory. By the early 1990s, the existence of prions was coming to be accepted in many quarters of the scientific community, but the mechanism by which normal

PrP was converted into the disease-causing form was still obscure. When Fred Cohen and I began to collaborate on PrP structural studies, I was again extremely fortunate. Fred brought an extraordinary set of skills in protein chemistry and computational biology to investigations of PrP structures.

As prions gained wider acceptance among scientists, I received many scientific prizes. The first major recognition of my work was accorded by neurologists with many other awards coming soon thereafter. But the most rewarding aspect of my work has been the numerous wonderful friends that I have made during an extensive series of collaborative studies. It has been a special privilege to work with so many talented scientists including numerous postdoctoral fellows and technical associates who have taught me so much. Besides the many collaborators who have contributed their scientific skills to advancing the study of prions, I have had many colleagues who have contributed indirectly to my work by being supportive of the special needs that such a project has demanded.

PRIONS

Nobel Lecture, December 8, 1997

by

STANLEY B. PRUSINER

Departments of Neurology and of Biochemistry and Biophysics, University of California, San Francisco, California 94103-0518, USA

ABSTRACT

Prions are unprecedented infectious pathogens that cause a group of invariably fatal neurodegenerative diseases by an entirely novel mechanism. Prion diseases may present as genetic, infectious, or sporadic disorders, all of which involve modification of the prion protein (PrP). Bovine spongiform encephalopathy (BSE), scrapie of sheep, and Creutzfeldt-Jakob disease (CJD) of humans are among the most notable prion diseases. Prions are transmissible particles that are devoid of nucleic acid and seem to be composed exclusively of a modified protein (PrP^{Sc}). The normal, cellular PrP (PrP^{C}) is converted into PrP^{Sc} through a posttranslational process during which it acquires a high β-sheet content. The species of a particular prion is encoded by the sequence of the chromosomal PrP gene of the mammals in which it last replicated. In contrast to pathogens carrying a nucleic acid genome, prions appear to encipher strain-specific properties in the tertiary structure of PrP^{Sc}. Transgenetic studies argue that PrP^{Sc} acts as a template upon which PrP^{C} is refolded into a nascent PrP^{Sc} molecule through a process facilitated by another protein. Miniprions generated in transgenic mice expressing PrP, in which nearly half of the residues were deleted, exhibit unique biological properties and should facilitate structural studies of PrP^{Sc}. While knowledge about prions has profound implications for studies of the structural plasticity of proteins, investigations of prion diseases suggest that new strategies for the prevention and treatment of these disorders may also find application in the more common degenerative diseases.

The tortuous path of the scientific investigation that led to an understanding of familial Creutzfeldt-Jakob disease (CJD) chronicles a remarkable scientific odyssey. By 1930, the high incidence of familial (f) CJD in some families was known [1,2]. Almost 60 years were to pass before the significance of this finding could be appreciated [3–5]. CJD remained a curious, rare neurodegenerative disease of unknown etiology throughout this period of three score years [6]. Only with transmission of disease to apes following inoculation of brain extracts prepared from patients, who died of CJD, did the story begin to unravel [7].

Once CJD was shown to be an infectious disease, relatively little attention was paid to the familial form of the disease since most cases were not found in families. It is interesting to speculate how the course of scientific investigation might have proceeded had transmission studies not been performed until after the molecular genetic lesion had been identified. Had that sequence of events transpired, then the prion concept, which readily explains how a single disease can have a genetic or infectious etiology, might have been greeted with much less skepticism [8].

Epidemiologic studies designed to identify the source of the CJD infection were unable to identify any predisposing risk factors although some geographic clusters were found [9–12]. Libyan Jews living in Israel developed CJD about 30 times more frequently than other Israelis [13]. This finding prompted some investigators to propose that the Libyan Jews had contracted CJD by eating lightly cooked brain from scrapie-infected sheep when they lived in Tripoli prior to emigration. Subsequently, the Libyan Jewish patients were all found to carry a mutation at codon 200 in their prion protein (PrP) gene [14–16].

My own interest in the subject began with a patient dying of CJD in the fall of 1972. At that time, I was beginning a residency in Neurology and was most impressed by a disease process that could kill my patient in two months by destroying her brain while her body remained unaffected by this process. No febrile response, no leucocytosis or pleocytosis, no humoral immune response, and yet I was told that she was infected with a "slow virus."

Slow viruses. The term "slow virus" had been coined by Bjorn Sigurdsson in 1954 while he was working in Iceland on scrapie and visna of sheep [17]. Five years later, William Hadlow had suggested that kuru, a disease of New Guinea highlanders, was similar to scrapie and thus, it, too, was caused by a slow virus [18]. Seven more years were to pass before the transmissibility of kuru was established by passaging the disease to chimpanzees inoculated intracerebrally [19]. Just as Hadlow had made the intellectual leap between scrapie and kuru, Igor Klatzo made a similar connection between kuru and CJD [20]. In both instances, these neuropathologists were struck by the similarities in light microscopic pathology of the CNS that kuru exhibited with scrapie or CJD. In 1968, the transmission of CJD to chimpanzees after intracerebral inoculation was reported [7].

In scrapie, kuru, CJD, and all of the other disorders now referred to as prion diseases (Table 1), spongiform degeneration and astrocytic gliosis is found upon microscopic examination of the CNS (Figure 1) [21]. The degree of spongiform degeneration is quite variable while the extent of reactive gliosis correlates with the degree of neuron loss [22].

Prions–a brief overview. Before proceeding with a detailed discussion of our current understanding of prions causing scrapie and CJD, I provide a brief overview of prion biology. Prions are unprecedented infectious pathogens that cause a group of invariably fatal neurodegenerative diseases mediated by an entirely novel mechanism. Prion diseases may present as genetic, infectious, or sporadic disorders, all of which involve modification of the prion protein (PrP), a constituent of normal mammalian cells [23]. CJD generally presents as progressive dementia while scrapie of sheep and bovine spongiform encephalopathy (BSE) are generally manifest as ataxic illnesses (Table 1) [24].

Table 1. The prion diseases

	Disease	Host	Mechanism of Pathogenesis
A.	Kuru	Fore people	Infection through ritualistic cannibalism
	Iatrogenic CJD*	Humans	Infection from prion-contaminated HGH, dura mater grafts, etc.
	Variant CJD	Humans	Infection from bovine prions?
	Familial CJD	Humans	Germline mutations in PrP gene
	GSS	Humans	Germline mutations in PrP gene
	FFI	Humans	Germline mutation in PrP gene (D178N, M129)
	Sporadic CJD	Humans	Somatic mutation or spontaneous conversion of PrP^{C} into PrP^{Sc}?
	FSI	Humans	Somatic mutation or spontaneous conversion of PrP^{C} into PrP^{Sc}?
B.	Scrapie	Sheep	Infection in genetically susceptible sheep
	BSE	Cattle	Infection with prion-contaminated MBM
	TME	Mink	Infection with prions from sheep or cattle
	CWD	Mule deer, elk	Unknown
	FSE	Cats	Infection with prion-contaminated bovine tissues or MBM
	exotic ungulate encephalopathy	Greater kudu, nyala, onyx	Infection with prion-contaminated MBM

*Abbreviations defined in Footnote a.

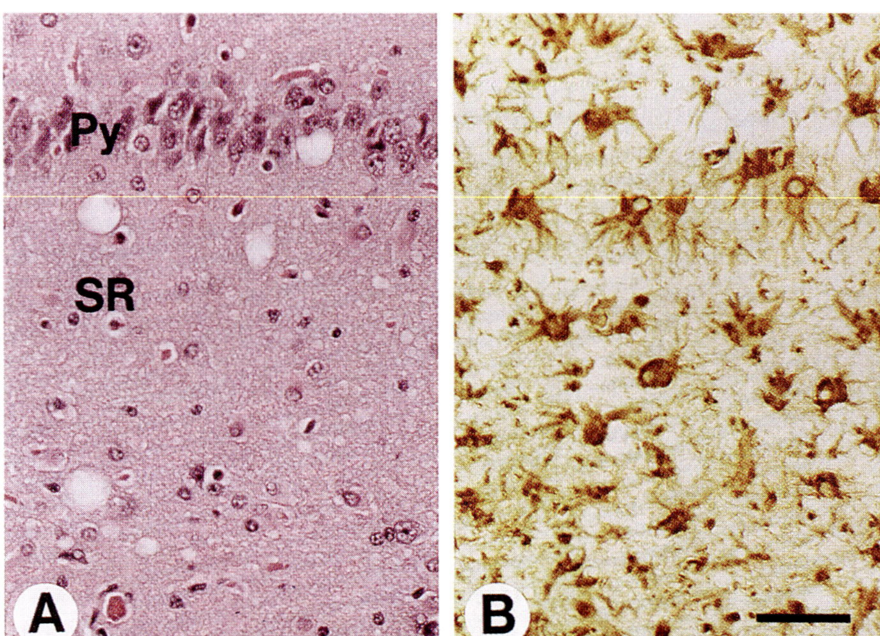

Figure 1. Neuropathologic changes in Swiss mice following inoculation with RML scrapie prions. **(A)** Hematoxylin and eosin stain of a serial section of the hippocampus shows spongiform degeneration of the neuropil, with vacuoles 10–30 μm in diameter. Brain tissue was immersion fixed in 10% buffered formalin solution after sacrificing the animals and embedded in paraffin. **(B)** GFAP immunohistochemistry of a serial section of the hippocampus shows numerous reactive astrocytes. Bar in B = 50 μm and also applies to A. Photomicrographs prepared by Stephen J. DeArmond.

Prions are devoid of nucleic acid and seem to be composed exclusively of a modified isoform of PrP designated PrP^{Sc}.[(a)] The normal, cellular PrP, denoted PrP^{C}, is converted into PrP^{Sc} through a process whereby a portion of its α-helical and coil structure is refolded into β-sheet [25]. This structural transition is accompanied by profound changes in the physicochemical properties of the PrP. The amino acid sequence of PrP^{Sc} corresponds to that encoded by the PrP gene of the mammalian host in which it last replicated. In contrast to pathogens with a nucleic acid genome that encode strain-specific properties in genes, prions encipher these properties in the tertiary structure of PrP^{Sc} [26–28]. Transgenetic studies argue that PrP^{Sc} acts as a template upon which PrP^{C} is refolded into a nascent PrP^{Sc} molecule through a process facilitated by another protein.

More than 20 mutations of the PrP gene are now known to cause the inherited human prion diseases and significant genetic linkage has been established for five of these mutations [4, 16, 29–31]. The prion concept readily explains how a disease can be manifest as a heritable as well as an infectious illness.

Resistance of scrapie agent to radiation. My fascination with CJD quickly shifted to scrapie once I learned of the remarkable radiobiological data that Tikvah Alper and her colleagues had collected on the scrapie agent [32–34]. The scrapie agent had been found to be extremely resistant to inactivation by UV and ionizing radiation as was later shown for the CJD agent [35]. It seemed to me that the most intriguing question was the chemical nature of the scrapie agent; Alper's data had evoked a torrent of hypotheses concerning its composition. Suggestions as to the nature of the scrapie agent ranged from small DNA viruses to membrane fragments to polysaccharides to proteins, the last of which eventually proved to be correct [36–42].

Scrapie of sheep and goats possesses a history no less fascinating than that of CJD. The resistance of the scrapie agent to inactivation by formalin and heat treatments [43], which were commonly used to produce vaccines against viral illnesses, suggested that the scrapie agent might be different from viruses, but it came at a time before the structure of viruses was understood. Later, this resistance was dismissed as an interesting observation but of little importance since some viruses can survive such treatments; indeed, this was not an unreasonable viewpoint. More than two decades were to pass before reports of the extreme resistance of the scrapie agent to inactivation by radiation again trumpeted the novelty of this infectious pathogen. Interestingly, British scientists had argued for many years about whether natural scrapie was a genetic or an infectious disease [44–46]. Scrapie, like kuru and CJD, produced

[a] Prions are defined as proteinaceous infectious particles that lack nucleic acid. PrP^{C} is the cellular prion protein; PrP^{Sc} is the pathologic isoform. NH_2-terminal truncation during limited proteolysis of PrP^{Sc} produces PrP 27-30. BSE, bovine spongiform encephalopathy; CJD, Creutzfeldt-Jakob disease; sCJD, sporadic CJD; fCJD, familial CJD; iCJD, iatrogenic CJD; vCJD, (new) variant CJD; CWD, chronic wasting disease; FFI, fatal familial insomnia; FSE, feline spongiform encephalopathy; FSI, fatal sporadic insomnia; GSS, Gerstmann-Sträussler-Scheinker disease; HGH, human growth hormone; MBM, meat and bone meal; SHa, Syrian hamster; Tg, transgenic; TME, transmissible mink encephalopathy; wt, wild-type.

death of the host without any sign of an immune response to a "foreign infectious agent."

My initial studies focused on the sedimentation properties of scrapie infectivity in mouse spleens and brains. From these studies, I concluded that hydrophobic interactions were responsible for the non-ideal physical behavior of the scrapie particle [47, 48]. Indeed, the scrapie agent presented a biochemical nightmare: infectivity was spread from one end to the other of a sucrose gradient and from the void volume to fractions eluting at 5–10 times the included volume of chromatographic columns. Such results demanded new approaches and better assays [49].

Bioassays. As the number of hypotheses about the molecular nature of the scrapie agent began to exceed the number of laboratories working on this problem, the need for new experimental approaches became evident. Much of the available data on the properties of the scrapie agent had been gathered on brain homogenates prepared from mice with clinical signs of scrapie. These mice had been inoculated 4–5 months earlier with the scrapie agent which originated in sheep but had been passaged multiple times in mice. Once an experiment was completed on these homogenates, an additional 12 months was required using an endpoint titration in mice [50]. Typically, 60 mice were required to determine the titer of a single sample. This slow, tedious, and expensive system discouraged systematic investigation.

Although the transmission of scrapie to mice had ushered in a new era of research, the 1.5–2 year intervals between designing experiments and obtaining results discouraged sequential studies. Infrequently, the results of one set of experiments were used as a foundation for the next and so on. Moreover, the large number of mice needed to measure the infectivity in a single sample prevented studies where many experiments were performed in parallel. These problems encouraged publication of inconclusive experimental results.

In 1972, when I became fascinated by the enigmatic nature of the scrapie agent, I thought that the most direct path to determining the molecular structure of the scrapie agent was purification. Fortunately, I did not appreciate the magnitude of that task although I had considerable experience and training in the purification of enzymes [51]. While many studies had been performed to probe the physicochemical nature of the scrapie agent using the mouse endpoint titration system, few systematic investigations had been performed on the fundamental characteristics of the infectious scrapie particle [42]. In fact, 12 years after introduction of the mouse bioassay, there were few data on the sedimentation behavior of the scrapie particle. Since differential centrifugation is frequently a useful initial step in the purification of many macromolecules, some knowledge of the sedimentation properties of the scrapie agent under defined conditions seemed mandatory. To perform such studies, Swiss mice were inoculated intracerebrally with the Chandler isolate of scrapie prions and the mice sacrificed about 30 and 150 days later when the titers in their spleens and brains, respectively, were expected to be at maximal levels. The two tissues were homogenized, extracted with detergent, and centrifuged for increasing times and speeds [47, 52]. The disappearance of scrapie infectivity was measured in supernatant fractions by endpoint titration which required one year to score.

Incubation time assays in hamsters. In view of these daunting logistical problems, the identification of an inoculum that produced scrapie in the golden Syrian hamster in ~70 days after intracerebral inoculation proved to be an important advance [53, 54] once an incubation time assay was developed [55, 56]. In earlier studies, Syrian hamsters (SHa) had been inoculated with prions, but serial passage with short incubation times was not reported [57]. Development of the incubation time bioassay reduced the time required to measure prions in samples with high titers by a factor of five: only 70 days were required instead of the 360 days previously needed. Equally important, 4 animals could be used in place of the 60 that were required for endpoint titrations, making possible a large number of parallel experiments. With this bioassay, research on the nature of the scrapie agent was accelerated nearly 100-fold and the hamster with high prion titers in its brain became the experimental animal of choice for biochemical studies.

The incubation time assay enabled development of effective purification schemes for enriching fractions for scrapie infectivity. It provided a means to assess quantitatively those fractions that were enriched for infectivity and those that were not. Such studies led rather rapidly to the development of a protocol for separating scrapie infectivity from most proteins and nucleic acids. With a ~100-fold purification of infectivity relative to protein, >98% of the proteins and polynucleotides were eliminated permitting more reliable probing of the constituents of these enriched fractions.

The prion concept. As reproducible data began to accumulate indicating that scrapie infectivity could be reduced by procedures that hydrolyze or modify proteins but was resistant to procedures that alter nucleic acids, a family of hypotheses about the molecular architecture of the scrapie agent began to emerge [58]. These data established, for the first time, that a particular macromolecule was required for infectivity and that this macromolecule was a protein. The experimental findings extended earlier observations on resistance of scrapie infectivity to UV irradiation at 250 nm [33] in that the four different procedures used to probe for a nucleic acid are based on physical principles that are independent of UV radiation damage.

Once the requirement of protein for infectivity was established, I thought that it was appropriate to give the infectious pathogen of scrapie a provisional name that would distinguish it from both viruses and viroids. After some contemplation, I suggested the term "prion" derived from **pro**teinaceous and **in**fectious [58]. At that time, I defined prions as proteinaceous infectious particles that resist inactivation by procedures that modify nucleic acids. I never imagined the irate reaction of some scientists to the word prion – it was truly remarkable!

Current definitions. Perhaps, the best current working definition of a prion is a proteinaceous infectious particle that lacks nucleic acid [28]. Because a wealth of data supports the contention that scrapie prions are composed entirely of a protein that adopts an abnormal conformation, it is not unreasonable to define prions as infectious proteins [25, 27, 59, 60]. But I hasten to add that we still cannot eliminate a small ligand bound to PrP^{Sc} as an essential component of the infectious prion particle. Learning how to renature PrP^{Sc} accompanied by restoration of prion infectivity or to generate de novo prion infectivity using a synthetic polypeptide should help address this as yet unresolved issue [61]. From a broader perspective, prions are elements that impart and propagate conformational variability.

Although PrP^{Sc} is the only *known* component of the infectious prion particles, these unique pathogens share several phenotypic traits with other infectious entities such as viruses. Because some features of the diseases caused by prions and viruses are similar, some scientists have difficulty accepting the existence of prions despite a wealth of scientific data supporting this concept [62–67].

Families of hypotheses. Once the requirement for a protein was established, it was possible to revisit the long list of hypothetical structures that had been proposed for the scrapie agent and to eliminate carbohydrates, lipids, and nucleic acids as the infective elements within a scrapie agent devoid of protein [58]. No longer could structures such as a viroid-like nucleic acid, a replicating polysaccharide, or a small polynucleotide surrounded by a carbohydrate be entertained as reasonable candidates to explain the puzzling properties of the scrapie agent [58, 68].

The family of hypotheses that remained after identifying a protein component was still large and required a continued consideration of all possibilities in which a protein was a critical element [49]. The prion concept evolved from a family of hypotheses in which an infectious protein was only one of several possibilities. With the accumulation of experimental data on the molecular properties of the prion, it became possible to discard an increasing number of hypothetical structures. In prion research, as in many other areas of scientific investigation, a single hypothesis is all too often championed at the expense of a reasoned approach which requires entertaining a series of complex arguments until one or more can be discarded on the basis of experimental data [69].

Genes and DNA. In some respects, the early development of the prion concept mirrors the story of DNA [70–72]. Prior to the acceptance of DNA as the genetic material of life [73, 74], many scientists asserted that the DNA preparations must be contaminated with protein which is the true genetic material [75]. The prejudices of these scientists were similar in some ways to those of investigators who have disputed the prion concept. But the scientists who attacked the hypothesis that genes are composed of DNA had no well proven alternative; they had only a set of feelings derived from poorly substantiated data sets that genes are made of protein. In contrast, those who attacked the hypothesis that the prion is composed only of protein had more than 30 years of cumulative evidence showing that genetic information in all organisms on our planet is encoded in DNA and that biological diversity resides in DNA. Studies of viruses and eventually viroids extended this concept to these small infectious pathogens [76] and showed that genes could also be composed of RNA [77, 78].

Discovery of the Prion Protein. The discovery of the prion protein transformed research on scrapie and related diseases [79, 80]. It provided a molecular marker that was subsequently shown to be specific for these illnesses as well as the major, and very likely the only, constituent of the infectious prion.

PrP 27–30 was discovered by enriching fractions from SHa brain for scrapie infectivity [79, 80]. This protein is the protease-resistant core of PrP^{Sc} and has an apparent molecular weight (M_r) of 27–30 kD [81, 82]. Although resistance to limited proteolysis proved to be a convenient tool for many but not all studies, use of proteases to enrich fractions for scrapie infectivity created a problem when the NH_2-terminal sequence of PrP 27–30 was determined [81]. The ragged NH_2-terminus of PrP 27–30 yielded three sets of signals in almost every cycle of the Edman degradation. Only once these signals were properly interpreted and placed in correct register could a unique sequence be assigned for the NH_2-terminus of PrP 27–30. The determination of amino acid sequence of the NH_2-terminus of PrP 27–30 made subsequent molecular cloning studies of the PrP gene possible [83, 84].

The finding that PrP mRNA levels were similar in normal uninfected and scrapie-infected tissues caused some investigators to argue that PrP 27–30 was not related to the infectious prion particle [83]. An alternate interpretation prompted a search for a prion protein in uninfected animals which was found to be protease sensitive and soluble in non-denaturing detergents, unlike PrP 27–30. This isoform was designated PrP^C (Figure 2) [84, 85]. Deduced amino acid sequences from PrP cDNA as well as immunoblotting studies revealed that PrP 27–30 was NH_2-terminally truncated and was derived from a larger molecule, designated PrP^{Sc}, that was unique to infected animals [81, 82, 84–86].

With the discovery of PrP 27–30 and the production of antiserum [87], brains from humans and animals with putative prion diseases were examined for the presence of this protein. In each case, PrP 27–30 was found and it was absent in other neurodegenerative disorders such as Alzheimer's disease, Parkinson's disease, and amyotrophic lateral sclerosis [88–91]. The extreme specificity of PrP^{Sc} for prion disease is an important feature of the protein and is consistent with the postulated role of PrP^{Sc} in both the transmission and pathogenesis of these illnesses (Table 2) [92].

The accumulation of PrP^{Sc} contrasts markedly with that of glial fibrillary acid protein (GFAP) in prion disease. In scrapie, GFAP mRNA and protein levels rise as the disease progresses [93] but the accumulation of GFAP is neither specific nor necessary for either the transmission or pathogenesis of disease. Mice deficient for GFAP show no alteration in their incubation times [94, 95].

Except for PrP^{Sc}, no macromolecule has been found in tissues of patients dying of the prion diseases that is specific for these encephalopathies. In searches for a scrapie-specific nucleic acid, cDNAs have been identified that are complementary to mRNAs encoding other proteins with increased expression in prion disease [96–98]. Yet, none of the proteins has been found to be specific for prion disease.

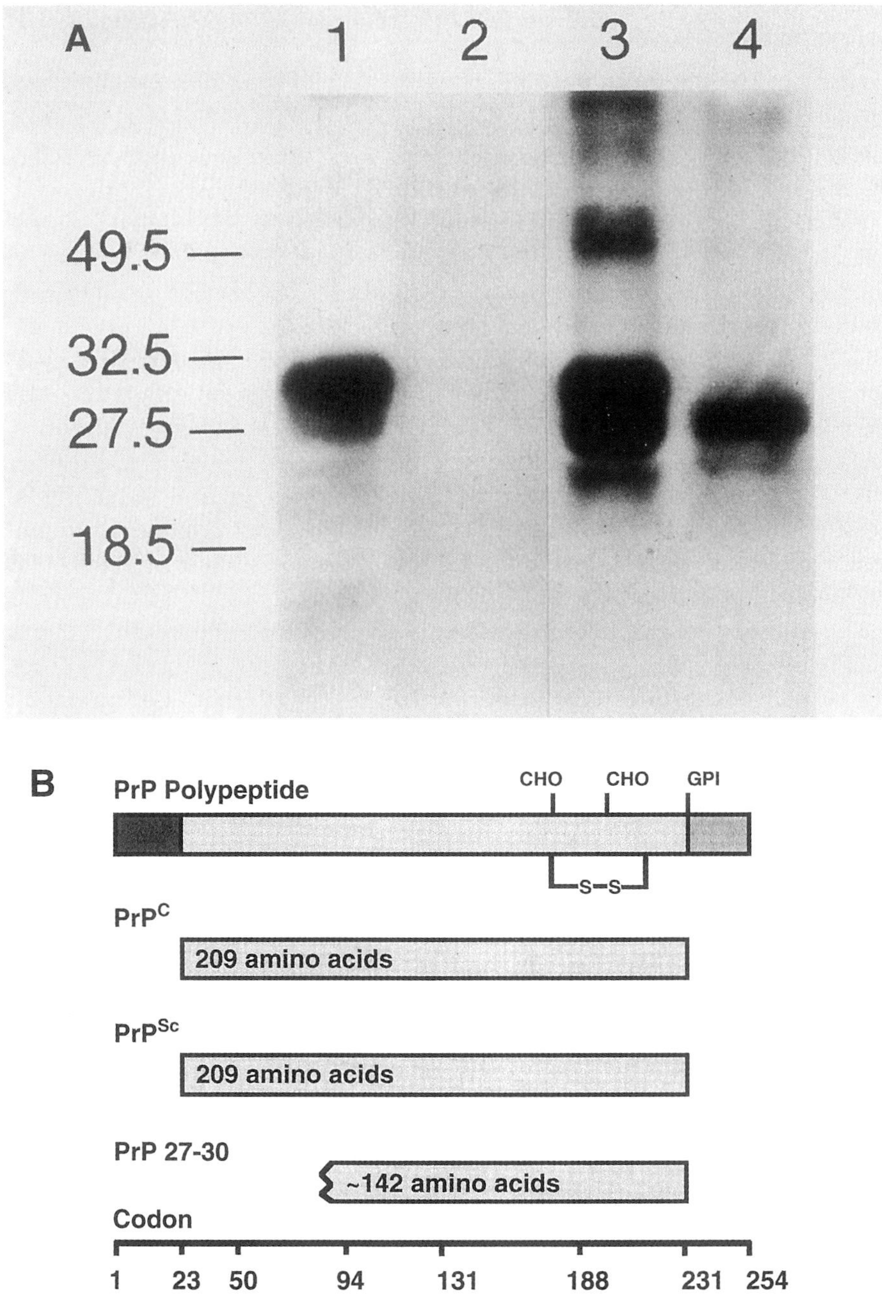

Figure 2. Prion protein isoforms. **(A)** Western immunoblot of brain homogenates from uninfected (lanes 1 and 2) and prion-infected (lanes 3 and 4) Syrian hamsters. Samples in lanes 2 and 4 were digested with 50 μg/ml of proteinase K for 30 min at 37°C. PrP^C in lanes 2 and 4 was completely hydrolyzed under these conditions, whereas approximately 67 amino acids were digested from the NH_2-terminus of PrP^{Sc} to generate PrP 27–30. After polyacrylamide gel electrophoresis (PAGE) and electrotransfer, the blot was developed with anti-PrP R073 polyclonal rabbit antiserum. Molecular size markers are in kilodaltons (kD). **(B)** Bar diagram of SHaPrP which consists of 254 amino acids. After processing of the NH_2- and COOH- termini, both PrP^C and PrP^{Sc} consist of 209 residues. After limited proteolysis, the NH_2-terminus of PrP^{Sc} is truncated to form PrP 27–30, which is composed of approximately 142 amino acids.

Table 2. Arguments for prions being composed largely, if not entirely, of PrP^{Sc} molecules and devoid of nucleic acid.

1. PrP^{Sc} and scrapie infectivity copurify using biochemical and immunologic procedures.
2. The unusual properties of PrP^{Sc} mimic those of prions. Many different procedures that modify or hydrolyze PrP^{Sc} inactivate prions.
3. Levels of PrP^{Sc} are directly proportional to prion titers. Non-denatured PrP^{Sc} has not been separated from scrapie infectivity.
4. No evidence for a virus-like particle or a nucleic acid genome.
5. Accumulation of PrP^{Sc} invariably associated with the pathology of prion diseases including PrP amyloid plaques that are pathognomonic.
6. PrP gene mutations are genetically linked to inherited prion disease and cause formation of PrP^{Sc}.
7. Overexpression of PrP^{C} increases the rate of PrP^{Sc} formation, which shortens the incubation time. Knock out of the PrP gene eliminates the substrate necessary for PrP^{Sc} formation and prevents both prion disease and prion replication.
8. Species variations in the PrP sequence are responsible, at least in part, for the species barrier which is found when prions are passaged from one host to another.
9. PrP^{Sc} preferentially binds to homologous PrP^{C} resulting in formation of nascent PrP^{Sc} and prion infectivity.
10. Chimeric and partially deleted PrP genes change susceptibility to prions from different species and support production of artificial prions with novel properties that are not found in nature.
11. Prion diversity is enciphered within the conformation of PrP^{Sc}. Strains can be generated by passage through hosts with different PrP genes. Prion strains are maintained by PrP^{C}/PrP^{Sc} interactions.
12. Human prions from fCJD(E200K) and FFI patients impart different properties to chimeric MHu2M PrP in transgenic mice, which provides a mechanism for strain propagation.

Attempts to falsify the prion hypothesis. Numerous attempts to disprove the prion hypothesis over the past 15 years have failed. Such studies have tried unsuccessfully to separate scrapie infectivity from protein and more specifically from PrP^{Sc}. No preparations of purified prions containing less than one PrP^{Sc} molecule per ID_{50} unit have been reported [99] and no replication of prions in PrP deficient ($Prnp^{0/0}$) mice was found [100–104].

Copurification of PrP 27–30 and scrapie infectivity demands that the physicochemical properties as well as antigenicity of these two entities be similar [105] (Table 2). The results of a wide array of inactivation experiments demonstrated the similarities in the properties of PrP 27–30 and scrapie infectivity [61, 106–109]. To explain these findings in terms of the virus hypothesis, it is necessary to postulate either a virus that has a coat protein which is highly homologous with PrP or a virus that binds tightly to PrP^{Sc}. In either case, the PrP-like coat protein or the PrP^{Sc}/virus complex must display properties indistinguishable from PrP^{Sc} alone. With each species that the putative virus invades, it must incorporate a new PrP sequence during replication.

Search for a scrapie-specific nucleic acid. The inability to inactivate preparations highly enriched for scrapie infectivity by procedures that modify nucleic acids militates against the existence of a scrapie-specific nucleic acid [58, 110, 111]. To explain the findings in terms of a virus, one must argue that PrP^{Sc} or an as yet undetected PrP-like protein of viral origin protects the viral genome from inactivation. The notion that the putative scrapie virus encodes a PrP-like protein was refuted by nucleic acid hybridization studies using a PrP cDNA probe. Less than 0.002 nucleic acid molecules encoding PrP per ID_{50} unit were found in purified preparations of SHa prions [84]. To circumvent this finding, it could be hypothesized that the genetic code used by the PrP gene differs so greatly from that found in the cell that a PrP cDNA probe failed to detect it in highly purified preparations.

If prions contained a genome with a unique genetic code, then it is likely that this genome would encode some specialized proteins required for replication as well as some unique tRNAs. But both UV and ionizing radiation inactivation studies as well as physical studies have eliminated the possibility of a large nucleic acid hiding within purified preparations of prions [110–112]. Only oligonucleotides of fewer than 50 bases were found at a concentration of one molecule per ID_{50} unit in prion preparations highly enriched for scrapie infectivity [113, 114]. These small nucleic acids were of variable length and are thought to be degradation byproducts generated during purification of prions. Failure to find a bona fide genome was attributed to the unusual properties of the putative viral nucleic acid or technical incompetence on the part of the investigators who were unable to find it [63, 115].

PrP Amyloid. In preparations highly enriched for scrapie infectivity and containing only PrP 27–30 by silver staining of gels after SDS-PAGE, numerous rod-shaped particles were seen by electron microscopy after negative staining (Figure 3) [107]. Each of the rods was slightly different, in contrast to viruses which exhibit extremely uniform structures [116]. These irregular rods

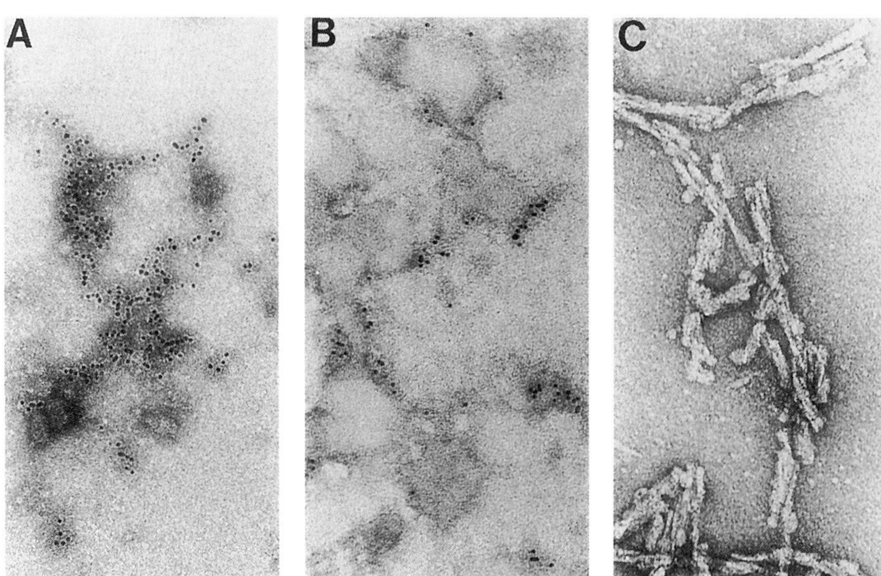

Figure 3. Electron micrographs of negatively stained and immunogold-labeled prion proteins: **(A)** PrP^C and **(B)** PrP^{Sc}. Neither PrP^C nor PrP^{Sc} form recognizable, ordered polymers. **(C)** Prion rods composed of PrP 27–30 were negatively stained. The prion rods are indistinguishable from many purified amyloids. Bar = 100 nm.

composed largely, if not entirely, of PrP 27–30, were indistinguishable morphologically from many other purified amyloids [117]. Studies of the prion rods with Congo red dye demonstrated that the rods also fulfilled the tinctorial criteria for amyloid [107] and immunostaining later showed that PrP is a major component of amyloid plaques in some animals and humans with prion disease [118–120]. Subsequently, it was recognized that the prion rods were not required for scrapie infectivity [121]; furthermore, the rods were shown to be an artifact of purification during which limited proteolysis of PrP^{Sc} generated PrP 27–30 that polymerized spontaneously in the presence of detergent (Figure 3) [122].

The idea that scrapie prions were composed of an amyloidogenic protein was truly heretical when it was introduced [107]. Since the prevailing view at the time was that scrapie is caused by an atypical virus, many argued that amyloid proteins are mammalian polypeptides and not viral proteins!

Some investigators have argued that the prion rods are synonymous with scrapie-associated fibrils [123–125] even though morphologic and tinctorial features of these fibrils clearly differentiated them from amyloid and as such from the prion rods [126, 127]. The scrapie-associated fibrils were identified by their unique ultrastructure in which two or four subfilaments were helically wound around each other [126] and were proposed to represent the first example of a filamentous animal virus [128]. After the argument for a filamentous animal virus causing scrapie faded, it was hypothesized that a virus induces the formation of PrP amyloid in order to explain the accumulation of PrP^{Sc} in prion diseases [129]. Besides the lack of evidence for a virus of any shape, no compelling data has been offered in support of the idea that prion diseases are caused by a filamentous bacterium called a spiroplasma [130].

Search for the ubiquitous "scrapie virus." When PrP gene mutations were discovered to cause familial prion diseases [4], it was postulated that PrP^C is a receptor for the ubiquitous scrapie virus that binds more tightly to mutant than to wt PrP^C [131]. A similar hypothesis was proposed to explain why the length of the scrapie incubation time was found to be inversely proportional to the level of PrP expression in Tg mice and why $Prnp^{0/0}$ mice are resistant to scrapie [132]. The higher the level of PrP expression, the faster the spread of the putative virus which results in shorter incubation times; conversely, mice deficient for PrP lack the receptor required for spread of the virus [63]. The inability to find virus-like particles in purified preparations of PrP^{Sc} was attributed to these particles being hidden [115] even though tobacco mosaic viruses could be detected when one virion was added per ID_{50} unit of scrapie prions [121].

Recent studies on the transmission of mutant prions from FFI and fCJD(E200K) to Tg(MHu2M) mice which results in the formation of two different PrP^{Sc} molecules [27] has forced a corollary to the ubiquitous virus postulate. To accommodate this result, at least two different viruses must reside worldwide, each of which binds to a different mutant $HuPrP^C$ and each of which induces a different MHu2M PrP^{Sc} conformer when transferred to Tg mice. Even more difficult to imagine is how one ubiquitous virus might acquire different mutant PrP^{Sc} molecules corresponding to FFI or fCJD(E200K) and then induce different MHu2M PrP^{Sc} conformers upon transmission to Tg mice.

Artificial prions. To explain the production of artificial prions from chimeric or mutant PrP transgenes in terms of a virus [133–135], mutated PrP^{Sc} molecules must be incorporated into the virus. In the case of mice expressing chimeric PrP transgenes, artificial prions are produced with host ranges not previously found in nature. Similarly, deleting specific regions of PrP resulted in the formation of "miniprions" with a unique host range and neuropathology as described below. The production of artificial prions that were generated by modifying the PrP gene sequence and exhibit unique biological properties is another compelling argument against the proposition that scrapie and CJD are caused by viruses.

Skepticism once well-justified. While the skepticism about prions was once well-justified and formed the basis for a vigorous scientific debate, the wealth of available data now renders such arguments moot. In summary, no single hypothesis involving a virus can explain the findings summarized above (Table 2); instead, a series of ad hoc hypotheses, virtually all of which can be refuted by experimental data, must be constructed to accommodate a steadily enlarging body of data.

It is notable that the search for an infectious pathogen with a nucleic acid genome as the cause of scrapie and CJD has done little to advance our understanding of these diseases. Instead, studies of PrP have created a wealth of data that now explain almost every aspect of these fascinating disorders. While no single experiment can refute the existence of the "scrapie virus", all of the data taken together from numerous experimental studies present an impressive edifice which argues that the 50 year quest for a virus has failed because it does not exist!

Prions defy rules of protein structure. Once cDNA probes for PrP became available, the PrP gene was found to be constitutively expressed in adult, uninfected brain [83, 84]. This finding eliminated the possibility that PrP^{Sc} stimulated production of more of itself by initiating transcription of the PrP gene as proposed nearly two decades earlier [37]. Determination of the structure of the PrP gene eliminated a second possible mechanism that might explain the appearance of PrP^{Sc} in brains already synthesizing PrP^{C}. Since the entire protein coding region was contained within a single exon, there was no possibility for the two PrP isoforms to be the products of alternatively spliced mRNAs [82]. Next, a posttranslational chemical modification that distinguishes PrP^{Sc} from PrP^{C} was considered but none was found in an exhaustive study [59] and we considered it likely that PrP^{C} and PrP^{Sc} differed only in their conformation, a hypothesis also proposed earlier [37]. However, this idea was no less heretical than that of an infectious protein.

For more than 25 years, it had been widely accepted that the amino acid sequence specifies one biologically active conformation of a protein [136]. Yet, in scrapie we were faced with the possibility that one primary structure for PrP might adopt at least two different conformations in order to explain the existence of both PrP^{C} and PrP^{Sc}. When the secondary structures of the PrP isoforms were compared by optical spectroscopy, they were found to be markedly different [25]. Fourier transform infrared (FTIR) and circular dichroism (CD) studies showed that PrP^C contains about 40% α-helix and little β-sheet while PrP^{Sc} is composed of about 30% α-helix and 45% β-sheet [25, 137]. Nevertheless, these two proteins have the same amino acid sequence!

Prior to comparative studies on the structures of PrP^C and PrP^{Sc}, we found by metabolic labeling studies that the acquisition of PrP^{Sc} protease-resistance is a posttranslational process [138]. In our quest for a chemical difference that would distinguish PrP^{Sc} from PrP^C, we found ethanolamine in hydrolysates of PrP 27–30 which signalled the possibility that PrP might contain a glycosylphosphatidyl inositol (GPI) anchor [139]. Both PrP isoforms were found to carry GPI anchors and PrP^C was found on the surface of cells where it could be released by cleavage of the anchor. Subsequent studies showed that PrP^{Sc} formation occurred after PrP^C reaches the cell surface [140, 141] and is localized to caveolae-like domains [142–145].

Modeling PrP structures. Molecular modeling studies predicted that PrP^C is a four-helix bundle protein containing four regions of secondary structure denoted H1, H2, H3, and H4 (Figure 4) [146, 147]. Subsequent NMR studies of a synthetic PrP peptide containing residues 90–145 provided good evidence for H1 [148]. This peptide contains the residues 113–128 that are most highly conserved among all species studied (Figure 4A) [147, 149, 150] and correspond to a transmembrane region of PrP which was delineated in cell-free translation studies [151, 152]. Recent studies show that a transmembrane form of PrP accumulates in GSS caused by the A117V mutation and in Tg mice overexpressing either mutant or wtPrP [153]. The paradoxical lack of evidence for an α-helix in this region from NMR studies of recombinant PrP in aqueous buffers [154–156] could be explained if the recombinant PrPs correspond to the secreted form of PrP that was also identified in cell-free translation studies. This contention is supported by studies with recombinant antibody fragments (Fabs) showing the GPI-anchored PrP^C on the surface of cells exhibits an immunoreactivity similar to that of recombinant PrP that was prepared with an α-helical conformation [157, 158]. GPI-anchored PrP^C is synthesized within the secretory pathway and transported to surface of the cell [139, 159].

Optical spectroscopic measurements of PrP^C provided the necessary background for more detailed structural studies [25]. Unable to produce crystals of PrP, we and others utilized NMR to determine the structure of an α-helical form of a recombinant PrP. The NMR structure of a COOH-terminal fragment of MoPrP consisting of 111 residues showed three helices, two of which corresponded to H3 and H4 in the PrP^C model, and two small β-strands each consisting of three residues [154]. How the structure of this MoPrP(121–231) fragment differs from PrP^C is of interest since this fragment is lethal when expressed in Tg mice [160]. Subsequently, structural studies were performed on a longer fragment of PrP containing residues 90–231 and corresponding to SHaPrP 27–30 [155, 161, 162]. Expression of PrP(90–231) in Tg mice did not produce spontaneous disease [163, 164]. More recently, NMR structures of recombinant full-length PrP have been reported [156, 165].

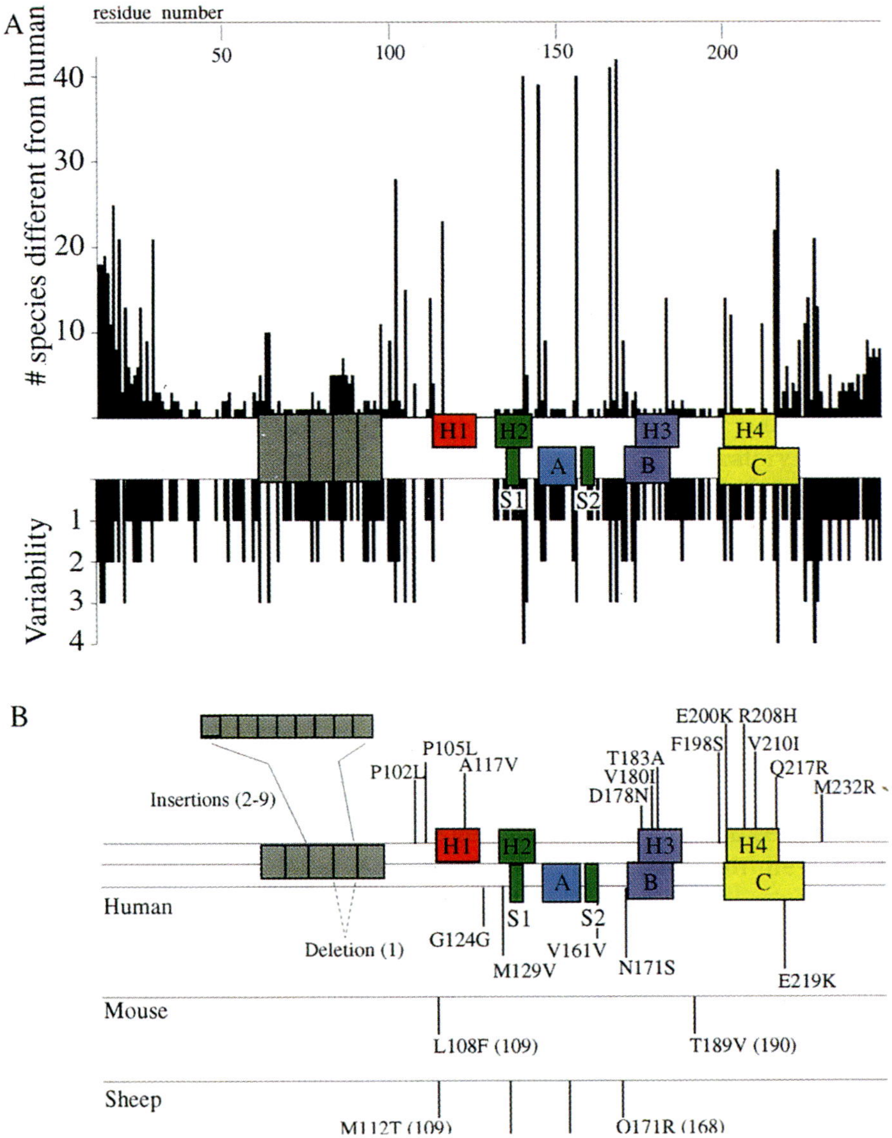

Figure 4. Species variations and mutations of the prion protein gene. (A) Species variations. The x-axis represents the human PrP sequence, with the five octarepeats and H1–H4 regions of putative secondary structure shown as well as the three α-helices A, B, and C and the two β-strands S1 and S2 as determined by NMR. The precise residues corresponding to each region of secondary structure are given in Figure 5. Vertical bars above the axis indicate the number of species that differ from the human sequence at each position. Below the axis, the length of the bars indicates the number of alternative amino acids at each position in the alignment. (B) Mutations causing inherited human prion disease and polymorphisms in human, mouse, and sheep. Above the line of the human sequence are mutations that cause prion disease. Below the lines are polymorphisms, some but not all of which are known to influence the onset as well as the phenotype of disease. Data were compiled by Paul Bamborough and Fred E. Cohen.

Models of PrP^{Sc} suggested that formation of the disease-causing isoform involves refolding of residues within the region between residues 90 and 140 into β-sheets [166]; the single disulfide bond joining COOH-terminal helices would remain intact since the disulfide is required for PrP^{Sc} formation (Figure 5E) [167, 168]. The high β-sheet content of PrP^{Sc} was predicted from the ability of PrP 27–30 to polymerize into amyloid fibrils [107]. Subsequent optical

Figure 5A

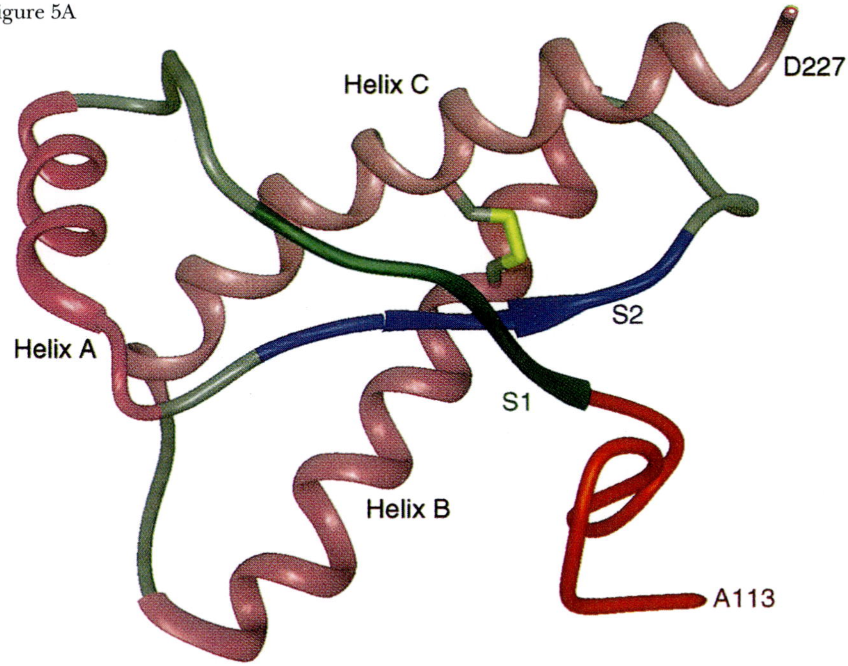

Figure 5. Structures of prion proteins. (A) NMR structure of Syrian hamster (SHa) recombinant (r) PrP(90–231). Presumably, the structure of the α-helical form of rPrP(90–231) resembles that of PrP^{C}. rPrP(90–231) is viewed from the interface where PrP^{Sc} is thought to bind to PrP^{C}. The color scheme is: α-helices A (residues 144–157), B (172–193), and C (200–227) in pink; disulfide between Cys179 and Cys214 in yellow; conserved hydrophobic region composed of residues 113–126 in red; loops in gray; residues 129–134 in green encompassing strand S1 and residues 159–165 in blue encompassing strand S2; the arrows span residues 129–131 and 161–163, as these show a closer resemblance to β–sheet [155]. (B) NMR structure of rPrP(90–231) is viewed from the interface where protein X is thought to bind to PrP^{C}. Protein X appears to bind to the side chains of residues that form a discontinuous epitope: some amino acids are in the loop composed of residues 165–171 and at the end of helix B (Gln168 and Gln172 with a low density van der Waals rendering) while others are on the surface of helix C (Thr215 and Gln219 with a high density van der Waals rendering) [178]. (C) PrP residues governing the transmission of prions [298]. NMR structure of recombinant SHaPrP region 121–231 [155] shown with the putative epitope formed by residues 184, 186, 203, and 205 highlighted in red. Residue numbers correspond to SHaPrP. Additional residues (138, 139, 143, 145, 148, and 155) that might participate in controlling the transmission of prions across species are depicted in green. Residues 168, 172, 215, and 219 that form the epitope for the binding of protein X are shown in blue. The three helices (A, B, and C) are highlighted in pink. (D) Schematic diagram showing the flexibility of the polypeptide chain for PrP(29–231) [156]. The structure of the portion of the protein representing residues 90–231 was taken from the coordinates of PrP(90–231) [155]. The remainder of the sequence was hand-built for illustration purposes only. The color scale corresponds to the heteronuclear $\{^{1}H\}$-^{15}N NOE data: red for the lowest (most negative) values, where the polypeptide is most flexible, to blue for the highest (most positive) values in the most structured and rigid regions of the protein. **(E)** Plausible model for the tertiary structure of human PrP^{Sc} [166]. Color scheme is: S1 β–strands are 108–113 and 116–122 in red; S2 β-strands are 128–135 and 138–144 in green; a-helices H3 (residues 178–191) and H4 (residues 202–218) in gray; loop (residues 142–177) in yellow. Four residues implicated in the species barrier are shown in ball-and-stick form (Asn 108, Met 112, Met 129, Ala 133).

spectroscopy confirmed the presence of β-sheet in both PrP^{Sc} and PrP 27–30 [25, 169–171]. Deletion of each of the regions of putative secondary structure in PrP, except for the NH_2-terminal 66 amino acids (residues 23–88) [163, 172] and a 36 amino acid region (Mo residues 141 -176) prevented formation of PrP^{Sc} as measured in scrapie–infected cultured neuroblastoma cells [168]. With α-PrP Fabs selected from phage display libraries [157] and two monoclonal antibodies (mAb) derived from hybridomas [173–175], the major conformational change that occurs during conversion of PrP^{C} into PrP^{Sc} has been localized largely, but not entirely, to a region bounded by residues 90 and 112 [158]. Similar conclusions were drawn from studies with an α-PrP IgM mAb [176]. While these results indicate that PrP^{Sc} formation involves primarily a conformational change in the domain comprised of residues 90–112, mutations causing inherited prion diseases have been found throughout the protein (Figure 4B). Interestingly, all of the known point mutations in PrP with biological significance occur either within or adjacent to regions of putative secondary structure in PrP and as such, appear to destabilize the structure of PrP [147, 148, 154].

NMR structure of recombinant PrP. The NMR structure of recombinant SHaPrP(90–231) derived from *E. coli* was determined after the protein was purified and refolded (Figure 5A). Residues 90–112 are not shown since marked conformational heterogeneity was found in this region while residues 113–126 constitute the conserved hydrophobic region that also displays some structural plasticity [162]. Although some features of the structure of rPrP(90–231) are similar to those reported earlier for the smaller recombinant MoPrP(121–231) fragment [154, 177], substantial differences were found.

Figure 5B

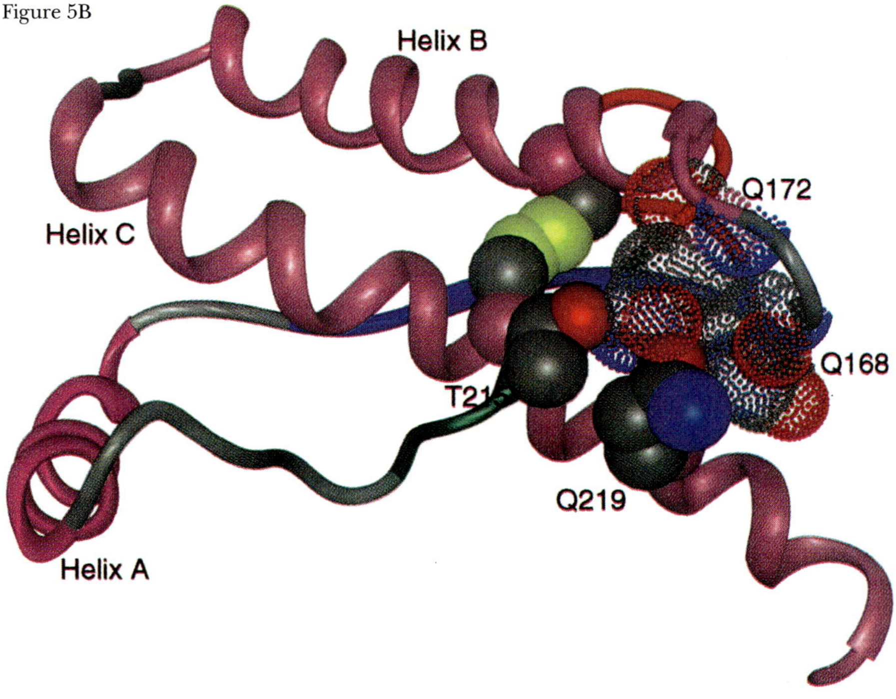

Figure 5C

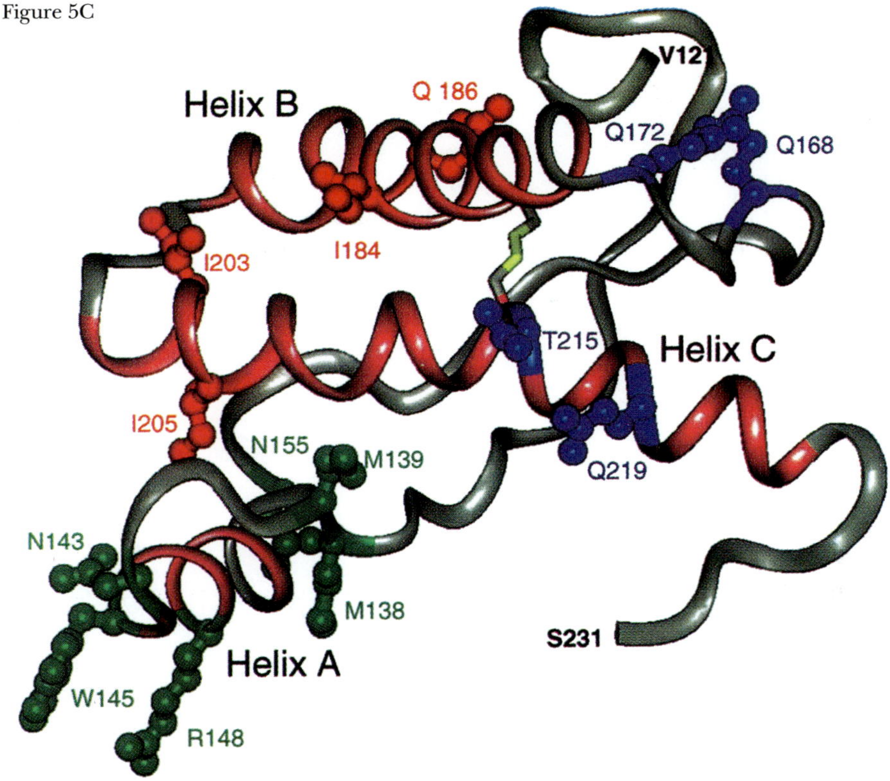

For example, the loop at the NH_2-terminus of helix B is well defined in rPrP(90–231) but is disordered in MoPrP(121–231); in addition, helix C is composed of residues 200–227 in rPrP(90–231) but extends only from 200–217 in MoPrP(121–231). The loop and the COOH-terminal portion of helix C are particularly important since they form the site to which protein X binds as described below (Figure 5B) [178]. Whether the differences between the two recombinant PrP fragments are due to (i) their different lengths, (ii) species specific differences in sequences, or (iii) the conditions used for solving the structures remains to be determined.

Studies of chimeric SHa/Mo and Hu/Mo PrP transgenes identified a domain comprised of residues 95 –170 where PrP^C binds to PrP^{Sc} [133, 179]. When chimeric Bo/Mo PrP transgenes failed to render mice sensitive to BSE prions, we examined the differences among the sequences in the chimeric and parent PrP genes [298]. The findings identified a second domain in PrP composed of residues 180–205 which seems to modulate the interaction between PrP^C and PrP^{Sc} (Figure 5C).

Recent NMR studies of full length MoPrP(23–231) and SHaPrP(29–231) have shown that the NH_2-termini are highly flexible and lack identifiable secondary structure under the experimental conditions employed (Figure 5D) [156, 165]. Studies of SHaPrP(29–231) indicate transient interactions between the COOH-terminal end of helix B and the highly flexible, NH_2-terminal ran-

Figure 5D

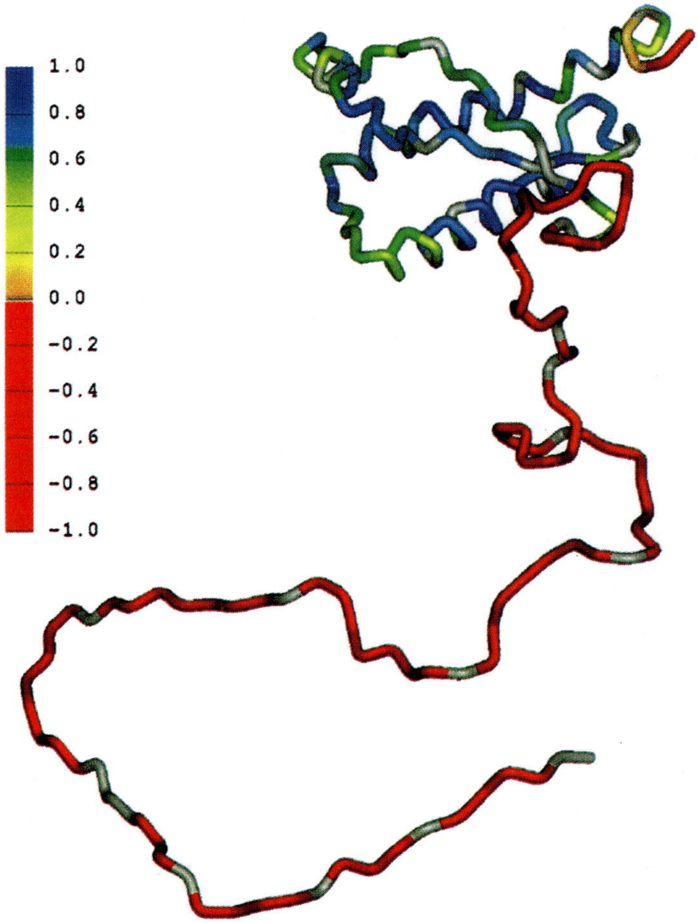

dom-coil containing the octareapeats (residues 29–125) [156]; such interactions were not reported for MoPrP(23–231) [165].

PrP appears to bind copper. The highly flexible NH_2-terminus of recombinant PrP may be more structured in the presence of copper. Each SHaPrP(29–231) molecule was found to bind two atoms of Cu^{2+}; other divalent cations did not bind to PrP [180]. Earlier studies with synthetic peptides corresponding to the octarepeat sequence demonstrated the binding of Cu^{2+} ions [181, 182] and optical spectroscopy showed that Cu^{2+} induced an α-helix formation in these peptides [183]. More recently, PrP-deficient ($Prnp^{0/0}$) mice were found to have lower levels of Zn/Cu superoxide dismutase (SOD) activity than controls [184]; SOD activity has been shown to mirror the state of copper metabolism [185]. Measurements of membrane extracts from brains of $Prnp^{0/0}$ mice showed low levels of Cu whereas Fe and Zn were unchanged suggesting PrP^C might function as a Cu^{2+} binding protein [186].

Disturbances in Cu^{2+} homeostasis leading to dysfunction of the CNS are well documented in humans and animals but are not known to be due to abnormalities in PrP metabolism: Menkes disease is manifest at birth and is due to a mutation of the MNK gene on the X chromosome while Wilson's disease

Figure 5E

appears in childhood and is due to a mutation of the WD gene on chromosome 13 [187–190]. Both the MNK and WD genes encode copper transporting ATPases. While both Menkes and Wilson's diseases are recessive disorders, only Wilson's disease can be treated with copper chelating reagents. Interestingly, cuprizone, a Cu^{2+} chelating reagent, has been used in mice to induce neuropathological changes similar to those found in the prion diseases [191, 192].

PrP gene structure and expression. The entire open reading frame (ORF) of all known mammalian and avian PrP genes resides within a single exon [4, 82, 193, 194]. The mouse (Mo), sheep, cattle, and rat PrP genes contain three exons with the ORFs in exon 3 [195–199] which is analogous to exon 2 of the Syrian hamster (SHa) gene [82]. The two exons of the Syrian hamster (SHa) PrP gene are separated by a 10-kb intron: exon 1 encodes a portion of the 5' untranslated leader sequence while exon 2 encodes the ORF and 3' untranslated region [82]. Recently, a low abundance SHaPrP mRNA containing an additional small exon in the 5′ untranslated region was discovered that is encoded by the SHaPrP gene [200]. Comparative sequencing of sheep and human (Hu) cosmid clones containing PrP genes revealed an additional putative, small untranslated 5′ exon in the HuPrP gene [201]. The promoters of both the

SHa and Mo PrP genes contain multiple copies of G-C rich repeats and are devoid of TATA boxes. These G-C nonamers represent a motif that may function as a canonical binding site for the transcription factor Sp1 [202]. Mapping of PrP genes to the short arm of Hu chromosome 20 and to the homologous region of Mo chromosome 2 argues for the existence of PrP genes prior to the speciation of mammals [203, 204].

Although PrP mRNA is constitutively expressed in the brains of adult animals [83, 84], it is highly regulated during development. In the septum, levels of PrP mRNA and choline acetyltransferase were found to increase in parallel during development [205]. In other brain regions, PrP gene expression occurred at an earlier age. *In situ* hybridization studies show that the highest levels of PrP mRNA are found in neurons [206].

PrP^C expression in brain was defined by standard immunohistochemistry [207] and by histoblotting in the brains of uninfected controls [208]. Immunostaining of PrP^C in the SHa brain was most intense in the stratum radiatum and stratum oriens of the CA1 region of the hippocampus and was virtually absent from the granule cell layer of the dentate gyrus and the pyramidal cell layer throughout Ammon's horn. PrP^{Sc} staining was minimal in those regions which were intensely stained for PrP^C. A similar relationship between PrP^C and PrP^{Sc} was found in the amygdala. In contrast, PrP^{Sc} accumulated in the medial habenular nucleus, the medial septal nuclei, and the diagonal band of Broca; in contrast, these areas were virtually devoid of PrP^C. In the white matter, bundles of myelinated axons contained PrP^{Sc} but were devoid of PrP^C. These findings suggest that prions are transported along axons and are in agreement with earlier findings in which scrapie infectivity was found to migrate in a pattern consistent with retrograde transport [209–211].

Molecular genetics of prion diseases. Independent of enriching brain fractions for scrapie infectivity that led to the discovery of PrP^{Sc}, the PrP gene was shown to be genetically linked to a locus controlling scrapie incubation times [212]. Subsequently, mutation of the PrP gene was shown to be genetically linked to the development of familial prion disease [4]. At the same time, expression of a SHaPrP transgene in mice was shown to render the animals highly susceptible to SHa prions demonstrating that expression of a foreign PrP gene could abrogate the species barrier [213]. Later, PrP-deficient ($Prnp^{0/0}$) mice were found to be resistant to prion infection and failed to replicate prions, as expected [100, 101]. The results of these studies indicated PrP must play a central role in the transmission and pathogenesis of prion disease but equally important, they argued that the abnormal isoform is an essential component of the prion particle [23].

PrP gene dosage controls length of incubation time. Scrapie incubation times in mice were used to distinguish prion strains and to identify a gene controlling its length [135, 214]. This gene was initially called *Sinc* based on genetic crosses between C57Bl and VM mice that exhibited short and long incubation times, respectively [214]. Because of the restricted distribution of VM mice, we searched for another mouse with long incubation times. I/Ln mice proved to be a suitable substitute for VM mice [215]; eventually, I/Ln and VM mice were found to be derived from a common ancestor [216]. With a PrP cDNA probe, we demonstrated genetic linkage between the PrP gene and a locus controlling the incubation time in crosses between NZW/LacJ and I/Ln mice [212]. We provisionally designated these genes as components of the *Prn* complex but eventually found that the incubation time gene, *Prn-i*, is either congruent with or closely linked to the PrP gene, *Prnp* [194].

Although the amino acid substitutions in PrP that distinguish NZW *(Prnp^a^)* from I/Ln *(Prnp^b^)* mice argued for congruency of *Prnp* and *Prn-i*, experiments with *Prnp^a^* mice expressing *Prnp^b^* transgenes demonstrated a "paradoxical" shortening of incubation times [195]. We had predicted that these Tg mice would exhibit a prolongation of the incubation time after inoculation with RML prions based on (*Prnp^a^* x*Prnp^b^*) F1 mice which do exhibit long incubation times. We described those findings as "paradoxical shortening" because we and others had believed for many years that long incubation times are dominant traits [212, 214]. From studies of congenic and transgenic mice expressing different numbers of the *a* and *b* alleles of *Prnp*, we learned that these findings were not paradoxical; indeed, they resulted from increased PrP gene dosage [217]. When the RML isolate was inoculated into congenic and transgenic mice, increasing the number of copies of the *a* allele was found to be the major determinant in reducing the incubation time; however, increasing the number of copies of the *b* allele also reduced the incubation time, but not to the same extent as that seen with the *a* allele. From the foregoing investigations, we concluded that both *Sinc* and *Prn-i* are congruent with PrP [217], and recent gene targeting studies have confirmed this view [218].

Overexpression of wtPrP transgenes. Mice were constructed expressing different levels of the wild-type (wt) SHaPrP transgene [213]. Inoculation of these Tg(SHaPrP) mice with SHa prions demonstrated abrogation of the species barrier resulting in abbreviated incubation times [219]. The length of the incubation time after inoculation with SHa prions was inversely proportional to the level of SHaPrP^{C} in the brains of Tg(SHaPrP) mice [219]. Bioassays of brain extracts from clinically ill Tg(SHaPrP) mice inoculated with Mo prions revealed that only Mo prions but no SHa prions were produced. Conversely, inoculation of Tg(SHaPrP) mice with SHa prions led only to the synthesis of SHa prions. Although the rate of PrP^{Sc} synthesis appears to be a function of the level of PrP^{C} expression in Tg mice, the level to which PrP^{Sc} finally accumulates seems to be independent of PrP^{C} concentration [219].

During transgenetic studies, we discovered that uninoculated older mice harboring numerous copies of wtPrP transgenes derived from Syrian hamsters, sheep, and *Prnp^b^* mice spontaneously developed truncal ataxia, hindlimb paralysis, and tremors [197]. These Tg mice exhibited a profound necrotizing myopathy involving skeletal muscle, a demyelinating polyneuropathy, and focal vacuolation of the CNS. Development of disease was dependent on transgene dosage. For example, Tg(SHaPrP$^{+/+}$)7 mice homozygous for the SHaPrP transgene array regularly developed disease between 400 and 600 days of age, while hemizygous Tg(SHaPrP$^{+/0}$)7 mice also developed disease, but only after >650 days.

PrP-deficient mice. The development and lifespan of two lines of PrP-deficient ($Prnp^{0/0}$) mice were indistinguishable from controls [220, 221] while two other lines exhibited ataxia and Purkinje cell degeneration at ~70 weeks of age [222] (R. Moore and D. Melton, in preparation). In the former two lines with normal development, altered sleep wake cycles have been reported [223] and altered synaptic behavior in brain slices was reported [224, 225] but could not be confirmed by others [226, 227].

$Prnp^{0/0}$ mice are resistant to prions [100, 101]. $Prnp^{0/0}$ mice were sacrificed 5, 60, 120, and 315 days after inoculation with RML prions and brain extracts bioassayed in CD-1 Swiss mice. Except for residual infectivity from the inoculum detected at 5 days after inoculation, no infectivity was detected in the brains of $Prnp^{0/0}$ mice [101]. One group of investigators found that $Prnp^{0/0}$ mice inoculated with RML prions and sacrificed 20 weeks later had $10^{3.6}$ ID_{50} units/ml of homogenate by bioassay [100]. Others have used this report to argue that prion infectivity replicates in the absence of PrP [67, 132]. Neither we nor the authors of the initial report could confirm the finding of prion replication in $Prnp^{0/0}$ mice [101, 103].

Prion species barrier and protein X. The passage of prions between species is often a stochastic process, almost always characterized by prolonged incubation times during the first passage in the new host [36]. This prolongation is often referred to as the "species barrier" [36, 228]. Prions synthesized *de novo* reflect the sequence of the host PrP gene and not that of the PrP^{Sc} molecules in the inoculum derived from the donor [90]. On subsequent passage in a homologous host, the incubation time shortens to that recorded for all subsequent passages. From studies with Tg mice, three factors have been identified that contribute to the species barrier: (i) the difference in PrP sequences between the prion donor and recipient, (ii) the strain of prion, and (iii) the species specificity of protein X, a factor defined by molecular genetic studies that binds to PrP^C and facilitates PrP^{Sc} formation. This factor is likely to be a protein, hence the provisional designation protein X [134, 178]. The prion donor is the last mammal in which the prion was passaged and its PrP sequence represents the "species" of the prion. The strain of prion, which seems to be enciphered in the conformation of PrP^{Sc}, conspires with the PrP sequence, which is specified by the recipient, to determine the tertiary structure of nascent PrP^{Sc}. These principles are demonstrated by studies on the transmission of Syrian hamster (SHa) prions to mice showing that expression of a SHaPrP transgene in mice abrogated the species barrier (Table 3) [213]. Besides the PrP sequence, the strain of prion modified transmission of SHa prions to mice (Table 3) [135, 229, 230].

Transmission of human prions. Protein X was postulated to explain the results on the transmission of human (Hu) prions to Tg mice (Table 4) [134, 179]. Mice expressing both Mo and HuPrP were resistant to Hu prions while those expressing only HuPrP were susceptible. These results argue that $MoPrP^C$ inhibited transmission of Hu prions i.e. the formation of nascent $HuPrP^{Sc}$. In contrast to the foregoing studies, mice expressing both MoPrP and chimeric MHu2M PrP were susceptible to Hu prions and mice expressing MHu2MPrP

Table 3. Influence of prion species and strains on transmission across a species barrier*

Inoculum	Host	Sc237		139H	
		Incubation time / [days ± SEM] (n/n_o)			
SHa	SHa	77 ± 1	(48/48)	167 ± 1	(94/94)
SHa	non-Tg mice	>700	(0/9)	499 ± 15	(11/11)
SHa	Tg(SHaPrP)81/ FVB mice	75 ± 2	(22/22)	110 ± 2	(19/19)
SHa	Tg(SHaPrP)81/ $Prnp^{0/0}$ mice	54 ± 1	(9/9)	65 ± 1	(15/15)

*data from references [101, 219, 230] (D. Groth and S.B. Prusiner, unpublished data). n/n_o = number of diseased animals / number of injected animals.

alone were only slightly more susceptible. These findings contend that $MoPrP^{C}$ has only a minimal effect on the formation of chimeric $MHu2MPrP^{Sc}$.

Table 4. Evidence for protein X from transmission studies of human prions†

Inoculum	Host	MoPrP gene	Incubation Time	
			[days ± SEM]	(n/no)
sCJD	Tg(HuPrP)	$Prnp^{+/+}$	721	(1/10)
sCJD	Tg(HuPrP)$Prnp^{0/0}$	$Prnp^{0/0}$	263 ± 2	(6/6)
sCJD	Tg(MHu2M)	$Prnp^{+/+}$	238 ± 3	(8/8)
sCJD	Tg(MHu2M)$Prnp^{0/0}$	$Prnp^{0/0}$	191 ± 3	(10/10)

†data with inoculum RG [134].

Genetic evidence for protein X. When the data on Hu prion transmission to Tg mice were considered together, they suggested that $MoPrP^{C}$ prevented the conversion of $HuPrP^{C}$ into PrP^{Sc} but had little effect on the conversion of MHu2M into PrP^{Sc} by binding to another Mo protein. We interpreted these results in terms of $MoPrP^{C}$ binding to this Mo protein with a higher affinity than does $HuPrP^{C}$. We postulated that $MoPrP^{C}$ had little effect on the formation of PrP^{Sc} from MHu2M (Table 4) because MoPrP and MHu2M share the same amino acid sequence at the COOH-terminus. We hypothesized that $MoPrP^{C}$ only weakly inhibited transmission of SHa prions to Tg(SHaPrP) mice (Table 3) because SHaPrP is more closely related to MoPrP than is HuPrP.

Using scrapie-infected mouse (Mo) neuroblastoma cells transfected with chimeric Hu/Mo PrP genes, we extended our studies of protein X. Substitution of a Hu residue at position 214 or 218 prevented PrP^{Sc} formation (Figure 5B) [178]. The side chains of these residues protrude from the same surface of the COOH-terminal α-helix forming a discontinuous epitope with residues 167 and 171 in an adjacent loop. Substitution of a basic residue at positions 167, 171, or 218 prevented PrP^{Sc} formation; these mutant PrPs appear to act as "dominant negatives" by binding protein X and rendering it unavailable for prion propagation. Our findings within the context of protein X explain the protective effects of basic polymorphic residues in PrP of humans and sheep [198, 231, 232].

Is protein X a molecular chaperone? Since PrP undergoes a profound structural transition during prion propagation, it seems likely that other proteins such as chaperones participate in this process. Whether protein X functions as a molecular chaperone is unknown. Interestingly, scrapie-infected cells in culture display marked differences in the induction of heat-shock proteins [233, 234] and Hsp70 mRNA has been reported to increase in scrapie of mice [235]. While attempts to isolate specific proteins that bind to PrP have been disappointing [236], PrP has been shown to interact with Bcl-2 and Hsp60 by two-hybrid analysis in yeast [237, 238]. Although these studies are suggestive, no molecular chaperone involved in prion formation in mammalian cells has been identified.

Miniprions. Using the four-helix bundle model of PrP^C (Figure 4A) [147], each region of proposed secondary structure was systematically deleted and the mutant constructs expressed in scrapie-infected neuroblastoma (ScN2a) cells and Tg mice [164, 168]. Deletion of any of the four putative helical regions prevented PrP^{Sc} formation while deletion of the NH_2-terminal region containing residues 23–89 did not affect the yield of PrP^{Sc}. In addition to the 67 residues at the NH_2-terminus, 36 residues from position 141 to 176 could be deleted without altering PrP^{Sc} formation (Figures 6 and 7). The resulting PrP molecule of 106 amino acids was designated PrP106. In this mutant PrP, helix A as well as the S2 β-strand were removed. When PrP106 was expressed in ScN2a cells, $PrP^{Sc}106$ was soluble in 1% Sarkosyl. Whether the structure of $PrP^{Sc}106$ can be more readily determined than that of full-length PrP^{Sc} remains uncertain.

Transgene-specified susceptibility. Tg(PrP106)$Prnp^{0/0}$ mice that expressed PrP106 developed neurological dysfunction ~300 days after inoculation with RML prions previously passaged in CD-1 Swiss mice (S. Supattapone, T. Muramoto, D. Peretz, S. J. DeArmond, A. Wallace, F. E. Cohen, S. B. Prusiner and M. R. Scott, in preparation). The resulting prions containing $PrP^{Sc}106$

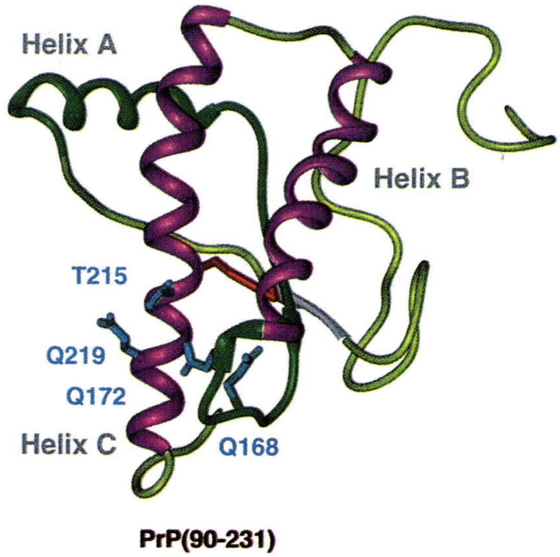

Figure 6. Miniprions produced by deleting PrP residues 23–89 and 141–176. The deletion of residues 141–176 (green) containing helix A and the S2 β-strand is shown.

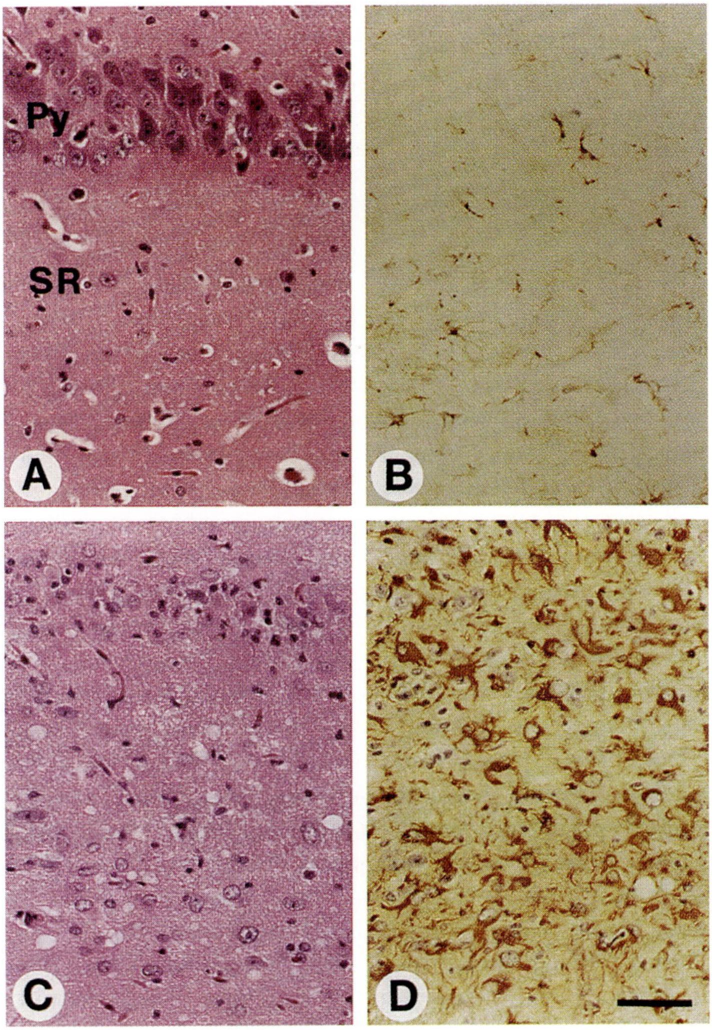

Figure 7. Tg(PrP106) $Prnp^{0/0}$ mice were inoculated with RML106 prions containing $PrP^{Sc}106$. Sections of the hippocampus stained with hematoxylin (A and C) and eosin (B and D). (A and B) Control Tg(PrP106) mouse uninoculated and without neurologic deficits. (C and D) Tg(PrP106) $Prnp^{0/0}$ mouse inoculated with RML106 prions and sacrificed after signs of neurologic dysfunction observed. Photomicrographs prepared by Stephen J. DeArmond.

produced CNS disease in ~66 days upon subsequent passage in Tg(PrP106)$Prnp^{0/0}$ mice. (Table 5). Besides widespread spongiform degeneration and PrP deposits, the pyramidal cells of the hippocampus comprising the CA-1, CA-2, and CA-3 fields disappeared in Tg(PrP106)$Prnp^{0/0}$ mice inoculated with prions containing $PrP^{Sc}106$ (Figure 7A–D). In no previous study of Tg mice have we seen similar neuropathological lesions. The Tg(MoPrP-A) mice overexpressing MoPrP are resistant to RML106 miniprions, but are highly susceptible to RML prions. These mice require more than 250 days to produce illness after inoculation with miniprions but develop disease in ~50 days when inoculated with RML prions containing full-length $MoPrP^{Sc}$.

Table 5. Susceptibility and resistance of transgenic mice to artificial miniprions*

Host	RML106 miniprions		RML prions	
	Incubation time [days ± SEM] (n/n_0)			
Tg(PrP106)Prnp$^{0/0}$ mice	66 ± 3	(10/10)	300 ± 2	(9/10)
Tg(MoPrP-A) mice	>250	(0/11)	50 ± 2	(16/16)

*data from references [217] (S. Supattapone, T. Muramoto, D. Peretz, S. J. DeArmond, A. Wallace, F. E. Cohen, S. B. Prusiner and M. R. Scott, in preparation).

Smaller prions and mythical viruses. The unique incubation times and neuropathogy in Tg mice caused by miniprions are difficult to reconcile with the notion that scrapie is caused by an as yet unidentified virus. When the mutant or wt PrP^{C} of the host matched PrP^{Sc} in the inoculum, the mice were highly susceptible (Table 5). However, when there was a mismatch between PrP^{C} and PrP^{Sc}, the mice were resistant to the prions. This principle of homologous PrP interactions which underlies the species barrier (Table 3) is recapitulated in studies of PrP106 where the amino acid sequence has been drastically changed by deleting nearly 50% of the residues. Indeed, the unique properties of the miniprions provide another persuasive argument supporting the contention that prions are infectious proteins.

Human prion diseases. Most humans afflicted with prion disease present with rapidly progressive dementia, but some manifest cerebellar ataxia. Although the brains of patients appear grossly normal upon post-mortem examination, they usually show spongiform degeneration and astrocytic gliosis under the microscope. The human prion diseases can present as sporadic, genetic, or infectious disorders [5] (Table 1A).

Sporadic CJD. Sporadic forms of prion disease comprise most cases of CJD and possibly a few cases of Gerstmann-Sträussler-Scheinker disease (GSS) (Table 1A) [4, 239, 240]. In these patients, mutations of the PrP gene are not found. How prions causing disease arise in patients with sporadic forms is unknown; hypotheses include horizontal transmission of prions from humans or animals [241], somatic mutation of the PrP gene, and spontaneous conversion of PrP^{C} into PrP^{Sc} [5, 15]. Since numerous attempts to establish an infectious link between sporadic CJD and a preexisting prion disease in animals or humans have been unrewarding, it seems unlikely that transmission features in the pathogenesis of sporadic prion disease [9–12, 242].

Inherited prion diseases. To date, 20 different mutations in the human PrP gene resulting in nonconservative substitutions have been found that segregate with the inherited prion diseases (Figure 4B). Familial CJD cases suggested that genetic factors might influence pathogenesis [1, 2, 243], but this was difficult to reconcile with the transmissibility of fCJD and GSS [3]. The discovery of genetic linkage between the PrP gene and scrapie incubation times in mice [212] raised the possibility that mutation might feature in the hereditary human prion diseases. The P102L mutation was the first PrP mutation to be genetically linked to CNS dysfunction in GSS (Figure 4B)[4] and has since been found in many GSS families throughout the world [244-246]. Indeed, a mutation in the protein coding region of the PrP gene has been found in all reported kindred with familial human prion disease; besides the P102L mutation, genetic linkage has been established for four other mutations [16, 29–31].

Tg mouse studies confirmed that mutations of the PrP gene can cause neurodegeneration. The P102L mutation of GSS was introduced into the MoPrP transgene and five lines of Tg(MoPrP-P101L) mice expressing high levels of mutant PrP developed spontaneous CNS degeneration consisting of widespread vacuolation of the neuropil, astrocytic gliosis, and numerous PrP amyloid plaques similar to those seen in the brains of humans who die from GSS(P102L) [247–249]. Brain extracts prepared from spontaneously ill Tg(MoPrP-P101L) mice transmitted CNS degeneration to Tg196 mice but contained no protease-resistant PrP [248, 249]. The Tg196 mice do not develop spontaneous disease but express low levels of the mutant transgene MoPrP-P101L and are deficient for mouse PrP ($Prnp^{0/0}$) [220]. These studies, combined with the transmission of prions from patients who died of GSS to apes and monkeys [3] or to Tg(MHu2M-P101L) mice [134], demonstrate that prions are generated de novo by mutations in PrP. Additionally, brain extracts from patients with some other inherited prion diseases, fCJD(E200K) or FFI, transmit disease to Tg(MHu2M) mice [27]. An artificial set of mutations in a PrP transgene consisting of A113V, A115V, and A118V produced neurodegeneration in neonatal mice; these Val substitutions were selected for their propensity to form β sheets [153, 250]. In preliminary studies, brain extracts from two of these mice transmitted disease to hamsters and to Tg mice expressing a chimeric SHa/Mo PrP.

Genetic disease that is transmissible. Had the PrP gene been identified in families with prion disease by positional cloning or through the purification and sequencing of PrP in amyloid plaques before brain extracts were shown to be transmissible, the prion concept might have been more readily accepted. Within that scenario, it seems likely that we would have explored the possibility that the mutant protein, upon inoculation into a susceptible host, stimulated production of more of a similar protein. Postulating an infectious pathogen with a foreign genome would have been the least likely candidate to explain how a genetic disease could be experimentally transmissible.

Infectious prion diseases. The infectious prion diseases include kuru of the Fore people in New Guinea where prions were transmitted by ritualistic cannibalism [241, 251, 252]. With the cessation of cannibalism at the urging of missionaries, kuru began to decline long before it was known to be transmissible (Figure 8). Sources of prions causing infectious CJD on several different continents include improperly sterilized depth electrodes, transplanted corneas, human growth hormone (HGH) and gonadotrophin derived from cadaveric pituitaries, and dura mater grafts [253]. Over 90 young adults have developed CJD after treatment with cadaveric HGH; the incubation periods range from 3 to more than 20 years [254, 255]. Dura mater grafts implanted during neurosurgical procedures seem to have caused more than 60 cases of CJD; these incubation periods range from one to more than 14 years [256–258].

Prion diversity. The existence of prion strains raises the question of how

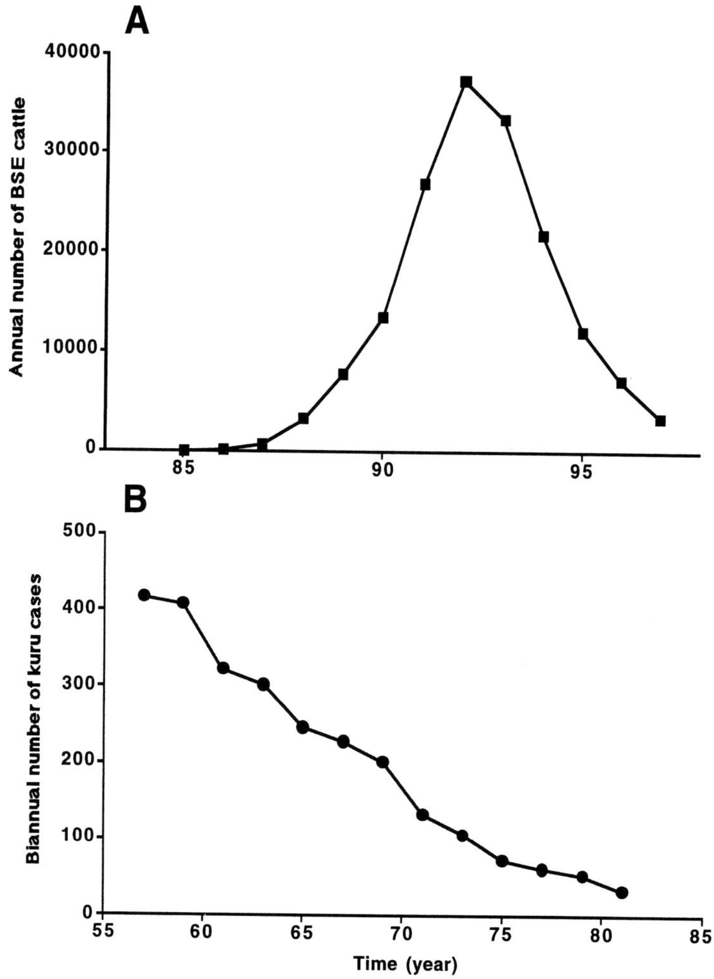

Figure 8. Disappearance of the kuru and BSE epidemics. (A) Number of annual cases of BSE in cattle in Great Britain. (B) Number of biannual cases of kuru in Papua New Guinea. Data compiled for BSE by John Wilesmith and for kuru by Michael Alpers.

heritable biological information can be enciphered in a molecule other than nucleic acid [131, 214, 259-263]. Strains or varieties of prions have been defined by incubation times and the distribution of neuronal vacuolation [214, 264]. Subsequently, the patterns of PrP^{Sc} deposition were found to correlate with vacuolation profiles and these patterns were also used to characterize strains of prions [230, 265, 266]

The typing of prion strains in C57Bl, VM, and F1(C57Bl x VM) inbred mice began with isolates from sheep with scrapie. The prototypic strains called Me7 and 22A gave incubation times of ~150 and ~400 days in C57Bl mice, respectively [214, 267, 268]. The PrP genes of C57Bl and I/Ln (and later VM) mice encode proteins differing at two residues and control scrapie incubation times [194, 212, 216–218, 269].

Until recently, support for the hypothesis that the tertiary structure of PrP^{Sc}

enciphers strain-specific information [23] was minimal except for the DY strain isolated from mink with transmissible encephalopathy by passage in Syrian hamsters [26, 270, 271]. PrP^{Sc} in DY prions showed diminished resistance to proteinase K digestion as well as a peculiar site of cleavage. The DY strain presented a puzzling anomaly since other prion strains exhibiting similar incubation times did not show this altered susceptibility to proteinase K digestion of PrP^{Sc} [135]. Also notable was the generation of new strains during passage of prions through animals with different PrP genes [135, 229].

PrP^{Sc} conformation enciphers diversity. Persuasive evidence that strain-specific information is enciphered in the tertiary structure of PrP^{Sc} comes from transmission of two different inherited human prion diseases to mice expressing a chimeric MHu2M PrP transgene [27]. In fatal familial insomnia (FFI), the protease-resistant fragment of PrP^{Sc} after deglycosylation has an M_r of 19 kD; whereas in fCJD(E200K) and most sporadic prion diseases, it is 21 kD (Table 6) [272, 273]. This difference in molecular size was shown to be due to different sites of proteolytic cleavage at the NH_2-termini of the two human PrP^{Sc} molecules reflecting different tertiary structures [272]. These distinct conformations were understandable since the amino acid sequences of the PrPs differ.

Table 6. Distinct prion strains generated in humans with inherited prion diseases and transmitted to transgenic mice*

Inoculum	Host Species	Host PrP Genotype	Incubation time [days ± SEM]	(n/n_o)	PrP^{Sc}(kD)
None	Human	FFI(D178N,M129)			19
FFI	Mouse	Tg(MHu2M)	206 ± 7	(7/7)	19
FFI → Tg(MHu2M)	Mouse	Tg(MHu2M)	136 ± 1	(6/6)	19
None	Human	fCJD(E200K)			21
fCJD	Mouse	Tg(MHu2M)	170 ± 2	(10/10)	21
fCJD → Tg(MHu2M)	Mouse	Tg(MHu2M)	167 ± 3	(15/15)	21

*data from reference [27] (G. Telling *et al.*, in preparation).

Extracts from the brains of FFI patients transmitted disease to mice expressing a chimeric MHu2M PrP gene about 200 days after inoculation and induced formation of the 19 kD PrP^{Sc}; whereas fCJD(E200K) and sCJD produced the 21 kD PrP^{Sc} in mice expressing the same transgene [27]. On second passage, Tg(MHu2M) mice inoculated with FFI prions showed an incubation time of ~130 days and a 19 kD PrP^{Sc} while those inoculated with fCJD(E200K) prions exhibited an incubation time of ~170 days and a 21 kD PrP^{Sc} [28]. The experimental data demonstrate that $MHu2MPrP^{Sc}$ can exist in two different conformations based on the sizes of the protease-resistant fragments; yet, the amino acid sequence of $MHu2MPrP^{Sc}$ is invariant.

The results of our studies argue that PrP^{Sc} acts as a template for the conversion of PrP^C into nascent PrP^{Sc}. Imparting the size of the protease-resistant fragment of PrP^{Sc} through conformational templating provides a mechanism for both the generation and propagation of prion strains.

Interestingly, the protease-resistant fragment of PrP^{Sc} after deglycosylation with an M_r of 19 kD has been found in a patient who developed a sporadic case of prion disease similar to FFI but with no family history. Since both PrP alleles encoded the wt sequence and a Met at position 129, we labeled this case, fatal sporadic insomnia (FSI). At autopsy, the spongiform degeneration, reactive astrogliosis, and PrP^{Sc} deposition were confined to the thalamus [274]. These findings argue that the clinicopathologic phenotype is determined by the conformation of PrP^{Sc} in accord with the results of the transmission of human prions from patients with FFI to Tg mice [27].

Selective neuronal targeting. Besides incubation times, profiles of spongiform change (Figure 1) have been used to characterize prion strains [275], but recent studies argue that such profiles are not an intrinsic feature of strains [276, 277]. The mechanism by which prion strains modify the pattern of spongiform degeneration was perplexing since earlier investigations had shown that PrP^{Sc} deposition precedes neuronal vacuolation and reactive gliosis [211, 230]. When FFI prions were inoculated into Tg(MHu2M) mice, PrP^{Sc} was confined largely to the thalamus (Figure 9A) as is the case for FFI in humans [27, 278]. In contrast, fCJD(E200K) prions inoculated into Tg(MHu2M) mice produced widespread deposition of PrP^{Sc} throughout the cortical mantle and many of the deep structures of the CNS (Figure 9B) as is seen in fCJD(E200K) of humans. To examine whether the diverse patterns of PrP^{Sc} deposition are influenced by Asn-linked glycosylation of PrP^{C}, we constructed Tg mice expressing PrPs mutated at one or both of the Asn-linked glycosylation consensus sites [277]. These mutations resulted in aberrant neuroanatomic topologies of PrP^{C} within the CNS, whereas pathologic point mutations adjacent to the consensus sites did not alter the distribution of PrP^{C}. Tg mice with mutation of the second PrP glycosylation site exhibited prion incubation times of >500 days and unusual patterns of PrP^{Sc} deposition. These findings raise two possible scenarios. First, glycosylation can modify the conformation of PrP^{C} and affect its affinity for a particular conformer of PrP^{Sc}, which results in specific patterns of PrP^{Sc} deposition; such interactions between PrP^{Sc} and PrP^{C} are likely to determine the rate of nascent PrP^{Sc} formation. Second, glycosylation modifies the stability of PrP^{Sc} and hence, the rate of PrP^{Sc} clearance. This latter explanation is consistent with the proposal that the binding of PrP^{C} to protein X is the rate-limiting step in PrP^{Sc} formation under most circumstances [255].

Bovine spongiform encephalopathy. Prion strains and the species barrier are of paramount importance in understanding the BSE epidemic in Great Britain, in which it is estimated that almost one million cattle were infected with prions [279, 280]. The mean incubation time for BSE is about 5 years. Most cattle therefore did not manifest disease since they were slaughtered between 2 and 3 years of age [281]. Nevertheless, more than 160,000 cattle, primarily dairy cows, have died of BSE over the past decade (Figure 5A) [279]. BSE is a massive common source epidemic caused by meat and bone meal (MBM) fed primarily to dairy cows [280, 282]. The MBM was prepared from the offal of sheep, cattle, pigs, and chickens as a high protein nutritional supplement. In the late 1970s, the hydrocarbon-solvent extraction method used in the rendering of offal began to be abandoned resulting in MBM with a much higher

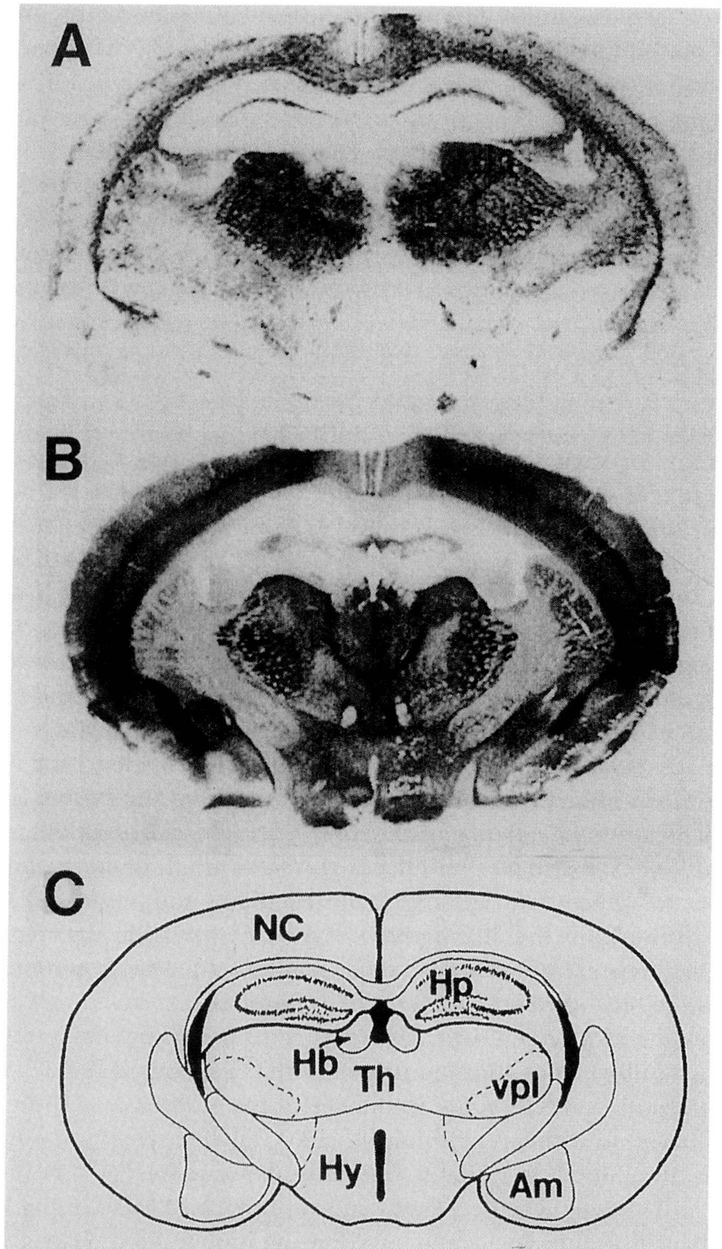

Figure 9. Regional distribution of PrP^{Sc} deposition in Tg(MHu2M)$Prnp^{0/0}$ mice inoculated with prions from humans who died of inherited prion diseases (Table 5). Histoblot of PrP^{Sc} deposition in a coronal section a Tg(MHu2M)$Prnp^{0/0}$ mouse through the hippocampus and thalamus [27]. (A) The Tg mouse was inoculated with brain extract prepared from a patient who died of FFI. (B) The Tg mouse was inoculated with extract from a patient with fCJD(E200K). Cryostat sections were mounted on nitrocellulose and treated with proteinase K to eliminate PrP^{C} [208]. To enhance the antigenicity of PrP^{Sc}, the histoblots were exposed to 3 guanidinium isothiocyanate before immunostaining using anti-PrP 3F4 mAb [174]. **(C)** Labelled diagram of a coronal sections of the hippocampus/thalamus region. NC, neocortex; Hp, hippocampus; Hb, habenula; Th, thalamus; vpl, ventral posterior lateral thalamic nucleus; Hy, hypothalamus; Am, amygdala. Photomicrographs prepared by Stephen J. DeArmond.

fat content [282]. It is now thought that this change in the rendering process allowed scrapie prions from sheep to survive rendering and to be passed into cattle. Alternatively, bovine prions were present at low levels prior to modification of the rendering process and with the processing change survived in sufficient numbers to initiate the BSE epidemic when inoculated back into cattle orally through MBM. Against the latter hypothesis is the widespread geographical distribution throughout England of the initial 17 cases of BSE, which occurred almost simultaneously [280, 283, 284]. Furthermore, there is no evidence of a pre-existing prion disease of cattle, either in Great Britain or elsewhere.

Origin of BSE prions? The origin of the bovine prions causing BSE cannot be determined by examining the amino acid sequence of PrP^{Sc} in cattle with BSE since the PrP^{Sc} in these animals has the bovine sequence whether the initial prions in MBM came from cattle or sheep. The bovine PrP sequence differs from that of sheep at 7 or 8 positions [285–287]. In contrast to the many PrP polymorphisms found in sheep, only one PrP polymorphism has been found in cattle. Though most bovine PrP alleles encode five octarepeats, some encode six. PrP alleles encoding six octarepeats do not seem to be overrepresented in BSE (Figure 4B) [288].

Brain extracts from BSE cattle cause disease in cattle, sheep, mice, pigs, and mink after intracerebral inoculation [289–293], but prions in brain extracts from sheep with scrapie fed to cattle produced illness substantially different from BSE [294]. However, no exhaustive effort has been made to test different strains of sheep prions or to examine the disease following bovine to bovine passage. The annual incidence of sheep with scrapie in Britain over the past two decades has remained relatively low (J. Wilesmith, unpublished data). In July 1988, the practice of feeding MBM to sheep and cattle was banned. Recent statistics argue that the epidemic is now disappearing as a result of this ruminant feed ban (Figure 8A) [279], reminiscent of the disappearance of kuru in the Fore people of New Guinea [241, 252] (Figure 8B).

Monitoring cattle for BSE prions. Although many plans have been offered for the culling of older cattle in order to minimize the spread of BSE [279], it seems more important to monitor the frequency of prion disease in cattle as they are slaughtered for human consumption. No reliable, specific test for prion disease in live animals is available but immunoassays for PrP^{Sc} in the brainstems of cattle might provide a reasonable approach to establishing the incidence of subclinical BSE in cattle entering the human food chain [176, 208, 287, 295–297]. Determining how early in the incubation period PrP^{Sc} can be detected by immunological methods is now possible since a reliable bioassay has been created by expressing the BoPrP gene in Tg mice [298]. Prior to development of Tg(BoPrP)$Prnp^{0/0}$ mice, non-Tg mice inoculated intracerebrally with BSE brain extracts required more than 300 days to develop disease [67, 293, 299, 300]. Depending on the titer of the inoculum, the structures of PrP^{C} and PrP^{Sc}, and the structure of protein X, the number of inoculated animals developing disease can vary over a wide range (Table 3). Some investigators have stated that transmission of BSE to mice is quite variable with incubation periods exceeding one year [67] while others report low prion titers in BSE brain homogenates [299, 300] compared to rodent brain scrapie [54, 56, 301, 302].

Have bovine prions been transmitted to humans? In 1994, the first cases of CJD in teenagers and young adults that were eventually labelled new variant (v) CJD occurred in Britain [303]. Besides the young age of these patients [304, 305], the brains of these patients showed numerous PrP amyloid plaques surrounded by a halo of intense spongiform degeneration (Figure 10A and B) [306]. Later, one French case meeting these criteria followed [307]. These unusual neuropathologic changes have not been seen in CJD cases in the United States, Australia, or Japan [306, 308]. Both macaque monkeys and marmosets developed neurologic disease several years after inoculation with bovine

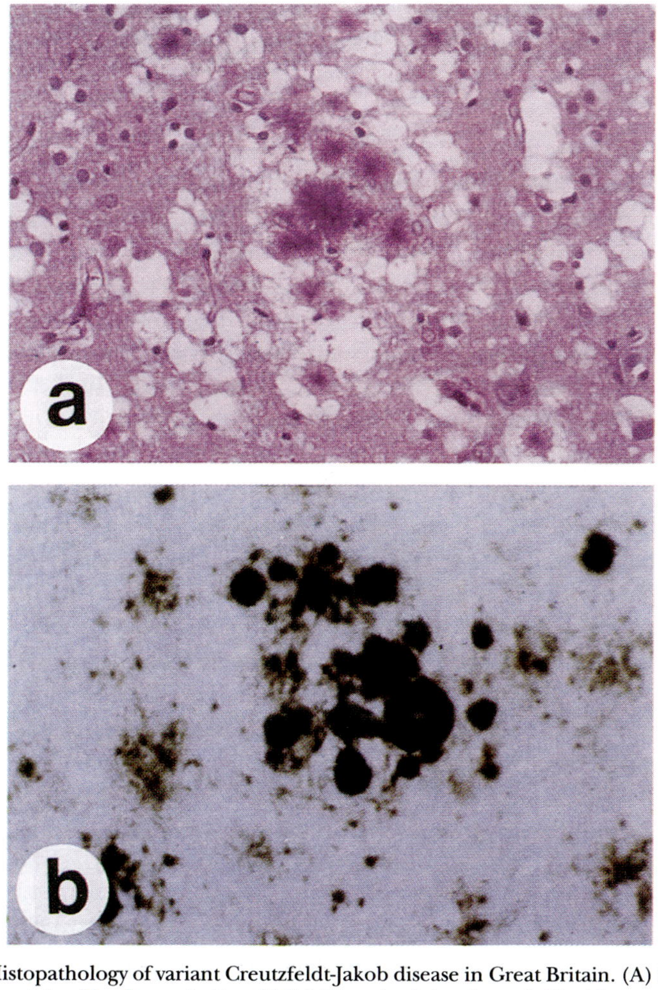

Figure 10. Histopathology of variant Creutzfeldt-Jakob disease in Great Britain. (A) Section from frontal cortex stained by the periodic acid Schiff method showing a field with aggregates of plaques surrounded by spongiform degeneration. Magnification X 100. (B) Multiple plaques and amorphous deposits are PrP immunopositive. Magnification X 500. Specimens provided by James Ironside, Jeanne Bell, and Robert Will; photomicrographs prepared by Stephen J. DeArmond.

prions [309], but only the macaques exhibited numerous PrP plaques similar to those found in vCJD [310] (R. Ridley and H. Baker, unpublished data).

The restricted geographical occurrence and chronology of vCJD have raised the possibility that BSE prions have been transmitted to humans. That only ~25 vCJD cases have been recorded and the incidence has remained relatively constant makes establishing the origin of vCJD difficult. No set of dietary habits distinguishes vCJD patients from apparently healthy people. Moreover, there is no explanation for the predilection of vCJD for teenagers and young adults. Why have older individuals not developed vCJD-based neuropathologic criteria? It is noteworthy that epidemiological studies over the past three decades have failed to find evidence for transmission of sheep prions to humans [9–12]. Attempts to predict the future number of cases of vCJD, assuming exposure to bovine prions prior to the offal ban, have been uninformative because so few cases of vCJD have occurred [311–313]. Are we at the beginning of a human prion disease epidemic in Britain like those seen for BSE and kuru (Figure 8) or will the number of vCJD cases remain small as seen with iCJD caused by cadaveric HGH [254, 255]?

Strain of BSE prions. Was a particular conformation of bovine PrP^{Sc} selected for heat-resistance during the rendering process and then reselected multiple times as cattle infected by ingesting prion-contaminated MBM were slaughtered and their offal rendered into more MBM? Recent studies of PrP^{Sc} from brains of patients who died of vCJD show a pattern of PrP glycoforms different from those found for sCJD or iCJD [314, 315]. But the utility of measuring PrP glycoforms is questionable in trying to relate BSE to vCJD [316, 317] because PrP^{Sc} is formed after the protein is glycosylated [138, 140] and enzymatic deglycosylation of PrP^{Sc} requires denaturation [318, 319]. Alternatively, it may be possible to establish a relationship between the conformations of PrP^{Sc} from cattle with BSE and those from humans with vCJD by using Tg mice as was done for strains generated in the brains of patients with FFI or fCJD [27, 298]. A relationship between vCJD and BSE has been suggested by finding similar incubation times in non-Tg RIII mice of ~310 days after inoculation with Hu or Bo prions [293].

Yeast and other prions. Although prions were originally defined in the context of an infectious pathogen [58], it is now becoming widely accepted that prions are elements that impart and propagate variability through multiple conformers of a normal cellular protein. Such a mechanism must surely not be restricted to a single class of transmissible pathogens. Indeed, it is likely that the original definition will need to be extended to encompass other situations where a similar mechanism of information transfer occurs.

Two notable prion-like determinants, [URE3] and [PSI], have already been described in yeast and one in another fungus denoted [Het-s*] [320–325]. Studies of candidate prion proteins in yeast may prove particularly helpful in the dissection of some of the events that feature in PrP^{Sc} formation. Interestingly, different strains of yeast prions have been identified [326]. Conversion to the prion-like [PSI] state in yeast requires the molecular chaperone Hsp104; however, no homolog of Hsp104 has been found in mammals [321, 327]. The NH_2-terminal prion domains of Ure2p and Sup35 that are responsible for the [URE3] and [PSI] phenotypes in yeast have been identified. In contrast to PrP, which is a GPI-anchored membrane protein, both Ure2p and Sup35 are cytosolic proteins [328]. When the prion domains of these yeast proteins were expressed in *E. coli*, the proteins were found to polymerize into fibrils with properties similar to those of PrP and other amyloids [322–324].

Whether prions explain some other examples of acquired inheritance in lower organisms is unclear [329, 330]. For example, studies on the inheritance of positional order and cellular handedness on the surface of small organisms have demonstrated the epigenetic nature of these phenomena but the mechanism remains unclear [331, 332].

Prevention of and therapeutics for prion diseases. As our understanding of prion propagation increases, it should be possible to design effective therapeutics. Because people at risk for inherited prion diseases can now be identified decades before neurologic dysfunction is evident, the development of an effective therapy for these fully penetrant disorders is imperative [333, 334]. Although we have no way of predicting the number of individuals who may develop neurologic dysfunction from bovine prions in the future [312], it would be prudent to seek an effective therapy now [28, 335]. Interfering with the conversion of PrP^C into PrP^{Sc} seems to be the most attractive therapeutic target [60]. Either stabilizing the structure of PrP^C by binding a drug or modifying the action of protein X, which might function as a molecular chaperone (Figure 5B), are reasonable strategies. Whether it is more efficacious to design a drug that binds to PrP^C at the protein X binding site or one that mimics the structure of PrP^C with basic polymorphic residues that seem to prevent scrapie and CJD remains to be determined [178, 232]. Since PrP^{Sc} formation seems limited to caveolae-like domains [142–145, 336], drugs designed to inhibit this process need not penetrate the cytosol of cells but they do need to be able to enter the CNS. Alternatively, drugs that destabilize the structure of PrP^{Sc} might also prove useful.

The production of domestic animals that do not replicate prions may also be important with respect to preventing prion disease. Sheep encoding the R/R polymorphism at position 171 seem to be resistant to scrapie [198, 231, 337–343]; presumably, this was the genetic basis of Parry's scrapie eradication program in Great Britain 30 years ago [44, 344]. A more effective approach using dominant negatives for producing prion-resistant domestic animals, including sheep and cattle, is probably the expression of PrP transgenes encoding R171 as well as additional basic residues at the putative protein X binding site (Figure 5B) [178]. Such an approach can be readily evaluated in Tg mice and once shown to be effective, it could be instituted by artificial insemination of sperm from males homozygous for the transgene. More difficult is the production of PrP-deficient cattle and sheep. Although such animals would not be susceptible to prion disease [100, 101], they might suffer some deleterious effects from ablation of the PrP gene [222–224, 227].

Whether gene therapy for the human prion diseases using the dominant negative approach described above for prion-resistant animals will prove feasible depends on the availability of efficient vectors for delivery of the transgene to the CNS.

Concluding remarks–looking to the future. Although the study of prions has taken several unexpected directions over the past three decades, a rather novel and fascinating story of prion biology is emerging. Investigations of prions have elucidated a previously unknown mechanism of disease in humans and animals. While learning the details of the structures of PrPs and deciphering the mechanism of PrP^C transformation into PrP^{Sc} will be important, the fundamental principles of prion biology have become reasonably clear. Though some investigators prefer to view the composition of the infectious prion particle as unresolved [335, 345], such a perspective denies an enlarging body of data, none of which refutes the prion concept. Moreover, the discovery of prion-like phenomena mediated by proteins unrelated to PrP in yeast and fungi serves not only to strengthen the prion concept but also to widen it [328].

Hallmark of prion diseases. The hallmark of all prion diseases–whether sporadic, dominantly inherited, or acquired by infection–is that they involve the aberrant metabolism and resulting accumulation of the prion protein (Table 1)[23]. The conversion of PrP^C into PrP^{Sc} involves a conformation change whereby the α-helical content diminishes and the amount of β-sheet increases [25]. These findings provide a reasonable mechanism to explain the conundrum presented by the three different manifestations of prion disease.

Understanding how PrP^C unfolds and refolds into PrP^{Sc} will be of paramount importance in transferring advances in the prion diseases to studies of other degenerative illnesses. The mechanism by which PrP^{Sc} is formed must involve a templating process whereby existing PrP^{Sc} directs the refolding of PrP^C into a nascent PrP^{Sc} with the same conformation. Not only will a knowledge of PrP^{Sc} formation help in the rational design of drugs that interrupt the pathogenesis of prion diseases, but it may also open new approaches to deciphering the causes of and to developing effective therapies for the more common neurodegenerative diseases including Alzheimer's disease, Parkinson's disease, and amyotrophic lateral sclerosis (ALS). Indeed, the expanding list of prion diseases and their novel modes of transmission and pathogenesis (Table 1), as well as the unprecedented mechanisms of prion propagation and information transfer (Tables 5 and 6), indicate that much more attention to these fatal disorders of protein conformation is urgently needed.

But prions may have even wider implications than those noted for the common neurodegenerative diseases. If we think of prion diseases as disorders of protein conformation and do not require the diseases to be transmissible, then what we have learned from the study of prions may reach far beyond these common illnesses.

Multiple conformers. The discovery that proteins may have multiple biologically active conformations may prove no less important than the implications of prions for diseases. How many different tertiary structures can PrP^{Sc} adopt? This query not only addresses the issue of the limits of prion diversity but also applies to proteins as they normally function within the cell or act to affect homeostasis in multicellular organisms. The expanding list of chaperones that assist the folding and unfolding of proteins promises much new knowledge about this process. For example, it is now clear that proproteases can carry their own chaperone activity where the *pro* portion of the protein functions as a chaperone in *cis* to guide the folding of the proteolytically active portion before it is cleaved [346]. Such a mechanism might well feature in the maturation of polypeptide hormones. Interestingly, mutation of the chaperone portion of prosubtilisin resulted in the folding of a subtilisin protease with different properties than the one folded by the wild-type chaperone. Such chaperones have also been shown to work in *trans* [346]. Besides transient metabolic regulation within the cell and hormonal regulation of multicellular organisms, it is not unreasonable to suggest that polymerization of proteins into multimeric structures such as intermediate filaments might be controlled at least in part by alternative conformations of proteins. Such regulation of multimeric protein assemblies might occur either in the proteins that form the polymers or the proteins that function to facilitate the polymerization process. Additionally, apoptosis during development and throughout adult life might also be regulated, at least in part, by alternative tertiary structures of proteins.

Shifting the debate. The debate about prions and the diseases that they cause has now shifted to such issues as how many biological processes are controlled by changes in protein conformation. Although the extreme radiation-resistance of the scrapie infectivity suggested that the pathogen causing this disease and related illnesses would be different from viruses, viroids, and bacteria [32, 33], few thought that alternative protein conformations might even remotely feature in the pathogenesis of the prion diseases [37]. Indeed, an unprecedented mechanism of disease has been revealed where an aberrant conformational change in a protein is propagated. The discovery of prions and their eventual acceptance by the community of scholars represents a triumph of the scientific process over prejudice. The future of this new and emerging area of biology should prove even more interesting and productive as a multitude of unpredicted discoveries emerge.

ACKNOWLEDGEMENT

I thank O. Abramsky, M. Baldwin, R. Barry, C. Bellinger, D. Borchelt, D. Burton, D. Butler, G. Carlson, J. Cleaver, F. Cohen, C. Cooper, S. DeArmond, T. Diener, J. Dyson, S. Farr-Jones, R. Fletterick, R. Gabizon, D. Groth, C. Heinrich, L. Hood, K. Hsiao, Z. Huang, T. James, K. Kaneko, R. Kohler, T. Kitamoto, H. Kretzschmar, V. Lingappa, H. Liu, D. Lowenstein, M. McKinley, Z. Meiner, B. Miller, W. Mobley, B. Oesch, K.-M. Pan, D. Peretz, D. Riesner, G. Roberts, J. Safar, M. Scott, A. Serban, N. Stahl, A. Taraboulos, J. Tateishi, M. Torchia, P. Tremblay, M. Vey, C. Weissmann, D. Westaway, A. Williamson, P. Wright, and L. Zullianello for their contributions to many phases of these studies, and H. Baron, W. Burke, I. Diamond, H. Fields, R. Fishman, D, Gilden, S. Hauser, H. Koprowski, J. Krevans, J. Martin, M. McCarty, R. C. Morris, N.

Nathanson, F. Seitz, and C. Yanofsky for support and encouragement. This research was supported by grants from the National Institute of Aging and the National Institute of Neurologic Diseases and Stroke of the National Institutes of Health, the National Science Foundation, International Human Frontiers of Science Program, and the American Health Assistance Foundation as well as by gifts from the Sherman Fairchild Foundation, the Keck Foundation, the G. Harold and Leila Y. Mathers Foundation, the Bernard Osher Foundation, the John D. French Foundation, the Howard Hughes Medical Institute, R. J. Reynolds, National Medical Enterprises, and Centeon.

REFERENCES

1. Meggendorfer F.: Klinische und genealogische Beobachtungen bei einem Fall von spastischer Pseudosklerose Jakobs. *Z. Gesamte Neurol. Psychiatr.* 128:337–341, 1930.
2. Stender A.: Weitere Beiträge zum Kapitel "Spastische Pseudosklerose Jakobs". *Z. Gesamte Neurol. Psychiatr.* 128:528–543, 1930.
3. Masters C.L., Gajdusek D.C., and Gibbs C.J., Jr.: Creutzfeldt-Jakob disease virus isolations from the Gerstmann-Sträussler syndrome. *Brain* 104:559–588, 1981.
4. Hsiao K., Baker H.F., Crow T.J., Poulter M., Owen F., Terwilliger J.D., Westaway D., Ott J., and Prusiner S.B.: Linkage of a prion protein missense variant to Gerstmann-Sträussler syndrome. *Nature* 338:342–345, 1989.
5. Prusiner S.B.: Scrapie prions. *Annu. Rev. Microbiol.* 43:345–374, 1989.
6. Kirschbaum W.R.: *Jakob-Creutzfeldt Disease,* Elsevier, Amsterdam, 1968.
7. Gibbs C.J., Jr., Gajdusek D.C., Asher D.M., Alpers M.P., Beck E., Daniel P.M., and Matthews W.B.: Creutzfeldt-Jakob disease (spongiform encephalopathy): transmission to the chimpanzee. *Science* 161:388–389, 1968.
8. Prusiner S.B.: The prion diseases. *Sci. Am.* 272:48–57, 1995.
9. Malmgren R., Kurland L., Mokri B., and Kurtzke J.: The epidemiology of Creutzfeldt-Jakob disease. *Slow Transmissible Diseases of the Nervous System, Vol. 1.* Prusiner S.B. and Hadlow W.J., Eds. Academic Press, New York, 1979 pp. 93–112.
10. Brown P., Cathala F., Raubertas R.F., Gajdusek D.C., and Castaigne P.: The epidemiology of Creutzfeldt-Jakob disease: conclusion of 15-year investigation in France and review of the world literature. *Neurology* 37:895–904, 1987.
11. Harries-Jones R., Knight R., Will R.G., Cousens S., Smith P.G., and Matthews W.B.: Creutzfeldt-Jakob disease in England and Wales, 1980–1984: a case-control study of potential risk factors. *J. Neurol. Neurosurg. Psychiatry* 51:1113–1119, 1988.
12. Cousens S.N., Harries-Jones R., Knight R., Will R.G., Smith P.G., and Matthews W.B.: Geographical distribution of cases of Creutzfeldt-Jakob disease in England and Wales 1970–84. *J. Neurol. Neurosurg. Psychiatry* 53:459–465, 1990.
13. Kahana E., Milton A., Braham J., and Sofer D.: Creutzfeldt-Jakob disease: focus among Libyan Jews in Israel. *Science* 183:90–91, 1974.
14. Goldfarb L., Korczyn A., Brown P., Chapman J., and Gajdusek D.C.: Mutation in codon 200 of scrapie amyloid precursor gene linked to Creutzfeldt-Jakob disease in Sephardic Jews of Libyan and non-Libyan origin. *Lancet* 336:637–638, 1990.
15. Hsiao K., Meiner Z., Kahana E., Cass C., Kahana I., Avrahami D., Scarlato G., Abramsky O., Prusiner S.B., and Gabizon R.: Mutation of the prion protein in Libyan Jews with Creutzfeldt-Jakob disease. *N. Engl. J. Med.* 324:1091–1097, 1991.
16. Gabizon R., Rosenmann H., Meiner Z., Kahana I., Kahana E., Shugart Y., Ott J., and Prusiner S.B.: Mutation and polymorphism of the prion protein gene in Libyan Jews with Creutzfeldt-Jakob disease. *Am. J. Hum. Genet.* 53:828–835, 1993.

17. Sigurdsson B.: Rida, a chronic encephalitis of sheep with general remarks on infections which develop slowly and some of their special characteristics. *Br. Vet. J.* 110:341–354, 1954.
18. Hadlow W.J.: Scrapie and kuru. *Lancet* 2:289–290, 1959.
19. Gajdusek D.C., Gibbs C.J., Jr., and Alpers M.: Experimental transmission of a kuru-like syndrome to chimpanzees. *Nature* 209:794–796, 1966.
20. Klatzo I., Gajdusek D.C., and Zigas V.: Pathology of kuru. *Lab. Invest.* 8:799–847, 1959.
21. Zlotnik I. and Stamp J.L.: Scrapie disease of sheep. *World Neurology* 2:895–907, 1961.
22. Masters C.L. and Richardson E.P., Jr.: Subacute spongiform encephalopathy Creutzfeldt-Jakob disease–the nature and progression of spongiform change. *Brain* 101:333–344, 1978.
23. Prusiner S.B.: Molecular biology of prion diseases. *Science* 252:1515–1522, 1991.
24. Wells G.A.H., Scott A.C., Johnson C.T., Gunning R.F., Hancock R.D., Jeffrey M., Dawson M., and Bradley R.: A novel progressive spongiform encephalopathy in cattle. *Vet. Rec.* 121:419–420, 1987.
25. Pan K.-M., Baldwin M., Nguyen J., Gasset M., Serban A., Groth D., Mehlhorn I., Huang Z., Fletterick R.J., Cohen F.E., and Prusiner S.B.: Conversion of α-helices into β-sheets features in the formation of the scrapie prion proteins. *Proc. Natl. Acad. Sci. USA* 90:10962–10966, 1993.
26. Bessen R.A. and Marsh R.F.: Distinct PrP properties suggest the molecular basis of strain variation in transmissible mink encephalopathy. *J. Virol.* 68:7859–7868, 1994.
27. Telling G.C., Parchi P., DeArmond S.J., Cortelli P., Montagna P., Gabizon R., Mastrianni J., Lugaresi E., Gambetti P., and Prusiner S.B.: Evidence for the conformation of the pathologic isoform of the prion protein enciphering and propagating prion diversity. *Science* 274:2079–2082, 1996.
28. Prusiner S.B.: Prion diseases and the BSE crisis. *Science* 278:245–251, 1997.
29. Dlouhy S.R., Hsiao K., Farlow M.R., Foroud T., Conneally P.M., Johnson P., Prusiner S.B., Hodes M.E., and Ghetti B.: Linkage of the Indiana kindred of Gerstmann-Sträussler-Scheinker disease to the prion protein gene. *Nat. Genet.* 1:64–67, 1992.
30. Petersen R.B., Tabaton M., Berg L., Schrank B., Torack R.M., Leal S., Julien J., Vital C., Deleplanque B., Pendlebury W.W., Drachman D., Smith T.W., Martin J.J., Oda M., Montagna P., Ott J., Autilio-Gambetti L., Lugaresi E., and Gambetti P.: Analysis of the prion protein gene in thalamic dementia. *Neurology* 42:1859–1863, 1992.
31. Poulter M., Baker H.F., Frith C.D., Leach M., Lofthouse R., Ridley R.M., Shah T., Owen F., Collinge J., Brown G., Hardy J., Mullan M.J., Harding A.E., Bennett C., Doshi R., and Crow T.J.: Inherited prion disease with 144 base pair gene insertion. 1. Genealogical and molecular studies. *Brain* 115:675–685, 1992.
32. Alper T., Haig D.A., and Clarke M.C.: The exceptionally small size of the scrapie agent. *Biochem. Biophys. Res. Commun.* 22:278–284, 1966.
33. Alper T., Cramp W.A., Haig D.A., and Clarke M.C.: Does the agent of scrapie replicate without nucleic acid? *Nature* 214:764–766, 1967.
34. Latarjet R., Muel B., Haig D.A., Clarke M.C., and Alper T.: Inactivation of the scrapie agent by near monochromatic ultraviolet light. *Nature* 227:1341–1343, 1970.
35. Gibbs C.J., Jr., Gajdusek D.C., and Latarjet R.: Unusual resistance to ionizing radiation of the viruses of kuru, Creutzfeldt-Jakob diseas. *Proc. Natl. Acad. Sci. USA* 75:6268–6270, 1978.
36. Pattison I.H.: Experiments with scrapie with special reference to the nature of the agent and the pathology of the disease. *Slow, Latent and Temperate Virus Infections, NINDB Monograph 2.* Gajdusek D.C., Gibbs C.J., Jr., and Alpers M.P., Eds. U.S. Government Printing, Washington, D.C., 1965 pp. 249–257.
37. Griffith J.S.: Self-replication and scrapie. *Nature* 215:1043–1044, 1967.
38. Pattison I.H. and Jones K.M.: The possible nature of the transmissible agent of scrapie. *Vet. Res.* 80:1–8, 1967.
39. Gibbons R.A. and Hunter G.D.: Nature of the scrapie agent. *Nature* 215:1041–1043, 1967.

40. Hunter G.D., Kimberlin R.H., and Gibbons R.A.: Scrapie: a modified membrane hypothesis. *J. Theor. Biol.* 20:355–357, 1968.
41. Field E.J., Farmer F., Caspary E.A., and Joyce G.: Susceptibility of scrapie agent to ionizing radiation. *Nature* 222:90–91, 1969.
42. Hunter G.D.: Scrapie: a prototype slow infection. *J. Infect. Dis.* 125:427–440, 1972.
43. Gordon W.S.: Advances in veterinary research. *Vet. Res.* 58:516–520, 1946.
44. Parry H.B.: Scrapie: a transmissible and hereditary disease of sheep. *Heredity* 17:75–105, 1962.
45. Dickinson A.G., Young G.B., Stamp J.T., and Renwick C.C.: An analysis of natural scrapie in Suffolk sheep. *Heredity* 20:485–503, 1965.
46. Parry H.B.: Recorded occurrences of scrapie from 1750. *Scrapie Disease in Sheep.* Oppenheimer D.R., Ed. Academic Press, New York, 1983 pp. 31–59.
47. Prusiner S.B.: An approach to the isolation of biological particles using sedimentation analysis. *J. Biol. Chem.* 253:916–921, 1978.
48. Prusiner S.B., Garfin D.E., Baringer J.R., Cochran S.P., Hadlow W.J., Race R.E., and Eklund C.M.: Evidence for multiple molecular forms of the scrapie agent. *Persistent Viruses.* Stevens J., Todaro G., and Fox C.F., Eds. Academic Press, New York, 1978 pp. 591–613.
49. Prusiner S.B.: Prions. *Sci. Am.* 251:50–59, 1984.
50. Chandler R.L.: Experimental scrapie in the mouse. *Res. Vet. Sci.* 4:276–285, 1963.
51. Prusiner S., Davis J.N., and Stadtman E.R.: Regulation of glutaminase B in *Escherichia coli.* I. Purification, properties, and cold liability. *J. Biol. Chem.* 251:3447–3456, 1976.
52. Prusiner S.B., Hadlow W.J., Eklund C.M., and Race R.E.: Sedimentation properties of the scrapie agent. *Proc. Natl. Acad. Sci. USA* 74:4656–4660, 1977.
53. Marsh R.F. and Kimberlin R.H.: Comparison of scrapie and transmissible mink encephalopathy in hamsters. II. Clinical signs, pathology and pathogenesis. *J. Infect. Dis.* 131:104–110, 1975.
54. Kimberlin R. and Walker C.: Characteristics of a short incubation model of scrapie in the golden hamster. *J. Gen. Virol.* 34:295–304, 1977.
55. Prusiner S.B., Groth D.F., Cochran S.P., Masiarz F.R., McKinley M.P., and Martinez H.M.: Molecular properties, partial purification, and assay by incubation period measurements of the hamster scrapie agent. *Biochemistry* 19:4883–4891, 1980.
56. Prusiner S.B., Cochran S.P., Groth D.F., Downey D.E., Bowman K.A., and Martinez H.M.: Measurement of the scrapie agent using an incubation time interval assay. *Ann. Neurol.* 11:353–358, 1982.
57. Zlotnik I.: Experimental transmission of scrapie to golden hamsters. *Lancet* 2:1072, 1963.
58. Prusiner S.B.: Novel proteinaceous infectious particles cause scrapie. *Science* 216:136–144, 1982.
59. Stahl N., Baldwin M.A., Teplow D.B., Hood L., Gibson B.W., Burlingame A.L., and Prusiner S.B.: Structural analysis of the scrapie prion protein using mass spectrometry and amino acid sequencing. *Biochemistry* 32:1991–2002, 1993.
60. Cohen F.E., Pan K.-M., Huang Z., Baldwin M., Fletterick R.J., and Prusiner S.B.: Structural clues to prion replication. *Science* 264:530–531, 1994.
61. Prusiner S.B., Groth D., Serban A., Stahl N., and Gabizon R.: Attempts to restore scrapie prion infectivity after exposure to protein denaturants. *Proc. Natl. Acad. Sci. USA* 90:2793–2797, 1993.
62. Carp R.I., Kascsak R.J., Rubenstein R., and Merz P.A.: The puzzle of PrP^{Sc} and infectivity–do the pieces fit? *Trends Neurosci.* 17:148–149, 1994.
63. Manuelidis L. and Fritch W.: Infectivity and host responses in Creutzfeldt-Jakob disease. *Virology* 216:46–59, 1996.
64. Narang H.: The nature of the scrapie agent: the virus theory. *Proc. Soc. Exp. Biol. Med.* 212:208–224, 1996.
65. Özel M., Baldauf E., Beekes M., and Diringer H.: Small virus-like particles in transmissible spongiform encephalopathies. *Transmissible Subacute Spongiform*

Encephalopathies: Prion Diseases. Court L. and Dodet B., Eds. Elsevier, Paris, 1996 pp. 369–373.

66. Caughey B. and Chesebro B.: Prion protein and the transmissible spongiform encephalopathies. *Trends Cell Biol.* 7:56–62, 1997.
67. Lasmézas C.I., Deslys J.-P., Robain O., Jaegly A., Beringue V., Peyrin J.-M., Fournier J.-G., Hauw J.-J., Rossier J., and Dormont D.: Transmission of the BSE agent to mice in the absence of detectable abnormal prion protein. *Science* 275:402–405, 1997.
68. Diener T.O., McKinley M.P., and Prusiner S.B.: Viroids and prions. *Proc. Natl. Acad. Sci. USA* 79:5220–5224, 1982.
69. Chamberlin T.C.: The method of multiple working hypotheses. *Science [Old Series]* 15:92–97, 1890.
70. Avery O.T., MacLeod C.M., and McCarty M.: Studies on the chemical nature of the substance inducing transformation of pneumococcal types. Induction of transformation by a deoxyribonucleic acid fraction isolated from pneumococcus type III. *J. Exp. Med.* 79:137–157, 1944.
71. Stanley W.M.: The "undiscovered" discovery. *Arch. Environ. Health* 21:256–262, 1970.
72. McCarty M.: *The Transforming Principle: Discovering that Genes Are Made of DNA,* W. W. Norton & Co., New York, 1985.
73. Hershey A.D. and Chase M.: Independent functions of viral protein and nucleic acid in growth of bacteriophage. *J. Gen. Physiol.* 36:39–56, 1952.
74. Watson J.D. and Crick F.H.C.: Genetical implication of the structure of deoxyribose nucleic acid. *Nature* 171:964–967, 1953.
75. Mirsky A.E. and Pollister A.W.: Chromosin, a desoxyribose nucleoprotein complex of the cell nucleus. *J. Gen. Physiol.* 30:134–135, 1946.
76. Diener T.O.: *Viroids and Viroid Diseases,* John Wiley, New York, 1979.
77. Fraenkel-Conrat H. and Williams R.C.: Reconstitution of active tobacco virus from the inactive protein and nucleic acid components. *Proc. Natl. Acad. Sci. USA* 41:690–698, 1955.
78. Gierer A. and Schramm G.: Infectivity of ribonucleic acid from tobacco mosaic virus. *Nature* 177:702–703, 1956.
79. Bolton D.C., McKinley M.P., and Prusiner S.B.: Identification of a protein that purifies with the scrapie prion. *Science* 218:1309–1311, 1982.
80. Prusiner S.B., Bolton D.C., Groth D.F., Bowman K.A., Cochran S.P., and McKinley M.P.: Further purification and characterization of scrapie prions. *Biochemistry* 21:6942–6950, 1982.
81. Prusiner S.B., Groth D.F., Bolton D.C., Kent S.B., and Hood L.E.: Purification and structural studies of a major scrapie prion protein. *Cell* 38:127–134, 1984.
82. Basler K., Oesch B., Scott M., Westaway D., Wälchli M., Groth D.F., McKinley M.P., Prusiner S.B., and Weissmann C.: Scrapie and cellular PrP isoforms are encoded by the same chromosomal gene. *Cell* 46:417–428, 1986.
83. Chesebro B., Race R., Wehrly K., Nishio J., Bloom M., Lechner D., Bergstrom S., Robbins K., Mayer L., Keith J.M., Garon C., and Haase A.: Identification of scrapie prion protein-specific mRNA in scrapie-infected and uninfected brain. *Nature* 315:331–333, 1985.
84. Oesch B., Westaway D., Wälchli M., McKinley M.P., Kent S.B.H., Aebersold R., Barry R.A., Tempst P., Teplow D.B., Hood L.E., Prusiner S.B., and Weissmann C.: A cellular gene encodes scrapie PrP 27–30 protein. *Cell* 40:735–746, 1985.
85. Meyer R.K., McKinley M.P., Bowman K.A., Braunfeld M.B., Barry R.A., and Prusiner S.B.: Separation and properties of cellular and scrapie prion proteins. *Proc. Natl. Acad. Sci. USA* 83:2310–2314, 1986.
86. Locht C., Chesebro B., Race R., and Keith J.M.: Molecular cloning and complete sequence of prion protein cDNA from mouse brain infected with the scrapie agent. *Proc. Natl. Acad. Sci. USA* 83:6372–6376, 1986.
87. Bendheim P.E., Barry R.A., DeArmond S.J., Stites D.P., and Prusiner S.B.: Antibodies to a scrapie prion protein. *Nature* 310:418–421, 1984.

88. Bockman J.M., Kingsbury D.T., McKinley M.P., Bendheim P.E., and Prusiner S.B.: Creutzfeldt-Jakob disease prion proteins in human brains. *N. Engl. J. Med.* 312:73–78, 1985.
89. Brown P., Coker-Vann M., Pomeroy K., Franko M., Asher D.M., Gibbs C.J., Jr., and Gajdusek D.C.: Diagnosis of Creutzfeldt-Jakob disease by Western blot identification of marker protein in human brain tissue. *N. Engl. J. Med.* 314:547–551, 1986.
90. Bockman J.M., Prusiner S.B., Tateishi J., and Kingsbury D.T.: Immunoblotting of Creutzfeldt-Jakob disease prion proteins: host species-specific epitopes. *Ann. Neurol.* 21:589–595, 1987.
91. Manuelidis L., Valley S., and Manuelidis E.E.: Specific proteins associated with Creutzfeldt-Jakob disease and scrapie share antigenic and carbohydrate determinants. *Proc. Natl. Acad. Sci. USA* 82:4263–4267, 1985.
92. Prusiner S.B.: Prions and neurodegenerative diseases. *N. Engl. J. Med.* 317:1571–1581, 1987.
93. Manuelidis L., Tesin D.M., Sklaviadis T., and Manuelidis E.E.: Astrocyte gene expression in Creutzfeldt-Jakob disease. *Proc. Natl. Acad. Sci. USA* 84:5937–5941, 1987.
94. Gomi H., Yokoyama T., Fujimoto K., Ikeda T., Katoh A., Itoh T., and Itohara S.: Mice devoid of the glial fibrillary acidic protein develop normally and are susceptible to scrapie prions. *Neuron* 14:29–41, 1995.
95. Tatzelt J., Maeda N., Pekny M., Yang S.-L., Betsholtz C., Eliasson C., Cayetano J., Camerino A.P., DeArmond S.J., and Prusiner S.B.: Scrapie in mice deficient in apolipoprotein E or glial fibrillary acidic protein. *Neurology* 47:449–453, 1996.
96. Duguid J.R., Rohwer R.G., and Seed B.: Isolation of cDNAs of scrapie-modulated RNAs by subtractive hybridization of a cDNA library. *Proc. Natl. Acad. Sci. USA* 85:5738–5742, 1988.
97. Duguid J.R., Bohmont C.W., Liu N., and Tourtellotte W.W.: Changes in brain gene expression shared by scrapie and Alzheimer disease. *Proc. Natl. Acad. Sci. USA* 86:7260–7264, 1989.
98. Diedrich J.F., Carp R.I., and Haase A.T.: Increased expression of heat shock protein, transferrin, and b_2-microglobulin in astrocytes during scrapie. *Microbial. Pathogenesis* 15:1–6, 1993.
99. Akowitz A., Sklaviadis T., and Manuelidis L.: Endogenous viral complexes with long RNA cosediment with the agent of Creutzfeldt-Jakob disease. *Nucleic Acids Res.* 22:1101–1107, 1994.
100. Büeler H., Aguzzi A., Sailer A., Greiner R.-A., Autenried P., Aguet M., and Weissmann C.: Mice devoid of PrP are resistant to scrapie. *Cell* 73:1339–1347, 1993.
101. Prusiner S.B., Groth D., Serban A., Koehler R., Foster D., Torchia M., Burton D., Yang S.-L., and DeArmond S.J.: Ablation of the prion protein (PrP) gene in mice prevents scrapie and facilitates production of anti-PrP antibodies. *Proc. Natl. Acad. Sci. USA* 90:10608–10612, 1993.
102. Manson J.C., Clarke A.R., McBride P.A., McConnell I., and Hope J.: PrP gene dosage determines the timing but not the final intensity or distribution of lesions in scrapie pathology. *Neurodegeneration* 3:331–340, 1994.
103. Sailer A., Büeler H., Fischer M., Aguzzi A., and Weissmann C.: No propagation of prions in mice devoid of PrP. *Cell* 77:967–968, 1994.
104. Sakaguchi S., Katamine S., Shigematsu K., Nakatani A., Moriuchi R., Nishida N., Kurokawa K., Nakaoke R., Sato H., Jishage K., Kuno J., Noda T., and Miyamoto T.: Accumulation of proteinase K-resistant prion protein (PrP) is restricted by the expression level of normal PrP in mice inoculated with a mouse-adapted strain of the Creutzfeldt–Jakob disease agent. *J. Virol.* 69:7586–7592, 1995.
105. Gabizon R., McKinley M.P., Groth D.F., and Prusiner S.B.: Immunoaffinity purification and neutralization of scrapie prion infectivity. *Proc. Natl. Acad. Sci. USA* 85:6617–6621, 1988.
106. McKinley M.P., Bolton D.C., and Prusiner S.B.: A protease-resistant protein is a structural component of the scrapie prion. *Cell* 35:57–62, 1983.

107. Prusiner S.B., McKinley M.P., Bowman K.A., Bolton D.C., Bendheim P.E., Groth D.F., and Glenner G.G.: Scrapie prions aggregate to form amyloid-like birefringent rods. *Cell* 35:349–358, 1983.
108. Bolton D.C., McKinley M.P., and Prusiner S.B.: Molecular characteristics of the major scrapie prion protein. *Biochemistry* 23:5898–5906, 1984.
109. Riesner D., Kellings K., Post K., Wille H., Serban H., Groth D., Baldwin M.A., and Prusiner S.B.: Disruption of prion rods generates 10-nm spherical particles having high α-helical content and lacking scrapie infectivity. *J. Virol.* 70:1714–1722, 1996.
110. Bellinger-Kawahara C., Cleaver J.E., Diener T.O., and Prusiner S.B.: Purified scrapie prions resist inactivation by UV irradiation. *J. Virol.* 61:159–166, 1987.
111. Bellinger-Kawahara C., Diener T.O., McKinley M.P., Groth D.F., Smith D.R., and Prusiner S.B.: Purified scrapie prions resist inactivation by procedures that hydrolyze, modify, or shear nucleic acids. *Virology* 160:271–274, 1987.
112. Bellinger-Kawahara C.G., Kempner E., Groth D.F., Gabizon R., and Prusiner S.B.: Scrapie prion liposomes and rods exhibit target sizes of 55,000 Da. *Virology* 164:537–541, 1988.
113. Kellings K., Meyer N., Mirenda C., Prusiner S.B., and Riesner D.: Further analysis of nucleic acids in purified scrapie prion preparations by improved return refocussing gel electrophoresis (RRGE). *J. Gen. Virol.* 73:1025–1029, 1992.
114. Kellings K., Prusiner S.B., and Riesner D.: Nucleic acids in prion preparations: unspecific background or essential component? *Phil. Trans. R. Soc. Lond. B* 343:425–430, 1994.
115. Manuelidis L., Sklaviadis T., Akowitz A., and Fritch W.: Viral particles are required for infection in neurodegenerative Creutzfeldt-Jakob disease. *Proc. Natl. Acad. Sci. USA* 92:5124–5128, 1995.
116. Williams R.C.: Electron microscopy of viruses. *Adv. Virus Res.* 2:183–239, 1954.
117. Cohen A.S., Shirahama T., and Skinner M.: Electron microscopy of amyloid. *Electron Microscopy of Proteins, Vol. 3.* Harris J.R., Ed. Academic Press, New York, 1982 pp. 165–206.
118. DeArmond S.J., McKinley M.P., Barry R.A., Braunfeld M.B., McColloch J.R., and Prusiner S.B.: Identification of prion amyloid filaments in scrapie-infected brain. *Cell* 41:221–235, 1985.
119. Kitamoto T., Tateishi J., Tashima I., Takeshita I., Barry R.A., DeArmond S.J., and Prusiner S.B.: Amyloid plaques in Creutzfeldt-Jakob disease stain with prion protein antibodies. *Ann. Neurol.* 20:204–208, 1986.
120. Roberts G.W., Lofthouse R., Brown R., Crow T.J., Barry R.A., and Prusiner S.B.: Prion-protein immunoreactivity in human transmissible dementias. *N. Engl. J. Med.* 315:1231–1233, 1986.
121. Gabizon R., McKinley M.P., and Prusiner S.B.: Purified prion proteins and scrapie infectivity copartition into liposomes. *Proc. Natl. Acad. Sci. USA* 84:4017–4021, 1987.
122. McKinley M.P., Meyer R., Kenaga L., Rahbar F., Cotter R., Serban A., and Prusiner S.B.: Scrapie prion rod formation *in vitro* requires both detergent extraction and limited proteolysis. *J. Virol.* 65:1440–1449, 1991.
123. Diringer H., Gelderblom H., Hilmert H., Ozel M., Edelbluth C., and Kimberlin R.H.: Scrapie infectivity, fibrils and low molecular weight protein. *Nature* 306:476–478, 1983.
124. Merz P.A., Kascsak R.J., Rubenstein R., Carp R.I., and Wisniewski H.M.: Antisera to scrapie-associated fibril protein and prion protein decorate scrapie-associated fibrils. *J. Virol.* 61:42–49, 1987.
125. Diener T.O.: PrP and the nature of the scrapie agent. *Cell* 49:719–721, 1987.
126. Merz P.A., Somerville R.A., Wisniewski H.M., and Iqbal K.: Abnormal fibrils from scrapie-infected brain. *Acta Neuropathol. (Berl.)* 54:63–74, 1981.
127. Merz P.A., Wisniewski H.M., Somerville R.A., Bobin S.A., Masters C.L., and Iqbal K.: Ultrastructural morphology of amyloid fibrils from neuritic and amyloid plaques. *Acta Neuropathol. (Berl.)* 60:113–124, 1983.
128. Merz P.A., Rohwer R.G., Kascsak R., Wisniewski H.M., Somerville R.A., Gibbs C.J., Jr.,

and Gajdusek D.C.: Infection-specific particle from the unconventional slow virus diseases. *Science* 225:437–440, 1984.
129. Diringer H.: Transmissible spongiform encephalopathies (TSE) virus-induced amyloidoses of the central nervous system (CNS). *Eur. J. Epidemiol.* 7:562–566, 1991.
130. Bastian F.O.: Bovine spongiform encephalopathy: relationship to human disease and nature of the agent. *ASM News* 59:235–240, 1993.
131. Kimberlin R.H.: Scrapie and possible relationships with viroids. *Semin. Virol.* 1:153–162, 1990.
132. Chesebro B. and Caughey B.: Scrapie agent replication without the prion protein? *Curr. Biol.* 3:696–698, 1993.
133. Scott M., Groth D., Foster D., Torchia M., Yang S.-L., DeArmond S.J., and Prusiner S.B.: Propagation of prions with artificial properties in transgenic mice expressing chimeric PrP genes. *Cell* 73:979–988, 1993.
134. Telling G.C., Scott M., Mastrianni J., Gabizon R., Torchia M., Cohen F.E., DeArmond S.J., and Prusiner S.B.: Prion propagation in mice expressing human and chimeric PrP transgenes implicates the interaction of cellular PrP with another protein. *Cell* 83:79–90, 1995.
135. Scott M.R., Groth D., Tatzelt J., Torchia M., Tremblay P., DeArmond S.J., and Prusiner S.B.: Propagation of prion strains through specific conformers of the prion protein. *J. Virol.* 71:9032–9044, 1997.
136. Anfinsen C.B.: Principles that govern the folding of protein chains. *Science* 181:223–230, 1973.
137. Pergami P., Jaffe H., and Safar J.: Semipreparative chromatographic method to purify the normal cellular isoform of the prion protein in nondenatured form. *Anal. Biochem.* 236:63–73, 1996.
138. Borchelt D.R., Scott M., Taraboulos A., Stahl N., and Prusiner S.B.: Scrapie and cellular prion proteins differ in their kinetics of synthesis and topology in cultured cells. *J. Cell Biol.* 110:743–752, 1990.
139. Stahl N., Borchelt D.R., Hsiao K., and Prusiner S.B.: Scrapie prion protein contains a phosphatidylinositol glycolipid. *Cell* 51:229–240, 1987.
140. Caughey B. and Raymond G.J.: The scrapie-associated form of PrP is made from a cell surface precursor that is both protease- and phospholipase-sensitive. *J. Biol. Chem.* 266:18217–18223, 1991.
141. Borchelt D.R., Taraboulos A., and Prusiner S.B.: Evidence for synthesis of scrapie prion proteins in the endocytic pathway. *J. Biol. Chem.* 267:16188–16199, 1992.
142. Taraboulos A., Scott M., Semenov A., Avrahami D., Laszlo L., and Prusiner S.B.: Cholesterol depletion and modification of COOH-terminal targeting sequence of the prion protein inhibits formation of the scrapie isoform. *J. Cell Biol.* 129:121–132, 1995.
143. Vey M., Pilkuhn S., Wille H., Nixon R., DeArmond S.J., Smart E.J., Anderson R.G., Taraboulos A., and Prusiner S.B.: Subcellular colocalization of the cellular and scrapie prion proteins in caveolae-like membranous domains. *Proc. Natl. Acad. Sci. USA* 93:14945–14949, 1996.
144. Kaneko K., Vey M., Scott M., Pilkuhn S., Cohen F.E., and Prusiner S.B.: COOH-terminal sequence of the cellular prion protein directs subcellular trafficking and controls conversion into the scrapie isoform. *Proc. Natl. Acad. Sci. USA* 94:2333–2338, 1997.
145. Naslavsky N., Stein R., Yanai A., Friedlander G., and Taraboulos A.: Characterization of detergent-insoluble complexes containing the cellular prion protein and its scrapie isoform. *J. Biol. Chem.* 272:6324–6331, 1997.
146. Gasset M., Baldwin M.A., Lloyd D., Gabriel J.-M., Holtzman D.M., Cohen F., Fletterick R., and Prusiner S.B.: Predicted α-helical regions of the prion protein when synthesized as peptides form amyloid. *Proc. Natl. Acad. Sci. USA* 89:10940–10944, 1992.
147. Huang Z., Gabriel J.-M., Baldwin M.A., Fletterick R.J., Prusiner S.B., and Cohen F.E.: Proposed three-dimensional structure for the cellular prion protein. *Proc. Natl. Acad. Sci. USA* 91:7139–7143, 1994.
148. Zhang H., Kaneko K., Nguyen J.T., Livshits T.L., Baldwin M.A., Cohen F.E., James T.L.,

and Prusiner S.B.: Conformational transitions in peptides containing two putative α-helices of the prion protein. *J. Mol. Biol.* 250:514–526, 1995.

149. Schätzl H.M., Da Costa M., Taylor L., Cohen F.E., and Prusiner S.B.: Prion protein gene variation among primates. *J. Mol. Biol.* 245:362–374, 1995.
150. Bamborough P., Wille H., Telling G.C., Yehiely F., Prusiner S.B., and Cohen F.E.: Prion protein structure and scrapie replication: theoretical, spectroscopic and genetic investigations. *Cold Spring Harb. Symp. Quant. Biol.* 61:495–509, 1996.
151. Hay B., Barry R.A., Lieberburg I., Prusiner S.B., and Lingappa V.R.: Biogenesis and transmembrane orientation of the cellular isoform of the scrapie prion protein. *Mol. Cell. Biol.* 7:914–920, 1987.
152. De Fea K.A., Nakahara D.H., Calayag M.C., Yost C.S., Mirels L.F., Prusiner S.B., and Lingappa V.R.: Determinants of carboxyl-terminal domain translocation during prion protein biogenesis. *J. Biol. Chem.* 269:16810–16820, 1994.
153. Hegde R.S., Mastrianni J.A., Scott M.R., DeFea K.A., Tremblay P., Torchia M., DeArmond S.J., Prusiner S.B., and Lingappa V.R.: A transmembrane form of the prion protein in neurodegenerative disease. *Science* 279: 827–834, 1998.
154. Riek R., Hornemann S., Wider G., Billeter M., Glockshuber R., and Wüthrich K.: NMR structure of the mouse prion protein domain PrP(121–231). *Nature* 382:180–182, 1996.
155. James T.L., Liu H., Ulyanov N.B., Farr-Jones S., Zhang H., Donne D.G., Kaneko K., Groth D., Mehlhorn I., Prusiner S.B., and Cohen F.E.: Solution structure of a 142-residue recombinant prion protein corresponding to the infectious fragment of the scrapie isoform. *Proc. Natl. Acad. Sci. USA* 94:10086–10091, 1997.
156. Donne D.G., Viles J.H., Groth D., Mehlhorn I., James T.L., Cohen F.E., Prusiner S.B., Wright P.E., and Dyson H.J.: Structure of the recombinant full-length hamster prion protein PrP(29–231): the N terminus is highly flexible. *Proc. Natl. Acad. Sci. USA* 94:13452–13457, 1997.
157. Williamson R.A., Peretz D., Smorodinsky N., Bastidas R., Serban H., Mehlhorn I., DeArmond S.J., Prusiner S.B., and Burton D.R.: Circumventing tolerance to generate autologous monoclonal antibodies to the prion protein. *Proc. Natl. Acad. Sci. USA* 93:7279–7282, 1996.
158. Peretz D., Williamson R.A., Matsunaga Y., Serban H., Pinilla C., Bastidas R., Rozenshteyn R., James T.L., Houghten R.A., Cohen F.E., Prusiner S.B., and Burton D.R.: A conformational transition at the N terminus of the prion protein features in formation of the scrapie isoform. *J. Mol. Biol.* 273:614–622, 1997.
159. Caughey B., Race R.E., Ernst D., Buchmeier M.J., and Chesebro B.: Prion protein biosynthesis in scrapie-infected and uninfected neuroblastoma cells. *J. Virol.* 63:175–181, 1989.
160. Shmerling D., Hegyi I., Fischer M., Blättler T., Brandner S., Götz J., Rülicke T., Flechsig E., Cozzio A., von Mering C., Hangartner C., Aguzzi A., and Weissmann C.: Expression of amino-terminally truncated PrP in the mouse leading to ataxia and specific cerebellar lesions. *Cell* 93:203–214, 1998.
161. Mehlhorn I., Groth D., Stöckel J., Moffat B., Reilly D., Yansura D., Willett W.S., Baldwin M., Fletterick R., Cohen F.E., Vandlen R., Henner D., and Prusiner S.B.: High-level expression and characterization of a purified 142-residue polypeptide of the prion protein. *Biochemistry* 35:5528–5537, 1996.
162. Zhang H., Stöckel J., Mehlhorn I., Groth D., Baldwin M.A., Prusiner S.B., James T.L., and Cohen F.E.: Physical studies of conformational plasticity in a recombinant prion protein. *Biochemistry* 36:3543–3553, 1997.
163. Fischer M., Rülicke T., Raeber A., Sailer A., Moser M., Oesch B., Brandner S., Aguzzi A., and Weissmann C.: Prion protein (PrP) with amino-proximal deletions restoring susceptibility of PrP knockout mice to scrapie. *EMBO J.* 15:1255–1264, 1996.
164. Muramoto T., DeArmond S.J., Scott M., Telling G.C., Cohen F.E., and Prusiner S.B.: Heritable disorder resembling neuronal storage disease in mice expressing prion protein with deletion of an α-helix. *Nat. Med.* 3:750–755, 1997.

165. Riek R., Hornemann S., Wider G., Glockshuber R., and Wüthrich K.: NMR characterization of the full-length recombinant murine prion protein, *m*PrP(23–231). *FEBS Lett* 413:282–288, 1997.
166. Huang Z., Prusiner S.B., and Cohen F.E.: Scrapie prions: a three-dimensional model of an infectious fragment. *Folding & Design* 1:13–19, 1996.
167. Turk E., Teplow D.B., Hood L.E., and Prusiner S.B.: Purification and properties of the cellular and scrapie hamster prion proteins. *Eur. J. Biochem.* 176:21–30, 1988.
168. Muramoto T., Scott M., Cohen F., and Prusiner S.B.: Recombinant scrapie-like prion protein of 106 amino acids is soluble. *Proc. Natl. Acad. Sci. USA* 93:15457–15462, 1996.
169. Caughey B.W., Dong A., Bhat K.S., Ernst D., Hayes S.F., and Caughey W.S.: Secondary structure analysis of the scrapie-associated protein PrP 27–30 in water by infrared spectroscopy. *Biochemistry* 30:7672–7680, 1991.
170. Gasset M., Baldwin M.A., Fletterick R.J., and Prusiner S.B.: Perturbation of the secondary structure of the scrapie prion protein under conditions associated with changes in infectivity. *Proc. Natl. Acad. Sci. USA* 90:1–5, 1993.
171. Safar J., Roller P.P., Gajdusek D.C., and Gibbs C.J., Jr.: Conformational transitions, dissociation, and unfolding of scrapie amyloid (prion) protein. *J. Biol. Chem.* 268:20276–20284, 1993.
172. Rogers M., Yehiely F., Scott M., and Prusiner S.B.: Conversion of truncated and elongated prion proteins into the scrapie isoform in cultured cells. *Proc. Natl. Acad. Sci. USA* 90:3182–3186, 1993.
173. Barry R.A. and Prusiner S.B.: Monoclonal antibodies to the cellular and scrapie prion proteins. *J. Infect. Dis.* 154:518–521, 1986.
174. Kascsak R.J., Rubenstein R., Merz P.A., Tonna-DeMasi M., Fersko R., Carp R.I., Wisniewski H.M., and Diringer H.: Mouse polyclonal and monoclonal antibody to scrapie-associated fibril proteins. *J. Virol.* 61:3688–3693, 1987.
175. Rogers M., Serban D., Gyuris T., Scott M., Torchia T., and Prusiner S.B.: Epitope mapping of the Syrian hamster prion protein utilizing chimeric and mutant genes in a vaccinia virus expression system. *J. Immunol.* 147:3568–3574, 1991.
176. Korth C., Stierli B., Streit P., Moser M., Schaller O., Fischer R., Schulz-Schaeffer W., Kretzschmar H., Raeber A., Braun U., Ehrensperger F., Hornemann S., Glockshuber R., Riek R., Billeter M., Wuthrick K., and Oesch B.: Prion (PrP^{Sc})-specific epitope defined by a monoclonal antibody. *Nature* 389:74–77, 1997.
177. Billeter M., Riek R., Wider G., Hornemann S., Glockshuber R., and Wüthrich K.: Prion protein NMR structure and species barrier for prion diseases. *Proc. Natl. Acad. Sci. USA* 94:7281–7285, 1997.
178. Kaneko K., Zulianello L., Scott M., Cooper C.M., Wallace A.C., James T.L., Cohen F.E., and Prusiner S.B.: Evidence for protein X binding to a discontinuous epitope on the cellular prion protein during scrapie prion propagation. *Proc. Natl. Acad. Sci. USA* 94:10069–10074, 1997.
179. Telling G.C., Scott M., Hsiao K.K., Foster D., Yang S.-L., Torchia M., Sidle K.C.L., Collinge J., DeArmond S.J., and Prusiner S.B.: Transmission of Creutzfeldt-Jakob disease from humans to transgenic mice expressing chimeric human-mouse prion protein. *Proc. Natl. Acad. Sci. USA* 91:9936–9940, 1994.
180. Stöckel J., Safar J., Wallace A.C., Cohen F.E., and Prusiner S.B.: Prion protein selectively binds copper-II-ions. *Biochemistry* 37: 7185–7193, 1998.
181. Hornshaw M.P., McDermott J.R., Candy J.M., and Lakey J.H.: Copper binding to the N-terminal tandem repeat region of mammalian and avian prion protein: structural studies using synthetic peptides. *Biochem. Biophys. Res. Commun.* 214:993–999, 1995.
182. Hornshaw M.P., McDermott J.R., and Candy J.M.: Copper binding to the N-terminal tandem repeat regions of mammalian and avian prion protein. *Biochem. Biophys. Res. Commun.* 207:621–629, 1995.
183. Miura T., Hori-i A., and Takeuchi H.: Metal-dependent α-helix formation promoted by the glycine-rich octapeptide region of prion protein. *FEBS Lett.* 396:248–252, 1996.
184. Brown D.R., Schulz-Schaeffer W.J., Schmidt B., and Kretzschmar H.A.: Prion protein-

deficient cells show altered response to oxidative stress due to decreased SOD-1 activity. *Exp. Neurol.* 146:104–112, 1997.

185. Harris E.D.: Copper as a cofactor and regulator of copper, zinc superoxide dismutase. *J. Nutr.* 122:636–640, 1992.
186. Brown D.R., Qin K., Herms J.W., Madlung A., Manson J., Strome R., Fraser P.E., Kruck T., von Bohlen A., Schulz-Schaeffer W., Giese A., Westaway D., and Kretzschmar H.: The cellular prion protein binds copper *in vivo*. *Nature* 390:684–687, 1997.
187. Vulpe C., Levinson B., Whitney S., Packman S., and Gitschier J.: Isolation of a candidate gene for Menkes disease and evidence that it encodes a copper-transporting ATPase. *Nat. Genet.* 3:7–13, 1993.
188. Chelly J., Tümer Z., Tønnesen T., Petterson A., Ishikawa-Brush Y., Tommerup N., Horn N., and Monaco A.P.: Isolation of a candidate gene for Menkes disease that encodes a potential heavy metal binding protein. *Nat. Genet.* 3:14–19, 1993.
189. Bull P.C., Thomas G.R., Rommens J.M., Forbes J.R., and Cox D.W.: The Wilson disease gene is a putative copper transporting P-type ATPase similar to the Menkes gene. *Nat. Genet.* 5:327–337, 1993.
190. Petrukhin K., Fischer S.G., Pirastu M., Tanzi R.E., Chernov I., Devoto M., Brzustowicz L.M., Cayanis E., Vitale E., Russo J.J., Matseoane D., Boukhgalter B., Wasco W., Figus A.L., Loudianos J., Cao A., Sternlieb I., Evgrafov O., Parano E., Pavone L., Warburton D., Ott J., Penchaszadeh G.K., Scheinberg I.H., and Gilliam T.C.: Mapping, cloning and genetic characterization of the region containing the Wilson disease gene. *Nat. Genet.* 5:338–343, 1993.
191. Pattison I.H. and Jebbett J.N.: Clinical and histological observations on cuprizone toxicity and scrapie in mice. *Res. Vet. Sci.* 12:378–380, 1971.
192. Kimberlin R.H., Millson G.C., Bountiff L., and Collis S.C.: A comparison of the biochemical changes induced in mouse brain by cuprizone toxicity and by scrapie infection. *J. Comp. Path.* 84:263–270, 1974.
193. Gabriel J.-M., Oesch B., Kretzschmar H., Scott M., and Prusiner S.B.: Molecular cloning of a candidate chicken prion protein. *Proc. Natl. Acad. Sci. USA* 89:9097–9101, 1992.
194. Westaway D., Goodman P.A., Mirenda C.A., McKinley M.P., Carlson G.A., and Prusiner S.B.: Distinct prion proteins in short and long scrapie incubation period mice. *Cell* 51:651–662, 1987.
195. Westaway D., Mirenda C.A., Foster D., Zebarjadian Y., Scott M., Torchia M., Yang S.-L., Serban H., DeArmond S.J., Ebeling C., Prusiner S.B., and Carlson G.A.: Paradoxical shortening of scrapie incubation times by expression of prion protein transgenes derived from long incubation period mice. *Neuron* 7:59–68, 1991.
196. Yoshimoto J., Iinuma T., Ishiguro N., Horiuchi M., Imamura M., and Shinagawa M.: Comparative sequence analysis and expression of bovine PrP gene in mouse L-929 cells. *Virus Genes* 6:343–356, 1992.
197. Westaway D., Cooper C., Turner S., Da Costa M., Carlson G.A., and Prusiner S.B.: Structure and polymorphism of the mouse prion protein gene. *Proc. Natl. Acad. Sci. USA* 91:6418–6422, 1994.
198. Westaway D., Zuliani V., Cooper C.M., Da Costa M., Neuman S., Jenny A.L., Detwiler L., and Prusiner S.B.: Homozygosity for prion protein alleles encoding glutamine-171 renders sheep susceptible to natural scrapie. *Genes Dev.* 8:959–969, 1994.
199. Saeki K., Matsumoto Y., Hirota Y., Matsumoto Y., and Onodera T.: Three-exon structure of the gene encoding the rat prion protein and its expression in tissues. *Virus Genes* 12:15–20, 1996.
200. Li G. and Bolton D.C.: A novel hamster prion protein mRNA contains an extra exon: increased expression in scrapie. *Brain Res.* 751:265–274, 1997.
201. Lee I., Westaway D., Smit A., Wang K., Cooper C., Yao H., Prusiner S.B., and Hood L.: in preparation.
202. McKnight S. and Tjian R.: Transcriptional selectivity of viral genes in mammalian cells. *Cell* 46:795–805, 1986.

203. Robakis N.K., Devine-Gage E.A., Kascsak R.J., Brown W.T., Krawczun C., and Silverman W.P.: Localization of a human gene homologous to the PrP gene on the p arm of chromosome 20 and detection of PrP-related antigens in normal human brain. *Biochem. Biophys. Res. Commun.* 140:758–765, 1986.
204. Sparkes R.S., Simon M., Cohn V.H., Fournier R.E.K., Lem J., Klisak I., Heinzmann C., Blatt C., Lucero M., Mohandas T., DeArmond S.J., Westaway D., Prusiner S.B., and Weiner L.P.: Assignment of the human and mouse prion protein genes to homologous chromosomes. *Proc. Natl. Acad. Sci. USA* 83:7358–7362, 1986.
205. Mobley W.C., Neve R.L., Prusiner S.B., and McKinley M.P.: Nerve growth factor increases mRNA levels for the prion protein and the beta-amyloid protein precursor in developing hamster brain. *Proc. Natl. Acad. Sci. USA* 85:9811–9815, 1988.
206. Kretzschmar H.A., Prusiner S.B., Stowring L.E., and DeArmond S.J.: Scrapie prion proteins are synthesized in neurons. *Am. J. Pathol.* 122:1–5, 1986.
207. DeArmond S.J., Mobley W.C., DeMott D.L., Barry R.A., Beckstead J.H., and Prusiner S.B.: Changes in the localization of brain prion proteins during scrapie infection. *Neurology* 37:1271–1280, 1987.
208. Taraboulos A., Jendroska K., Serban D., Yang S.-L., DeArmond S.J., and Prusiner S.B.: Regional mapping of prion proteins in brains. *Proc. Natl. Acad. Sci. USA* 89:7620–7624, 1992.
209. Kimberlin R.H., Field H.J., and Walker C.A.: Pathogenesis of mouse scrapie: evidence for spread of infection from central to peripheral nervous system. *J. Gen. Virol.* 64:713–716, 1983.
210. Fraser H. and Dickinson A.G.: Targeting of scrapie lesions and spread of agent via the retino-tectal projection. *Brain Res.* 346:32–41, 1985.
211. Jendroska K., Heinzel F.P., Torchia M., Stowring L., Kretzschmar H.A., Kon A., Stern A., Prusiner S.B., and DeArmond S.J.: Proteinase-resistant prion protein accumulation in Syrian hamster brain correlates with regional pathology and scrapie infectivity. *Neurology* 41:1482–1490, 1991.
212. Carlson G.A., Kingsbury D.T., Goodman P.A., Coleman S., Marshall S.T., DeArmond S.J., Westaway D., and Prusiner S.B.: Linkage of prion protein and scrapie incubation time genes. *Cell* 46:503–511, 1986.
213. Scott M., Foster D., Mirenda C., Serban D., Coufal F., Wälchli M., Torchia M., Groth D., Carlson G., DeArmond S.J., Westaway D., and Prusiner S.B.: Transgenic mice expressing hamster prion protein produce species-specific scrapie infectivity and amyloid plaques. *Cell* 59:847–857, 1989.
214. Dickinson A.G., Meikle V.M.H., and Fraser H.: Identification of a gene which controls the incubation period of some strains of scrapie agent in mice. *J. Comp. Pathol.* 78:293–299, 1968.
215. Kingsbury D.T., Kasper K.C., Stites D.P., Watson J.D., Hogan R.N., and Prusiner S.B.: Genetic control of scrapie and Creutzfeldt-Jakob disease in mice. *J. Immunol.* 131:491–496, 1983.
216. Carlson G.A., Goodman P.A., Lovett M., Taylor B.A., Marshall S.T., Peterson-Torchia M., Westaway D., and Prusiner S.B.: Genetics and polymorphism of the mouse prion gene complex: the control of scrapie incubation time. *Mol. Cell. Biol.* 8:5528–5540, 1988.
217. Carlson G.A., Ebeling C., Yang S.-L., Telling G., Torchia M., Groth D., Westaway D., DeArmond S.J., and Prusiner S.B.: Prion isolate specified allotypic interactions between the cellular and scrapie prion proteins in congenic and transgenic mice. *Proc. Natl. Acad. Sci. USA* 91:5690–5694, 1994.
218. Moore R.C., Hope J., McBride P.A., McConnell I., Selfridge J., and Melton D.W.: Mice with gene targeted prion protein alterations show that *Prn-p*, *Sinc* and *Prn-i* are congruent. *Nat. Genet.* 18:118–125, 1998.
219. Prusiner S.B., Scott M., Foster D., Pan K.-M., Groth D., Mirenda C., Torchia M., Yang S.-L., Serban D., Carlson G.A., Hoppe P.C., Westaway D., and DeArmond S.J.: Transgenetic studies implicate interactions between homologous PrP isoforms in scrapie prion replication. *Cell* 63:673–686, 1990.

220. Büeler H., Fischer M., Lang Y., Bluethmann H., Lipp H.-P., DeArmond S.J., Prusiner S.B., Aguet M., and Weissmann C.: Normal development and behaviour of mice lacking the neuronal cell-surface PrP protein. *Nature* 356:577–582, 1992.
221. Manson J.C., Clarke A.R., Hooper M.L., Aitchison L., McConnell I., and Hope J.: 129/Ola mice carrying a null mutation in PrP that abolishes mRNA production are developmentally normal. *Mol. Neurobiol.* 8:121–127, 1994.
222. Sakaguchi S., Katamine S., Nishida N., Moriuchi R., Shigematsu K., Sugimoto T., Nakatani A., Kataoka Y., Houtani T., Shirabe S., Okada H., Hasegawa S., Miyamoto T., and Noda T.: Loss of cerebellar Purkinje cells in aged mice homozygous for a disrupted PrP gene. *Nature* 380:528–531, 1996.
223. Tobler I., Gaus S.E., Deboer T., Achermann P., Fischer M., Rülicke T., Moser M., Oesch B., McBride P.A., and Manson J.C.: Altered circadian activity rhythms and sleep in mice devoid of prion protein. *Nature* 380:639–642, 1996.
224. Collinge J., Whittington M.A., Sidle K.C., Smith C.J., Palmer M.S., Clarke A.R., and Jefferys J.G.R.: Prion protein is necessary for normal synaptic function. *Nature* 370:295–297, 1994.
225. Whittington M.A., Sidle K.C.L., Gowland I., Meads J., Hill A.F., Palmer M.S., Jefferys J.G.R., and Collinge J.: Rescue of neurophysiological phenotype seen in PrP null mice by transgene encoding human prion protein. *Nat. Genet.* 9:197–201, 1995.
226. Herms J.W., Kretzschmar H.A., Titz S., and Keller B.U.: Patch-clamp analysis of synaptic transmission to cerebellar purkinje cells of prion protein knockout mice. *Eur. J. Neurosci.* 7:2508–2512, 1995.
227. Lledo P.-M., Tremblay P., DeArmond S.J., Prusiner S.B., and Nicoll R.A.: Mice deficient for prion protein exhibit normal neuronal excitability and synaptic transmission in the hippocampus. *Proc. Natl. Acad. Sci. USA* 93:2403–2407, 1996.
228. Pattison I.H. and Jones K.M.: Modification of a strain of mouse-adapted scrapie by passage through rats. *Res. Vet. Sci.* 9:408–410, 1968.
229. Kimberlin R.H., Walker C.A., and Fraser H.: The genomic identity of different strains of mouse scrapie is expressed in hamsters and preserved on reisolation in mice. *J. Gen. Virol.* 70:2017–2025, 1989.
230. Hecker R., Taraboulos A., Scott M., Pan K.-M., Torchia M., Jendroska K., DeArmond S.J., and Prusiner S.B.: Replication of distinct prion isolates is region specific in brains of transgenic mice and hamsters. *Genes Dev.* 6:1213–1228, 1992.
231. Hunter N., Goldmann W., Benson G., Foster J.D., and Hope J.: Swaledale sheep affected by natural scrapie differ significantly in PrP genotype frequencies from healthy sheep and those selected for reduced incidence of scrapie. *J. Gen. Virol.* 74:1025–1031, 1993.
232. Shibuya S., Higuchi J., Shin R.-W., Tateishi J., and Kitamoto T.: Protective prion protein polymorphisms against sporadic Creutzfeldt-Jakob disease. *Lancet* 351:419, 1998.
233. Tatzelt J., Zuo J., Voellmy R., Scott M., Hartl U., Prusiner S.B., and Welch W.J.: Scrapie prions selectively modify the stress response in neuroblastoma cells. *Proc. Natl. Acad. Sci. USA* 92:2944–2948, 1995.
234. Tatzelt J., Prusiner S.B., and Welch W.J.: Chemical chaperones interfere with the formation of scrapie prion protein. *EMBO J.* 15:6363–6373, 1996.
235. Kenward N., Hope J., Landon M., and Mayer R.J.: Expression of polyubiquitin and heat-shock protein 70 genes increases in the later stages of disease progression in scrapie-infected mouse brain. *J. Neurochem.* 62:1870–1877, 1994.
236. Oesch B., Teplow D.B., Stahl N., Serban D., Hood L.E., and Prusiner S.B.: Identification of cellular proteins binding to the scrapie prion protein. *Biochemistry* 29:5848–5855, 1990.
237. Kurschner C. and Morgan J.I.: Analysis of interaction sites in homo- and heteromeric complexes containing Bcl-2 family members and the cellular prion protein. *Mol. Brain Res.* 37:249–258, 1996.

238. Edenhofer F., Rieger R., Famulok M., Wendler W., Weiss S., and Winnacker E.-L.: Prion protein PrP^{C} interacts with molecular chaperones of the Hsp60 family. *J. Virol.* 70:4724–4728, 1996.
239. Gerstmann J., Sträussler E., and Scheinker I.: Über eine eigenartige hereditär-familiäre Erkrankung des Zentralnervensystems zugleich ein Beitrag zur frage des vorzeitigen lokalen Alterns. *Z. Neurol.* 154:736–762, 1936.
240. Masters C.L., Harris J.O., Gajdusek D.C., Gibbs C.J., Jr., Bernouilli C., and Asher D.M.: Creutzfeldt-Jakob disease: patterns of worldwide occurrence and the significance of familal and sporadic clustering. *Ann. Neurol.* 5:177–188, 1978.
241. Gajdusek D.C.: Unconventional viruses and the origin and disappearance of kuru. *Science* 197:943–960, 1977.
242. Bobowick A.R., Brody J.A., Matthews M.R., Roos R., and Gajdusek D.C.: Creutzfeldt-Jakob disease: a case-control study. *Am. J. Epidemiol.* 98:381–394, 1973.
243. Rosenthal N.P., Keesey J., Crandall B., and Brown W.J.: Familial neurological disease associated with spongiform encephalopathy. *Arch. Neurol.* 33:252–259, 1976.
244. Doh-ura K., Tateishi J., Sasaki H., Kitamoto T., and Sakaki Y.: Pro->Leu change at position 102 of prion protein is the most common but not the sole mutation related to Gerstmann-Sträussler syndrome. *Biochem. Biophys. Res. Commun.* 163:974–979, 1989.
245. Goldgaber D., Goldfarb L.G., Brown P., Asher D.M., Brown W.T., Lin S., Teener J.W., Feinstone S.M., Rubenstein R., Kascsak R.J., Boellaard J.W., and Gajdusek D.C.: Mutations in familial Creutzfeldt-Jakob disease and Gerstmann-Sträussler-Scheinker's syndrome. *Exp. Neurol.* 106:204–206, 1989.
246. Kretzschmar H.A., Honold G., Seitelberger F., Feucht M., Wessely P., Mehraein P., and Budka H.: Prion protein mutation in family first reported by Gerstmann, Sträussler, and Scheinker. *Lancet* 337:1160, 1991.
247. Hsiao K.K., Scott M., Foster D., Groth D.F., DeArmond S.J., and Prusiner S.B.: Spontaneous neurodegeneration in transgenic mice with mutant prion protein. *Science* 250:1587–1590, 1990.
248. Hsiao K.K., Groth D., Scott M., Yang S.-L., Serban H., Rapp D., Foster D., Torchia M., DeArmond S.J., and Prusiner S.B.: Serial transmission in rodents of neurodegeneration from transgenic mice expressing mutant prion protein. *Proc. Natl. Acad. Sci. USA* 91:9126–9130, 1994.
249. Telling G.C., Haga T., Torchia M., Tremblay P., DeArmond S.J., and Prusiner S.B.: Interactions between wild-type and mutant prion proteins modulate neurodegeneration in transgenic mice. *Genes & Dev.* 10:1736–1750, 1996.
250. Scott M.R., Nguyen O., Stöckel J., Tatzelt J., DeArmond S.J., Cohen F.E., and Prusiner S.B.: Designer mutations in the prion protein promote β-sheet formation *in vitro* and cause neurodegeneration in transgenic mice (Abstr.). *Protein Sci.* 6 [Suppl. 1]:84, 1997.
251. Alpers M.P.: Kuru: implications of its transmissibility for interpretation of its changing epidemiological pattern. *The Central Nervous System, Some Experimental Models of Neurological Diseases.* Bailey O.T. and Smith D.E., Eds. Williams and Wilkins, Baltimore, 1968 pp. 234–251.
252. Alpers M.: Epidemiology and clinical aspects of kuru. *Prions–Novel Infectious Pathogens Causing Scrapie and Creutzfeldt-Jakob Disease.* Prusiner S.B. and McKinley M.P., Eds. Academic Press, Orlando, 1987 pp. 451–465.
253. Brown P., Preece M.A., and Will R.G.: "Friendly fire" in medicine: hormones, homografts, and Creutzfeldt-Jakob disease. *Lancet* 340:24–27, 1992.
254. Billette de Villemeur T., Deslys J.-P., Pradel A., Soubrié C., Alpérovitch A., Tardieu M., Chaussain J.-L., Hauw J.-J., Dormont D., Ruberg M., and Agid Y.: Creutzfeldt-Jakob disease from contaminated growth hormone extracts in France. *Neurology* 47:690–695, 1996.
255. Prusiner S. B., Scott M. R., Dearmond S. J., and Cohen F. E. Prion protein biology. *Cell* 93:337–348, 1998.
256. Esmonde T., Lueck C.J., Symon L., Duchen L.W., and Will R.G.: Creutzfeldt-Jakob dis-

ease and lyophilised dura mater grafts: report of two cases. *J. Neurol. Neurosurg. Psychiatry* 56:999–1000, 1993.
257. Lane K.L., Brown P., Howell D.N., Crain B.J., Hulette C.M., Burger P.C., and DeArmond S.J.: Creutzfeldt-Jakob disease in a pregnant woman with an implanted dura mater graft. *Neurosurgery* 34:737–740, 1994.
258. CDC: Creutzfeldt-Jakob disease associated with cadaveric dura mater grafts–Japan, January 1979–May 1996. *MMWR* 46:1066–1069, 1997.
259. Kimberlin R.H.: Reflections on the nature of the scrapie agent. *Trends Biochem. Sci.* 7:392–394, 1982.
260. Dickinson A.G. and Outram G.W.: Genetic aspects of unconventional virus infections: the basis of the virino hypothesis. *Novel Infectious Agents and the Central Nervous System. Ciba Foundation Symposium 135.* Bock G. and Marsh J., Eds. John Wiley and Sons, Chichester, UK, 1988 pp. 63–83.
261. Bruce M.E., McConnell I., Fraser H., and Dickinson A.G.: The disease characteristics of different strains of scrapie in *Sinc* congenic mouse lines: implications for the nature of the agent and host control of pathogenesis. *J. Gen. Virol.* 72:595–603, 1991.
262. Weissmann C.: A "unified theory" of prion propagation. *Nature* 352:679–683, 1991.
263. Ridley R.M. and Baker H.F.: To what extent is strain variation evidence for an independent genome in the agent of the transmissible spongiform encephalopathies? *Neurodegeneration* 5:219–231, 1996.
264. Fraser H. and Dickinson A.G.: Scrapie in mice. Agent-strain differences in the distribution and intensity of grey matter vacuolation. *J. Comp. Pathol.* 83:29–40, 1973.
265. Bruce M.E., McBride P.A., and Farquhar C.F.: Precise targeting of the pathology of the sialoglycoprotein, PrP, and vacuolar degeneration in mouse scrapie. *Neurosci. Lett.* 102:1–6, 1989.
266. DeArmond S.J., Yang S.-L., Lee A., Bowler R., Taraboulos A., Groth D., and Prusiner S.B.: Three scrapie prion isolates exhibit different accumulation patterns of the prion protein scrapie isoform. *Proc. Natl. Acad. Sci. USA* 90:6449–6453, 1993.
267. Dickinson A.G. and Meikle V.M.: A comparison of some biological characteristics of the mouse-passaged scrapie agents, 22A and ME7. *Genet. Res.* 13:213–225, 1969.
268. Bruce M.E. and Dickinson A.G.: Biological evidence that the scrapie agent has an independent genome. *J. Gen. Virol.* 68:79–89, 1987.
269. Hunter N., Hope J., McConnell I., and Dickinson A.G.: Linkage of the scrapie-associated fibril protein (PrP) gene and Sinc using congenic mice and restriction fragment length polymorphism analysis. *J. Gen. Virol.* 68:2711–2716, 1987.
270. Bessen R.A. and Marsh R.F.: Identification of two biologically distinct strains of transmissible mink encephalopathy in hamsters. *J. Gen. Virol.* 73:329–334, 1992.
271. Bessen R.A., Kocisko D.A., Raymond G.J., Nandan S., Lansbury P.T., and Caughey B.: Non-genetic propagation of strain-specific properties of scrapie prion protein. *Nature* 375:698–700, 1995.
272. Monari L., Chen S.G., Brown P., Parchi P., Petersen R.B., Mikol J., Gray F., Cortelli P., Montagna P., Ghetti B., Goldfarb L.G., Gajdusek D.C., Lugaresi E., Gambetti P., and Autilio-Gambetti L.: Fatal familial insomnia and familial Creutzfeldt-Jakob disease: different prion proteins determined by a DNA polymorphism. *Proc. Natl. Acad. Sci. USA* 91:2839–2842, 1994.
273. Parchi P., Castellani R., Capellari S., Ghetti B., Young K., Chen S.G., Farlow M., Dickson D.W., Sima A.A.F., Trojanowski J.Q., Petersen R.B., and Gambetti P.: Molecular basis of phenotypic variability in sporadic Creutzfeldt-Jakob disease. *Ann. Neurol.* 39:767–778, 1996.
274. Mastrianni J., Nixon F., Layzer R., DeArmond S.J., and Prusiner S.B.: Fatal sporadic insomnia: fatal familial insomnia phenotype without a mutation of the prion protein gene. *Neurology* 48 [Suppl.]:A296, 1997.
275. Fraser H. and Dickinson A.G.: The sequential development of the brain lesions of scrapie in three strains of mice. *J. Comp. Pathol.* 78:301–311, 1968.

276. Carp R.I., Meeker H., and Sersen E.: Scrapie strains retain their distinctive characteristics following passages of homogenates from different brain regions and spleen. *J. Gen. Virol.* 78:283–290, 1997.
277. DeArmond S.J., Sánchez H., Yehiely F., Qiu Y., Ninchak-Casey A., Daggett V., Camerino A.P., Cayetano J., Rogers M., Groth D., Torchia M., Tremblay P., Scott M.R., Cohen F.E., and Prusiner S.B.: Selective neuronal targeting in prion disease. *Neuron* 19:1337–1348, 1997.
278. Medori R., Tritschler H.-J., LeBlanc A., Villare F., Manetto V., Chen H.Y., Xue R., Leal S., Montagna P., Cortelli P., Tinuper P., Avoni P., Mochi M., Baruzzi A., Hauw J.J., Ott J., Lugaresi E., Autilio-Gambetti L., and Gambetti P.: Fatal familial insomnia, a prion disease with a mutation at codon 178 of the prion protein gene. *N. Engl. J. Med.* 326:444–449, 1992.
279. Anderson R.M., Donnelly C.A., Ferguson N.M., Woolhouse M.E.J., Watt C.J., Udy H.J., MaWhinney S., Dunstan S.P., Southwood T.R.E., Wilesmith J.W., Ryan J.B.M., Hoinville L.J., Hillerton J.E., Austin A.R., and Wells G.A.H.: Transmission dynamics and epidemiology of BSE in British cattle. *Nature* 382:779–788, 1996.
280. Nathanson N., Wilesmith J., and Griot C.: Bovine spongiform encephalopathy (BSE): cause and consequences of a common source epidemic. *Am. J. Epidemiol.* 145:959–969, 1997.
281. Stekel D.J., Nowak M.A., and Southwood T.R.E.: Prediction of future BSE spread. *Nature* 381:119, 1996.
282. Wilesmith J.W., Ryan J.B.M., and Atkinson M.J.: Bovine spongiform encephalopathy–Epidemiologic studies on the origin. *Vet. Rec.* 128:199–203, 1991.
283. Wilesmith J.W.: The epidemiology of bovine spongiform encephalopathy. *Semin. Virol.* 2:239–245, 1991.
284. Kimberlin R.H.: Speculations on the origin of BSE and the epidemiology of CJD. *Bovine Spongiform Encephalopathy: The BSE Dilemma.* Gibbs C.J., Jr., Ed. Springer, New York, 1996 pp. 155–175.
285. Goldmann W., Hunter N., Manson J., and Hope J.: The PrP gene of the sheep, a natural host of scrapie. *Proceedings of the VIIIth International Congress of Virology, Berlin, Aug. 26–31* 284, 1990.
286. Goldmann W., Hunter N., Martin T., Dawson M., and Hope J.: Different forms of the bovine PrP gene have five or six copies of a short, G-C-rich element within the protein-coding exon. *J. Gen. Virol.* 72:201–204, 1991.
287. Prusiner S.B., Fuzi M., Scott M., Serban D., Serban H., Taraboulos A., Gabriel J.-M., Wells G., Wilesmith J., Bradley R., DeArmond S.J., and Kristensson K.: Immunologic and molecular biological studies of prion proteins in bovine spongiform encephalopathy. *J. Infect. Dis.* 167:602–613, 1993.
288. Hunter N., Goldmann W., Smith G., and Hope J.: Frequencies of PrP gene variants in healthy cattle and cattle with BSE in Scotland. *Vet. Rec.* 135:400–403, 1994.
289. Fraser H., McConnell I., Wells G.A.H., and Dawson M.: Transmission of bovine spongiform encephalopathy to mice. *Vet. Rec.* 123:472, 1988.
290. Dawson M., Wells G.A.H., and Parker B.N.J.: Preliminary evidence of the experimental transmissibility of bovine spongiform encephalopathy to cattle. *Vet. Rec.* 126:112–113, 1990.
291. Dawson M., Wells G.A.H., Parker B.N.J., and Scott A.C.: Primary parenteral transmission of bovine spongiform encephalopathy to the pig. *Vet. Rec.* 127:338, 1990.
292. Bruce M., Chree A., McConnell I., Foster J., and Fraser H.: Transmissions of BSE, scrapie and related diseases to mice (Abstr.). *Proceedings of the IXth International Congress of Virology* 93, 1993.
293. Bruce M.E., Will R.G., Ironside J.W., McConnell I., Drummond D., Suttie A., McCardle L., Chree A., Hope J., Birkett C., Cousens S., Fraser H., and Bostock C.J.: Transmissions to mice indicate that 'new variant' CJD is caused by the BSE agent. *Nature* 389:498–501, 1997.
294. Robinson M.M., Hadlow W.J., Knowles D.P., Huff T.P., Lacy P.A., Marsh R.F., and

Gorham J.R.: Experimental infection of cattle with the agents of transmissible mink encephalopathy and scrapie. *J. Comp. Path.* 113:241–251, 1995.

295. Hope J., Reekie L.J.D., Hunter N., Multhaup G., Beyreuther K., White H., Scott A.C., Stack M.J., Dawson M., and Wells G.A.H.: Fibrils from brains of cows with new cattle disease contain scrapie-associated protein. *Nature* 336:390–392, 1988.
296. Serban D., Taraboulos A., DeArmond S.J., and Prusiner S.B.: Rapid detection of Creutzfeldt-Jakob disease and scrapie prion proteins. *Neurology* 40:110–117, 1990.
297. Grathwohl K.-U.D., Horiuchi M., Ishiguro N., and Shinagawa M.: Sensitive enzyme-linked immunosorbent assay for detection of PrP^{Sc} in crude tissue extracts from scrapie-affected mice. *J. Virol. Methods* 64:205–216, 1997.
298. Scott M.R., Safar J., Telling G., Nguyen O., Groth D., Torchia M., Koehler R., Tremblay P., Walther D., Cohen F.E., DeArmond S.J., and Prusiner S.B.: Identification of a prion protein epitope modulating transmission of bovine spongiform encephalopathy prions to transgenic mice. *Proc. Natl. Acad. Sci. USA* 94:14279–14284, 1997.
299. Taylor K.C.: The control of bovine spongiform encephalopathy in Great Britain. *Vet. Rec.* 129:522–526, 1991.
300. Fraser H., Bruce M.E., Chree A., McConnell I., and Wells G.A.H.: Transmission of bovine spongiform encephalopathy and scrapie to mice. *J. Gen. Virol.* 73:1891–1897, 1992.
301. Hunter G.D., Millson G.C., and Chandler R.L.: Observations on the comparative infectivity of cellular fractions derived from homogenates of mouse-scrapie brain. *Res. Vet. Sci.* 4:543–549, 1963.
302. Eklund C.M., Kennedy R.C., and Hadlow W.J.: Pathogenesis of scrapie virus infection in the mouse. *J. Infect. Dis.* 117:15–22, 1967.
303. Will R.G., Ironside J.W., Zeidler M., Cousens S.N., Estibeiro K., Alperovitch A., Poser S., Pocchiari M., Hofman A., and Smith P.G.: A new variant of Creutzfeldt-Jakob disease in the UK. *Lancet* 347:921–925, 1996.
304. Bateman D., Hilton D., Love S., Zeidler M., Beck J., and Collinge J.: Sporadic Creutzfeldt-Jakob disease in a 18-year-old in the UK (Lett.). *Lancet* 346:1155–1156, 1995.
305. Britton T.C., Al-Sarraj S., Shaw C., Campbell T., and Collinge J.: Sporadic CreutzfeldtJakob disease in a 16-year-old in the UK (Lett.). *Lancet* 346:1155, 1995.
306. Ironside J.W.: The new variant form of Creutzfeldt-Jakob disease: a novel prion protein amyloid disorder (Editorial). *Amyloid: Int. J. Exp. Clin. Invest.* 4:66–69, 1997.
307. Chazot G., Broussolle E., Lapras C.I., Blättler T., Aguzzi A., and Kopp N.: New variant of Creutzfeldt-Jakob disease in a 26-year-old French man. *Lancet* 347:1181, 1996.
308. CDC: Surveillance for Creutzfeldt-Jakob Disease–United States. *MMWR* 45:665–668, 1996.
309. Baker H.F., Ridley R.M., and Wells G.A.H.: Experimental transmission of BSE and scrapie to the common marmoset. *Vet. Rec.* 132:403–406, 1993.
310. Lasmézas C.I., Deslys J.-P., Demaimay R., Adjou K.T., Lamoury F., Dormont D., Robain O., Ironside J., and Hauw J.-J.: BSE transmission to macaques. *Nature* 381:743–744, 1996.
311. Collinge J., Palmer M.S., Sidle K.C., Hill A.F., Gowland I., Meads J., Asante E., Bradley R., Doey L.J., and Lantos P.L.: Unaltered susceptibility to BSE in transgenic mice expressing human prion protein. *Nature* 378:779–783, 1995.
312. Cousens S.N., Vynnycky E., Zeidler M., Will R.G., and Smith P.G.: Predicting the CJD epidemic in humans. *Nature* 385:197–198, 1997.
313. Raymond G.J., Hope J., Kocisko D.A., Priola S.A., Raymond L.D., Bossers A., Ironside J., Will R.G., Chen S.G., Petersen R.B., Gambetti P., Rubenstein R., Smits M.A., Lansbury P.T., Jr., and Caughey B.: Molecular assessment of the potential transmissibilities of BSE and scrapie to humans. *Nature* 388:285–288, 1997.
314. Collinge J., Sidle K.C.L., Meads J., Ironside J., and Hill A.F.: Molecular analysis of prion strain variation and the aetiology of "new variant" CJD. *Nature* 383:685–690, 1996.

315. Hill A.F., Desbruslais M., Joiner S., Sidle K.C.L., Gowland I., Collinge J., Doey L.J., and Lantos P.: The same prion strain causes vCJD and BSE. *Nature* 389:448–450, 1997.
316. Parchi P., Capellari S., Chen S.G., Petersen R.B., Gambetti P., Kopp P., Brown P., Kitamoto T., Tateishi J., Giese A., and Kretzschmar H.: Typing prion isoforms (Lett.). *Nature* 386:232–233, 1997.
317. Somerville R.A., Chong A., Mulqueen O.U., Birkett C.R., Wood S.C.E.R., and Hope J.: Biochemical typing of scrapie strains. *Nature* 386:564, 1997.
318. Haraguchi T., Fisher S., Olofsson S., Endo T., Groth D., Tarantino A., Borchelt D.R., Teplow D., Hood L., Burlingame A., Lycke E., Kobata A., and Prusiner S.B.: Asparagine-linked glycosylation of the scrapie and cellular prion proteins. *Arch. Biochem. Biophys.* 274:1–13, 1989.
319. Endo T., Groth D., Prusiner S.B., and Kobata A.: Diversity of oligosaccharide structures linked to asparagines of the scrapie prion protein. *Biochemistry* 28:8380–8388, 1989.
320. Wickner R.B.: [URE3] as an altered URE2 protein: evidence for a prion analog in *Saccharomyces cerevisiae. Science* 264:566–569, 1994.
321. Chernoff Y.O., Lindquist S.L., Ono B., Inge-Vechtomov S.G., and Liebman S.W.: Role of the chaperone protein Hsp104 in propagation of the yeast prion-like factor [psi^+]. *Science* 268:880–884, 1995.
322. Glover J.R., Kowal A.S., Schirmer E.C., Patino M.M., Liu J.-J., and Lindquist S.: Self-seeded fibers formed by Sup35, the protein determinant of [PSI^+], a heritable prion-like factor of S. cerevisiae. *Cell* 89:811–819, 1997.
323. King C.-Y., Tittman P., Gross H., Gebert R., Aebi M., and Wüthrich K.: Prion-inducing domain 2–114 of yeast Sup35 protein transforms *in vitro* into amyloid-like filaments. *Proc. Natl. Acad. Sci. USA* 94:6618–6622, 1997.
324. Paushkin S.V., Kushnirov V.V., Smirnov V.N., and Ter-Avanesyan M.D.: In vitro propagation of the prion-like state of yeast Sup35 protein. *Science* 277:381–383, 1997.
325. Coustou V., Deleu C., Saupe S., and Begueret J.: The protein product of the *het-s* heterokaryon incompatibility gene of the fungus *Podospora anserina* behaves as a prion analog. *Proc. Natl. Acad. Sci. USA* 94:9773–9778, 1997.
326. Derkatch I.L., Chernoff Y.O., Kushnirov V.V., Inge-Vechtomov S.G., and Liebman S.W.: Genesis and variability of [*PSI*] prion factors in *Saccharomyces cerevisiae. Genetics* 144:1375–1386, 1996.
327. Patino M.M., Liu J.-J., Glover J.R., and Lindquist S.: Support for the prion hypothesis for inheritance of a phenotypic trait in yeast. *Science* 273:622–626, 1996.
328. Wickner R.B.: A new prion controls fungal cell fusion incompatibility [Commentary]. *Proc. Natl. Acad. Sci. USA* 94:10012–10014, 1997.
329. Sonneborn T.M.: The determination of hereditary antigenic differences in genically identical Paramecium cells. *Proc. Natl. Acad. Sci. USA* 34:413–418, 1948.
330. Landman O.E.: The inheritance of acquired characteristics. *Annu. Rev. Genetics* 25:1–20, 1991.
331. Beisson J. and Sonneborn T.M.: Cytoplasmid inheritance of the organization of the cell cortex of *Paramecium aurelia. Proc. Natl. Acad. Sci. USA* 53:275–282, 1965.
332. Frankel J.: Positional order and cellular handedness. *J. Cell Sci.* 97:205–211, 1990.
333. Chapman J., Ben-Israel J., Goldhammer Y., and Korczyn A.D.: The risk of developing Creutzfeldt-Jakob disease in subjects with the *PRNP* gene codon 200 point mutation. *Neurology* 44:1683–1686, 1994.
334. Spudich S., Mastrianni J.A., Wrensch M., Gabizon R., Meiner Z., Kahana I., Rosenmann H., Kahana E., and Prusiner S.B.: Complete penetrance of Creutzfeldt-Jakob disease in Libyan Jews carrying the E200K mutation in the prion protein gene. *Mol. Med.* 1:607–613, 1995.
335. Aguzzi A. and Weissmann C.: Prion research: the next frontiers. *Nature* 389:795–798, 1997.
336. Gorodinsky A. and Harris D.A.: Glycolipid-anchored proteins in neuroblastoma cells form detergent-resistant complexes without caveolin. *J. Cell Biol.* 129:619–627, 1995.

337. Goldmann W., Hunter N., Smith G., Foster J., and Hope J.: PrP genotype and agent effects in scrapie: change in allelic interaction with different isolates of agent in sheep, a natural host of scrapie. *J. Gen. Virol.* 75:989–995, 1994.
338. Belt P.B.G.M., Muileman I.H., Schreuder B.E.C., Ruijter J.B., Gielkens A.L.J., and Smits M.A.: Identification of five allelic variants of the sheep PrP gene and their association with natural scrapie. *J. Gen. Virol.* 76:509–517, 1995.
339. Clousard C., Beaudry P., Elsen J.M., Milan D., Dussaucy M., Bounneau C., Schelcher F., Chatelain J., Launay J.M., and Laplanche J.L.: Different allelic effects of the codons 136 and 171 of the prion protein gene in sheep with natural scrapie. *J. Gen. Virol.* 76:2097–2101, 1995.
340. Ikeda T., Horiuchi M., Ishiguro N., Muramatsu Y., Kai-Uwe G.D., and Shinagawa M.: Amino acid polymorphisms of PrP with reference to onset of scrapie in Suffolk and Corriedale sheep in Japan. *J. Gen. Virol.* 76:2577–2581, 1995.
341. Hunter N., Moore L., Hosie B.D., Dingwall W.S., and Greig A.: Association between natural scrapie and PrP genotype in a flock of Suffolk sheep in Scotland. *Vet. Rec.* 140:59–63, 1997.
342. Hunter N., Cairns D., Foster J.D., Smith G., Goldmann W., and Donnelly K.: Is scrapie solely a genetic disease? *Nature* 386:137, 1997.
343. O'Rourke K.I., Holyoak G.R., Clark W.W., Mickelson J.R., Wang S., Melco R.P., Besser T.E., and Foote W.C.: PrP genotypes and experimental scrapie in orally inoculated Suffolk sheep in the United States. *J. Gen. Virol.* 78:975–978, 1997.
344. Parry H.B.: *Scrapie Disease in Sheep,* Academic Press, New York, 1983.
345. Chesebro B.: Prion diseases: BSE and prions: uncertainties about the agent. *Science* 279:42–43, 1998.
346. Shinde U.P., Liu J.J., and Inouye M.: Protein memory through altered folding mediated by intramolecular chaperones. *Nature* 389:520-522, 1997.

Physiology or Medicine 1998

ROBERT F. FURCHGOTT, LOUIS J. IGNARRO and FERID MURAD

"for their discoveries concerning nitric oxide as a signalling molecule in the cardiovascular system"

THE NOBEL PRIZE IN PHYSIOLOGY OR MEDICINE

Speech by Professor Sten Lindahl of the Nobel Committee at the Karolinska Institute.
Translation of the Swedish text.

Your Majesties, Your Royal Highness, Ladies and Gentlemen,

When Robert Furchgott, Louis Ignarro and Ferid Murad independently discovered that a short-lived gas, nitric oxide, NO, was endogenously produced and acted as a signaling molecule between cells – it was unexpected and unique. It initiated a new chapter in biomedical research and new horizons were anticipated.

It was Robert Furchgott who opened up the field in 1980. During the 1970s researchers realized that the innermost cell layer of blood vessels, the endothelium, not only had passive, protective properties. Furchgott demonstrated, quite unexpectedly, that contraction and relaxation of blood vessels were dependent on whether the endothelium was present or not. In a brilliant experiment – the so-called sandwich experiment – he made a key discovery which set the stage for future scientific developments. In his sandwich study, responses from different pieces of aorta were investigated. One piece had the endothelial layer intact while in the other it was removed. In the absence of endothelium he recorded a contraction at stimulation. The piece with endothelium was prepared so that neither contractions nor relaxations were possible. When he put the two pieces together in a sandwich model he found that the same stimulation no longer resulted in a contraction, it relaxed! Furchgott concluded that an unknown substance, a factor, was produced in the endothelium, was transported to the piece of aorta without endothelium, and caused relaxation.

This was a great discovery. It became the starting-shot of a scientific hunt for the identity of the endothelial factor. A search that went on for six years. Different hypotheses were launched. One of them was that nitro-compounds were involved. It was within this field of research that Ferid Murad was active. He knew that nitroglycerine activated an enzyme in aortic muscle cells, guanylylcyclase, which increased cyclic GMP and caused relaxation. Ferid Murad now asked an important question. Did nitroglycerine act via release of nitric oxide, NO? He tested his hypothesis by simply allowing NO gas to bubble through a tissue preparation containing guanylylcyclase. The production of cGMP increased. A new mode of action for drugs to activate enzyme function was discovered. The working principle of nitroglycerine, up to this point unknown in spite of more than 100 years of successful treatment for angina, was now revealed. These experiments by Ferid Murad, performed a few years ahead of Furchgott's discovery of the endothelial factor, created new knowledge which later became the "trail" for identification of the endothelial factor.

It was on this "trail" that the third Nobel Laureate in Physiology or Medicine of this year, Louis Ignarro, also had his scientific experience and activities. Inspired by Murad's findings he also reported that NO relaxed blood vessels. Simultaneously with, but independently from Robert Furchgott, he also added, during the first half of the 1980s, knowledge about the factor. Its identity was becoming clearer. The hunt for Furchgott's endothelial factor came to an end during a scientific meeting at the Mayo Clinic, Rochester, Minnesota, USA in the summer of 1986. At the meeting Furchgott concluded, based on several findings, that the factor was identical with NO. Ignarro supported this at the same meeting. Ignarro also went one step further in an interesting experiment. He used spectral analysis, which means that each substance emits a specific, unique, spectrum. He found identical spectra when reduced hemoglobin reacted with the endothelial factor and with NO, and concluded that the factor was NO.

The hunt was over. The riddle concerning the endothelial factor was finally solved. An endogenous short-lived gas had the capacity to act as a signaling molecule between cells in the body. This was a new phenomenon. The discovery explained the mode of action for nitroglycerine when used to treat high blood pressure and angina, a disease that Alfred Nobel also suffered from. He wrote: "It sounds like the Fate of irony that I have been prescribed nitroglycerine – internally. They have named it Trinitrin in order not to upset pharmacists and the public." Alfred Nobel knew that nitroglycerine could also cause headache. He did not follow his doctor's orders. The discovery that the endothelial factor was nitric oxide also initiated new therapies in clinical medicine, improved diagnostic capabilities for serious inflammatory diseases and opened up new possibilities for drug development. The continued research within the NO field, from 1986 till today, has been enormous.

Professors Robert Furchgott, Louis Ignarro and Ferid Murad,

Your discoveries concerning nitric oxide as a signaling molecule in the cardiovascular system have not only explained the working principle of an old important group of drugs, the nitrovasodilators, they have also opened new avenues for patient treatment and diagnoses of various diseases. Your discoveries have lifted medical research into a new era.

On behalf of the Nobel Assembly at the Karolinska Institute I wish to convey to you our warmest congratulations and I now ask you to step forward to receive your Nobel Prize from the hands of His Majesty the King.

ROBERT F. FURCHGOTT

EARLY EDUCATION

I was born in the lovely coastal city of Charleston, S.C. in 1916 and lived there until I was thirteen. In Charleston I first became enamored of "natural history" when I attended nature study classes and field trips to nearby beaches, marshes and woods, sponsored by the Charleston Museum. I became an avid shell collector and bird watcher (that was before the term "birder" was coined), and I still enjoy these hobbies. In 1929, my family moved from Charleston to Orangeburg, S.C., an inland, rural town of about 8,000 inhabitants, where my mother had grown up and still had some family. The reason for the move was that the Furchgott department store in Charleston, which had been started by my grandfather and was being run by my father and his two brothers, was unable to survive in the midst of the Depression, and my father decided to open a women's clothing store in Orangeburg. So I spent my high school years in Orangeburg, enjoying small town life and competing with my first cousin Edwin Moseley for the highest grades in our class. He won.

Within the first couple of years of high school, I knew that I would like to be a scientist. My parents were encouraging: they gave me chemistry sets and a small microscope as presents. I liked to read popular books about scientists, although there were not many available at that time. My father subscribed to the Sunday New York Times, in which there was often a column on science that I found very exciting.

During the four years that I was in high school, my older brother Arthur was at the University of North Carolina at Chapel Hill. I wanted to attend college there also, but that was not possible when I finished high school in 1933 because tuition for me, as an out-of-state resident, was more than my father could afford at that time. So I spent my freshman year at the University of South Carolina, where my tuition was much less. However, by the summer of 1934, my father moved his business from Orangeburg to Goldsboro, N.C., where he felt that the local economy was better. So now, as a resident of North Carolina, I was able to register at the University at Chapel Hill as a sophomore majoring in chemistry.

At Chapel Hill, I had a number of excellent teachers in chemistry. During my junior and senior years, I had a small amount of financial support from an NYA job (NYA being the initials of the National Youth Administration set up by the federal government to help students during the Depression). In that job, I was a lab assistant in research to a junior faculty member working on the physical chemistry of solutions of cellulose. I had decided early in my college years that I would go on to graduate work in some branch of chemistry.

My preference by the time I was a senior was physical organic chemistry. I sent letters to dozens of chemistry departments applying for a graduate fellowship or teaching assistantship. I had an excellent academic record, but by graduation time I still had no definite offer of a position for graduate training. I was almost resigned to taking a job in chemical industry, when around the middle of June while I was in Florence, S.C., where my parents now lived, an unexpected offer of a teaching assistantship came to me from the Physiological Chemistry Department of Northwestern University Medical School in Chicago. I was to be a graduate student of Dr. Henry Bull, who had recently come to Northwestern, and whose research interests were physical chemical aspects of biochemistry.

NORTHWESTERN AND COLD SPRING HARBOR (1937–1940)

Before I went to Chicago, I worked for two summer months in 1937 for Eastern Airlines at the Philadelphia airport – a job which my older brother Arthur, who was employed by that airline, helped me obtain. The job allowed me to save some money and also allowed me free air travel to Chicago. That helped a lot since my stipend as a teaching assistant at Northwestern was only $50 a month for a nine-month academic year. When I arrived in Chicago, it had already been arranged for me to share a room with two more advanced graduate students. Living in Chicago was quite a change from living in the Carolinas. When I would walk to work in the winter from our rooming house, which was about a mile from the medical school, the chill wind whipping in from Lake Michigan along Chicago Avenue was quite an experience for a Southern boy.

My course work at Northwestern was partly at the medical school , and partly at the Evanston campus to which I would travel via the El. At the Evanston campus, my courses were mainly in physical chemistry under Dr. Malcolm Dole, who was also on my Ph.D. advisory committee. At the Chicago campus, I had to take physiology and bacteriology (along with medical students), Henry Bull's course on physical chemistry in biochemistry, and some assorted graduate courses in physiology and biochemistry. The physiology course was under the direction of Dr. Andrew Ivy, who had built up a sizeable physiology department faculty for those times. In contrast, the biochemistry faculty consisted only of the chairman, Dr. Chester Farmer, Dr. Bull and two part-time lecturers.

My laboratory work with Bull started out with the preparation of purified egg albumin. He was studying physical chemical changes in this protein after different methods of denaturation. He had begun to involve me in some of his studies when the summer of 1938 came along, and that turned out to be a special summer for me. Bull had been invited to present a paper on his work at the sixth Cold Spring Harbor Symposium on Quantitative Biology which was to take place at the Cold Spring Harbor Biological Laboratory of the Long Island Biological Association. The theme of the symposium, which was to run for five weeks in a leisurely fashion was the structure and function of proteins. Bull had obtained permission from the director of the Cold Spring Harbor Laboratory, Dr. Eric Ponder, for me to attend the symposium, while earning my room and board by running the lantern slide projector at the lectures. The symposium was very exciting. I met a many distinguished scientists. Ponder and a physician-scientist, Harold Abramson, arranged to have me assist in a research project at the laboratory for the rest of the summer after the symposium was over. The project was on the electrophoretic mobility of rabbit erythrocytes and ghosts, measured with the use of a microelectrophoresis cell and light- and dark-field microscopy.

By the end of the summer, I had become very interested in the physical chemistry of the red blood cell membrane. When I returned to Northwestern in the fall of 1938, Bull approved continuation of my research on red blood cells as a Ph.D. thesis project. In particular, I was fascinated by the unexplained phenomenon of the transformation of mammalian red blood cells, suspended in unbuffered isotonic saline from discs to perfect spheres when a small drop of the suspension was placed between slide and coverglass. I discovered that the disc-sphere transformation depended on two factors. The first was a rise in pH to over 9.0 in the unbuffered suspension, as a result of the alkaline nature of the glass surfaces (pH being measured with a semimicro glass electrode that I constructed). The second factor was the removal from the suspension of the red blood cells by adsorption onto the glass surfaces of the slide and coverglass of a substance in the suspension that prevented sphering on elevation of pH of the suspension. I demonstrated that this substance, which I termed the anti-sphering factor, was serum albumin which could not be effectively removed from the red cells simply by multiple washing and centrifuging. In addition to the work on shape changes in erythrocytes, my Ph.D. thesis work involved additional studies on the electrophesis of the cells under various conditions and on other aspects of the physical chemistry of erythrocyte membranes.

In the summer of 1939 at the invitation of Ponder, with whom I had extensive correspondence during the year and who had become in effect the major advisor for my Ph.D. thesis research, I returned to Cold Spring Harbor to continue research on red blood cells. To earn my room and board, I waited on tables in the communal dining room. I also was able to attend the symposium talks of that year, which were on the subject of biological oxidations. There I first became aware of the new developments in oxidative energy metabolism and the importance of high energy phosphate compounds. Among the many outstanding biochemists attending were L. Michaelis, Fritz Lipmann and Carl Cori. Ponder and his young wife Ruth were very hospitable to me. I was much impressed with his skill in applying mathematics in his research, his facility in scientific writing, and his large collection of records of classical music.

I was able to complete and defend my thesis in time to receive the Ph.D. degree in June of 1940. Earlier that spring I had attended the annual meeting of the Federation of American Societies for Experimental Biology (FASEB) in New Orleans. I had fortunately been asked by Henry Tauber, an Austrian biochemist working for a pharmaceutical firm in Chicago, to share the driving in his car on the round trip to New Orleans as well as his room in a rundown hotel in New Orleans. Thus, I was able to attend this meeting at very little expense. At the FASEB meeting in New Orleans, where gatherings of participants were still called "smokers" and even a fancy meal was not more than two dollars, I had some interviews with persons about possible post-doctoral jobs. One of the interviews was with Dr. Ephraim Shorr, an Associate Professor of Medicine at Cornell University Medical School in New York City, whom I had met at Cold Spring Harbor the summer before. A few weeks later Shorr offered me a postdoctoral position in his laboratory. Although I was hoping to get a position which would allow me to continue work on physical chemistry of proteins or cell membranes, none came through, and I accepted the position with Shorr, with the understanding that I would begin in September.

The reason for waiting until September to begin work at Cornell was because I wanted to spend one more summer at the Biological Laboratory at Cold Spring Harbor. This time, however, I went there as an invited speaker at the symposium which that summer was on the topic of permeability of cell membranes. My talk was entitled "Observations on the structure of red cell ghosts." At that symposium, there were again a number of established distinguished scientists like K.S. Cole, Robert Chambers and F.O. Schmitt; and in addition, a number of bright young scientists like Hans Neurath, who had also been at the 1938 symposium, Hugh Davson, who with Danielli had developed the lipid bilayer membrane model, and Benjamin Zweifach, with whom I was to collaborate later in research.

CORNELL UNIVERSITY MEDICAL COLLEGE (1940–1949)

I stayed at Cornell University Medical College working in the laboratory of Ephraim Shorr for nine years. When I arrived, Sam Barker, a young research associate, was there to instruct me in methods and procedures they were using to study tissue metabolism (largely using Warburg manometers) and the turnover of rather ill-defined tissue organic phosphate fractions from canine cardiac muscle during incubations *in vitro*. For such studies the lab was one of the first to use radioactive phosphate, which we obtained from the cyclotron laboratory at Berkeley. Barker left toward the end of my first year at Cornell, and I was then responsible for running the laboratory for Shorr. Shorr himself, would sometimes take part in preparing tissue for the Warburg experiments. He was quite capable in the laboratory in addition to being a busy and excellent clinician.

During my first two years at Cornell, my major project was on phosphate exchange and turnover, using radioactive phosphate and slices of dog left ventricular muscle. A full paper on the work was published in the Journal of Biological Chemistry in 1943. The methods and equipment we used in that work have long been superseded, but we did manage with chemical and some early enzymatic methods to show the extremely fast turnover of creatinine phosphate and the terminal phosphate of ATP in resting cardiac muscle.

The 1943 paper was my first full publication after three years of work at Cornell. One likely reason for sparse output was that the United States had entered World War II in December of 1941, and Shorr, like many others, began to undertake research that had more relevance to the war effort. With government and other support, he shifted the major research in the lab to circulatory shock – first on changes in tissue energy metabolism resulting from hypoxia associated with hemorrhagic shock, and then mainly on factors that might account for "irreversible" shock, the condition in which restoration of blood volume is no longer able to raise pressure and sustain life in the animal subjected to maintained low blood pressure as a result of controlled hemorrhage. To help in this new line of research, Shorr recruited Benjamin Zweifach, then a bright young physiologist who had trained with Robert Chambers and had developed a beautiful method for microscopic observation of blood flow in part of the mesentery (the "mesoappendix" area) of the anesthetized rat. In brief, the "rat mesoappendix test", conducted by Zweifach and technicians whom he trained, produced evidence by 1944 for two vasoactive factors in circulatory shock. The first factor appeared in the plasma of dogs in the early reversible (by transfusion) stage of hemorrhage. Intravenous injections of this plasma increased the sensitivity of the small arterioles and pre-capillary sphincters to topically applied epinephrine in the mesoappendix test. This factor was termed VEM (for vasoexcitatory materials). As the irreversible stage of circulatory shock developed, VEM activity disappeared from the plasma and a new factor appeared which markedly decreased the sensitivity to epinephrine in the mesoappendix test. This factor was termed VDM (for vasodepressor material). We developed evidence, in part from *in vitro* experiments with tissue slices, that hypoxic kidney was the probable source of VEM and that hypoxic liver was the probable source of VDM. By late 1945, these developments led to a lead article in the journal *Science* by Shorr, Zweifach and myself.

During the war years, I was not solely involved in research on tissue metabolism and circulatory shock. In 1943, Eugene DuBois, chairman of the Department of Physiology at Cornell, arranged that I join his department as an instructor in order to replace a staff member lost to military service. Although I was teaching in physiology, I still spent most of my time in research in Shorr's lab, which was partially funded by the federal Office of Scientific Research and Development. The work on VEM and VDM continued after the war ended. I had attempted to isolate the VEM-like material that accumulated in incubation fluid when kidney slices were incubated anaerobically. I was able to concentrate it somewhat and it appeared to be a labile dialyzable peptide, but I failed to isolate it. On the other hand, Abraham Mazur, a professor of biochemistry at the City College of New York who worked part time with us, purified a VDM-like material from liver which appeared to be ferritin. (Ferritin or not, we might now wonder whether VDM could somehow be related to nitric oxide!)

Unfortunately, the only bioassay procedure for detecting VEM and VDM activity was that involving changes in sensitivity to epinephrine in the rat mesoappendix test. Intravenous injections of solutions containing high levels of impure VEM or purified ferritin did not effect blood pressure in experimental animals. Attempts to develop an *in vitro* bioassay system also failed. These failures tempered my enthusiasm, and I think that of Zweifach, for the significance of VEM and VDM in the regulation of circulation. However, the failed attempts to develop an *in vitro* bioassay for VEM and VDM were very important for me for they introduced me to the pharmacology of smooth muscle, a subject that has been a major interest of mine ever since.

Two of the isolated smooth muscle preparations that I unsuccessfully tested for bioassay of VEM and VDM were a helically-cut strip of rabbit aorta, which responded with contraction to epinephrine, and a longitudinal segment of rabbit duodenum, which exhibited spontaneous rhythmic contractions that were inhibited by epinephrine and stimulated by acetylcholine. At that time, contractions of such smooth muscle preparations mounted in organ baths were recorded with isotonic levers on kymographs. One day in the course of making tests on segments of rabbit duodenum mounted in oxygenated Krebs solution, I was surprised to see that during the first hours of the experiment, contraction amplitude did not stabilize as usual but declined gradually and markedly even though the rhythmic frequency remained unchanged. I suspected that my technician had forgotten to add glucose to the Krebs solution. Adding glucose now quickly increased contraction amplitude to the normal level. This finding led to a simple procedure for finding out what sugars and fatty acids could be utilized for energy for contraction in the intestinal smooth muscle under aerobic and anaerobic conditions and to analyze the sites of action of metabolic inhibitors.

In the spring of 1949, I had two interesting offers at the assistant professorship level – one in physiology at Duke and one in pharmacology at Washington University School of Medicine. I decided on Washington University, partly because the new chairman there, Oliver Lowry, was someone I had known in the Enzyme Club in New York City and partly because I had begun to be very interested in pharmacology as a discipline. This was partly because of the studies I had begun on the effects of drugs and other agents on smooth muscle preparations *in vitro*, but also in large part because of my close friendship with Walter Riker, who was then a junior member in the Pharmacology Department at Cornell at the beginning of a distinguished career. His enthusiasm for research in pharmacology was contagious.

In the summer of 1949, my family and I drove from New York to St. Louis. My wife, Lenore, a native New Yorker, said she felt like she was going West in a covered wagon. By that time we had two daughters, ages four and one. Later we had a third daughter born in St. Louis. It might be noted here that none of my daughters became scientists. Instead, they all went into art (like my younger brother, Max). It might also be noted here that my wife Lenore died in 1983; and that now I have a new wife, Margaret (Maggie). I have been very fortunate in having wives who encouraged my work, even though it often reduced the time I could give to family matters.

WASHINGTON UNIVERSITY (1949–1956)

My seven years in the Pharmacology Department at Washington University were enjoyable ones. Oliver (Ollie) Lowry had been appointed chairman of that department a year or so before I came. He was already well recognized for his ingenuous methods involving enzymology, spectrometry and fluorometry in the quantitative analysis of important enzymes, substrates and products in extremely small amounts of tissue. He was very helpful in introducing me to enzymatic-spectroscopic methods (as developed by Kalckar) for analysis of ATP, ADP and AMP. As a new chairman, Lowry inherited two faculty members, Helen Graham and Edward Hunter, and recruited two new ones, namely myself and Morris (Morrie) Friedkin. I had never had a course in pharmacology as a student, much less taught in one, and so I had to spend a lot of time during my first year in St. Louis keeping ahead of the medical students. Later, when I set up my own department in Brooklyn, I adopted for the pharmacology course there much of the lecture, laboratory and conference program that I had participated in at St. Louis.

Lowry's department was a stimulating place for research. Over the years I was there, the departmental staff grew steadily. Lowry attracted outstanding postdoctoral fellows, such as Eli Robbins and Jack Strominger. We often joined the members of Carl Cori's Biochemistry Department for seminars and journal club meetings.

My first research project at Washington University was a continuation of the work I had begun at Cornell on energy-metabolism and function of rabbit intestinal smooth muscle. I was able to obtain a small grant to support my research on smooth muscle, and to hire a technician, Marilyn (Wales) McCaman, who later became my first graduate student. By the middle of 1951, my favorite *in vitro* smooth muscle preparation had shifted from the rabbit duodenum to the rabbit thoracic aorta. I had found that the helical (spiral) strip of that vessel, properly cut and mounted in organ chambers for isotonic recording, gave very reproducible contractions to epinephrine and norepinephrine after equilibration in oxygenated Krebs bicarbonate solution. I had at first planned to study the effects of disturbances in energy-metabolism on these contractions, but I became much more interested in using the aortic strip for studies on drug-receptor interactions.

By 1953, I had published a paper entitled "Reactions of strips of rabbit aorta to epinephrine, isoproterenol, sodium nitrite and other drugs". Among the other drugs was acetylcholine. I found that it only produced contractions, whether it was added to resting strips or strips precontracted with some other agent. That was a paradoxical response since acetylcholine was known to be a very potent vasodilator *in vivo.* Little did I suspect then what I was able to show many years later – namely, that relaxation of arteries by acetylcholine is strictly endothelium-dependent, and that my method of preparing the strips inadvertently resulted in the mechanical removal of all the endothelial cells.

In 1954, I published a paper on the use of dibenamine in differentiating receptors in the aortic strip, and in 1955 a review in Pharmacological Reviews on the pharmacology of vascular smooth muscle. In that review, I tried to develop receptor theory as a logical base for interpreting the responses of vascular smooth muscle to many neurotransmitters, hormones and drugs. In order to derive equations to account for the very slow onset and offset kinetics of competitive antagonists as compared to the fast kinetics of agonists, I developed a biophase model in which the agents moved between an aqueous extracellular phase and a lipid membrane phase containing the receptors. Although I paid homage in my review to A.J. Clark for his pioneering work in developing receptor theory, I took issue with his hypothesis that response of a tissue to an agonist is proportional to the fraction of receptors occupied by the agonist. Our results with dibenamine, which behaved as an irreversible competitive blocking agent of adrenergic α-receptors, had indicated that with a strong agonist like epinephrine, one could still achieve well over half of the maximum contraction when only a small fraction of receptors were still active. This was the beginning of my interest in the concept of "receptor reserve" or "spare receptors." (A year later, R.P. Stephenson published his classic paper on the subject in which he introduced the concepts of efficacy, full agonist and partial agonist.)

In the review of 1955, I also briefly reported on a newly discovered phenomenon – namely, that strips of rabbit aorta undergo reversible relaxation when exposed to light of proper wavelength and intensity. This photorelaxation was an accidental discovery that came from the observation that in one experiment active contractile tone of two strips in one pair of organ chambers fluctuated with time, whereas that of two strips in another pair of chambers remained steady. The two strips showing fluctuations did so synchronously. Those two strips, but not the other two, were in organ chambers near a window through which they were exposed to skylight. Suspecting that the fluctuations in tone were due to fluctuations in light intensity on the strips near the window (it was a cloudy-bright day), I closed the shade on the window and both strips increased in tone. I opened the shade and both decreased in tone. From that point on, we never allowed our strips to be exposed to direct skylight. (The usual overhead fluorescent lights do not produce photorelaxation.) Some studies on the characteristics of photorelaxation were begun in St. Louis, and then extended when I moved to Brooklyn.

In addition to working on *in vitro* smooth muscle preparations at Washington University, I also began what became many years of research on the pharmacology of an *in vitro* cardiac muscle preparation – namely the isolated electrically-driven right atrium of the guinea pig. In starting that work, I had the assistance of a very able technician, Taisija De Gubareff. Using chemical and enzymatic methods for analysis of creatinine phosphate, ATP, ADP, and AMP, we showed that neither development of "experimental failure" *in vitro* (a steady loss of contractile force over hours) nor recovery from failure on addition of a cardiac glycoside was due to changes in concentration of these high-energy phosphates. We also reported on the effects of anaerobiosis and of a number of positive and negative inotropic agents. We collaborated with my good friend William Sleator of the Physiology Department in the study of changes in cellular action potentials (measured with intracellular microelectrodes) associated with the changes in contractility of the guinea-pig atrium in response to epinephrine and acetylcholine, and a number of other inotropic agents.

SUNY MEDICAL CENTER IN BROOKLYN (1956–)

In 1956, I accepted the position of chairman of the new Department of Pharmacology at the State University of New York (SUNY) College of Medicine at New York City (actually in Brooklyn, and later changed in name to SUNY Downstate Medical Center and more recently to SUNY Health Science Center at Brooklyn). The department had previously been part of a joint physiology and pharmacology department headed by Chandler Brooks but with the opening of a new, relatively huge (for the time) basic science building for the medical school and with good financial support from the State University, there was ample room and resources for a separate department. From the former joint department, I inherited Julius Belford as an associate professor and Bernard Mirkin as an assistant professor. For additional faculty, I recruited Kwang Soo Lee, Leonard Procita, Lowell Greenbaum, Walter Wosilait and Arthur Zimmerman, all in time for them to teach our first course for medical students. The following year C.Y. Kao joined the staff. Also during the first year, we accepted our first graduate students, namely Maurice Feinstein, who worked with me, and Arnold Schwartz, who worked with Lee. During that year I didn't do much bench work in the research lab since most of my time was spent organizing the department and learning how to be a chairman. (I never became a well-organized administrator and was always poor at delegating authority.)

In Brooklyn, I continued research on photorelaxation of blood vessels, factors influencing contractility of cardiac muscle, peripheral adrenergic mechanisms, and receptor theory and mechanisms. Then, about twenty-three years after moving to Brooklyn, the research in my laboratory largely shifted to endothelium-dependent relaxation of blood vessels. For convenience, I shall divide the discussion of research in Brooklyn into subsections corresponding to the areas that I have listed.

Photorelaxation of Blood Vessels

Helping with this research were Eugene Greenblatt, my first postdoctoral fellow, and Stuart Ehrreich, my third graduate student. Among other things, we were able to obtain an accurate action spectrum (with a peak at 310 nm) for the photorelaxation. Later we observed that addition of sodium nitrite to the bathing medium greatly sensitized the rabbit aortic strip to photorelaxation and shifted the peak of the action spectrum to about 355 nm. Ehrreich and I found that many other smooth muscle preparations (from stomach, intestine and uterus) which did not ordinarily relax in response to radiation did so in the presence of inorganic nitrite. Percy Lindgren, a visiting faculty member from the Karolinska Institute, also worked with us for a while on photosensitization by nitrite.

Many years later in the early 1980's, after the discovery of endothelium-derived relaxing factor (EDRF), I again began research on photorelaxation. Although photorelaxation did not depend on the presence of endothelium on the strip or ring of rabbit aorta, we found many similarities between it and endothelium-dependent relaxation (as produced by acetylcholine or A23187). Not only was photorelaxation, like endothelium-dependent relaxation, causally dependent on the elevation of cyclic GMP as a result of stimulation of guanylate cyclase, but both were inhibited by hemoglobin and by methylene blue. This work was carried out with Desingarao Jothianandan, who has been a most helpful research associate in my lab over the past seventeen years. Then, after EDRF was identified in 1986 as nitric oxide, Kazuki Matsunaga (a postdoctoral fellow) and I reinvestigated the potentiation of photorelaxation by sodium nitrite. Using a cleverly designed perfusion-bioassay type apparatus, Matsunaga clearly demonstrated that the potentiation was due to the photoactivated release of NO from nitrite. It is tempting to hypothesize that light (in the absence of added nitrite) produces relaxation of vascular smooth muscle by photoactivating the release of NO from some endogenous compound in the muscle cell.

Factors Influencing Contractility of Cardiac Muscle

My first graduate student in Brooklyn, Maurice Feinstein, did his Ph.D. thesis research on the effects of experimental congestive heart failure, asphyxia and ouabain on high energy phosphates and creatine content of the guinea pig heart. My second graduate student, Albert Grossman, who began work in 1957, did his thesis research on the effects of frequency of stimulation, extracellular calcium concentration and various drugs on calcium exchange and contractility of the guinea-pig left atrium. Grossman and I published three papers based on his thesis research, which was one of the first attempts to determine the rates of exchange of calcium (using ^{45}Ca) between extracellular fluid and various intracellular "pools" of calcium in cardiac muscle under various conditions affecting contractility. We showed that the positive inotropic effects of norepinephrine and strophanthin-K were correlated with an increase in rate of exchange of calcium in an intracellular pool associated with the contractile process and that the negative inotropic effects of acetylcholine and adenosine were correlated with a decrease in rate of exchange in that pool.

We also continued work with ryanodine, which produced a negative inotropic effect on the guinea-pig atrium and actually changed the force-frequency effect from a positive to negative staircase (mimicking the normal staircase in frog heart). Sleator, De Gubareff and I had shown that the decrease in force with ryanodine (unlike that with acetylcholine or adenosine) was not associated with a decrease in duration of the action potential. The thesis research of Grossman and a few years later that of another graduate student, Peter Wolf, also using ^{45}Ca to measure effects of ryanodine on calcium exchange, led to a hypothetical model that fits fairly well with more recent work of others on the reactions of ryanodine with "receptors" involved with calcium transport in the sarcoplasmic reticulum.

Peripheral Adrenergic Mechanisms

In writing the 1955 review on the "Pharmacology of vascular smooth muscle," I had become very interested in the mechanisms by which sympathetic postganglionic denervation and certain drugs like cocaine markedly potentiate the response of effector organs to epinephrine and norepinephrine, yet markedly reduce the response to the sympathomimetic tyramine. My second postdoctoral fellow, Sadashiv (Sada) Kirpekar, was assigned to work in this area. He proved to be a gifted investigator, and we published a number of papers together on work carried out between 1959 and 1962. In one paper, with the running page heading of "the cocaine paradox," we presented evidence that in aortic strips of rabbit and isolated electrically-driven atria from guinea pig and cat, cocaine potentiated responses to norepinephrine and inhibited those to tyramine by blocking one and the same site on adrenergic nerve terminals. Blockade of this site inhibited the neuronal uptake of norepinephrine from the region of the adrenergic receptors, thus potentiating its action; however, blockade of the site also inhibited uptake of tyramine, whose sympathomimetic action depends on release of norepinephrine from neuronal storage sites, thus inhibiting its action. The site, which we called the "transfer site" later became known as the uptake-1 (U1) site. In the same paper we showed that reserpine, which depleted neuronal storage granules of norepinephrine, did not interfere with activity of the uptake site. In addition to Kirpekar, Peter Cervoni came in as a postdoctoral fellow to work on peripheral adrenergic mechanisms. Both he and Kirpekar later became faculty members in the department, with Kirpekar staying on and becoming a stellar figure in the field of adrenergic mechanisms before his untimely death in 1983.

In 1960, I was invited to present a paper on some of my studies on receptors for sympathomimetic amines at a CIBA Foundation conference on Adrenergic Mechanisms held at CIBA House in London. It was the occasion for my first trip abroad and was very exciting. Among the many distinguished pharmacologists at the conference were Sir Henry Dale, Sir John Gaddum and J.H. Burn. Burn at that time was pushing his "cholinergic-link" hypothesis for norepinephrine release at adrenergic nerve terminals. I felt strongly that he had misinterpreted the experimental results which had led to the hypothesis and in the discussion sessions I presented our own results with isolated atria which indicated that there were nicotinic cholinergic receptors on adrenergic nerve terminals which when stimulated by nicotine or acetylcholine triggered a transient release of norepinephrine, but which played no role in release of norepinephrine on electrical stimulation of the nerve.

In 1962–63, I spent a sabbatical year in the Department of Physiology of the

University of Geneva, where Jean Posternak was chairman. Although I did some research and teaching there, I spent most of my time writing papers on research that my colleagues and I had completed during the preceding few years and on a review on receptor mechanisms (see below). I also visited a number of laboratories in Europe where outstanding research on adrenergic mechanisms was in progress. Among these were the laboratories of U.S. von Euler in Stockholm, E. Muscholl in Mainz and John Gillespie in Glasgow.

Between 1965 and 1970 I was fortunate in having a number of very competent coworkers in research on peripheral adrenergic mechanisms. In addition to Kirpekar, there were Pedro Sanchez-Garcia, (a visiting research associate who later became a leading pharmacologist in his native Spain), Jerome Levin (a postdoctoral fellow) and Arun Wakade (a graduate student who later became a faculty member).

In early 1971, I began my second sabbatical leave, this time at the relatively new medical school of the University of California at San Diego (located in La Jolla). I became a visiting professor in Steve Mayer's Pharmacology Division of the Department of Medicine. One reason for this choice of a sabbatical site was that I wanted to learn the method for analysis of cyclic AMP that Mayer had developed (this was before the development of radioimmunoassays for cyclic nucleotides). However, I did not do a lot of research at La Jolla, partly because a fair amount of my time that year was devoted to duties as president of the American Society for Pharmacology and Experimental Therapeutics.

On returning from La Jolla to Brooklyn in 1972, I continued research on the role of receptors located on prejunctional terminals (varicosities) of adrenergic nerves. I collaborated with Kirpekar in an attempt to characterize the inhibitory prejunctional α-adrenergic receptors on the nerve terminals in cat spleen. At the same time, one of my graduate students, Odd Steinsland, was conducting a very exciting thesis project on cholinergic receptors on prejunctional adrenergic nerve terminals in the isolated, perfused central ear artery of the rabbit. He first pharmacologically characterized with the use of various muscarinic agonists and antagonists the prejunctional receptor through which acetylcholine produces a marked inhibition of norepinephrine release (monitored by both the degree of vasoconstriction and [^{3}H]norepinephrine release). He then went on to study the release of norepinephrine from the adrenergic neurons in the ear artery by cholinergic agonists acting on prejunctional nicotinic receptors. At the same time I was continuing studies, with the assistance of Taruna Wakade, on the pharmacology of cholinergic nicotinic receptors on adrenergic prejunctional terminals in the guinea-pig left atrium.

Receptor Theory and Mechanisms

When I first gave a course on receptor theory and mechanisms to graduate students in 1957–1958, the literature on the subject was relatively sparse: papers by Clark, Gaddum, Schild, Ariëns, Stephenson, Nickerson and myself.

I became interested in developing suitable theory (occupation theory) and *in vitro* procedures for differentiating and characterizing receptors. In particular, I concentrated on receptors for adrenergic and cholinergic agents using as test tissues the rabbit aortic strip, duodenal segment, and stomach fundus muscle, and the guinea-pig electrically driven left atrium and tracheal ring.

In 1963, toward the end of my sabbatical year at the University of Geneva, I completed a review on "Receptor Mechanisms" for Volume 4 of the Annual Review of Pharmacology. In it, I took the opportunity to stress the importance of Stephenson's ideas on efficacy and spare receptors. In 1965 at a symposium on receptor mechanisms at Chelsea College in London, I presented a paper on the use of β-haloalkylamines, as irreversible receptor antagonists, in the differentiation of receptors and in the determination of dissociation constants of receptor-agonist complexes. Using a slightly modified form of Stephenson's equations and introducing a term, ϵ, for intrinsic efficacy, I derived a simple equation that predicted that the slope and ordinate intercept of a double reciprocal plot of equiactive concentrations of an agonist before and after irreversible inactivation of a fraction of its receptors, could permit the determination of both the fraction of receptors still active as well as the dissociation constant (K_A) of the agonist-receptor complex. For different agonists acting on the same receptor, one could calculate from the K_A values the fractional occupation by each to obtain the same standard response before receptor inactivation, and thus obtain relative efficacies. Using this approach, Paula (Bursztyn) Goldberg (a graduate student) and I compared the dissociation constants and relative efficacies of agonists acting on muscarinic cholinergic receptors of isolated strips of rabbit stomach fundus muscle; and later John Besse (a postdoctoral fellow) and I compared the dissociation constants and relative efficacies of agonists acting on α_1-adrenergic receptors of rabbit aorta. In light of what is now known about receptor signalling pathways through G-proteins, it is probably better to admit that the pharmacological procedure which we developed for obtaining agonist-receptor dissociation constants can only give approximate relative values. Nevertheless, the procedure has proven useful in a number of studies.

In 1972, I published a review entitled "The classification of adrenoceptors (adrenergic receptors). An evaluation from the standpoint of receptor theory". In it I attempted to formulate the methods and necessary conditions for the classification and differentiation of receptors by pharmacological procedures designed to give accurate dissociation constants of competitive antagonists acting on a given receptor, and accurate relative potencies and, if possible, dissociation constants of agonists acting on the same receptor. In particular, I attempted to point out pitfalls in such procedures and how to avoid them. For example, I derived theoretical equations to illustrate how removal of the agonist from the region of the receptor by active uptake or enzymatic destruction could markedly alter the slope of a Schild plot for competitive antagonism from the theoretical slope of 1. Later, Aaron Jurkiewicz, a visiting research associate from Sao Paulo, Niede Jurkiewicz and I successfully used these theoretical equations in the analysis of propranolol antagonism to isoproterenol in guinea-pig tracheal strips before and after blockade of removal of the agonist by active uptake.

In 1977, I organized for the annual FASEB meeting a symposium on receptors. By then binding of radioligands (usually ^{3}H-labelled competitive antagonists) had been used for several years for quantifying specific receptors in membranes from homogenized cells and for determining the dissociation constants of competitive antagonists and agonists for those receptors. Most of the papers at the symposium were reports of studies with radioligands (e.g., R.J. Lefkowitz on both α- and β-adrenergic receptors; P. Seeman on dopamine receptors; S. Snyder and colleagues on serotonin receptors and opiate receptors). My paper at the symposium was partly a discussion of how pharmacological procedures for differentiating and characterizing receptors based on occupation theory were still very useful in conjunction with the exciting new developments in receptor research being made with specific radioligands.

Also, I reviewed work that had been carried out in my laboratory on β-adrenergic receptors mediating relaxation of guinea-pig tracheal smooth muscle, and presented results of pharmacological experiments that showed that this smooth muscle did not have exclusively the β_2-type of the β-adrenergic receptor, as dogma of that time would have it, but had an admixture of the β_1-type as well – usually as a small fraction of the total of β-receptors, but, depending on the guinea-pig used, sometimes much more.

EndoThelium-dependent Relaxation

Having obtained pharmacological evidence that guinea-pig tracheal smooth muscle sometimes has a sizeable fraction of the β_1-type adrenergic receptor along with the β_2-type (see above), I decided that it would be well to reexamine the smooth muscle of rabbit thoracic aorta to see if it also might have varying amounts of the β_1-type receptor mixed with the β_2-type. However, in the very first experiment designed for this new study in May, 1978, an accidental finding as a result of a technician's error completely changed the course of research in my laboratory. The accidental finding was that on the preparation of rabbit aorta being used in the experiment, the muscarinic agents acetylcholine and carbachol induced relaxation rather than the expected contraction. Why this accidental finding was so exciting, how it led to our discovery of the endothelium-derived relaxing factor (EDRF), and how that factor was eventually identified as nitric oxide will not be discussed here since those matters will be considered in detail in my Nobel Lecture.

In 1982, I resigned from the chairmanship of the Department of Pharmacology at the SUNY Downstate Medical Center, but continued as a professor. In 1989, I retired from my professorship (receiving emeritus status), so that I now was free of teaching duties and committee work related to the medical curriculum but could still continue research in the department. My retirement also now allowed me to spend about three and a half months each winter as an adjunct professor in the Department of Molecular and Cellular Pharmacology of the University of Miami School of Medicine. Most of my time there I have spent trying to catch up on the writing of manuscripts and on the reading of the burgeoning literature in the field of nitric oxide research – an impossible task these days! During the winter sojourns in Miami, I keep in touch with what is going on in my research laboratory in Brooklyn by means of an occasional visit, but mainly by frequent fax and telephone communications with my one or two coworkers there. I consider myself very fortunate in having this Brooklyn-Miami arrangement. Of course, an additional advantage for my wife Maggie and me is that the arrangement allows us to enjoy the very pleasant winter weather in Miami and some of the outdoor activities that it fosters (golf, for instance, in my case).

From 1982 until the present writing, I have been the recipient of a number of honors and awards for my research. Naturally, I have been very pleased to be the recipient. Yet, in thinking back about what aspects of my research have given me the greatest pleasure, I would not place the honors and awards first. I think that my greatest pleasure has come from each first demonstration in my laboratory that experiments designed to test a new hypothesis developed to explain some earlier, often puzzling or paradoxical finding, have given results consistent with the hypothesis. It is not just the immediate pleasure of obtaining such results but also the anticipated pleasure of discussing the results with others doing research in the same area – obviously an ego supportive aspect.

I still enjoy doing bench work in the laboratory with my co-workers. The research still is rather "old fashion" pharmacological research. I was very lucky to stumble on unexpected results in 1978 that led to the finding of endothelium-dependent relaxation and EDRF, and eventually to NO; for if I had not, I would probably have still concentrated on research on receptor theory and mechanisms, and been left far behind by others in that field who have so brilliantly and successfully developed and used molecular biological and other advanced methodologies in their research.

ENDOTHELIUM-DERIVED RELAXING FACTOR: DISCOVERY, EARLY STUDIES, AND IDENTIFICATION AS NITRIC OXIDE

Nobel Lecture, December 8, 1998

by

ROBERT F. FURCHGOTT

Department of Pharmacology, State University of New York, Health Science Center at Brooklyn, Brooklyn, NY 11203, USA

AN ACCIDENTAL FINDING

In a paper published forty-five years ago, in which I introduced the helical strip of the rabbit aorta as a useful preparation for quantitative pharmacological studies on vasoactive drugs and drug-receptor interactions, I reported on the responses of this preparation to a number of catecholamines, sodium nitrite, histamine and acetylcholine (ACh). One somewhat surprising finding (Furchgott and Bhadrakom, 1953) was that ACh, well recognized as a very potent vasodilator in whole animal and perfused organ studies, elicited no relaxation but only contraction of the aortic strip, whether or not the strip was tested at rest or precontracted with a vasocontrictor like norepinephrine. Somewhat later, we used ACh as the contracting agent when determining the relative potencies of norepinephrine, epinephrine and isoproterenol as relaxing agents acting via β-adrenoceptors on helical strips in which α-adrenoceptors had been irreversibly blocked by pretreatment with dibenamine (Furchgott, 1954). The relative potencies of these catecholamines as relaxants (approximately 1:50:200) indicated a β_2-type adrenoceptor.

In the mid-1970s, in pharmacological studies on isolated strips of guinea-pig tracheal smooth muscle, we obtained evidence that the β-adrenoceptors of that muscle, which according to dogma were pure β_2-type, frequently had a fairly high admixture of the β_1 type with the β_2-type. In view of this unexpected finding, I decided that it would be well to reexamine the type of β-receptor in the smooth muscle of rabbit thoracic aorta. The plan was first to determine anew in a series of experiments the relative potencies of norepinephrine, epinephrine and isoproterenol in producing relaxation of preparations of rabbit aorta after dibenamine pretreatment. To produce contraction after the pretreatment, we planned to use a muscarinic agonist like carbachol or ACh, as in our earlier studies. However, in the very first experiment of this planned series in May 1978, my technician did not follow my directions correctly: early in the experiment before blocking the α-adrenergic receptors with dibenamine, he tested carbachol for its contracting activity before rather than after (as prescribed in the directions) washout of a previous test dose of norepinephrine. The response to carbachol was not a contraction, but was a partial relaxation of the norepinephrine-induced contraction. ACh was then tested. It too produced relaxation of the aortic preparation precontracted with norepinephrine (Fig. 1). This was the first time that I had ever observed relaxation of rabbit aorta in response to muscarinic agonists over the many years that I had been using this blood vessel for *in vitro* studies.The unexpected relaxation of rabbit aorta *in vitro* by muscarinic agonists was very exciting, for it was an effect that was in accord with the potent vasodilating action of these agonists *in vivo.*

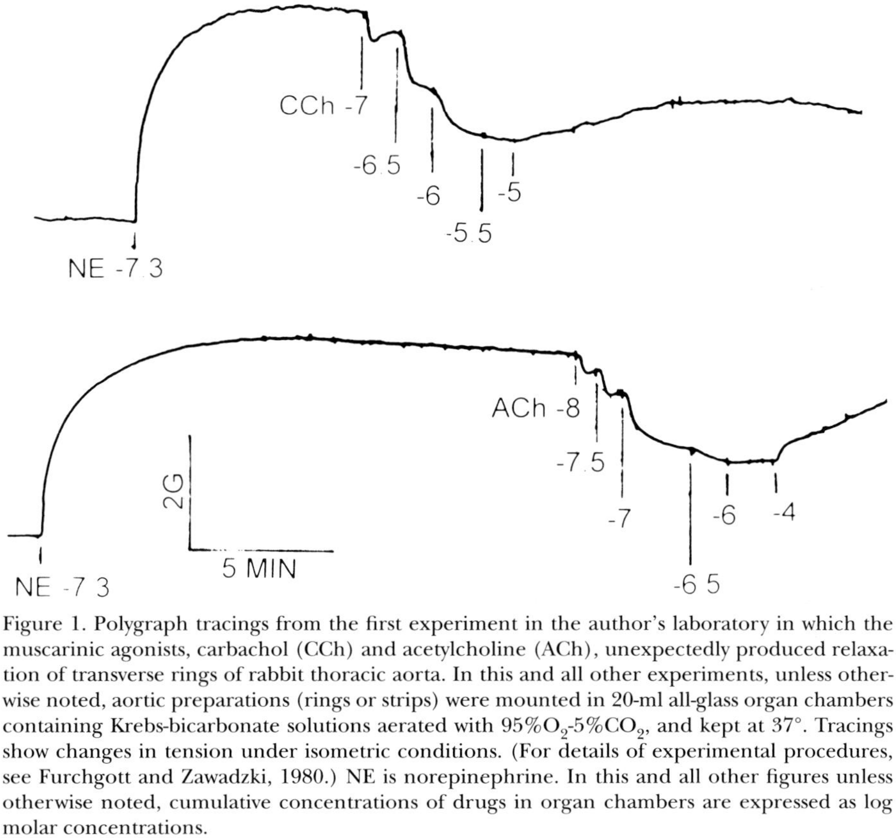

Figure 1. Polygraph tracings from the first experiment in the author's laboratory in which the muscarinic agonists, carbachol (CCh) and acetylcholine (ACh), unexpectedly produced relaxation of transverse rings of rabbit thoracic aorta. In this and all other experiments, unless otherwise noted, aortic preparations (rings or strips) were mounted in 20-ml all-glass organ chambers containing Krebs-bicarbonate solutions aerated with 95%O_2-5%CO_2, and kept at 37°. Tracings show changes in tension under isometric conditions. (For details of experimental procedures, see Furchgott and Zawadzki, 1980.) NE is norepinephrine. In this and all other figures unless otherwise noted, cumulative concentrations of drugs in organ chambers are expressed as log molar concentrations.

DISCOVERY AND EARLY STUDIES ON EDRF

Why had we failed to observe ACh-induced relaxation or aortic preparations in the past, and why did it now become manifest? A detailed account of our investigation of this matter has been published (Furchgott, 1993; 1996), so it will suffice here to outline briefly the sequence of findings. First, we were aware that the kind of preparation of rabbit thoracic aorta that we were now using was the transverse ring, rather than the helical strip which we had used in all of the earlier experiments with muscarinic agonists.

Several experiments were therefore carried out in which we compared the responses to cumulative concentrations of carbachol or ACh on helical strips and transverse rings prepared from the same aorta. Only the rings responded with relaxation to these muscarinic agonists, while the strips responded with contraction (Fig. 2). After several weeks of work, we recognized that in any

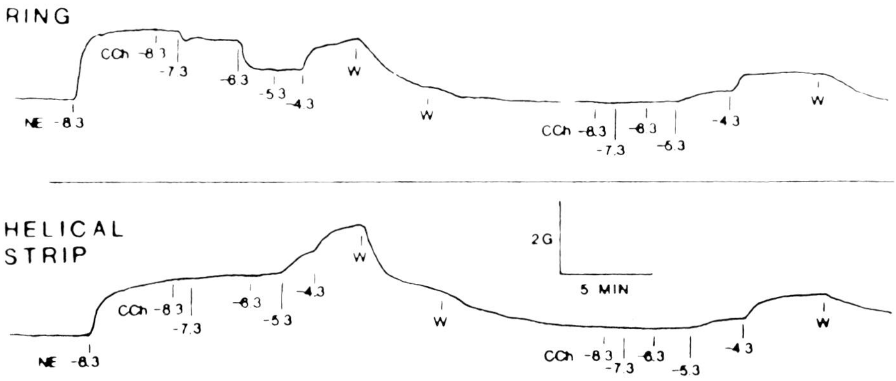

Figure 2. Comparison of responses to carbachol (CCh) of a transverse ring and a helical strip prepared from the same rabbit thoracic aorta. The ring (2.5-mm wide) was mounted on a pair of L-shaped hooks for attachment to the base of the organ chamber and to the force transducer. The strip, prepared by our standard method, had thread ties at each end for attachment. Only the ring, when precontracted with norepinephrine (NE), relaxed in response to low concentrations of CCh. It was subsequently recognized that in our standard method of preparing strips,the endothelial cells were unintentionally rubbed off the intimal surface. (There is no relaxation of the ring in response to CCh added after washout (W) of NE because such preparations exhibit practically no active contractile tension in the absence of vasoconstrictor agents.) In this figure, concentrations of drugs are expressed as log of concentrations in g per ml. Reproduced with permission from Furchgott (1993).

kind of preparation (ring or strip), gentle rubbing of the intimal surface, whether intentionally or unintentionally (as had occurred, unbeknown to us, in our standard method of preparing helical strips), eliminated the relaxing response to Ach and other muscarinic agonists. Strips as well as rings gave excellent relaxation when care was taken not to rub the intimal surface. That the loss of the relaxation response to ACh after intimal rubbing was the result of mechanical removal of the endothelial cells was clearly demonstrated with the use of a silver staining procedure to allow *en face* microscopy of the intimal surfaces. It was also demonstrated that complete removal of endothelial cells of the aorta by pre-incubation of the intimal surface with collagenase also resulted in a complete loss of the relaxation response to ACh.

Figure 3 is a slightly modified figure from the first full paper on the obligatory role of endothelial cells in the relaxation of arterial smooth muscle by ACh (Furchgott and Zawadzki, 1980). In that paper, we also reported that in addition to the thoracic aorta, a number of other arteries from the rabbit and various arteries from other species of laboratory animals also exhibited relaxation in response to ACh only if endothelial cells were present. Cyclooxygenase inhibitors did not interfere with the endothelium-dependent relaxation. One hypothesis to explain the obligatory role of endothelial cells was that ACh acting on a muscarinic receptor of these cells stimulates them to release a non-prostanoid substance that diffuses to and activates relaxation of

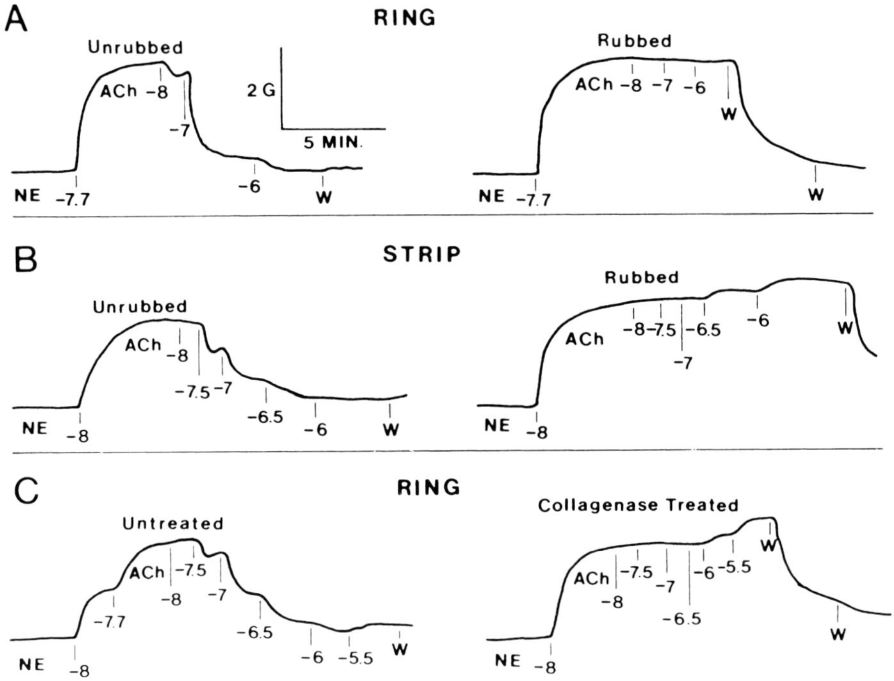

Figure 3. Recordings showing loss of relaxing response of rabbit aorta to ACh after complete removal of endothelial cells. In each panel, the recording on the left was made before removal and that on the right after removal of endothelial cells. The cells were removed from the ring in panel A by rubbing its intimal surface with a small wooden stick, from the transverse strip in panel B by rubbing its intimal surface on moistened filter paper, and from the ring in panel C by exposing its intimal surface to collagenase. After the procedure to remove endothelial cells, each preparation was allowed to re-equilibrate in its organ chamber for about one hour before being retested. Reproduced with permission from Furchgott *et al.* (1981).

the subjacent smooth muscle cells. Direct evidence for this hypothesis was obtained with the so-called "sandwich" procedure in which it was shown that a transverse strip of aorta freed of endothelial cells and therefore unable to relax on exposure to ACh, could be made to relax in response to ACh if it were remounted for recording with its endothelium-free intimal surface placed against the endothelium-containing intimal surface of a longitudinal strip of aorta (Fig. 4). The ACh-induced relaxation of the transverse strip in the "sandwich" demonstrated that ACh stimulated the release of a diffusible relaxing substance (or substances) from the endothelial cells of the longitudinal strip (Furchgott and Zawadzki, 1980). The relaxing substance was later referred to as EDRF (for endothelium-derived relaxing factor) (Cherry *et al.*, 1982).

It should be noted here that although we had never observed relaxation of rabbit aortic preparations by ACh in my laboratory prior to our accidental finding in 1978, there had been occasional reports from other investigators of relaxation of isolated preparations of arteries in response to that agent. For example, Jellife (1962) had reported that a chain of thread-tied rings of rabbit aorta precontracted with serotonin, relaxed in response to low concentra-

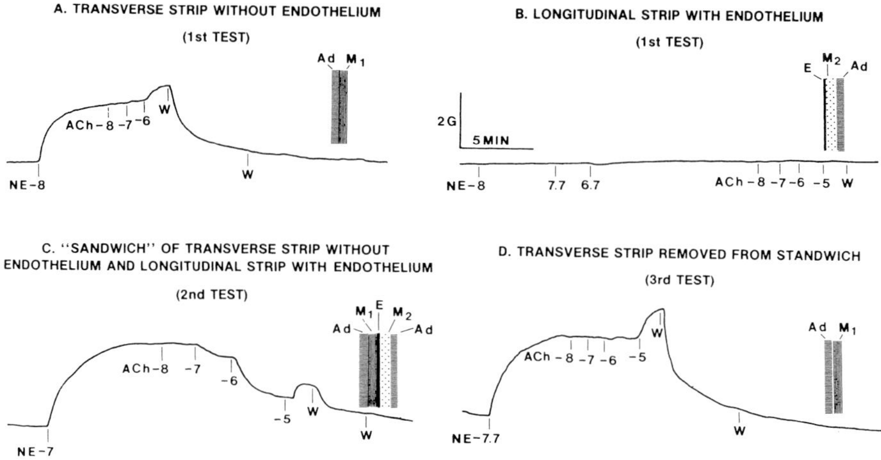

Figure 4. "Sandwich" experiment demonstrating release of a relaxing substance from endothelial cells in response to ACh. **A:** On a transverse strip of rabbit aorta with endothelial cells removed by rubbing, ACh failed to relax the NE-induced tone. **B:** On a longitudinal strip from the same aorta with endothelium present, NE and ACh produced only trivial changes in tension because of the orientation of the smooth muscle cells. **C:** On a "sandwich" made from the transverse strip without endothelium and the longitudinal strip with endothelium mounted together in clips with their entire intimal surfaces apposed, and allowed to re-equilibrate, ACh now produced relaxation of the NE-induced tone. **D:** After removing the longitudinal strip with endothelium from the endothelium-denuded transverse strip, the latter after re-equilibration, again gave no relaxation in response to ACh. In the diagrams of the vertical cross-sections of the preparations, Ad is adventitia; E is endothelial cell layer; M_1 and M_2, respectively, are medial layers of smooth muscle cells oriented vertically and horizontally. Reproduced with permission from Furchgott *et al.* (1981).

tions of ACh; and Vanhoutte (1974) and Toda (1974) had found that on a number of canine arteries prepared as strips, ACh elicited significant relaxation of spontaneous or drug-induced contractions. After our findings on the obligatory role of endothelial cells in ACh-induced relaxation of arteries, it became apparent that these earlier examples of ACh-induced relaxation were dependent on the presence of endothelial cells in the preparations of the arteries being studied.

It is of some interest that an unintentional removal of endothelial cells in another vascular preparation used in my laboratory may have prevented the discovery of endothelium-dependent vasodilation at an earlier date. That preparation was the perfused rabbit central ear artery which Odd Steinsland, while a graduate student in my laboratory in the early 1970s, used for investigating the inhibition by ACh and other muscarinic agonists of adrenergic neurotransmission by their action on prejunctional muscarinic inhibitory receptors. He readily demonstrated that ACh was a potent inhibitor of neurotransmission, as evidenced by its inhibition of both the release of norepinephrine and the vasoconstriction in response to nerve stimulation (Steinsland *et al.,* 1973). To make sure that the inhibition of the vasoconstriction was only due to inhibition of transmitter release, he tested whether infused ACh could inhibit vasoconstriction produced by infused norepinephrine, and found that it did not (Steinsland *et al.*, 1973). About ten years later, after endothelium-dependent relaxation by ACh had been found in a large variety of arterial preparations, we were puzzled by the earlier finding that the perfused ear artery constricted with infused norepinephrine, had failed to show any vasodilation in response to ACh. Dr. Steinsland provided us with the answer. In his earlier work, he had not taken precautions to prevent air bubbles from entering the perfusion line and passing through the ear artery. When care was now taken to avoid the occurrence of bubbles in the perfusion stream through the artery, ACh elicited good vasodilation (relaxation) of norepinephrine-induced vasoconstriction (Furchgott *et al.*, 1990b). Apparently, in the earlier work, the bubbles had mechanically removed the endothelial cells and we had been unaware of this action.

EARLY STUDIES ON AGENTS INDUCING AND INHIBITING ENDOTHELIUM-DEPENDENT RELAXATION

Within a few years after the discovery that ACh-induced relaxation of isolated arteries was endothelium-dependent, a number of other endothelium-dependent relaxing agents were identified in organ chamber experiments using rings or strips of blood vessels from a variety of species, including man. Among these were the calcium ionophore A23187, ATP and ADP, substance P, bradykinin, histamine, thrombin, serotonin and vasopressin. [For references to first reports on endothelium-dependent relaxation by these and other agents, see Furchgott and Vanhoutte (1989) and Furchgott (1990)]. The early finding that the calcium ionophore A23187 was a very potent endothelium-dependent relaxing agent on all vessels tested suggested a central role for calcium in the release, or synthesis and release, of EDRF by endothelium-dependent relaxing agents (Furchgott 1981).

It was soon recognized in this early work that, depending on the agent, endothelium-dependent relaxation might be limited to certain species, or even to specific arteries from a given species. For example, whereas ACh relaxed by this mechanism practically all systemic arteries from all mammalian species tested, histamine, another potent vasodilator, produced endothelium-dependent relaxation of rat (van de Voorde and Leusen, 1983) but not of rabbit arteries (Furchgott, 1983). Another example was bradykinin, which gave endothelium-dependent relaxation of isolated canine and human arteries but not of rabbit or cat arteries (Cherry *et al.*, 1982). As examples of vessel specificity, it was found that noradrenaline activated endothelium-dependent relaxation in isolated canine coronary arteries by acting on α_2-adrenoceptors of the endothelial cells, but failed to activate this kind of relaxation in canine systemic arteries (Cocks and Angus, 1983), and that serotonin activated endothelium-dependent relaxation in canine coronary arteries but not in canine systemic arteries (Cocks and Angus, 1983; Cohen *et al.*, 1983).

Among the conditions and agents which were found to inhibit endothelium-dependent relaxation of arteries in the early organ chamber experiments were anoxia and some agents that were recognized as inhibitors in the metabolic pathway of arachidonic acid (AA) from phospholipids through lipoxygenase (Furchgott and Zawadzki, 1980; Furchgott *et al.*, 1981). Because it had already been reported that in certain smooth muscles there was a positive relationship between an increase in cyclic guanosine monophosphate (cGMP) and relaxation (Katsuki and Murad, 1977; Böhme *et al.*, 1978; Murad *et al.*, 1978, 1979) and that guanylate cyclase (G-cyclase) was markedly stimulated by hydroperoxides of AA (Hidaka and Asano, 1977; Goldberg *et al.*, 1978), I speculated that (1) EDRF is either a short-lived hydroperoxide or free radical arising as an intermediate product in the oxidation of liberated AA by the lipoxygenase pathway, and (2) that EDRF stimulates the G-cyclase of the vascular smooth muscle, causing an increase in cGMP which then somehow activates relaxation (Furchgott *et al.*, 1981). As it turned out, this early speculation about the nature of EDRF was wrong, but that about the role of cGMP proved to be right.

CYCLIC GMP AS A MEDIATOR FOR EDRF AND NITROVASODILATORS

The speculation that EDRF released by ACh would stimulate an increase in cGMP was soon proven correct in a number of independent studies – e.g., Rapoport and Murad (1983) using rat aorta, Diamond and Chu (1983) and Furchgott *et al.* (1984) using rabbit aorta, Holzmann (1982) using bovine coronary artery, and Ignarro *et al.* (1984) using bovine pulmonary artery. All of these investigators showed that a rise in cGMP accompanied the ACh-induced relaxation of the blood vessel, consistent with the proposal that the increase in cGMP in the muscle has a causal role in the relaxation (Fig. 5).

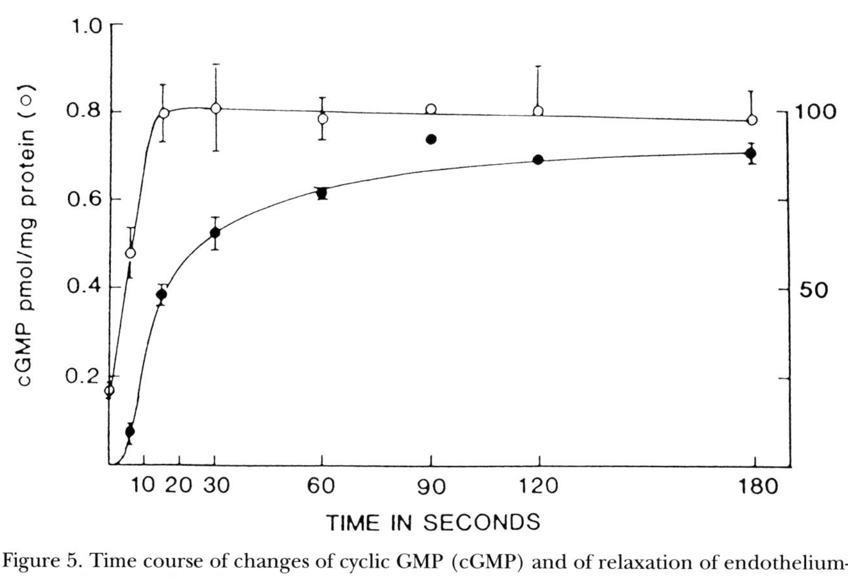

Figure 5. Time course of changes of cyclic GMP (cGMP) and of relaxation of endothelium-containing rings of rabbit thoracic aorta after addition of ACh. Norepinephrine (0.1 μM) was first added to produce a steady contraction. ACh (1 μM) was then added, and the tissue was instantly frozen in liquid nitrogen after the desired period of exposure to ACh. The level of cyclic AMP (not shown) did not change significantly during the endothelium-dependent relaxation. Reproduced with permission from Furchgott *et al.* (1984).

Endothelium-dependent relaxation of rat aorta by A23187 and by histamine (Rapoport and Murad, 1983) and of rabbit aorta by A23187 (Furchgott *et al.*, 1984) was also accompanied by a rise in cGMP.

It is well to stress at this point that Murad and coworkers (see Murad *et al.*, 1978 and 1979 for reviews) had previously found that nitric oxide (NO) is a very potent stimulus of G-cyclase, and had proposed that a number of agents (e.g. glyceryl trinitrate, nitroprusside, azide) which they and others had found to increase cGMP and produce relaxation of certain non-vascular smooth muscle preparations, do so by generating NO as the proximal activator of G-cyclase. Also, Ignarro and his coworkers (Gruetter *et al.*, 1979) had found that the relaxation of a vascular preparation (bovine coronary artery) by NO as well as by nitroprusside was associated with an increase in cGMP. However, despite these early findings with NO, none of the investigators who several years later showed that endothelium-dependent relaxation of arteries in organ chamber experiments was also correlated with an increase in cGMP (see above) hypothesized that EDRF might be the free radical NO. Certain key findings had yet to be made to foster this novel proposal.

FURTHER STUDIES ON EDRF, INCLUDING THOSE WITH PERFUSION-BIOASSAY PROCEDURES

Hemoglobin and Methylene Blue as Inhibitors

Two agents that became widely used as inhibitors of endothelium-dependent relaxation were hemoglobin (Hb) and methylene blue (MB). Hb had been found by Murad *et al.* (1978) to inhibit guanylate cyclase in cellular extracts, and by workers at Glasgow University (Bowman *et al.*, 1982) to inhibit the relaxation of strips of bovine retractor penis (BRP) muscle in response either to nonadrenergic, noncholinergic (NANC) nerve stimulation or to an acid-activated extract of that tissue. Gruetter *et al.* (1981) had reported that methemoglobin (metHb) interfered with NO-induced increases in vascular cGMP. In our laboratory, work with Hb began in 1983 when William Martin, who had studied the properties of extracts of the BRP muscle while a graduate student at Glasgow University (Gillespie *et al.*, 1981; Bowman *et al.*, 1981), came to my laboratory as a postdoctoral fellow (see below). It was soon found that Hb was an extremely potent and fast acting inhibitor of endothelium-dependent relaxation and of the associated rise in cGMP induced by ACh, A23187 and other agents on rabbit and rat aorta (Furchgott *et al.*, 1984; Martin *et al.*, 1985a,1986a). Myoglobin was as effective as Hb, but metHb only inhibited very weakly (Martin *et al.*, 1985b) (Fig. 6).

The degree of potentiation by Hb of contractions of arterial rings by vasoconstrictors such as phenylephrine also provided a measure of the relaxing activity of basally released EDRF (Martin *et al.*, 1986a). Using this measure, the relaxing activity of basally released EDRF in the case of rings of rat aorta appeared to be considerably greater than in the case of rings of rabbit aorta. With rings from both species, addition of phosphodiesterase inhibitors (such as zaprinast and IBMX) produced relaxation associated with increases in

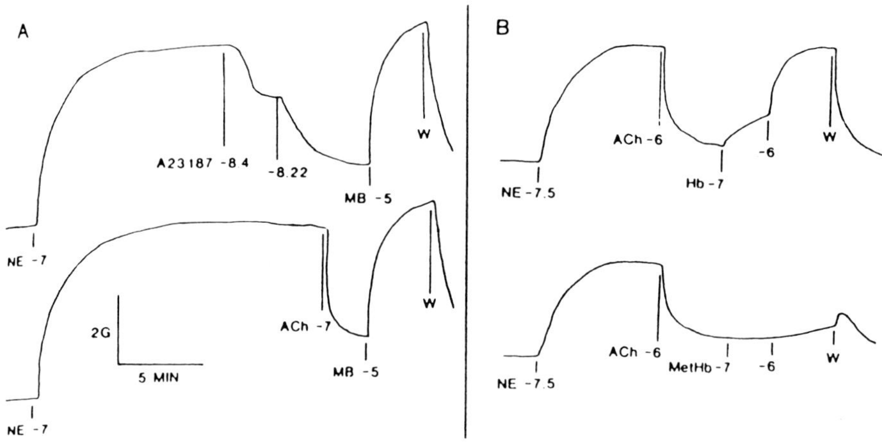

Figure 6. Examples of inhibition by methylene blue (MB) and by hemoglobin (Hb) of endothelium-dependent relaxation in rings of rabbit aorta. **A.** Addition of MB (10 μM) rapidly and completely inhibited the ongoing relaxation produced by the addition of ACh and that produced by addition of the calcium ionophore A23187. **B.** Addition of Hb at 1 μM rapidly and almost completely inhibited the relaxation produced by addition of ACh; however, methemoglobin (MetHb) at the same concentration inhibited ACh-induced relaxation only to a very small extent.

cGMP (Martin *et al.*, 1986b). The responses to these inhibitors of the enzymes which hydrolyze cGMP to GMP were more pronounced in the case of rat aorta than in rabbit aorta, consistent with a higher basal release of EDRF and higher basal stimulation of G-cyclase in the former. The relaxation of endothelium-containing rings of both species produced by phosphodiesterase inhibitors could be completely blocked by Hb.

It was originally proposed that Hb inhibits relaxation and the rise in cGMP by binding EDRF as it diffuses from the endothelial cells, thus preventing it from reaching and activating the G-cyclase in the muscle cells (Martin *et al.*, 1985a). However, after the identification of EDRF as NO, it became apparent that the scavenging of EDRF by Hb under the usual aerobic conditions was the result of the very rapid reaction of HbO_2 with NO to form NO_3^- and metHb, as first reported by Doyle and Hoekstra (1981).

MB had first been shown by Gruetter *et al.* (1981) to inhibit stimulation of G-cyclase by nitrovasodilators in cell-free systems and to inhibit both the relaxation and increase in cGMP elicited by these agents in vascular smooth muscle. Subsequently, MB was found to inhibit both the endothelium-dependent relaxation and accompanying increase in cGMP produced by ACh in bovine arteries (Holzman, 1982; Ignarro *et al.*, 1984) and by ACh and A23187 in rabbit aorta (Furchgott *et al.*, 1984; Martin *et al.*, 1985a). Holzman (1982) and Ignarro *et al.* (1984) concluded that the inhibition of relaxation was the result of the inhibition of G-cyclase by the MB. On the other hand, the present author and his coworkers, although agreeing that such enzyme inhibition accounted for the sustained inhibition by MB developed during prolonged exposure of the rabbit aorta, proposed that the extremely fast inhibition of endothelium-dependent relaxation immediately after addition of MB to an organ chamber (as in Fig. 6) resulted from some more direct inactivation of EDRF by the MB (Furchgott *et al.*, 1984; Martin *et al.*, 1985a). It now appears that this more direct inactivation is due to superoxide anions, generated by reactions of the MB in the tissue (also see Wolin *et al.*, 1990).

Results with Perfusion-Bioassay Procedures

For more direct studies on the properties of EDRF than were possible in organ chamber experiments, perfusion-bioassay procedures were developed. These procedures, although first developed in laboratories other than my own, should be considered here, because they produced results which contributed much to the identification of EDRF. In such procedures, endothelial cells are upstream in the perfusion (or superfusion) cascade and an endothelium-denuded arterial preparation (usually a ring or one or more strips) whose contractile state is monitored, is downstream to bioassay EDRF in the perfusion fluid. In some procedures, the endothelial cells upstream were native cells on the luminal surface of a perfused artery (e.g. Forstermann *et al.*, 1984, Griffith *et al.*, 1984, Rubanyi *et al.*, 1985, Khan and Furchgott, 1987); in other procedures, they were cultured cells (usually from bovine or porcine aortas) on the surface of microcarrier beads contained in a perfused column (e.g. Cocks *et al.*, 1985, Gryglewski *et al.*, 1986).

Perfusion-bioassay procedures allowed one to determine clearly whether an inhibitor of endothelium-dependent relaxation interfered with the synthesis/release of EDRF or inactivated the EDRF after its release. In addition, by altering the transit time of the perfusate between the endothelial cells and the bioassay preparation, it was possible to study the rate of decay of EDRF. Surprisingly, estimated half-life values for EDRF ranged from as low as 4-6 s to as high as 50 s depending on the biological preparations and the experimental conditions. The likely explanation of these discrepant values came with the important finding that superoxide dismutase (SOD) markedly stabilizes EDRF released from either endothelial cells on arteries (Rubanyi and Vanhoutte 1986) or cultured endothelial cells on microcarrier beads (Gryglewski *et al.*, 1986), clearly showing that the superoxide anion (O_2^-) very rapidly inactivates EDRF. Thus, the marked differences in half-life of EDRF reported from different laboratories could largely be attributed to differences in concentrations of O_2^- in the perfusion fluid leaving the endothelial cells under the different experimental conditions used. Subsequently, Moncada *et al.* (1986) were able to show with their perfusion-bioassay procedure that inhibition of endothelium-dependent relaxation by certain reducing agents (e.g. hydroquinone, phenidone and pyrogallol) originally demonstrated in organ chamber experiments, was principally the result of inactivation of EDRF by O_2^- formed in a single-electron oxidation of the reducing agent by O_2.

Development of perfusion-bioassay procedures also led to the important finding that increases in shear stress on endothelial cells of the perfused artery with increases in rate of flow of the perfusion fluid resulted in increases in the rate of release of EDRF (e.g. Holtz *et al.*, 1984; Rubanyi *et al.*, 1986). Increase in flow through arteries both *in situ* (Holtz *et al.* 1984; Kaiser and Sparks 1986; Pohl *et al.*, 1986) and *in vitro* produced vasodilation that could not be blocked by inhibitors of cyclooxygenase but was abolished by removal of the endothelium. These findings concerning increased EDRF release with increased shear stress were important for they provided evidence that EDRF has a continuous physiological role in the control of regional blood flow and regulation of blood pressure.

FINDINGS THAT LED TO THE IDENTIFICATION OF EDRF AS NO

While Billy Martin was working in my laboratory (1983-1985), I became interested in the work that he and others had done in John Gillespie's laboratory in Glasgow in trying to identify the neurotransmitter released from NANC (non-adrenergic, non-cholinergic) nerves that elicit relaxation of bovine retractor penis (BRP) muscle and certain other smooth muscle preparations. I learned that both relaxation produced by NANC nerve stimulation and relaxation produced by a partly purified extract made by Martin from the BRP muscle (referred to as BRP inhibitory factor or BRPIF) were both inhibited by Hb (Bowman *et al.*, 1982) and were both associated with increases in cGMP in the muscle (Bowman and Drummond, 1984). One important aspect of Martin's work with BRPIF extract was that it had to be acidified for a short period prior to testing in order for it to have a relaxing effect on smooth muscle preparations of retractor penis and arteries in organ baths (Gillespie *et al.*, 1981). Once activated by acidification, the BRPIF extract gave good relaxations of the smooth muscle preparations, but the relaxations were transient, lasting no more than a few minutes.

As discussed in the paper that I presented at a symposium in Rochester, MN, in July, 1986 (Furchgott, 1988), I was struck by the similarities of the transient relaxations produced by the acid-activated BRPIF extract of Martin and the transient relaxations of rabbit aorta produced by acidified solutions of sodium nitrite ($NaNO_2$). That acidified, but not neutral solutions of $NaNO_2$, produce strong transient relaxations of rabbit aorta had been found about fifteen years earlier in my laboratory when a new postdoctoral fellow, assigned to obtain cumulative dose-response curves for $NaNO_2$, accidentally made his standard solutions of that salt with acidified diluting fluid (used in my laboratory for making stabile standard solutions of catecholamines) rather than with neutral diluting fluid (Fig. 7). At that time, I became aware that the HNO_2 (pK_a of 3.2) produced by acidification of $NaNO_2$ solutions immediately generates low concentrations of NO and NO_2 as a result of a reversible dismutation.

$$NO_2^- + H^+ \longleftrightarrow HNO_2$$
$$2HNO_2 \longleftrightarrow NO + NO_2 + H_2O$$

However, when a solution of acidified nitrite is added to the oxygenated, well-buffered physiological salt solution used in our organ chambers, any NO present will be removed very rapidly – thus accounting for the transiency of the relaxation of the aortic rings receiving acidified nitrite.

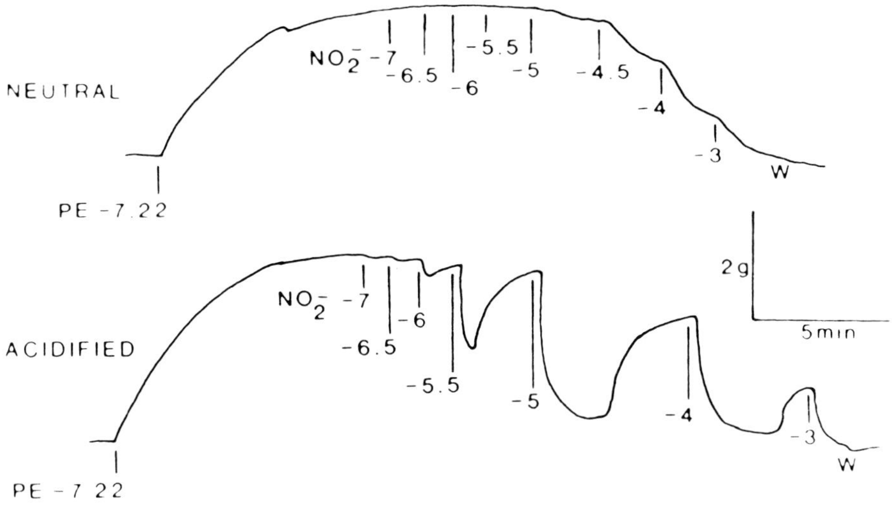

Figure 7. Comparison of the relaxing effects of cumulative additions of neutral and acidified solutions of sodium nitrite on rings of rabbit aorta. In this typical experiment, two rings from the same aorta after contraction with phenylephrine (PE), were exposed, respectively, to additions of $NaNO_2$ diluted with neutral physiological saline and to additions of $NaNO_2$ diluted with the saline acidified with 11 mM HCl. The concentrations shown are the final nitrite concentrations (log molar) in the organ chambers. Note the striking transient relaxations in the case of the acidified nitrite solutions. The rings in this experiment were denuded of endothelial cells, but results were essentially the same when endothelial cells were present. Reproduced with permission from Furchgott (1988).

In the spring of 1986, recalling the much earlier accidental finding in my laboratory with acidified sodium nitrite solutions, I began to consider the possibility that Martin's acid-activatable BRPIF was inorganic nitrite. An encouraging early finding was that the degree of transient relaxation as a function of pH of the added $NaNO_2$ solution in the case of rabbit aorta was similar to that found by Martin for his added extract in the case of the BRP – maximal transient relaxation at a pH of 2 or lower, and approximately half-maximal at a pH of 3.

By the time I had begun to use acidified nitrite solutions as a source of NO in experiments with rabbit aorta in 1986, I had also become aware of the recent findings by Gryglewski, Palmer and Moncada (1986) and by Rubanyi and Vanhoutte (1986) demonstrating that the superoxide anion (O_2^-) rapidly inactivates EDRF and that superoxide dismutase (SOD) protects against this inactivation. Assuming that the free radical NO might rapidly react with the free radical O_2^-, to form inactive NO_3^-, I began to speculate on the possibility that EDRF is NO. [Later, it was pointed out by Beckman *et al.* (1990) that the immediate product of the reaction between NO and O_2^- is peroxynitrite ($ONOO^-$), which is protonated to HONOO ($pK_a = 6.8$), which decays within seconds at physiological pH.]

My speculation in the spring of 1986 that EDRF might be NO led to a large number of experiments on rabbit aorta, carried out with the assistance of D. Jothianandan and M.T. Khan, comparing the characteristics of relaxation by

NO (present in acidified solutions of $NaNO_2$) with the characteristics of relaxation by EDRF (released by ACh). It was found that relaxation by NO like that produced by EDRF was strongly inhibited by Hb, strongly inhibited by MB, strongly inhibited by superoxide generators (e.g., hydroquinone, phenidone, ferrous sulfate), and potentiated by SOD when superoxide was limiting the degree of relaxation (Fig. 8). It was already known that relaxation of

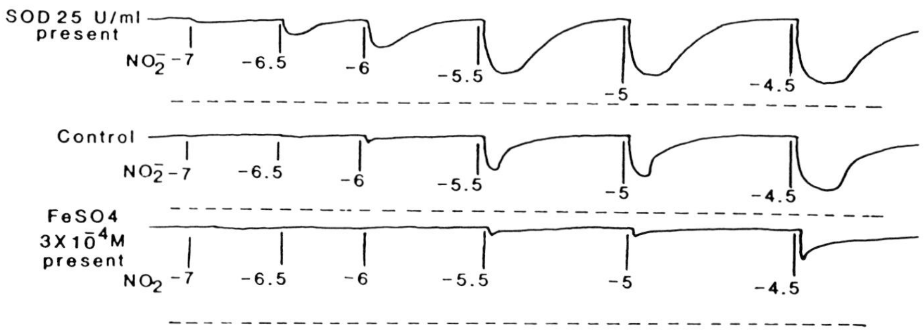

Figure 8. Transient relaxations of endothelium-free rings of rabbit aorta by nitric oxide present in acidified solutions (pH 2) of $NaNO_2$, potentiation of the relaxation by superoxide dismutase (SOD), and inhibition by $FeSO_4$ (a superoxide generator). Three matched rings from one aorta were contracted above basal tone (dashed line) with phenylephrine. SOD and $FeSO_4$ were added about five min before first additions of the acidified nitrite. Reproduced with permission from Furchgott (1988).

vascular smooth muscle by NO, like that by EDRF, is accompanied by an increase in cGMP. On finding the characteristics of NO and EDRF so similar, I had no hesitancy in proposing in the paper which I delivered at the July, 1986, symposium that EDRF is NO (Furchgott, 1988). [Shortly after the symposium, we began using solutions of NO gas rather than acidified nitrite as a source of NO in our experiments and obtained identical results (Fig. 9).]

In my symposium presentation, I also proposed that the acid-activatable inhibitory factor in the BRP extract of Martin and Gillespie is inorganic nitrite, and that the possibility had to be considered that the neurotransmitter of NANC nerves might also be NO (Furchgott, 1988).

At the same symposium at which I proposed that EDRF is NO, Ignarro, on the basis of studies in his laboratory on isolated bovine pulmonary arteries, independently made the same proposal (Ignarro *et al.*, 1988). (Unfortunately, the papers presented at that symposium were delayed in publication until 1988.) Soon after the 1986 symposium, three laboratories utilized perfusion-bioassay procedures to compare more accurately the biological and chemical characteristics of EDRF and NO. Ignarro and coworkers using bovine pulmonary vessels (Ignarro *et al.* 1987), Moncada and coworkers using cultured porcine aortic endothelial cells (Palmer *et al.*, 1987), and Khan and myself using rabbit aorta (Khan and Furchgott, 1987) as a source of EDRF, all found EDRF released upstream and NO infused upstream to have similar rates of decay, similar susceptibility to inhibitors like Hb and superoxide generators, similar stabilization by SOD, etc. In addition, Ignarro's laboratory presented

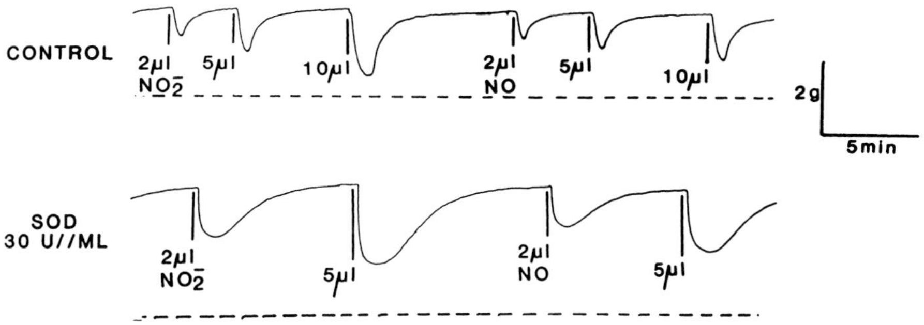

Figure 9. Similarity of the relaxing effects of a solution of acidified sodium nitrite and a solution of nitric oxide gas on rings of rabbit aorta in the absence and presence of superoxide dismutase (SOD). The acidified nitrite was 2.5 mM at pH 2.0. The NO solution was 15 μM. (See Furchgott *et al.* (1990a) for method for making solutions of NO.) Additions were to 20 ml of Krebs solution in each organ chamber, so that 2 μl of the NO solution gave an initial concentration of 1.5 nM. Both rings were precontracted with phenylephrine. The potentiation by SOD of the transient relaxations produced by the NO in the acidified nitrite and in the NO solution indicates that there was a considerable endogenous production of the superoxide anion O_2^- in the organ chambers.

spectroscopic evidence that the product of the reaction of released EDRF and Hb was the same as that of NO and Hb, and Moncada's laboratory showed that the amount of NO released by bradykinin (as determined by a chemiluminescence assay) could fully account for the relaxation of the bioassay strip by the EDRF released by the peptide. A year late, Moncada and coworkers made the major finding that the source of endothelial NO was a guanidinium nitrogen of L-arginine and that the enzyme responsible for its formation was an oxygenase (now called endothelial nitric oxide synthase or eNOS) (Palmer *et al.*, 1988). These findings marked the beginning of a major worldwide expansion of research on the role of NO in vascular physiology and pathophysiology.

Despite the evidence for the identity of EDRF and NO, some results obtained in various laboratories in the late 1980s did not appear to be completely consistent with that conclusion. [For a review of these apparent or real inconsistencies, see Furchgott *et al.*, 1990a.] Nevertheless, as of now, it has become generally accepted that EDRF is either NO, or some adduct that readily releases NO, or perhaps a mixture of NO and some adduct of NO. This does not preclude a contribution to endothelium-dependent relaxation in some blood vessels from the release of a non-nitric oxide endothelium-derived hyperpolarizing factor (EDHF) along with EDRF (for review see Garland *et al.*, 1995).

ACKNOWLEDGEMENT

The research in the author's laboratory referred to in this article was supported by grants from the National Heart, Lung, and Blood Institute Of the National Institutes of Health.

REFERENCES

Beckman, J.S., Beckman, T.W., Chen, J., Marshall, P.A. and Freeman, B.A. (1990) Apparent hydroxyl radical production by peroxynitrite: implications for endothelial injury from nitric oxide and superoxide. *Proc. Natl. Acad. Sci. U.S.A.* **87**, 1620–1624.

Böhme, E, Graf, H. and Schultz, G. (1978) Effects of sodium nitroprusside and other smooth muscle relaxants on cyclic GMP formation in smooth muscle and platelets. *Adv. Cyclic Nucleotide Res.* **9**, 131–143.

Bowman, A., and Drummond, A.H. (1984) Cyclic GMP mediates neurogenic relaxation in the bovine retractor penis muscle. *Br. J. Pharmacol.*, **81**, 665–674.

Bowman, A., Gillespie, J.S., and Martin, W. (1981) Actions on the cardiovascular system of an inhibitory material extracted from the bovine retractor penis. *Br. J. Pharmacol.* **72**, 365–372.

Bowman, A., Gillespie, J.S., and Pollock, D. (1982) Oxyhaemoglobin blocks nonadrenergic inhibition in the bovine retractor penis muscle. *Eur. J. Pharmacol.* **85**, 221–224.

Cherry, P.D., Furchgott, R.F., Zawadzki, J.V., and Jothianandan, D. (1982) The role of endothelial cells in the relaxation of isolated arteries by bradykinin. *Proc. Nat. Acad. Sci.* **79**, 2106–2110.

Cocks, T.M. and Angus, J.A. (1983) Endothelium-dependent relaxation of coronary arteries by noradrenaline and serotonin. *Nature* **305**, 627–630.

Cocks, T.M., Angus, J.A., Campbell, J.H. and Campbell, G.R. (1985) Release and properties of endothelium-derived relaxing factor (EDRF) from endothelial cells in culture. *J. Cell Physiol.* **123**, 310–320.

Cohen, R.A., Shepherd, J.T. and Vanhoutte, P.M. (1983) 5-Hydroxytryptamine can mediate endothelium-dependent relaxation of coronary arteries. *Am. J. Physiol.* **245**, H1077–1080.

Diamond, J. and Chu, E.B. (1983) Possible role for cyclic GMP in endothelium-dependent relaxation of rabbit aorta by ACh. Comparison with nitroglycerin. *Res. Comm. Chem. Pathol. Pharmacol.* **41**, 369–381.

Doyle, M.P., and Hoekstra, J.W. (1981) Oxidation of nitrogen oxides by bound dioxygen in hemoproteins. *J. Inorg. Biochem.* **14**, 351–358.

Forstermann, U., Trogisch, G. and Busse, R. (1984) Species-dependent differences in the value of endothelium-derived vascular relaxing factor. *Eur. J. Pharmacol.* **106**, 639–643.

Furchgott, R.F. (1954) Dibenamine blockade in strips of rabbit aorta and its use in differentiating receptors. *J. Pharmacol. Exp. Ther.* **111**, 265–284.

Furchgott, R.F. (1981) The requirement for endothelial cells in the relaxation of arteries by ACh and some other vasodilators. *Trends Pharmacol. Sci.* **2**, 173–176.

Furchgott, R.F. (1983) Role of endothelium in the responses of vascular smooth muscle. *Circ. Res.* **53**, 557–573.

Furchgott, R.F. (1988) Studies on relaxation of rabbit aorta by sodium nitrite: the basis for the proposal that the acid-activatable inhibitory factor from retractor penis is inorganic nitrite and the endothelium-derived relaxing factor is nitric oxide. In: *Vasodilatation: Vascular Smooth Muscle, Peptides, and Endothelium,* edited by P.M. Vanhoutte, pp. 401–414, Raven Press, New York.

Furchgott, R.F. (1990) The 1989 Ulf von Euler Lecture: Studies on endothelium-dependent vasodilation and the endothelium-derived relaxing factor. *Acta Physiol. Scand.* **139**, 257–270.

Furchgott, R.F. (1993) The discovery of endothelium-dependent relaxation. *Circulation* **87**(Suppl V) V3–V8.

Furchgott, R.F. (1996) The discovery of endothelium-derived relaxing factor and its importance in the identification of nitric oxide. J.A.M.A. **276**, 1186–1188.

Furchgott, R.F. and Bhadrakom, S. (1953) Reactions of strips of rabbit aorta to epinephrine, isopropylarterenol, sodium nitrite and other drugs. *J. Pharmacol. Exp. Therap.*, **108**, 129–143.

Furchgott, R.F.and Vanhoutte, P.M. (1989) Endothelium-derived relaxing and contracting factors. *FASEB J.* **3**, 2007–2018.

Furchgott, R.F., and Zawadzki, J.V. (1980) The obligatory role of endothelial cells in the relaxation of arterial smooth muscle by ACh. *Nature* **288**, 373–376.

Furchgott, R.F., Zawadzki, J.V., and Cherry, P.D. (1981) Role of endothelium in the vasodilator response to ACh. In: *Vasodilatation*, edited by P. Vanhoutte and I. Leusen, pp. 49–66. Raven Press, New York.

Furchgott, R.F., Cherry, P.D., Zawadzki, J.V., Jothianandan, D. (1984) Endothelial cells as mediators of vasodilation of arteries. *J. Cardiovasc. Pharmacol.* **6**, Suppl. 2, S336–S344.

Furchgott, R.F., Jothianandan, D. and Freay, A.D. (1990b) Endothelium-derived relaxing factor: some old and new findings. In *Nitric Oxide from L-Arginine: A Bioregulatory System*, edited by S. Moncada and E.A. Higgs, pp. 5–17, Elsevier, Amsterdam.

Furchgott, R.F., Khan, M.T. and Jothianandan, D. (1990a) Comparison of properties of nitric oxide and endothelium-derived relaxing factor: Some cautionary findings. In *Endothelium-derived Relaxing Factors*, edited by G.M. Rubanyi and P.M. Vanhoutte, pp. 8–21, Karger, Basel.

Garland, C.J., Plane, F., Kemp, B.K., and Cocks, T.M. (1995) Endothelium-dependent hyperpolarization: a role in the control of vascular tone. *Trends Pharmacol. Sci.* **16**, 23–30.

Gillespie, J.S., Hunter, J.C., and Martin, W. (1981) Some physical and chemical properties of the smooth muscle inhibitory factor in extracts of the bovine retractor penis muscle. *J. Physiol.* **315**, 111–125.

Goldberg, N.D., Graff, G., Haddox, M.K., Stephenson, J.H., Glass, D.B., and Moser, M.E. (1978) Redox modulation of splenic cell soluble guanylate cyclase activity: Activation by hydrophilic and hydrophobic antioxidants represented by ascorbic and dehydroascorbic acids, fatty acid hydroperoxides, and prostaglandin endoperoxides. *Adv. Cyclic Nucleotide Res.* **9**, 101–130.

Griffith, T.M., Edwards, D.H., Newby, A.C., Lewis, M.J., Newby, A.C. and Henderson, A.H. (1984) The nature of the endothelium-derived vascular relaxant factor. *Nature* **305**, 645–647.

Gruetter, C.A., Barry, B.K., McNamara, D.B., Gruetter, D.Y., Kadowitz, P.J. and Ignarro, L.J. (1979) Relaxation of bovine coronary artery and activation of coronary arterial guanylate cyclase by nitric oxide, nitroprusside and carcinogenic nitrosoamine. *J. Cyclic Nucleotide Res.* **5**, 211–214.

Gruetter, C.A., Gruetter, D.Y., Lyon, J.E., Kadowitz, P.J. and Ignarro, E.F. (1981) Relationship between cyclic guanosine 3',5'-monophosphate formation and relaxation of coronary arterial smooth muscle by glyceryl trinitrate, nitroprusside, nitrite and nitric oxide: Effects of methylene blue and methemoglobin. *J. Pharmacol. Exp. Ther.* **219**, 181–186.

Gryglewski, P.J., Palmer, R.M. and Moncada, S.A. (1986) Superoxide anion is involved in the breakdown of endothelium-derived relaxing factor. *Nature* **320**, 454–456.

Hidaka, H., and Asano, T. (1977) Stimulation of platelet guanylate cyclase by unsaturated fatty acid peroxides. *Proc. Natl. Acad. Sci. USA* **74**, 3657–3661.

Holtz, J., Förstermann, U., Pohl, U., Giesler, M., and Bassenge, E. (1984) Flow-dependent, endothelium-mediated dilation of epicardial coronary arteries in conscious dogs: effects of cyclooxygenase inhibition. *J. Cardiovasc. Pharmacol.* **6**, 1161–1169.

Holzmann, S. (1982) Endothelium-induced relaxation by ACh associated with larger rises in cyclic GMP in coronary arterial strips. *J. Cyclic Nucleotide Res.* **8**, 409–419.

Ignarro, L.J., Burke, T.M., Wood, K.S., Wolin, M.S., Kadowitz, P.J. (1984) Association between cyclic GMP accumulation and ACh-elicited relaxation of bovine intrapulmonary artery. *J. Pharmacol. Exp. Ther.* **228**, 682–690.

Ignarro, L.J., Buga, G.M., Wood, K.S., Byrns, R.E. and Chaudhuri, G. (1987) Endothelium–derived relaxing factor produced and released from artery and vein is nitric oxide. *Proc. Natl. Acad. Sci. U.S.A.* **84**, 9265–9269.

Ignarro, L.J., Byrns, R.E. and Wood, K.S. (1988) Biochemical and pharmacological properties of endothelium-derived relaxing factor and its similarity to nitric oxide radical. In: *Vasodilatation: Vascular Smooth Muscle, Peptides, and Endothelium*, edited by P.M. Vanhoutte, pp. 427–435, Raven press, new York.

Jelliffe, R.W. (1962) Dilator and constrictor effects of acetylcholine on isolated rabbit aortic chains. *J. Pharmacol. Exp. Ther.* **135**, 349–353.

Kaiser, L., and Sparks, H.V., Jr. (1986) Mediation of flow-dependent arterial dilation by endothelial cells. *Circulatory Shock* **18**, 109–114.

Katsuki, S., and Murad, F. (1977) Regulation of adenosine cyclic 3',5'-monophosphate levels and contractility in bovine tracheal smooth muscle. *Mol. Pharmacol.* **13,** 330–341.

Khan, M.T. and Furchgott, R.F. (1987) Additional evidence that endothelium-derived relaxing factor is nitric oxide. In: *Pharmacology*, edited by M.J. Rand and C. Raper, pp. 341–344 Elsevier, Amsterdam.

Martin, W., Villani, G.M., Jothianandan, D. and Furchgott, R.F. (1985a) Selective blockade of endothelium-dependent and glyceryl trinitrate-induced relaxation by hemoglobin and by methylene blue in the rabbit aorta. *J. Pharmacol. Exp. Ther.* **232**, 708–716.

Martin, W., Villani, G.M., Jothianandan, D. and Furchgott, R.F. (1985b) Blockade of endothelium-dependent and glyceryl trinitrate-induced relaxation of rabbit aorta by certain ferrous hemoproteins. *J. Pharmacol. Exp. Ther.* **233**, 679–685.

Martin, W., Furchgott, R.F., Villani, G.M., and Jothianandan, D. (1986a) Depression of contractile responses in rat aorta by spontaneously released endothelium-derived relaxing factor (EDRF). *J. Pharmacol. Exp. Ther.* **237**, 529–538.

Martin, W., Furchgott, R.F., Villani, G.M. and Jothianandan, D. (1986b) Phosphodiesterase inhibitors induce endothelium-dependent relaxation of rat and rabbit aorta by potentiating the effects of spontaneously released endothelium-derived relaxing factor (EDRF) *J. Pharmacol. Exp. Ther.* **237**, 539–547.

Moncada, S., Palmer, R.M.J. and Gryglewski, R.J. (1986) Mechanism of action of some inhibitors of endothelium-derived relaxing factor. *Proc. Natl. Acad. Sci. U.S.A.* **83**, 9164–9168.

Murad, F., Mittal, C.K., Arnold, W.P., Katsuki, S., Kimura, H. (1978) Guanylate cyclase activation by azide, nitro compounds, nitric oxide, and hydroxyl radical and inhibition by hemoglobin and myoglobin. *Adv. Cyclic Nucleotide Res.* **9**, 145–158.

Murad, F., Arnold, W.P., Mittal, C.K., Braughler, J.M. (1979) Properties and regulation of guanylate cyclase and some proposed functions of cyclic GMP. *Adv. Cyclic Nucleotide Res.* **11**: 175–204.

Palmer, R.M.J., Ferrige, A.G. and Moncada, S. (1987) Nitric oxide release accounts for the biological activity of endothelium-derived relaxing factor. *Nature (Lond.)* **327**, 524–526.

Palmer, R.M.J., Ashton, D.S. and Moncada, S. (1988a) Vascular endothelial cells synthesize nitric oxide from L-arginine. *Nature* **333**, 664–666.

Pohl, U., Holtz, J., Busse, R., Bassenge, E. (1986) Crucial role of endothelium in the vasodilator response to increased flow *in vivo*. *Hypertension* **8**, 37–44.

Rapoport, R.M. and Murad, F. (1983) Agonist-induced endothelium-dependent relaxations in rat thoracic aorta may be mediated through cGMP. *Circ. Res.* **52**, 352–357.

Rubanyi, G.M. and Vanhoutte, P.M. (1986) Superoxide anions and hyperoxia inactive endothelium-derived relaxing factor. *Am. J. Physiol.* **250**, H822–H827.

Rubanyi, G.M., Lorenz, R.R. and Vanhoutte, P.M. (1985) Bioassay of endothelium-derived relaxing factor. *Am. J. Physiol.* **215**, H1077–H1080.

Rubanyi, G.M., Romero, J.C., Vanhoutte, P.M. (1986) Flow-induced release of endothelium-derived relaxing factor. *Am. J. Physiol.* **250**, H1145–H1149.

Steinsland, O.S., Furchgott, R.F., Kirpekar, S.M. (1973) Inhibition of adrenergic neurotransmission by parasympathomimetics in the rabbit ear artery. *J. Pharmacol. Exp. Ther.* **184**, 346–356.

Toda, N. (1974) The action of vasodilating drugs on isolated basilar, coronary and mesenteric arteries of the dog. *J. Pharmacol. Exp. Ther.* **193**, 376–384.

Van de Voorde, J. and Leusen, I. (1983) Role of endothelium in the vasodilator responses of rat theriac aorta to histamine. *Eur. J. Pharmacol.* **87**, 113–120.

Vanhoutte, P.M. (1974) Inhibition by acetylcholine of adrenergic neurotransmission in vascular smooth muscle. *Circ. Res.* **34**, 317–326.

Wolin, M.S., Cherry, P.D., Rodenburg, J.M., Messina, E.J., and Kaley, G. (1990) Methylene blue inhibits vasodilation of skeletal muscle arterioles to ACh and nitric oxide via the extracellular generation of superoxide anion. *J. Pharmacol. Exp. Ther.* **254**, 872–876.

LOUIS J. IGNARRO

The first two decades of my life were spent in the New York City area, where the families of both my parents had settled in the 1920s after immigrating from Italy. My father had been a ship builder in Naples but my mother was still a young child when she came from Sicily. They met for the first time in Brooklyn, New York in the 1930s, were married, and then moved to the nearby coastal city of Long Beach. I was born on May 31, 1941 in Brooklyn and my brother, Angelo, followed on January 10, 1944. My father worked as a carpenter, whereas my mother elected to bring up her two sons at home.

Long Beach was a beautiful town, about 25 miles east of New York City located on the south shore of Long Island. We had a lovely home within walking distance of the beach along the Atlantic ocean. I can still recall walking to the beach and going for a swim nearly every day in the summer. My greatest joy each morning was building gigantic sand castles using dripping sand wetted by the incoming tide. All my friends believed and predicted that I would grow up to become an architect or engineer. This view was reinforced by my eagerness even as a young child to disassemble anything I could find and put it back together again. The joy of discovering that I could actually get the object to function again was quite rewarding and satisfying. But my greatest joy came when I was 8 years old. To my surprise and delight, mother and father finally responded favorably to my relentless request to have a chemistry set, and bought me one. I can recall vividly following every step of every experiment and becoming overjoyed at the success of each one. This was much more fun than building sand castles on the beach. My inquisitiveness drove me to the library to study more applied aspects of chemistry. Soon after completing dozens of additional experiments and going through several larger chemistry sets, I realized that what I really wanted to accomplish was to build a bomb and to send up a rocket. After about one year of experiments, I finally achieved those goals, albeit at the expense of numerous horrified reactions from the neighbors.

My interest in chemistry remained strong at Central Grade School and Long Beach High School, which led me to apply to Columbia University in New York City to study chemistry and pharmacy. I was especially pleased when I learned that I had been accepted to the freshman class at Columbia. I wanted to attend a university that was within commuting distance of home because I did not want to leave my family and friends in Long Beach. During my high school years I had developed a great interest in playing ball and racing cars, and I did not want that to come to an end, at least not just yet. My favorite sport was one-on-one stickball, the New York City sport of sports, where a "bouncy" rubber ball is thrown by the opponent pitcher against a brick or cement wall on which is drawn a "strike zone". The batter uses a stick conveniently detached from a suitable broom or mop to hit the fast pitched ball. When I was not playing stickball I was building and racing cars at the West Hampton Drag Raceway. I guess I could never get away from taking things apart and putting them back together again. Indeed, I spent many long hours thinking about whether I should study chemistry or open up my own drag racing shop out on Long Island. Well, chemistry it was. I took dozens of chemistry courses, but a course in pharmacology, although poorly taught, really caught my attention. I studied the subject well beyond the course requirements and tried to hang around the pharmacology laboratories as often as I could. The result of this was my application to graduate school in pharmacology upon graduation from Columbia University in 1962.

I was delighted to be admitted to the pharmacology program at the University of Minnesota in Minneapolis, which was considered to be one of the best departments of pharmacology in the nation at that time. Actually, I had applied to the University of Wisconsin in Madison, where the department was located when I first applied. But for one reason or another, the entire department was relocated from Madison to Minneapolis just after I had been accepted in Madison. A bit confused, I reported to Minneapolis in September of 1962 to study pharmacology. At first, things were difficult for me because I had left my family, friends, stickball, racing cars and the beach behind. And then things got even worse when I experienced my first winter season of–40°F with winds of 30 mph. But I survived my first winter and went on to enjoy the upper midwest and the "Big Ten" college football games.

My studies in graduate school involved developing a better understanding of why and how neurons of the sympathetic nervous system innervate the heart and produce and release norepinephrine. I spent three of the most intense years of my life in the laboratory, where I was determined to unravel every bit of information possible within the time frame allotted to me to satisfy the research requirements for the Ph.D. degree in pharmacology. My research was different from most in that it required, in addition to pharmacology, a great deal of knowledge in several other distinct disciplines such as physiology, biochemistry and anatomy. My major, of course, was pharmacology and I selected cardiovascular physiology as my minor. But that was insufficient, so I took several additional courses in biochemistry and anatomy. The most demanding course I took was enzymology, taught by Paul Boyer, who was awarded the Nobel Prize in Chemistry last year (1997). I have not stopped using enzymology in my research since taking that course. My research turned out to be acceptable to my committee, chaired by the late Frederick E. Shideman, M.D., Ph.D., who was also Chairman of the Department of Pharmacology at the University of Minnesota. He decided that I should write four separate manuscripts on my thesis research and that we should submit them to the Journal of Pharmacology and Experimental Therapeutics. The editors of the journal accepted all four papers and published them back-to-back in one issue of the journal, a feat never again repeated either by the journal or by me.

After Minneapolis, I accepted a postdoctoral position at the National Institutes of Health in the Laboratory of Chemical Pharmacology in the National Heart, Lung and Blood Institute. My mentor was Elwood Titus, a brilliant scientist who was able to mix chemistry and pharmacology with the greatest of ease. I tried to learn as much as I could from him in two years. Perhaps I tried a bit too hard. For example, he asked me to study the chemistry of beta adrenergic receptors and I decided that I was going to isolate, characterize and elucidate the chemical structure not only of beta but also of alpha adrenergic receptors, all in two years. Having published four consecutive papers in a distinguished journal on my first try, I thought that my research career was going to be a breeze. The N.I.H. proved to me that this was not going to be the case, and it was not. My work resulted in only one publication, but the agony of frustration caused me to mature quickly. The atmosphere of the N.I.H. was highly conducive to learning science and I had the opportunity to discuss my work and research in general with Bernard Brodie, Jim Gillette, Julie Axelrod and other distinguished scientists.

My first real job after my research training was with the drug industry. Geigy Pharmaceuticals recruited me in 1968 with an attractive package including the responsibility of heading the biochemical antiinflammatory program. Although this was an entirely new research topic for me, I accepted the position because of the enormous responsibility that would suddenly be mine. The work was quite satisfying in that I became a part of a larger group whose efforts led to the development and marketing of a new nonsteroidal antiinflammatory drug (diclofenac). About half way through my career at Geigy, my daughter, Heather, was born. I recall that day vividly (January 10, 1970) because I had to rush my wife to the nearby hospital in the midst of a snow storm. But all turned out well and I found myself devoting a great deal of time to something other than my own research. With the birth of Heather came a move from a small apartment in Hartsdale to a much larger unit in Irvington on the Hudson. This was a lovely neighborhood in which to raise a child.

In addition to my work on drug development, Geigy allowed me the freedom to pursue basic research in biochemical pharmacology, which led to my interest in studying the relatively new cyclic nucleotide, cyclic GMP. Although I enjoyed my work at Geigy Pharmaceuticals, when the company merged with Ciba Pharmaceuticals I decided to try my hand at academic research and teaching. In January of 1973, I accepted the position of Assistant Professor of pharmacology at Tulane University School of Medicine in New Orleans. I chose to go to Tulane because I wanted to continue my research on cyclic GMP, and there was a young pharmacologist at Tulane with the same interest. We moved to New Orleans, where we bought our first home in Terrytown, an attractive nearby suburb.

My interest and motivation in studying the possible physiological significance of cyclic GMP grew and grew during my first two years at Tulane.

Thanks to my own laboratory and those of other interested collaborators, we made many significant contributions to the field of cyclic GMP and cyclic nucleotide research in general. My early work with cyclic GMP involved leukocytes and the heart, but this eventually led to an interest in blood vessels. I recall reading an interesting paper by Ferid Murad's group in 1977, in which nitric oxide and various nitro compounds were shown to activate the cytosolic form of guanylate cyclase and to elevate cyclic GMP levels in various tissues. Nitroglycerin was one of those nitro compounds that Ferid had studied and speculated might release nitric oxide which then activated guanylate cyclase. It occurred to me that nitric oxide might account for the vascular smooth muscle relaxing action of nitroglycerin and that cyclic GMP might be the second messenger responsible for mediating the vasorelaxant effect of nitric oxide. In 1979 we published the first account of the capacity of nitric oxide to relax vascular smooth muscle. We purchased a small cylinder of nitric oxide gas, made a dilution in nitrogen (nitric oxide is very unstable in the presence of oxygen), and injected a fine stream of gas bubbles into an organ bath in which was mounted a strip of bovine coronary artery precontracted by addition of phenylephrine. The result was a rapid and profound relaxation of the coronary artery strip. This vasorelaxant effect of nitric oxide was blocked by addition of hemoglobin, which promotes oxidation of nitric oxide, and methylene blue, which had been known to inhibit guanylate cyclase. And so we knew right away that nitric oxide was probably responsible for the vasorelaxant effect of nitroglycerin and that cyclic GMP was the likely ultimate mediator of relaxation, just as Ferid Murad had predicted.

We wondered whether the platelet antiaggregatory action of certain nitrovasodilators could also be attributed to nitric oxide and cyclic GMP. A relatively straightforward experiment was conducted with human platelet-rich plasma, in which we examined the influence of added nitric oxide on ADP-induced platelet aggregation. The results were dramatic. Nitric oxide potently inhibited platelet aggregation and actually reversed aggregation once it had occurred. This effect was mediated by cyclic GMP. Thus, at least two biological actions of nitric oxide were clear from these early studies. Nitric oxide is a vasorelaxant and inhibitor of platelet aggregation, and both effects are mediated by cyclic GMP.

The next step was to elucidate the mechanism by which nitroglycerin is converted to nitric oxide by vascular smooth muscle. After reading nearly every paper in the field of organic nitrate esters and their vasodilator effects, I was motivated by the work of Phil Needleman, who showed that the vasodilator action of nitroglycerin and other organic nitrate esters was dependent somehow on the presence of thiols. A long and tedious series of experiments in my laboratory led to the discovery that thiols were required for the activation of guanylate cyclase by nitroglycerin and related nitrovasodilators. Interaction between thiols and nitro compounds led to the formation of intermediate S-nitrosothiols, which were chemically unstable and decomposed to liberate nitric oxide gas. Depletion of tissue thiols resulted in diminished vasorelaxation by nitroglycerin because nitric oxide could no longer be generated. Moreover, tolerance to the vasodilator action of nitroglycerin appeared to be due to thiol depletion, which could be reversed by adding back thiols in order to generate more nitric oxide. This work was published in 1981 in the Journal of Pharmacology and Experimental Therapeutics.

Having elucidated the mechanism of action of nitroglycerin as a vasodilator, the next step was to understand how nitric oxide activates guanylate cyclase. An elegant series of experiments was published in the late 1970s by Patricia Craven and Fred DeRubertis, showing that activation of guanylate cyclase by nitric oxide might require the presence of heme. This made sense to me because heme iron had long been known to have a high binding affinity for nitric oxide. Suppose guanylate cyclase had a heme prosthetic group that bound nitric oxide and somehow became activated to generate more cyclic GMP from GTP? In 1981 we set out to purify and characterize guanylate cyclase from bovine lung. A young biochemically trained postdoctoral fellow from Yale University, Mike Wolin, joined my laboratory to tackle this project. After an incredibly long and tedious series of experiments, each often lasting for 96 consecutive hours, we found the heme in purified guanylate cyclase. Subsequent experiments revealed that the presence of enzyme-bound heme was an absolute requirement for guanylate cyclase activation by nitric oxide. We went on to propose that nitric oxide reacts with heme iron to alter the configuration of the catalytic binding site for GTP and promote the conversion of GTP to cyclic GMP and pyrophosphate. In conducting these experiments, we discovered that the non-nitric oxide containing substance, protoporphyrin IX, activated heme-deficient guanylate cyclase by kinetic mechanisms that were indistinguishable from the mechanism by which nitric oxide activates heme-containing guanylate cyclase.

Although the above observations were exciting, they were also puzzling because it was unclear why mammalian cells were so sensitive to nitric oxide. Why do we have receptors for nitric oxide, an air pollutant and a metabolite of nitroglycerin? Was it possible that our own cells actually produced nitric oxide or nitroglycerin but we were unaware of it? In 1983, my laboratory set out to determine whether or not mammalian cells can produce either nitric oxide or a nitro compound that could be metabolized to nitric oxide. A separate project in the laboratory was to study endothelium-dependent vasorelaxation and to attempt to identify the mysterious "EDRF" (endothelium-derived relaxing factor) discovered three years earlier by Robert Furchgott. Both research projects came together in 1984 when we suddenly realized that EDRF and nitric oxide possessed similar pharmacological and biochemical properties. EDRF and nitric oxide were both chemically unstable and both activated guanylate cyclase and elevated tissue levels of cyclic GMP. The cyclic GMP elevating and vasorelaxant effects of both EDRF and nitric oxide were inhibited by addition of methylene blue to organ chambers. These findings, reported in 1984, prompted me to ascertain whether EDRF, like nitric oxide, required bound heme on guanylate cyclase in order to activate the enzyme and stimulate cyclic GMP formation. I can recall vividly the positive results of the first experiment, and I knew we had it. EDRF must be nitric oxide. I first reported these findings in the summer of 1986 at a vascular conference held at the Mayo Clinic in Rochester, Minnesota. Unexpectedly, at least to me, my colleague Robert Furchgott presented his own evidence that EDRF might be nitric oxide. I presented additional evidence a few months later at the fall American Heart Association meeting in Dallas and at the spring FASEB meeting in Washington, DC in 1987. So now it was clear why nitric oxide is such a potent vasorelaxant. This small lipophilic chemical is produced by vascular endothelial cells and functions to decrease vascular smooth muscle tone and to inhibit platelet aggregation.

The frenzy and excitement of these times in the mid-1980s was stalled at times by my divorce and my decision to leave Tulane University and begin a new personal life and academic career at UCLA School of Medicine. I moved to Los Angeles in May of 1985 and bought a small home in Encino, just 12 miles from the UCLA campus. My daughter, Heather, joined me in 1988 and attended California State University at Northridge. As a result of witnessing her dad's commitment to many long hours of research and teaching, Heather chose to major in radio, film and television. At first, her decision to shy away from a career in science concerned me, but then I realized how talented she was and how successful she would become.

The discovery that EDRF was nitric oxide led to an avalanche of studies that created an exciting new field in biological research. New physiological and pathophysiological roles for nitric oxide were being discovered on a weekly basis. In record time, several prominent laboratories elucidated the biochemical mechanisms involved in the synthesis of nitric oxide by various cell types. While studying the relaxant effects of nitric oxide on vascular and nonvascular smooth muscle from corpus cavernosum erectile tissue, we realized that the naturally occurring physiological neurotransmitter involved in the erectile response in mammals was unknown. John Garthwaite had just reported that nitric oxide was a neurotransmitter in the brain, and we wondered whether or not nitric oxide could be the neurotransmitter in the so called nonadrenergic noncholinergic neurons that were known to innervate the corpus cavernosum smooth muscle. After all, nitric oxide released from such nerves would be expected to diffuse into the nearby vascular and nonvascular smooth muscle and cause relaxation. Such an effect could account for the marked relaxation of both vascular and nonvascular smooth muscle that accompanies the erectile response and allows for the engorgement of blood in the sinusoidal or trabecular network of blood vessels in the corpus cavernosum. The first carefully designed experiment was successful. Electrical stimulation of strips of rabbit corpus cavernosum caused a transient but marked smooth muscle relaxation that was prevented by addition of a nitric oxide synthase inhibitor and enhanced by addition of a cyclic GMP phosphodiesterase inhibitor. Addition of authentic nitric oxide to organ chambers mimicked the effects of electrical stimulation. A subsequent experiment revealed that electrical stimulation results in the production of nitric oxide in the corpus cavernosum. Further studies using human tissue showed that patients with impotence suffer from an impaired nitric oxide-cyclic GMP pathway in the erectile tissue, and this work laid the foundation for the development by others of a drug that proved to be effective for the treatment of impotency in humans. Sildenafil (Viagra^{R}) promotes the erectile response by inhibiting a specific isoform of cyclic GMP phosphodiesterase and allowing cyclic GMP to accumulate when guanylate cyclase is activated by nitric oxide released from the nerves innervating the erectile tissue.

In the fall of 1994, I met Sharon Elizabeth Williams, a lovely and charming medical student here at UCLA. Sharon had been a nurse anesthetist for several years and then decided to obtain an M.D. degree in order to practice anesthesiology at a more professional level. After graduating from UCLA, Sharon moved to the east coast to begin her internship and residency at Johns Hopkins University. Shortly after her move, we started dating by long distance and were married in July of 1997. A year later, in the spring of 1998, Sharon transferred back to UCLA to continue her residency in anesthesiology. Finally, we were together. During the week we reside in an apartment adjacent to the UCLA campus in Westwood and we spend our weekends in my home in Malibu.

As a result of my work during the past decade, many investigators jumped in to extend our findings. This led to the development of close collaborations with numerous laboratories and the formation of close and genuine friendships in many different parts of the world. I treasure these friendships even more than the awards I have received for my research accomplishments. I also realize that these accomplishments would not have been possible without the interest, hard work, and commitment on the part of my technical assistants, graduate students, postdoctoral fellows, medical fellows, visiting scientists, research collaborators at home, and collaborators at other institutions.

Another rewarding development has been my discovery that I also have a real knack for and love of teaching what I know to medical and graduate students. I have consequently made teaching a regular part of my schedule since I came to UCLA and I cherish the Golden Apple teaching awards I have won from my classes. I trust that I have helped guide at least some of these young people toward careers that will be a blessing to them and to humanity. In my own case, the combination of biomedical research and teaching continues to provide me with an exciting and useful life, and I am exceedingly grateful.

NITRIC OXIDE: A UNIQUE ENDOGENOUS SIGNALING MOLECULE IN VASCULAR BIOLOGY

Nobel Lecture, December 8, 1998

by

LOUIS J. IGNARRO

Department of Molecular and Medical Pharmacology, UCLA School of Medicine, 10833 La Conte Avenue, Los Angeles, California 90095-1735, USA

INTRODUCTION

My early research career in graduate school focused on trying to develop a better understanding of catecholamine mechanisms during the embryological development of the sympathetic nervous system. This research required a strong background not only in pharmacology but also in physiology and biochemistry. I recall reading some interesting papers on cyclic AMP, but I never addressed this topic in my early research. At that time, there was no such thing as cyclic GMP, and nitric oxide was considered to be no more than a pollutant in the atmosphere around us. After completing my postdoctoral training at the National Institutes of Health in Bethesda, I accepted an industrial research position to learn more about biochemical aspects of the inflammatory process in order to develop novel antiinflammatory drugs. Although my research did not yet involve cyclic AMP, I continued to read the literature on cyclic AMP and witnessed the discovery of cyclic GMP. I then decided to ascertain whether or not cyclic AMP and cyclic GMP played any modulatory roles in the inflammatory process. We found that cyclic AMP stabilized lysosomal membranes, whereas cyclic GMP promoted instability of lysosomal membranes. Moreover, cyclic AMP production was associated with the inhibition of human neutrophil function such as phagocytosis and lysosomal enzyme secretion, whereas cyclic GMP was associated with increased neutrophil function. In a different research project, it appeared that this new cyclic nucleotide, cyclic GMP, was involved in mediating the negative effects of acetylcholine in the heart, whereas cyclic AMP was already well known to elicit the opposite effects in the heart. A theme was clearly developing that cyclic AMP and cyclic GMP mediated opposing influences on cell function.

In the midst of our studies on cyclic AMP and cyclic GMP, I recall reading two interesting papers on cyclic GMP that were published by Ferid Murad and colleagues (1,2). These studies revealed that nitric oxide and nitro compounds that might release nitric oxide all activated cytosolic guanylate cyclase. Nitroglycerin was one of the nitro compounds studied. Nitric oxide and nitro compounds also stimulated cyclic GMP production in isolated tissues *in vitro* (3). These observations suggested that nitroglycerin might activate guanylate cyclase and stimulate cyclic GMP formation by mechanisms involving nitric oxide. Additional studies revealed that nitric oxide might be involved in the nonvascular smooth muscle relaxant effects of nitroglycerin and other nitro compounds (4). These findings prompted me to conduct some experiments to ascertain whether nitroglycerin and related organic nitrate and nitrite esters could actually release nitric oxide gas in aqueous solution. After observing the generation of nitric oxide from a series of nitro compounds that were known to be smooth muscle relaxants, we thought that nitric oxide might also be responsible for the vasorelaxant effects of nitroglycerin and that cyclic GMP might be the intracellular second messenger mediating this effect of nitric oxide. Accordingly, a series of experiments was conducted to test the hypothesis forwarded by Ferid Murad that nitroglycerin, nitroprusside and nitroso compounds all cause relaxation of smooth muscle, including vascular smooth muscle, by liberating nitric oxide, which elicits its action by stimulating the production of cyclic GMP.

MECHANISM OF RELEASE OF NITRIC OXIDE FROM NITROGLYCERIN AND OTHER NITRO COMPOUNDS

The most important experiment that we conducted to test the hypothesis that nitric oxide is responsible for the vasorelaxant action of nitroglycerin was to determine whether nitric oxide itself causes vascular smooth muscle relaxation. In 1979, (5) we reported that nitric oxide gas is a potent relaxant of bovine coronary artery and activates guanylate cyclase isolated from this tissue (Figure 1). This observation confirmed what Ferid Murad had found and extended the hypothesis to include vascular smooth muscle. Therefore, nitroprusside, nitroglycerin and related nitro compounds appeared to relax vascular smooth muscle by liberating nitric oxide, which then stimulates cyclic GMP formation and results in vascular smooth muscle relaxation (Figure 2). The hypothesis that cyclic GMP is involved in smooth muscle relaxation was met with some controversy in the late 1970s because of the prevailing view that cyclic GMP and cyclic AMP mediate opposing biological actions. Cyclic AMP was known to mediate the smooth muscle relaxant effects of certain catecholamines and prostaglandins and cyclic GMP was thought to be involved in mediating smooth muscle contraction. However, the work conducted in several laboratories including those of Ferid Murad and myself indicated that cyclic GMP and cyclic AMP can mediate common biological actions in smooth muscle such as relaxation.

The next step in elucidating the precise mechanism of action of nitroglycerin was to ascertain how nitric oxide was released from the parent molecule in smooth muscle tissue. The spontaneous release of nitric oxide from nitroso compounds such as nitroprusside had been understood but nitroglycerin and other organic nitrate and nitrite esters are chemically stable compounds that would have to undergo enzymatic or chemical reactions to release nitric oxide. In early experiments, we found that thiols such as dithiothreitol enhanced the activation of guanylate cyclase by nitroprusside and N-nitroso

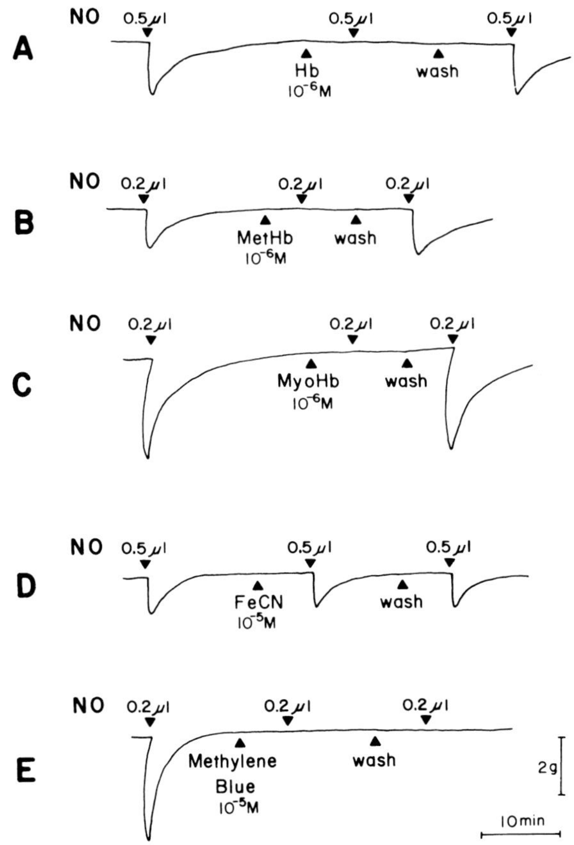

Figure 1. Relaxation of strips of bovine coronary artery by nitric oxide (NO). Strips were mounted, equilibrated and set up for recording changes in isometric force as described previously (5). Strips were submaximally precontracted by exposure to 30mM KCl. The concentrations of NO were estimated to be in the range of 0.01 to 0.05μM. Hemoglobin (Hb), methemoglobin (MetHb), myoglobin (MyoHb), potassium ferricyanide (FeCN) and methylene blue were added at the concentrations illustrated. Modified from Gruetter *et al.* (5) with permission.

compounds without affecting enzyme activation by nitric oxide itself (6–8). The mechanism of this enhancement was attributed to a reaction between thiol and nitroso compound to form an S-nitrosothiol which decomposed with the discharge of nitric oxide. A series of S-nitrosothiols were synthesized and found to be labile nitric oxide donor agents. In experiments using nitroglycerin, we found that nitroglycerin alone did not activate guanylate cyclase and that added dithiothreitol or glutathione failed to modify this lack of effect of nitroglycerin on guanylate cyclase. In the presence of cysteine, however, nitroglycerin caused a marked activation of guanylate cyclase (9). Sodium nitrite also failed to activate guanylate cyclase in the absence of added thiols but unlike nitroglycerin, a variety of thiols unmasked the enzyme activating action of sodium nitrite. Additional experiments revealed that cysteine reacts with nitroglycerin, whereas sodium nitrite reacts with any of a variety of thiols to generate the corresponding S-nitrosothiol (10). The S-nitrosothiols de-

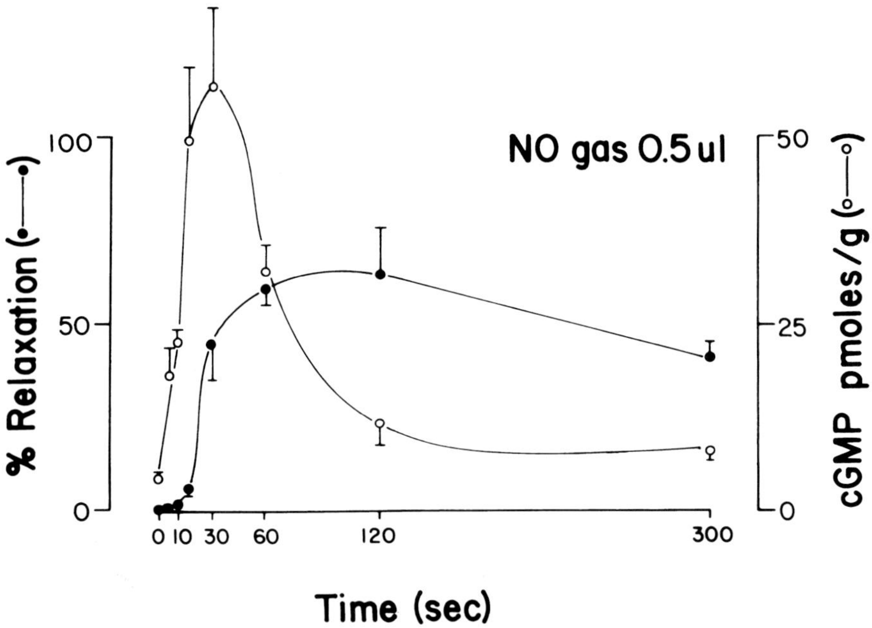

Figure 2. Time course of cyclic GMP accumulation in and relaxation of bovine coronary artery elicited by nitric oxide gas (NO). Strips were mounted, equilibrated and set up for recording changes in isometric force as described previously (5,36). Strips were submaximally precontracted by exposure to 30mM KCl. Relaxation and cyclic GMP (cGMP) levels were measured in the same arterial strips (10,36). Data are expressed as the mean ± SEM using 16 arterial strips from 4 separate experiments. Modified from Gruetter *et al.* (36) with permission.

compose readily with the release of nitric oxide, which activates guanylate cyclase. These concepts are illustrated schematically in Figure 3. Examples of such S-nitrosothiols are S-nitroso-N-acetylpenicillamine (SNAP), S-nitroso-glutathione (GSNO) and S-nitrosocysteine (CYSNO), each of which we synthesized and first examined their biological effects in mammalian systems (7–10). These S-nitrosothiols turned out to be useful nitric oxide donor agents that are now readily employed to test the effects of nitric oxide *in vitro* and *in vivo.*

INHIBITION OF PLATELET AGGREGATION BY NITRIC OXIDE

In reviewing the literature on nitroglycerin and nitroprusside, I came across a paper showing that nitroprusside inhibited human platelet aggregation (11). The obvious question raised was whether nitroprusside inhibited platelet aggregation by mechanisms involving nitric oxide and cyclic GMP just as had been shown for the smooth muscle relaxant effect of nitroprusside. Using a platelet-rich plasma fraction from human volunteers, we showed that nitric oxide was a potent inhibitor of platelet aggregation induced by ADP, thrombin, collagen and other aggregatory agents, and that the anti-platelet effect of nitric oxide was mediated by cyclic GMP (12). The anti-platelet action of nitroprusside appeared to involve the actions of S-nitrosothiols, which were

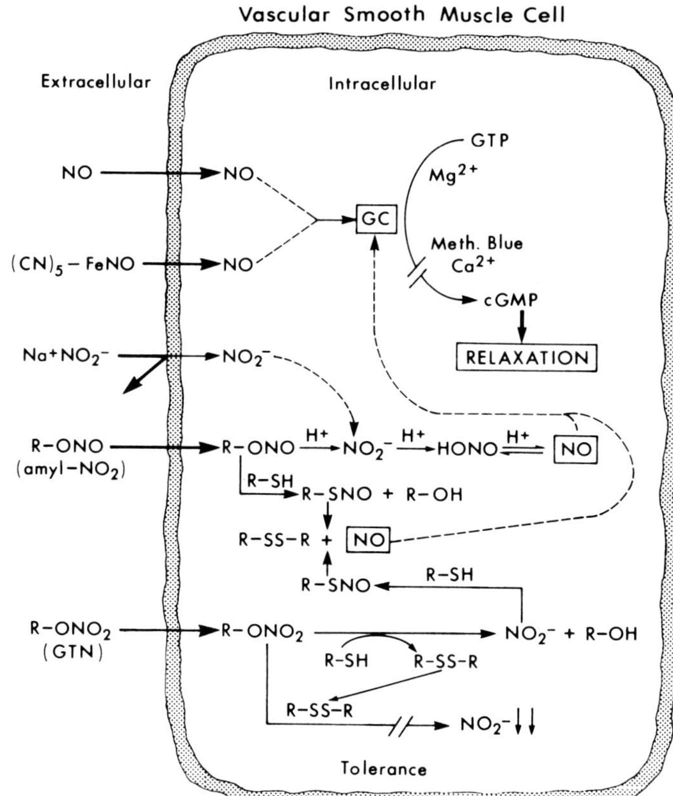

Figure 3. Schematic diagram of the proposed mechanisms by which organic nitrate and nitrite esters, inorganic nitrite and nitroso compounds relax vascular smooth muscle. Abbreviations: NO, nitric oxide; $(CN)_5$-FeNO, nitroprusside; R-ONO, organic nitrite ester; amyl-NO_2, amyl nitrite; R-ONO_2, organic nitrate ester, GTN, nitroglycerin; GC, cytosolic guanylate cyclase; cGMP, cyclic GMP; Meth. Blue, methylene blue; R-SH, thiol; R-SNO, S-nitrosothiol. Nitroprusside liberates NO in solution, whereas organic nitrite and nitrate esters require chemical interaction with thiols to generate S-nitrosothiols, when then decompose with the liberation of NO. NO activates guanylate cyclase and the resulting cyclic GMP causes vascular smooth muscle relaxation. Tolerance to nitroglycerin develops as a result of depletion of thiols and therefore production of S-nitrosothiols and NO. This is a schematic representation of data published by Ignarro *et al.* (10).

shown to be potent inhibitors of platelet aggregation (13). Human platelets were rich in cytoplasmic guanylate cyclase and nitric oxide activated partially purified guanylate cyclase from platelets by heme-dependent mechanisms.

Thus, nitric oxide not only caused vascular smooth muscle relaxation but also inhibited platelet aggregation, and both biological responses were mediated by cyclic GMP. It is perhaps of interest that these two biological actions of nitric oxide were discovered before or at the time of the discovery of endothelium-dependent vascular smooth muscle relaxation by Robert Furchgott in 1980 (14). In an attempt to better understand the signal transduction mechanisms involved in the stimulation of cyclic GMP formation by nitric oxide, experiments were designed to elucidate the biochemical mechanism by which nitric oxide activates guanylate cyclase.

MECHANISM OF ACTIVATION OF GUANYLATE CYCLASE BY NITRIC OXIDE

Some of the early mechanistic experiments on guanylate cyclase focused on the possible involvement of enzyme-bound thiols in the regulation of catalysis (15–18). We followed up on these studies and found that guanylate cyclase was very sensitive to oxygen, undergoing reversible inactivation with increasing oxygen tension (19). Sulfhydryl oxidizing agents produced identical effects which were prevented and reversed by addition of sulfhydryl reducing agents. Both basal enzymatic activity and activation of the enzyme by nitric oxide were affected similarly. Protection against enzyme inactivation was afforded by addition of either excess substrate (MgGTP) or thiol antioxidants. The thiol 2,3-dimercaprol, which possesses vicinal dithiols, but not dithiothreitol markedly inhibited guanylate cyclase activity, indicating that two closely juxtaposed -SH groups may be located at the active site. Additional studies suggested that enzyme -SH groups might bind to nitric oxide as part of the enzyme activation process. However, these studies were not definitive in terms of defining the mechanism of activation of guanylate cyclase by nitric oxide. After reading two papers published by Craven and DeRubertis (20,21), who suggested that heme may be required for guanylate cyclase activation by nitric oxide, we decided to purify cytosolic guanylate cyclase and look for enzyme-bound heme. Heme iron, when in the reduced ferrous state, has a high binding affinity for nitric oxide and this might explain why nitric oxide is such a potent activator of guanylate cyclase. Sure enough, cytosolic guanylate purified from bovine lung was found to contain one mole of heme bound per mole of enzyme dimer (22,23). Similar results were reported by another group (24).

We wanted to ascertain whether the heme bound to guanylate cyclase was indeed required for activation by nitric oxide and whether the heme played any role in the catalytic mechanism of the enzyme. Various procedures were attempted to remove the heme from guanylate cyclase without destroying the enzyme but this approach was unsuccessful at first. Accidentally, we found that passing relatively crude cytosolic fractions containing guanylate cyclase through ion exchange resins such as diethylaminoethyl cellulose at 48°C instead of 36°C (the cold room thermostat was not operating properly) caused the heme to detach from the enzyme without causing enzyme inactivation (23). That is, basal catalytic activity remained unaltered but guanylate cyclase was incapable of being activated by nitric oxide. Preformed nitrosyl-heme, however, markedly activated the heme-deficient enzyme. Addition of heme back to enzyme reaction mixtures completely restored activation of heme-deficient guanylate cyclase by nitric oxide. Eventually, we learned how to prepare heme-containing, heme-deficient and heme-reconstituted forms of guanylate cyclase from rat liver and bovine lung with relative ease (25). The interpretation of the data was clear (Figure 4). Cytosolic guanylate cyclase was a hemoprotein and the function of the heme prosthetic group was to bind to and react with nitric oxide to form the nitrosyl-heme complex and allow en-

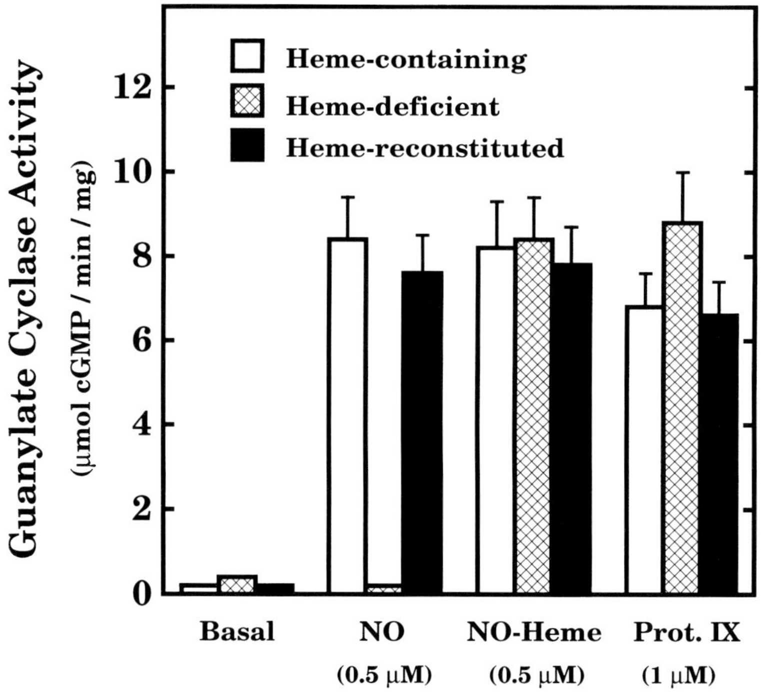

Figure 4. Heme-dependent activation of cytosolic guanylate cyclase by nitric oxide (NO). Guanylate cyclase was purified from bovine lung and assayed as described (22). The purified enzyme contained stoichiometric quantities of heme. Heme was removed from aliquots of native enzyme to yield heme-deficient guanylate cyclase, and the heme-deficient enzyme was reconstituted with stoichiometric quantities of heme as described (23,25). Enzyme reactions were conducted for 5 min under initial velocity conditions. NO-Heme, preformed nitrosyl-heme; Prot. IX, protoporphyrin IX. Data represent the mean ± SEM of 12 determinations from 3 separate experiments.

zyme activation by nitric oxide. In addition, guanylate cyclase possessed at least two reactive -SH groups at or near the catalytic site which had to be in the reduced state for the full expression of catalytic activity and activation by nitric oxide (19).

The next step was to elucidate the mechanism by which nitrosyl-heme activates guanylate cyclase. This was a difficult question to answer because the structure of guanylate cyclase was unknown. After studying the published interactions between hemoglobin or myoglobin and nitric oxide, we came up with a hypothesis to explain how nitrosyl-heme activates guanylate cyclase. We envisioned that heme (iron-protoporphyrin IX) was bound to guanylate cyclase at a site that was adjacent to the catalytic site that binds the substrate MgGTP. As in other hemoproteins, the axial ligand (bond between heme iron and protein) was thought to involve a histidine residue in guanylate cyclase, thereby yielding a 5-coordinate complex (where the heme iron and each of the four pyrrole nitrogen atoms form ligands with the enzyme protein). Removal of the heme group does not influence basal catalytic activity but abolishes enzyme activation by nitric oxide. We proposed that nitric oxide forms a covalent bond with the reduced heme iron (ferrous) to yield a 5-coordinate complex rather than a 6-coordinate complex. This means that the axial ligand would have to break and the resulting nitrosyl-heme complex would undergo a change in configuration while binding to the enzyme protein. This conformational change near the catalytic site could increase the affinity of guanylate cyclase for MgGTP and also increase the maximal velocity of catalysis (Figure 5). In order to test this hypothesis, protoporphyrin IX (heme without its iron) was tested to see if it could activate heme-deficient guanylate cyclase. The reasoning behind this experiment was that the axial

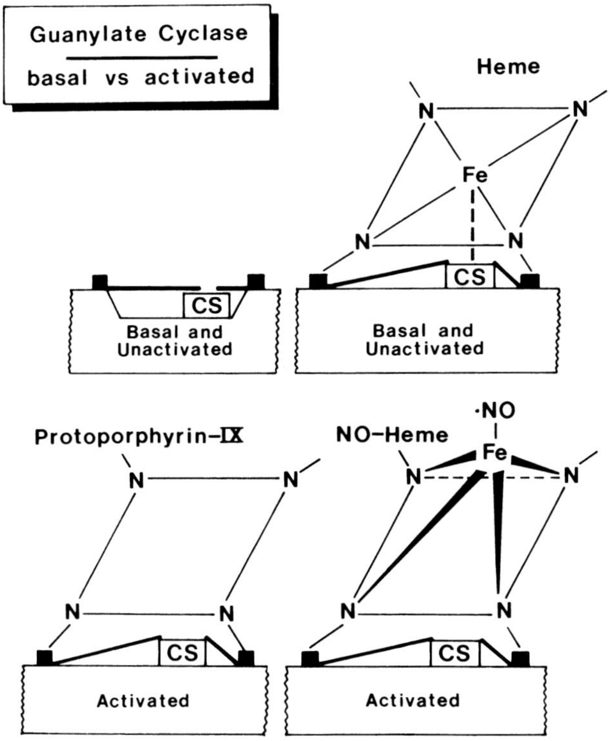

Figure 5. Schematic illustration of the mechanism of heme-dependent activation of cytosolic guanylate cyclase by nitric oxide (NO). This schematic is based on the knowledge that heme is required for the activation of guanylate cyclase by NO. In the basal or unactivated state and in the absence of the heme prosthetic group, the catalytic site (CS) in guanylate cyclase is only partially exposed so that substrate (MgGTP) availability is limited and cyclic GMP production is limited. In the presence of the heme prosthetic group (native enzyme) but in the absence of any NO (basal or unactivated state), catalytic activity is still minimal because of steric hindrance at the catalytic site. Enzyme-bound heme permits exposure of the catalytic site but the axial ligand (bond between heme iron (Fe) and enzyme protein near CS) limits access of substrate to CS, thereby limiting cyclic GMP production. NO binds to heme iron to form the nitrosyl-heme adduct as illustrated. This NO-heme complex remains as a 5-coordinate complex, thereby resulting in the breakage of the axial ligand and exposure of CS to enzyme substrate. The activated enzyme shows up to a 400-fold increase in Vmax and a 3-fold decrease in the Km for MgGTP. Addition of protoporphyrin IX to heme-deficient guanylate cyclase mimics the effect of NO-heme because protoporphyrin IX binding allows exposure of CS without any steric hindrance due to the absence of heme iron.

ligand between heme iron and guanylate cyclase might create steric hindrance to the binding of MgGTP to the catalytic site, whereas protoporphyrin IX should bind to the heme binding site but create no steric hindrance because of the absence of iron, and the result should be enzyme activation by protoporphyrin IX. That is exactly what we observed. Protoporphyrin IX was a potent activator of guanylate cyclase (22) and it mimicked the activating effect of nitrosyl-heme in that enzyme activation by protoporphyrin IX was heme-independent, like that for preformed nitrosyl-heme (26). A kinetic analysis revealed that protoporphyrin IX and nitrosyl-heme activated guanylate cyclase by kinetically indistinguishable mechanisms (26).

These observations revealed that guanylate cyclase possessed a porphyrin binding site that was responsible for enzyme activation by nitric oxide. More elaborate studies were later conducted by others (27–31) to verify our original hypothesis (32). Revelation of the crystal structure of cytosolic guanylate cyclase will undoubtedly explain the more precise mechanisms by which nitric oxide activates guanylate cyclase.

IDENTIFICATION OF ENDOTHELIUM-DERIVED RELAXING FACTOR AS NITRIC OXIDE

Most of the research discussed above was conducted before or shortly after the discovery of endothelium-dependent vasorelaxation by Robert Furchgott in 1980 (14). The knowledge that nitroglycerin was such a potent vasorelaxant and the more recent findings that nitric oxide was a potent vasorelaxant and inhibitor of platelet aggregation led to the question of why do mammals have receptors for such potent pharmacological compounds. This thought suggested that mammals might possess either an endogenous nitro compound that releases nitric oxide or nitric oxide itself. The early studies on endothelium-derived relaxing factor (EDRF) were conducted without any knowledge about its chemical structure. We started to study endothelium-dependent vasorelaxation in 1983, because we wanted to determine whether cyclic GMP was involved, and not because we thought that EDRF might be nitric oxide. Indeed, a separate project was also underway to find the endogenous nitric oxide. Just as we completed our experiments on EDRF and cyclic GMP, Ferid Murad's group published a paper showing that endothelium-dependent vasorelaxation by acetylcholine and other agents was associated with cyclic GMP production in vascular smooth muscle (33). We observed similar results and also found that methylene blue, an inhibitor of guanylate cyclase activity, prevented both the cyclic GMP accumulating effect and vasorelaxant effect of acetylcholine (34). At first, we entertained the idea that acetylcholine might be stimulating the formation of a metabolite of arachidonic acid, which then somehow activated guanylate cyclase to elevate cyclic GMP levels. This view was supported by studies from this and other laboratories showing that agents that interfere with the formation of arachidonic acid and certain arachidonic acid metabolites also interfere with acetylcholine-mediated vasorelaxation (33–35). But I was not satisfied with this view because experiments in my laboratory failed to reveal that anything other than nitric oxide could activate guanylate cyclase. After reviewing our data again, we suddenly realized that our finding that methylene blue prevented both the increase in cyclic GMP and vasorelaxation in response to acetylcholine was similar to our earlier finding on the prevention of both cyclic GMP formation and vasorelaxation in response to nitric oxide (36). This indicated that the vasorelaxant action of acetylcholine was pharmacologically similar to that for nitric oxide. This concept is indicated in Figure 6, which was published in 1984 (34). We were cautious, however, and did not suggest in writing that EDRF might be nitric oxide. Later, after other laboratories showed that EDRF was a very unstable molecule (37,38), we realized that the data from several laboratories were consistent with our view that EDRF and nitric oxide were closely similar. The two distinct projects in my laboratory, one focusing on EDRF/cyclic GMP and the other focusing on endogenous nitric oxide, converged but we were not yet ready to propose that EDRF was nitric oxide.

A series of experiments was designed to test our yet unpublished hypothesis that EDRF might be nitric oxide. The first experiment was to determine whether EDRF released from artery and vein could activate guanylate cyclase and thereby account for elevated tissue cyclic GMP levels in response to acetylcholine or bradykinin, and this study was published in 1986 (39). Isolated arterial and venous rings prepared from bovine intrapulmonary vessels were incubated in reaction mixtures containing cytosolic guanylate cyclase purified from bovine lung. Addition of acetylcholine to reaction mixtures containing

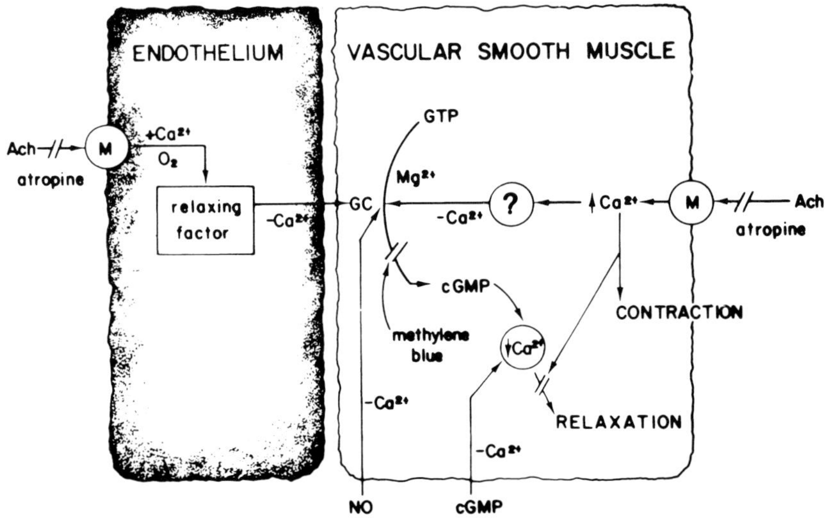

Figure 6. Schematic diagram of the similarities in the mechanisms of vascular smooth muscle relaxation elicited by acetylcholine and nitric oxide (NO). Abbreviations: Ach, acetylcholine; M, muscarinic receptor; GC, cytosolic guanylate cyclase; cGMP, cyclic GMP; -Ca^{2+}, calcium-independent mechanism; +Ca^{2+}, calcium-dependent mechanism. Acetylcholine and NO cause vasorelaxation by mechanisms involving cyclic GMP formation. Blockade of cyclic GMP formation by methylene blue inhibits vasorelaxation by both acetylcholine and NO, as illustrated. Modified from Ignarro *et al.* (34) with permission.

arterial rings resulted in the activation of guanylate cyclase, and bradykinin elicited the same effect in reaction mixtures containing either arterial or venous rings. Activation of guanylate cyclase by acetylcholine and bradykinin was dependent on the presence of an intact endothelial layer in arterial and venous rings, and the response was blocked by methylene blue and enhanced by antioxidants. These experiments revealed that EDRF from artery and vein activates guanylate cyclase by mechanisms that can be inhibited by methylene blue and enhanced by antioxidants. Clearly, the data were pointing toward the possibility that EDRF is nitric oxide or a chemically related substance such as a labile nitroso compound. The next experiment was to determine whether the activation of guanylate cyclase by EDRF was heme-dependent as is enzyme activation by nitric oxide. Experiments similar to those described above were repeated using guanylate cyclase that had been purified in both the heme-containing and heme-deficient forms. Sure enough, guanylate cyclase activation by EDRF, like enzyme activation by nitric oxide, was heme-dependent (Figure 7). At the end of the first experiment of this series, we

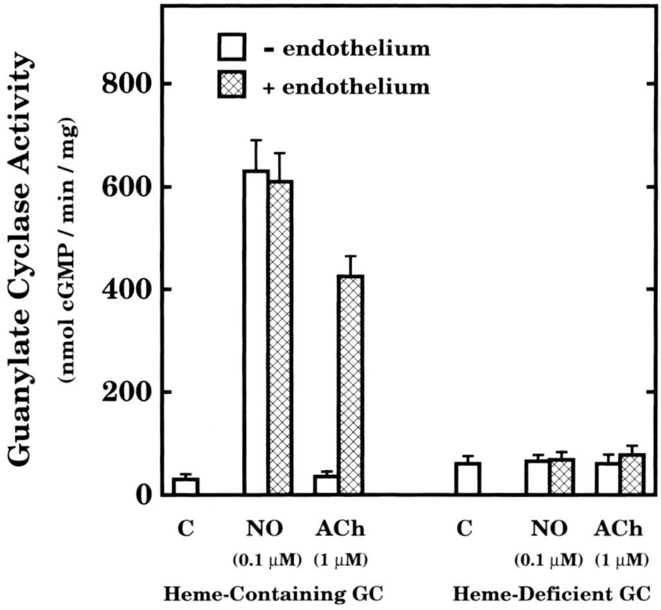

Figure 7. Endothelium-dependent and heme-dependent activation of cytosolic guanylate cyclase by endothelium-derived relaxing factor (EDRF) released from arterial rings. These experiments were based on techniques described previously (39). Bovine intrapulmonary arterial rings were prepared with and without intact endothelium and mounted in tissue baths for monitoring vasorelaxation. Acetylcholine (ACh) and nitric oxide (NO) were added to tissue baths to monitor relaxant responses. ACh elicited endothelium-dependent, whereas NO elicited endothelium-independent vasorelaxation. After the first round of additions to the tissue baths, the rings were rapidly detached and lowered into reaction vessels containing either heme-containing guanylate cyclase or heme-deficient guanylate cyclase, to which were added acetylcholine or NO. As illustrated, acetylcholine and NO caused heme-dependent activation of guanylate cyclase, but only NO caused endothelium-independent activation of guanylate cyclase. The conclusion was that acetylcholine causes the formation of NO from the vascular endothelium, and the NO causes heme-dependent activation of guanylate cyclase. Data represent the mean ± SEM of 12 determinations from 6 separate experiments.

knew that we had it. EDRF must be nitric oxide. But I wanted to conduct more definitive experiments before going out on a limb and advocating that EDRF is nitric oxide. A bioassay cascade apparatus was set up in order to study the chemical and pharmacological properties of EDRF released from artery and vein, and to compare these with the properties of nitric oxide. These experiments revealed the very close similarity of EDRF and nitric oxide, and were presented at several conferences in 1986 and early 1987 (40–42) before their publication in 1987 (43,44).

There was one experiment in particular that convinced me beyond any shadow of a doubt that EDRF was nitric oxide. We took advantage of a previous experiment in my laboratory on guanylate cyclase, in which we demonstrated that guanylate cyclase was a hemoprotein, like hemoglobin, that reacted with nitric oxide to form the nitrosyl-heme adduct by monitoring the spectral shift in the Soret region (23). This classical chemical analysis provides proof of the presence of heme iron, and established guanylate cyclase as a hemoprotein. Accordingly, we conducted similar experiments with EDRF released from cultured bovine aortic endothelial cells and showed that EDRF caused a spectral shift in hemoglobin that was identical to that obtained for authentic nitric oxide (43)(Figure 8). Suddenly, all previous data from our laboratory and several other laboratories made complete sense and were consistent with the view that EDRF was nitric oxide. EDRF is a small, labile and lipophilic molecule that is inactivated by oxygen and oxygen-derived radicals, activates cytosolic guanylate cyclase by heme-dependent mechanisms, causes smooth muscle relaxation and inhibition of platelet aggregation and adhesion by cyclic GMP-dependent mechanisms, and whose actions are blocked by hemoproteins or methylene blue but enhanced by antioxidants or oxygen radical scavengers.

At the time we were conducting some of the above experiments, other laboratories were conducting experiments that turned out to be consistent with the view that EDRF was nitric oxide. For example, hemoglobin was reported to interfere with acetylcholine-elicited, endothelium-dependent vasorelaxation (45,46), superoxide anion was shown to decrease, whereas superoxide dismutase increased the chemical stability of EDRF (37,38) and EDRF inhibited platelet aggregation (47). All of these effects on or actions of EDRF were identical to those for nitric oxide. Despite this obvious analogy, no one wanted to go out on a limb and propose that EDRF was nitric oxide, except for Robert Furchgott and myself , when we independently proposed this hypothesis at the same international conference in Rochester, Minnesota in June of 1986 (40).

THE PHYSIOLOGICAL SIGNIFICANCE OF NITRIC OXIDE IN VASCULAR BIOLOGY

Soon after the discovery that EDRF is nitric oxide, evidence was presented that vascular endothelial cells can produce nitric oxide and citrulline from arginine (48). This finding was remarkably similar to the observation in cyto-

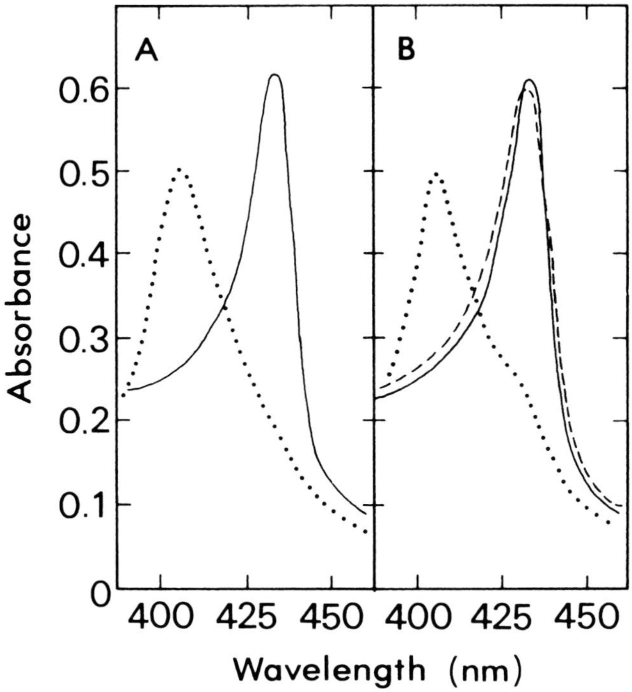

Figure 8. Identification of endothelium-derived relaxing factor (EDRF) as nitric oxide (NO) by spectrophotometric analysis of the reaction between deoxyhemoglobin and NO or EDRF. The EDRF was generated from freshly isolated bovine aortic endothelial cells stimulated by the calcium ionophore, A23187. Experiments were performed in quartz cuvettes as described previously (43). Panel A: solid curve signifies 5 μM deoxyhemoglobin alone; dotted curve signifies deoxyhemoglobin gassed with NO. Panel B: solid curve signifies 5 μM deoxyhemoglobin alone; dashed curve signifies deoxyhemoglobin after reaction with 10^6 endothelial cells; dotted curve signifies deoxyhemoglobin after reaction with endothelial cells in the presence of 1 μM A23187. Data illustrated are from one experiment representative of 3 separate experiments. Modified from Ignarro *et al.* (43) with permission.

kine-activated macrophages that nitrite (NO_2^-) and nitrate (NO_3^-) are produced from arginine (49–51). Indeed, these studies revealed that a structural analog of arginine, N^G-methylarginine, inhibited the formation of NO_2^- and NO_3^- from arginine in macrophages and also inhibited the production of nitric oxide from arginine in endothelial cells. Confirmation of these original observations and unequivocal demonstration of an enzyme that catalyzes the conversion of arginine to nitric oxide plus citrulline was made by Bredt and Snyder (52). The enzyme was termed nitric oxide synthase and this study started an avalanche of research leading to the isoform characterization, purification, elucidation of catalytic mechanisms and structure-function characterization of nitric oxide synthase (53–61).

During this period of time, incredible progress was also being made on the physiological and pathophysiological properties of endogenous nitric oxide. Vascular endothelial cells synthesize nitric oxide continuously for the purpose of causing vasodilation to limit any increases in systemic blood pressure (62,63). This conclusion was drawn from experiments in which N^G-methylarginine, a competitive inhibitor of nitric oxide synthase, caused a sustained increase in systemic blood pressure in animals after intravenous injection. The hypertensive response was completely reversed by intravenous injection of excess L-arginine, the substrate for nitric oxide synthase. The continuous generation of endothelium-derived nitric oxide appears to occur as a consequence of continuous blood flow or shear stress (64), which can trigger the calcium-dependent activation of endothelial nitric oxide synthase (65). Endothelium-derived nitric oxide also functions to inhibit platelet aggregation and adhesion (66), and to slow or prevent proliferation of the underlying vascular smooth muscle (67). Research continues on the elucidation of the protective role of nitric oxide in reperfusion or reoxygenation injury, restenosis, and atherosclerosis. The protective effects of nitric oxide may stem from its capacity to improve local blood flow, inhibit thrombosis or cell adhesion, interfere with cell proliferation, and inhibit key enzymes involved in mediating cellular modification or destruction.

The neurotransmitter function of nitric oxide was first shown in the brain (68). These observations were based on the findings that glutamate and nitric oxide stimulate the formation of cyclic GMP in the brain, and that arginine can also elevate cyclic GMP levels under certain conditions. The function of nitric oxide as a neurotransmitter in the central nervous system remains unknown but one possible hypothesis has been that nitric oxide modulates long-term potentiation in the control of learning and memory (66). The bad side of nitric oxide is that excessive production of nitric oxide in the brain stimulated by glutamate can lead to serious tissue injury in specialized regions of the brain (69). Moreover, excessive production of nitric oxide in close proximity to oligodendrocytes can lead to inhibition of myelin formation and perhaps multiple sclerosis (70). Evidence has amassed that nitric oxide is an important transmitter of the nonadrenergic-noncholinergic (NANC) neurons that innervate many smooth muscle tissues such as the airways, gastro-intestinal tract and genito-urinary tract (66).

We explored the possibility that nitric oxide might be the principal neurotransmitter that mediates penile erection. The reason for pursuing this project was that before 1990 the physiological mechanism by which nerve stimulation promotes erectile function was unknown. The common belief was that excitation of the parasympathetic nervous system caused acetylcholine-mediated erectile responses. The erectile response consists of neuronally mediated relaxation of both vascular and nonvascular smooth muscle associated with the cavernous or trabecular or sinusoidal arrangement of arterioles in the corpus cavernosum. The problem that this hypothesis presented to pharmacologists and physiologists is that the cholinergic neurotransmitter, acetylcholine, cannot relax nonvascular smooth muscle and can relax vascular smooth muscle only by endothelium-dependent mechanisms. It was viewed highly unlikely that acetylcholine, a quaternary ammonium compound (highly charged) would reach the vascular endothelial cells from the neuronal synapses at the vascular smooth muscle cells. However, excitation of any nitric oxide-containing NANC nerves that innervate the nonvascular and vascular smooth muscle in the corpus cavernosum would be expected to promote extensive smooth muscle relaxation and consequent penile erection. At the time of our first experiment, it was not known whether or not erectile tissue in mammals was innervated by NANC nerves, and we were not equipped to conduct such histochemical experiments. Instead, we used the classical tissue bath approach in which strips of corpus cavernosum isolated from rabbit and human penis were mounted in bath chambers, precontracted and then electrically stimulated in order to cause smooth muscle relaxation.

Our first experiment revealed that NANC stimulation of isolated strips of rabbit corpus cavernosum resulted in smooth muscle relaxation that was blocked by inhibitors of nitric oxide synthase, hemoglobin and methylene blue (71). Nitric oxide in the form of S-nitrosothiols also caused smooth muscle relaxation, and electrical stimulation of the tissue resulted in elevated levels of both nitrite (NO_2^-) and cyclic GMP (71). The conclusion was clear. Nitric oxide is the neurotransmitter in NANC nerves that innervate the corpus cavernosum and that nerve stimulation results in nitric oxide-mediated vascular and nonvascular smooth muscle relaxation. The action of the nitric oxide neurotransmitter is, in turn, mediated by cyclic GMP. Therefore, the signal transduction pathway involving nitric oxide as a neurotransmitter appears to be the same as that for endothelium-derived nitric oxide. Additional experiments with human corpus cavernosum yielded virtually identical data as the experiments using rabbit tissue (72,73). Figure 9 illustrates the typical marked but transient smooth muscle relaxation responses elicited by electrical stimulation of corpus cavernosum from both rabbit and human. We were struck with the very transient nature of the responses and that these electrically-induced responses could be mimicked by addition of authentic nitric oxide gas. Perhaps the most convincing experiment indicating that electrical stimulation triggers erectile function via mechanisms involving nitric oxide was the finding that N^G-nitroarginine, a competitive inhibitor of nitric oxide synthase, abolished the electrical responses, whereas excess enzyme substrate, L-arginine, overcame the inhibitory effect of N^G-nitroarginine (Figure 10). Our early studies also revealed that inhibition of the degradation of cyclic GMP through the use of phosphodiesterase inhibitors caused an enhancement of both electrically-induced smooth muscle relaxation and nitric oxide-elicited smooth muscle relaxation (72–74). The data for enhancement of nitric oxide-elicited relaxation of corpus cavernosum by the cyclic GMP phosphodiesterase inhibitor, M&B 22,948 (zaprinast), is illustrated in Figure 11. These observations indicated clearly that relaxation of corpus cavernosum smooth muscle by NANC nerve excitation or addition of nitric oxide is mediated by cyclic GMP.

These observations shed a great deal of light on the physiological mechanism of penile erection by revealing a critical signal transduction pathway that could be modulated by drugs. For example, drugs that interfere with the formation or action of nitric oxide or cyclic GMP could also interfere with the erectile response, and might be useful for treating priapism. But much more importantly, drugs that enhance the formation or action of nitric oxide or

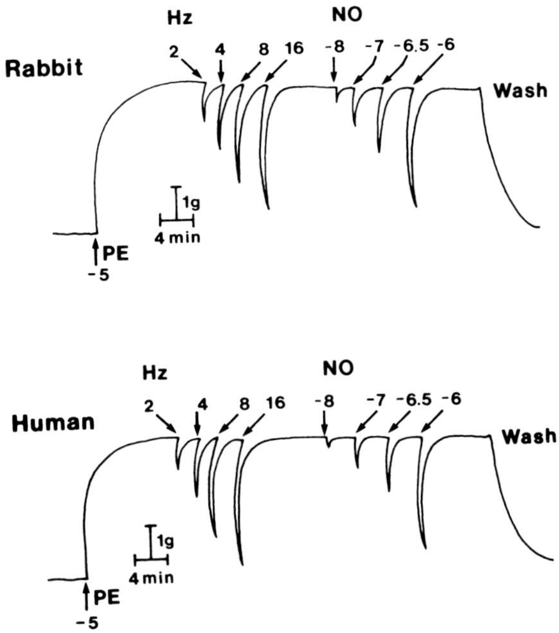

Figure 9. Relaxation of corpus cavernosum smooth muscle by nitric oxide (NO) closely resembles relaxation elicited by electrical field stimulation of NANC nerves. Strips of corpus cavernosum from rabbit and human penis were prepared, mounted in tissue baths and isometric tension recordings obtained (71,73). Smooth muscle was precontracted by addition of phenylephrine (PE) as shown. Electrical field stimulation was conducted at 2, 4, 8 and 16 hertz (Hz) as shown. Concentrations of NO and PE are expressed as Molar (ie. 10^{-8} M). Tracings are representative of 12–18 strips from 4 rabbits or 4 humans.

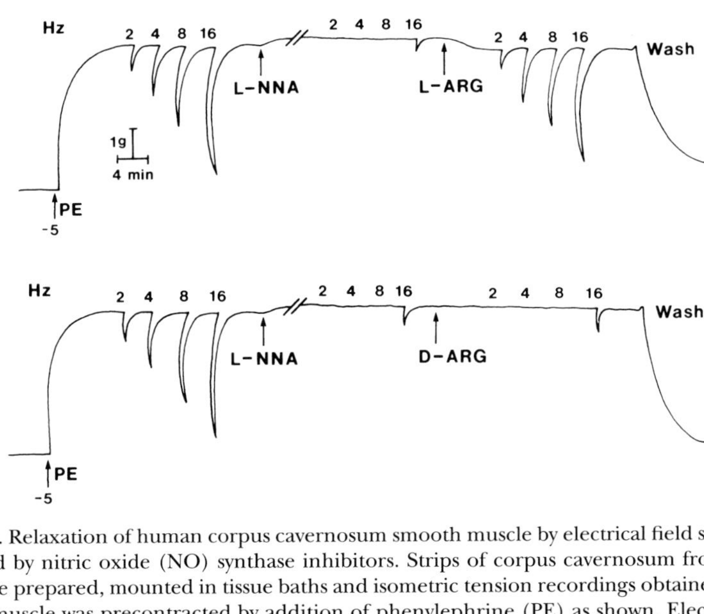

Figure 10. Relaxation of human corpus cavernosum smooth muscle by electrical field stimulation is blocked by nitric oxide (NO) synthase inhibitors. Strips of corpus cavernosum from human penis were prepared, mounted in tissue baths and isometric tension recordings obtained (71,73). Smooth muscle was precontracted by addition of phenylephrine (PE) as shown. Electrical field stimulation was conducted at 2, 4, 8 and 16 hertz (Hz) as shown. N^G-Nitroarginine (L-NNA) and arginine (L-ARG or D-ARG) were tested at 30 μM and 300μM, respectively. Tracings are representative of 12-18 strips from 4 humans.

cyclic GMP could also enhance the erectile response, and might be effective in the treatment of impotence, the most prevalent medical disorder in men. In 1998, the drug sildenafil (Viagra[R]) was marketed for the effective treatment of impotence. Sildenafil is a relatively selective inhibitor of one of the phosphodiesterase isoforms that degrades cyclic GMP and is present in relatively large quantities in the corpus cavernosum. Therefore, sildenafil allows intracellular cyclic GMP in corpus cavernosum smooth muscle to accumulate to high levels when ever there is a nitric oxide signal coming from the NANC nerves. Penile erection resulting from any form of excitation of the NANC nerves innervating the corpus cavernosum, whether mental or physical, would be enhanced by sildenafil, but the drug alone would not be expected to cause an erectile response in the absence of any NANC nerve excitation.

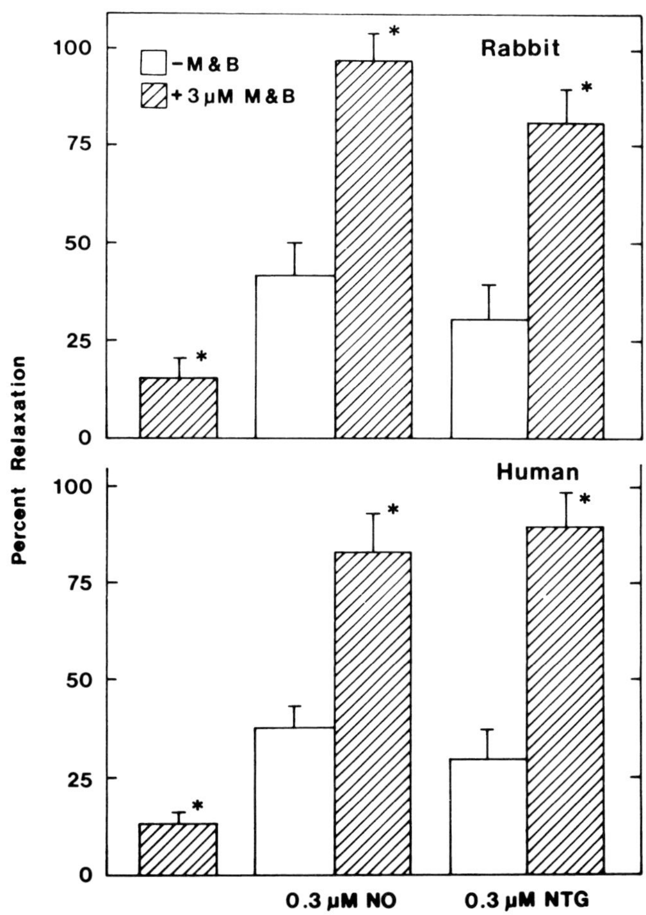

Figure 11. Enhancement of nitric oxide (NO)-elicited and nitroglycerin (NTG)-elicited relaxation of rabbit and human corpus cavernosum smooth muscle by M&B 22,948 (M&B). Strips of corpus cavernosum from rabbit and human penis were prepared, mounted in tissue baths and isometric tension recordings obtained (71,73). Smooth muscle was precontracted by addition of 10 μM phenylephrine. Each point represents the mean ± SEM of 9-12 strips from 4 rabbits or 4 humans. * All values in presence of M&B 22,948 are significantly different ($p < 0.001$) from corresponding values obtained in absence of M&B 22,948.

Sildenafil appears to be very effective in treating various types and degrees of impotence. Our improved understanding of the physiology of erectile function will enable scientists to unravel the pathophysiology and causes of impotence, thereby further opening the doors to develop even more promising drugs for the treatment of impotence.

FUTURE DIRECTIONS FOR NITRIC OXIDE RESEARCH IN VASCULAR BIOLOGY

The knowledge that vascular endothelium-derived nitric oxide and neuronally-generated nitric oxide promote the relaxation of vascular and nonvascular smooth muscle associated with the vascular system will undoubtedly afford scientists and clinicians the opportunity to make great advances in the diagnosis, treatment and prevention of life-threatening cardiovascular disorders such as coronary artery disease, essential hypertension, stroke, vascular complications of diabetes and related diseases. Already, it appears that nitric oxide plays many critical physiological roles in the regulation and protection of numerous tissue and organ functions. Defects or lesions in the nitric oxide-cyclic GMP pathway can probably account for much of the observed pathophysiology of cardiovascular disease. The development of a better understanding of basic physiological mechanisms involving the nitric oxide-cyclic GMP pathway should lead to the development of novel therapeutic strategies to reduce the morbidity and mortality attributed to cardiovascular disease.

ACKNOWLEDGMENTS

Many individuals have been responsible for my development as a scientist and to whom I am grateful. My mentors Fred Shideman and Elwood Titus taught me the meaning of concepts like the experimental approach, the establishment and testing of the hypothesis, the critical evaluation of experiments that not only work but also do not work as expected, and the importance of thinking. In my first job after my postdoctoral training, Barbara Petrack taught me the meaning of respect and hard work in science. In my early academic career, my colleagues Bill George and Phil Kadowitz provided strong encouragement and motivation. Jim Fisher consistently motivated me by example to develop into an effective teacher of both medical and graduate students. Perhaps I owe my most profound gratitude to Gautam Chaudhuri, who provided the greatest encouragement and motivation for me to continue my work and take it to new highs. All of my graduate students, postdoctoral fellows, clinical fellows and visiting scientists have made substantial contributions. These individuals include (in alphabetical order): John Adams, Ernesto Aeberhard, Nicole Arabolos, William Aronson, Bryan Ballot, Barbara Barry, Georgette Buga, Theresa Burke, Peggy Bush, Russell Byrns, Stella Cech, Gordon Cohen, Ronald Day, Jonathan Degnan, James Edwards, Richard Fitch, Michele Gold, Jeanette Griscavage, Rosemarie Gross, Carl Gruetter, Darlene Gruetter, Richard Harbison, Toshio Hayashi, Adrian Hobbs, Philip

Horwitz, Howard Lippton, Dennis McNamara, B. Theo Mellion, Steve Napoli, Eliot Ohlstein, Richard Paddock, Waldemar Radziszewski, Norma Rogers, Robert Smith, Hugo Vargas, Liu Hua Wei, Michael Wolin, and Keith Wood.

Collaborations with others on the faculty and elsewhere have been instrumental in our accomplishments. Particularly notable roles have been played by William Baricos, Gerald Buckberg, Steven Cederbaum, Gautam Chaudhuri, Jon Fukuto, William George, Scott Henderson, Albert Hyman, Philip Kadowitz, Jack Lancaster, Jean Merrill, William Pearce, Jacob Rajfer, Michael Sherman, Dennis Stuehr, and Sherwin Wilk.

Our research has been supported continuously since 1973 by grants from the National Heart, Lung and Blood Institute of the National Institutes of Health. Other invaluable sources of support have included the American Heart Association, the American Cancer Society, the Laubisch Fund for Cardiovascular Research, the Tobacco-Related Disease Research Program, The Council for Tobacco Research – USA, Inc., the Pharmaceutical Manufacturers Association, and the Schlieder Educational Foundation.

REFERENCES

1. Katsuki, S., Arnold, W., Mittal, C., and Murad, F. *J. Cyclic Nucleotide Res.* **3:** 23–35, 1977.
2. Arnold, W., Mittal, C., Katsuki, S., and Murad, F. *Proc. Natl. Acad. Sci. USA* **74:** 3203–3207, 1977.
3. Murad, F., Mittal, C., Arnold, W., Katsuki, S., and Kimura, H. *Adv. Cyclic Nucleotide Res.* **9:** 145–158, 1978.
4. Katsuki, S., and Murad, F. *Mol. Pharm.* **13:** 330–341, 1977.
5. Gruetter, C.A., Barry, B.K., McNamara, D.B., Gruetter, D.Y., Kadowitz, P.J., and Ignarro, L.J. *J. Cyclic Nucleotide Res.* **5:** 211–224, 1979.
6. Ohlstein, E.H., Barry, B.K., Gruetter, D.Y., and Ignarro, L.J. *FEBS Lett.* **102:** 316–320, 1979.
7. Ignarro, L.J., Edwards, J.C., Gruetter, D.Y., Barry, B.K., and Gruetter, C.A. *FEBS Lett.* **110:** 275–278, 1980.
8. Ignarro, L.J., Barry, B.K., Gruetter, D.Y., Edwards, J.C., Ohlstein, E.H., Gruetter, C.A., and Baricos, W.H. *Biochem. Biophys. Res. Commun.* **94:** 93–100, 1980.
9. Ignarro, L.J., and Gruetter, C.A. *Biochim. Biophys. Acta* **631:** 221–231, 1980.
10. Ignarro, L.J., Lippton, H., Edwards, J.C., Baricos, W.H., Hyman, A.L., Kadowitz, P.J., and Gruetter, C.A. *J. Pharmacol. Exp. Ther.* **218:** 739–749, 1981.
11. Saxon, A., and Kattlove, H.E. *Blood* **47:** 957–961, 1976.
12. Mellion, B.T., Ignarro, L.J., Ohlstein, E.H., Pontecorvo, E.G., Hyman, A.L., and Kadowitz, P.J. *Blood* **57:** 946–955, 1981.
13. Mellion, B.T., Ignarro, L.J., Myers, C.B., Ohlstein, E.H., Ballot, B.A., Hyman, A.L, and Kadowitz, P.J. *Mol. Pharm.* **23:** 653–664, 1983.
14. Furchgott, R.F., and Zawadzki, J.V. *Nature* **288:** 373–376, 1980.
15. White, A.A., Crawford, K.M., Patt, C.S., and Lad, P.J. *J. Biol. Chem.* **251:** 7304–7312, 1976.
16. Goldberg, N.D., and Haddox, M.K. *Annu. Rev. Biochem.* **46:** 823–896, 1977.
17. DeRubertis, F.R., Craven, P.A. *Biochim. Biophys. Acta* **499:** 337–351, 1977.
18. Craven, P.A., and DeRubertis, F.R. *Biochim. Biophys. Acta* **524:** 231–244, 1978.
19. Ignarro, L.J., Kadowitz, P.J., and Baricos, W.H. *Arch. Biochem. Biophys.* **208:** 75–86, 1981.
20. Craven, P.A., and DeRubertis, F.R. *J. Biol. Chem.* **253:** 8433–8443, 1978.
21. Craven, P.A., DeRubertis, F.R., and Pratt, D.W. *J. Biol. Chem.* **254:** 8213–8222, 1979.

22. Ignarro, L.J., Wood, K.S., and Wolin, M.S. *Proc. Nat. Acad. Sci. USA* **79:** 2870–2873, 1982.
23. Ignarro, L.J., Degnan, J.N., Baricos, W.H., Kadowitz, P.J., and Wolin, M.S. *Biochim. Biophys. Acta* **718:** 49–59, 1982.
24. Gerzer, R., Bohme, E., Hofmann, F., and Schultz, G. *FEBS Lett.* **132:** 71–74, 1981.
25. Ignarro, L.J., Wood, K.S., Ballot, B., and Wolin, M.S. *J. Biol. Chem.* **259:** 5923–5931, 1984.
26. Wolin, M.S., Wood, K.S., and Ignarro, L.J. *J. Biol. Chem.* **257:** 13312–13320, 1982.
27. Stone, J.R., and Marletta, M.A. *Biochemistry* **34:** 1468–1474, 1995.
28. Burstyn, J.N., Yu, A.E., Dierks, E.A., Hawkins, B.K., and Dawson, J.H. *Biochemistry* **34:** 5896–5903, 1995.
29. Dierks, E.A., and Burstyn, J.N. *Biochem. Pharmacol.* **51:** 1593–1600, 1996.
30. Stone, J.R., and Marletta, M.A. *Biochemistry* **35:** 1093–1099, 1996.
31. Zhao, Y., Schelvis, J.P., Babcock, G.T., and Marletta, M.A. *Biochemistry* **37:** 4502–4509, 1998.
32. Ignarro, L.J., Wood, K.S., and Wolin, M.S. *Adv. Cyclic Nucleotide Res.* **17:** 267–274, 1984.
33. Rapoport, R.M., and Murad, F. *Circ. Res.* **52:** 352–357, 1983.
34. Ignarro, L.J., Burke, T.M., Wood, K.S., Wolin, M.S., and Kadowitz, P.J. *J. Pharmacol. Exp. Ther.* **228:** 682–690, 1984.
35. Furchgott, R.F., Zawadzki, J.V., and Cherry, P.D. *In* Vasodilatation, ed. by P.M. Vanhoutte and I. Leusen, pp. 49–66, Raven Press, NY, 1981.
36. Gruetter, C.A., Gruetter, D.Y., Lyon, J.E., Kadowitz, P.J., and Ignarro, L.J. *J. Pharmacol. Exp. Ther.* **219:** 181–186, 1981.
37. Gryglewski, R.J., Moncada, S., and Palmer, R.M.J. *Br. J. Pharmacol.* **87:** 685–694, 1986.
38. Rubanyi, G.M., and Vanhoutte, P.M. *Am. J. Physiol.* **250:** H822–H827, 1986.
39. Ignarro, L.J., Harbison, R.G., Wood, K.S., and Kadowitz, P.J. *J. Pharmacol. Exp. Ther.* **237:** 893–900, 1986.
40. Ignarro, L.J. Invited talk presented at the IV International Symposium on Mechanisms of Vasodilatation, Rochester, Minnesota, 1986.
41. Ignarro, L.J., Byrns, R.E., and Wood K.S. *Circulation* **74:** II–287, 1986 (abstract).
42. Ignarro, L.J., Byrns, R.E., Buga, G.M., and Wood, K.S. *Fed. Proc.* **46:** 640, 1987 (abstract).
43. Ignarro, L.J., Buga, G.M., Wood, K.S., Byrns, R.E., and Chaudhuri, G. *Proc. Natl. Acad. Sci. USA* **84:** 9265–9269, 1987.
44. Ignarro, L.J., Byrns, R.E., Buga, G.M., and Wood, K.S. *Circ. Res.* **61:** 866–879, 1987.
45. Martin, W., Villani, G.M., Jothianandan, D., and Furchgott, R.F. *J. Pharmacol. Exp. Ther.* **232:** 708–716, 1985.
46. Martin, W., Villani, G.M., Jothianandan, D., and Furchgott, R.F. *J. Pharmacol. Exp. Ther.* **233:** 679–685, 1985.
47. Azuma, H., Ishikawa, M., and Sekizaki, S. *Br. J. Pharmacol.* **88:** 411–415, 1986.
48. Palmer, R.M.J., Ashton, D.S., and Moncada, S. *Nature* **333:** 664–666, 1988.
49. Iyengar, R., Stuehr, D.J., and Marletta, M.A. *Proc. Natl. Acad. Sci. USA* **84:** 6369–6373, 1987.
50. Hibbs, J.B., Vavrin, Z., and Taintor, R.R. *J. Immunol.* **138:** 550–565, 1987.
51. Hibbs, J.B., Taintor, R.R., and Vavrin, Z. *Science* **235:** 473–476, 1987.
52. Bredt, D.S., and Snyder, S.H. *Proc. Natl. Acad. Sci. USA* **87:** 682–685, 1990.
53. Stuehr, D.J., and Griffith, O.W. *Adv. Enzymol. Related Areas Mol. Biol.* **65:** 287–346, 1992.
54. Bredt, D.S., and Snyder, S.H. *Annu. Rev. Biochem.* **63:** 175–195, 1994.
55. Förstermann, U., Closs, E.I., Pollock, J.S., Nakane, M., Schwarz, P, Gath, I, and Kleinert, H. *Hypertension* **23:** 1121–1131, 1994.
56. Griffith, O.W., and Stuehr, D.J. *Annu. Rev. Physiol.* **57:** 707–736, 1995.
57. Presta, A., Siddhanta, U., Wu, C., Sennequier, N., Huang, L., Abu-Soud, H.M., Erzurum, S., and Stuehr, D.J. *Biochemistry* **37:** 298–310, 1998.
58. Crane, B.R., Arvai, A.S., Gachhui, R., Wu, C., Ghosh, D.K., Getzhoff, E.D., Stuehr, D.J., and Tainer, J.A. *Science* **278:** 425–431, 1997.

59. Salerno, J.C., McMillan, K., and Masters, B.S.S. *Biochemistry* **35:** 11839–11845, 1996.
60. Feron, O., Michel, J.B., Sase, K., and Michel, T. *Biochemistry* **37:** 193–200, 1998.
61. Garcia-Cardena, G., Martasek, P., Masters, B.S.S., Skidd, P.M., Couet, J., Li, S., Lisanti, M.P., and Sessa, W.C. *J. Biol. Chem.* **272:** 25437–25440, 1997.
62. Aisaka, K., Gross, S.S., Griffith, O.W., and Levi, R. *Biochem. Biophys. Res. Commun.* **160:** 881–886, 1989.
63. Rees, D.D., Palmer, R.M.J., and Moncada, S. *Proc. Natl. Acad. Sci. USA* **86:** 3375–3378, 1989.
64. Holz, J., Giesler, M., and Bassenge, E. *Z. Cardiol.* **72:** 98–106, 1983.
65. Buga, G.M., Gold, M.E., Fukuto, J.M., and Ignarro, L.J. *Hypertension* **17:** 187–193, 1991.
66. Moncada, S., Palmer, R.M.J., and Higgs, E.A. *Pharmacol. Rev.* **43:** 109–142, 1991.
67. Buga, G.M., Wei, L.H., Bauer, P.M., Fukuto, J.M., and Ignarro, L.J. *Am. J. Physiol.* **44:** R1256–R1264, 1998.
68. Garthwaite, J., and Garthwaite, G. *J. Neurochem.* **48:** 29–39, 1987.
69. Garthwaite, J. *In* The NMDA Receptor, ed. by J.C. Watkins and G.L. Collingridge, pp. 187–205, Oxford University Press, Oxford, England, 1989.
70. Merrill, J.E., Ignarro, L.J., Sherman, M.P., Melinek, J., Lane T.E. *J. Immunol.* **151:** 2132–2141, 1993.
71. Ignarro, L.J., Bush, P.A., Buga, G.M., Wood, K.S., Fukuto, J.M. and Rajfer, J. *Biochem. Biophys. Res. Commun.* **170:** 843–850, 1990.
72. Bush, P.A., Aronson, W.J., Buga, G.M., Rajfer, J., and Ignarro, L.J. *J. Urol.* **147:** 1650–1655, 1992.
73. Rajfer, J., Aronson, W.J., Bush, P.A., Dorey, F.J., and Ignarro, L.J. *N. Eng. J. Med.* **326:** 90–94, 1992.
74. Bush, P.A., Aronson, W.J., Rajfer, J., Buga, G.M., and Ignarro, L.J. *Int. J. Impotence Res.* **4:** 85–93, 1992.

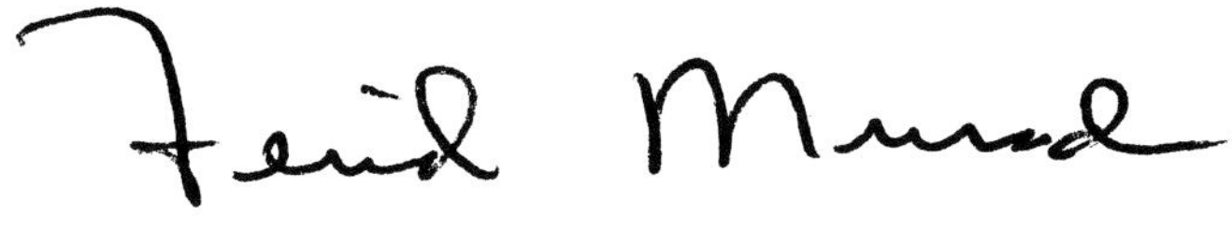

FERID MURAD

My father, Jabir Murat Ejupi, was born in Albania in 1892 and was the oldest of four children. His mother died when he was 13 years old. He and his family were shepherds and he subsequently ran away from home to sell candy in the Balkan countries as a teenager for several years. Although he had less than a year of education, he learned to speak seven languages before he died at the age of 84 in 1976. He met a group of other teenagers in Austria and they immigrated to the United States. The immigration officer at Ellis Island, August, 1913, asked his name, after which the officer declared him to be John Murad and stamped his papers. It was not uncommon to have names changed and abbreviated upon immigration. After working briefly in the steel mills and factories in Cleveland and Detroit, he settled in Chicago where he had several friends. His career was quite diverse and although he never admitted it, I learned subsequently from some of his colleagues that he was quite a playboy with fancy automobiles, perhaps the reason for my love of nice cars.

My mother, Henrietta Josephine Bowman, was born in 1918 in Alton, Illinois and was the third of six surviving children of Elizabeth Lillian and Andrew Orvie Bowman. My grandmother was a kind and wonderful woman. Only six of her eleven children survived due to stillbirths and some died of diseases and other conditions of poverty. My mother went to grade school for several years before she too quit to help her mother and younger siblings while her mother and two older sisters went to work. My grandfather was a carpenter who generally worked part-time and frequently spent his modest paycheck at the local bars before going home. The childhood poverty of both my parents and their minimal education did much to influence me and my two younger brothers in our education and career choices. One brother became a dentist and the other a professor of anthropology with a Ph.D. degree.

My mother also ran away from home at 17 in 1935 to marry my father who was 39. I was born September 14, 1936 at home in their hot and small apartment over a bakery in Whiting, Indiana. My brothers John Abderhaman and Turhon Allen were born in 1938 and 1944. We were raised in a four room apartment behind my parents' restaurant in Whiting, Indiana. This small apartment undoubtedly influenced my desire for large expensive homes.

The restaurant business had a profound effect on my future and that of my two brothers. When we were able to stand on a stool to reach the sink we washed dishes and later when we could see over the counter, we waited tables and managed the cash register. I did this throughout grade school and high school each evening and on weekends. I created a game from those chores and learned to memorize all of the customer's orders in our restaurant with a capacity of 28 customers and before they left I would tally their bills mentally and meet them at the cash register. I met a diverse and wonderful group of customers that ranged from laborers in the local refineries and steel mills to local bankers, businessmen, families and school teachers. My parents worked long hours as is typical of a family business, particularly a restaurant. My father worked 16 to 18 hours daily while my mother put in similar hours between the restaurant and raising three children. They owned the building that also included two other small apartments, another small business and 21 sleeping rooms upstairs. Many of the tenants were old and retired and my mother would often care for them and prepare their meals when they were sick. I learned from my mother and grandmother Bowman about compassion and generosity for people and this in turn influenced my career choice in medicine. My father taught me some business skills and how to repair numerous items that were continually breaking down in this old building. He was quite good at remembering how he took anything apart in order to repair it and reassemble the pieces as I stood at his side as a youngster passing him tools.

With this background I knew that I wanted considerable education so I wouldn't have to work as hard as my parents. Also, I knew at the age of 12 that I was going to become a doctor. My parents always encouraged us to get an education and establish a profession. However, my brothers and I grew up with considerable freedom whether it was saving or spending our tips from the restaurant or our career choices. This was also applied to our religious choices as my father was Muslim, my mother Baptist and we were raised in a Catholic community. Subsequently, my brothers became Catholic when they married Catholic wives and I was baptized Episcopalian in college. My wife of more than forty years is Presbyterian, two of our daughters married Jewish men and one married a Catholic man.

In eighth grade the class was asked to write an essay of our top three career choices. My choices were 1) physician, 2) teacher and 3) pharmacist (in 1948 clinical pharmacology was not yet a discipline in medicine). Today I do just that, as I am a board certified physician and internist doing both basic and clinical research with considerable teaching in medicine, pharmacology and clinical pharmacology and with a Ph.D. in pharmacology. While I am probably working much harder and longer hours than my parents, I certainly love my profession and have considerably more enjoyment and disposable income than they did. Until my graduation from high school only three of my cousins had finished high school and no relatives had ever gone to college. Grade school, middle school and high school were relatively easy for me and with little studying I was an honor student every semester graduating 5th in my high school class. Fortunately several high school teachers, some of whom frequented our restaurant, Jack Taylor in Spanish and history, LaDonna Thue Elson in art, Bernard Quebeck in music, Jesse Allen in math, and coach Peter Kovachic convinced me I had some potential and were wonderful counselors and advisers. I lettered in track and cross country as a distance runner in the one and two mile events and music. I also played football and basketball but spent most of my time keeping the bench warm. I played offense and defense left guard at 5' 11" and 140 pounds. After three monsters ran over the top of me I spent more of my energy with distance running in cross country. While I started to play golf in grade school, I stopped playing for many years during college and medical training and I continue to struggle with my game after I began playing again about 20 years ago.

There was one notable friend since kindergarten, Ronald Delismon, who influenced me considerably. We competed constantly with everything: grades, chess, fencing, sports, etc. Today he is an aeronautical engineer recently retired from Boeing. His projects were always top secret such as the stealth bomber and some of the star war defense projects. He would never discuss his work with me for security reasons and often joked with me by saying, "if I told you, I would then have to kill you". After 57 years we remain the best of friends and still compete, generally at golf, skiing and more pleasant encounters. His recent comment was, "one Nobel to zero".

The University of Chicago had a new program in the 1950s that accepted students after three years of high school and friends in the restaurant who were alumni from the University of Chicago encouraged me to apply. However, after considerable thought I decided not to enter college prematurely but rather completed my senior year in high school. In retrospect, this was the correct decision for me as my senior year in high school was wonderful. I coasted through the year with excellent grades and lots of fun participating in the school's chorus and took the lead in several operettas. This was probably the only year in school where I wasn't compulsive about grades and didn't study constantly.

Since my parents couldn't afford to help me with my college costs, I looked for a school that offered the best scholarship. I considered the military programs at the Naval Academy and Westpoint, but I knew I wouldn't have received the biology training for medical school since these were primarily engineering programs with a requisite four years of military duty afterwards. I competed successfully for a Rector Scholarship at DePauw University in Greencastle, Indiana, a small and excellent liberal arts university and went there from 1954 to 1958 on a tuition scholarship. The first year my grades were okay but not great with several A's, one C and the rest B's due to the hazing and distractions of being a pledge in the fraternity. In subsequent years my grades progressively improved as I was developing more self confidence and better study habits. I lived in "annexes", or small apartments with other fraternity brothers since the fraternity couldn't accommodate all of us and I generally chose other premeds as roommates. We often studied together and competed for grades. I was the scholarship chairman of the fraternity and remained a premed major with a second major in chemistry as I enjoyed both biology and chemistry. Throughout college I waited tables, taught the anatomy and embryology labs and worked one and sometimes two jobs during the summers to cover my expenses. If I had only one summer job I would take additional classes at one of the local extensions of Indiana University for additional math or literature classes in order to take more courses in biology, chemistry, physics or Greek and Latin at DePauw. The Greek and Latin courses in high school and college were of great value subsequently in learning the root derivatives of many scientific words.

In the spring of my junior year in 1957 on spring break in Fort Lauderdale, Florida, I met Carol Ann Leopold, my wife to be. She and her family were from St. Louis. We were at DePauw together where she was an English and Spanish major planning to become a teacher. Although she dated many of my fraternity brothers, I had not met her previously. After spring break we began to date and I gave her my fraternity pin a month later. Our dates were primarily "study dates" at the library (the only thing I could afford) and after mostly A's my senior year I was elected to Phi Beta Kappa. At Christmas we were engaged and married within several weeks of graduation from DePauw on June 21, 1958.

During my senior year of college I began to apply to medical schools and planned to go to Washington University Medical School in St. Louis. However, my faculty advisor Forst Fuller, a professor in the biology department and also my mentor during an elective research project to understand how fish managed calcium metabolism without parathyroid glands, suggested that I consider a new M.D.-Ph.D. program at Western Reserve University. A fraternity brother, Bill Sutherland, also advised that I consider this new combined degree program that his father Earl Sutherland, Jr. initiated in Cleveland in 1957. The program paid full tuition for both degrees and provided a modest stipend of $2000 per year. I quickly applied and was interviewed on a Saturday morning in February of 1958 by the entire Pharmacology Department. Needless to say, I was awed by the attention they gave me and decided immediately to accept their offer. Carol, my fiancé, was somewhat concerned that I was now planning seven more years of education but she has always been understanding and supportive of my training, career path and numerous moves around the country. The game plan was to have Carol teach high school English as I went through the combined degree program. These plans abruptly changed within three months when Carol became pregnant. After teaching for only one semester, she was asked to resign when the pregnancy "began to show". Subsequently, she was a substitute teacher, part time secretary and hospital clinic coordinator as we progressed with our family; four girls, including a set of identical twins before I finished medical school and graduate school in 1965. Number five, the first boy, was born as I finished my residency in 1967. Fortunately, we didn't stop as planned after number four was born.

As I entered the new combined degree program my mentors were Earl Sutherland, Jr. the chairman of the Pharmacology Department and Theodore Rall a new young assistant professor and collaborator of Sutherland's. The year before I arrived they had discovered cyclic AMP as a "second messenger" of epinephrine – and glucagon-mediated effects on glycogenolysis in liver preparations. My assignment was to show that the catecholamine effects on cyclic AMP formation were due to effects through the beta adrenergic receptor. Alquist had previously reported that adrenergic effects could be classified as alpha or beta depending on the relative potency of several catecholamines. The new and only beta adrenergic receptor antagonist, dichloroisoproterenol, had also been just described and was to become a useful antagonist in our work. We found that catecholamine effects on adenylyl cyclase activation in both heart and liver preparations were, indeed, due to beta adrenergic effects as shown by the relative potencies of *l*-isoproterenol, *l*-epinephrine and *l*-norepinephrine with inhibition by dichloroisoproterenol and failure of alpha blockers and agonists to have effects. I also found that acetylcholine and other cholinergic agents inhibited adenylyl cyclase preparations, the first description of hormones inhibiting cyclic AMP formation. I then became interested in agents that could block the effects of cyclic AMP on phosphorylase kinase and phosphorylase activation. This required some novel assays and an acquaintance with numerous cyclic AMP analogues and other nucleotides including cyclic GMP, cyclic IMP, cyclic CMP, etc. Many of these nucleotides and their analogues were synthesized by Theo Pasternak, a professor from Geneva who was on sabbatical collaborating with Sutherland and Rall. This work subsequently influenced my desire to work with cyclic GMP as described in my Nobel lecture. Later I again played organic chemist to make some nucleotides.

I was first in my class every year in medical school and graduate school. This was a wonderful and exciting time in my life working with these mentors, watching a new area of biology develop and actively participating in the work. I loved research as Earl Sutherland was quite a visionary who was able to bring together multiple disciplines and areas to apply to his work. Ted Rall taught how to do those fool proof "Sunday experiments" as we came to call them. It was on Sundays that I could design and conduct those large and complex experiments with all of Ted's required controls such that the data were "publishable". We and others in the department were able to determine that multiple hormones including catecholamines, cholinergics, ACTH, vasopressin, etc. could increase or decrease adenylyl cyclase activity and cyclic AMP formation. Prior to this the view of Sutherland was that receptors and adenylyl cyclase were a single macromolecule or a tightly associated complex in cell membranes. My work as a student and the work of others questioned this hypothesis and suggested that different receptors for this growing list of hormones must be coupled to adenylyl cyclase in yet to be determined complex ways (see Gilman's and Rodbells' Nobel lecture of 1994 for a greater description of their subsequent work).

I also enjoyed medical school and found myself learning everything presented before me. I knew that I couldn't determine what was to be true and important and many of our faculty acknowledged this as well. Since anything could be important, I began to learn everything taught. The new experimental integrated organ-system approach to medical education at Western Reserve permitted me to assimilate and integrate information more readily. I also thoroughly enjoyed my clinical rotations in medicine, surgery, OB-GYN, pediatrics, orthopedics, neurology, etc. There were few clinical rotations that I didn't think about as a possible discipline for my future academic career. I subsequently learned that I was at the top of the medical school and graduate school class each year and received prizes at graduation for both clinical medicine and research. I was in my element and loved it. There was no doubt in my mind about an academic career in medicine, research and teaching.

In order to supplement my stipend with so many children, I moonlighted at the Cleveland Clinic working one or two nights per week on the OB-GYN service to follow mothers with pelvic exams as they progressed through labor, assisted in deliveries and Caesarian sections and then scrubbed tables and floors after each delivery. All of this for $20.00 per night for 12 hours of work from 7:00 p.m. to 7:00 a.m. one or two nights per week for four years. On slow evenings I was able to study, analyze lab data and write research protocols. Some nights required that I work all night and then attend a full day of classes the next day. I continued this during my clinical clerkships requiring my absence from my family as often as 4 to 5 nights per week. However, I tried to have dinner with my family as often as my schedule permitted. My wife and children were very understanding. They grew up as wonderful children and adults in spite of my absence, obviously due to a devoted wife and mother. My current fetish is my 5 grandchildren who I try to spend as much time with as possible, undoubtedly due to my guilt as an absent father. I did manage to spend several weeks each summer with my family as we took them camping all over the U.S. to various scientific meetings. There are only a few states where we have not camped together as a family and they all became proficient swimmers at a young age.

I decided to go to Massachusetts General Hospital for my internship and residency in medicine (1965–67). What a wonderful experience this was with some of the worlds' leading scientists, teachers and clinicians. Our group of 14 housestaff included exciting bright minds such as Tom Smith, Tony Gotto, Jim Willerson, Ed Scolnik and others that had considerable influence on me. My attendings and chief residents included Alex Leaf, Dan Federman, Roman DeSanctis, Frank Austen, Sam Thier, Ken Shine and others. As a resident Joe Goldstein and Mike Brown were two of our interns. I couldn't have asked for a greater introduction to medicine in spite of being on call every other night and weekend. I did, however, miss the laboratory and each spring I found myself in the library reading many of the abstracts of the Federation meeting (currently FASEB meeting) to see what I was missing in "second messengers and hormone signaling". I generated a notebook that contained numerous "obvious experiments" to be done. When I subsequently went to NIH as a clinical associate in the Heart Institute I was able to do many of the planned experiments in Martha Vaughan's laboratory. She too was an excellent mentor with a style different from either Sutherland or Rall. She gave me considerable freedom to pursue a number of areas related to cyclic AMP and hormonal regulation. Her husband, the late Jack Orloff, while superficially a gruff and tough man, was a sensitive person and talented scientist. I was indeed fortunate that they and many others at NIH influenced my thinking and career planning. I soon learned that I had numerous role models and attempted to extract the best features of each as I planned my career path and future.

I remained at NIH for more than three years (1967–70) when the University of Virginia called to recruit me to develop a new Clinical Pharmacology Division in the Department of Medicine with an appointment as an Associate Professor in medicine and pharmacology. I couldn't resist the offer from Ed Hook, the new chairman of medicine and Joe Larner, the new chairman of pharmacology. Other faculty such as Tom Hunter, the Vice President of Medical Affairs, Ken Crispell the Dean, Bob Berne, Bob Haynes and others influenced my decision to leave NIH. I had known Larner, Berne and Haynes since they were faculty at Western Reserve when I was a student. Charlottesville was also an appealing place to raise my five children. Some colleagues around the country, particularly David Kipnis, another one of my role models, questioned me about going to Charlottesville. Just the previous year I called him to apply for a fellowship in endocrinology at Washington University. I was then 33 years old with 5 children and his advice was appropriate. He said, "Fred, time for you to get a job and support your family", and I took his advice to heart.

I joined the faculty at the University of Virginia, September 1, 1970 and nervously thought about how I could launch my own independent research career. I decided to work with cyclic GMP as it was beginning to emerge as a possible new "second messenger" to mediate hormone effects. This is detailed in my Nobel lecture. I remained at the University of Virginia from 1970 to 1981 where I was promoted as one of the youngest professors in 1975; I was also asked to become the Director of their Clinical Research Center in 1971 and the Director of Clinical Pharmacology in 1973. I built a research program with both clinical and basic studies and started to recruit many exciting students and fellows to work with me. Of the 82 fellows and students I have trained and collaborated with to date twenty are professors, chairmen, research directors and division chiefs around the world. I view them as offspring and keep in contact with most of them in my travels. There is no question that one of my greatest accomplishments is to have participated in the training of such successful scientists in my own laboratory and also influence the careers of many talented medical students, graduate students and housestaff.

After looking at many university positions around the country as a chair of medicine or pharmacology and industrial positions, I decided to go to Stanford in July 1981 as Chief of Medicine of the Palo Alto Veterans Hospital, a Stanford affiliated hospital. I was a professor of medicine and pharmacology and the associate chairman of medicine. While it was difficult to leave many friends and colleagues at the University of Virginia where we conducted the first experiments with the biological effects of nitric oxide, I couldn't turn down this exciting opportunity at Stanford. Ken Melmon was chairman of medicine and during our first three years together we recruited about 30 new young faculty. Inspite of the large administrative and clinical teaching demands, I continued to supervise a large and productive laboratory with about 15 students, fellows and staff. Trainees continued to come to our laboratory from all over the world. Some of my students and fellows subsequently went to medical school and after completing residencies have become very productive physician scientists at a number of institutions.

After a stint as Acting Chairman of Medicine at Stanford (1986–88), I left to become a Vice President at Abbott Laboratories as I was becoming concerned about managed health care on the horizon and its possible effects on patient care, research and education. After considering several industrial positions, I chose Abbott primarily because of its president Jack Schuler, a sales and marketing person with an MBA from Stanford who also had considerable vision. We worked well together as he taught me many business principles and I taught him about drug discovery and development. I enjoyed the access to all of Abbott's resources, scientific staff, instrumentation and what initially seemed like an unlimited research budget. I eventually learned that one can never have enough resources when one looks for novel therapies of major diseases; it's an expensive undertaking. Nevertheless, in four years of directing their pharmaceutical discovery and development programs we were able to discover many novel drug targets and we brought forward about 24 new compounds for clinical trials for various diseases. I continued to have a very productive lab with two NIH grants, some outside funding for fellows and about 20 scientists working with me on nitric oxide and cyclic GMP. The administrative demands and travel were considerable since I was a corporate officer, vice president and also overseeing many industrial collaborations around the world. When I left Abbott I was supervising about 1500 scientists and staff and probably earned the equivalent of an MBA from the experience on the job plus periodic management courses required by the company. Before my arrival at Abbott the company had no postdoctoral fellows or extramural funding. When I left we had about $3.5 mill. per year of extramural grant support and about 35 fellows in pharmaceutical research. Unfortunately, Abbott reorganized its senior management and my business roll models were asked to leave. As Abbott's senior scientist I found myself wedged between upper management, the marketing staff and the scientists and constantly was defending my decisions about the research programs. There were always considerable marketing pressures on me that in my opinion were often the wrong decisions to develop novel therapeutics for diseases without adequate therapy.

I left Abbott in 1993 to be a founder, President and CEO of a new biotech company, Molecular Geriatrics Corporation. The plan was to create another intensive research-based biotech company. Unfortunately, my investment banker never raised the amounts of money promised and he eventually lost a major personal fortune with his leveraging tactics. I found myself skipping around the world to find investors and partners to keep the company afloat and pay the bills. After a partnership with a major pharmaceutical company and some more financing as a private company, I left to rejoin academics, hopefully much wiser.

After considering a number of Vice President, Dean positions and Chairmanships, I realized that such positions would probably totally remove me from the laboratory, fellows and students, things I could not give up. In

April 1997, I became the University of Texas-Houston's first chairman of a newly combined basic science department, Integrative Biology, Pharmacology and Physiology. I am also creating a new Division of Clinical Pharmacology jointly between our department and medicine. I plan to continue an active basic and clinical research program and will participate in clinical medicine and teaching again. Thus, I have come full circle. I am back in my academic element again and I love it. I also expect to continue some business adventures and exercise my entrepreneurial skills, areas that I also enjoy and view as lucrative hobbies. The freedom and intellectual environment of academic medicine and bright young students and fellows are exciting and a daily joy for me. After all, I hope to tell Ron Delismon some day "Two Nobels to zero".

DISCOVERY OF SOME OF THE BIOLOGICAL EFFECTS OF NITRIC OXIDE AND ITS ROLE IN CELL SIGNALING

Nobel Lecture, December 8, 1998

by

FERID MURAD

Department of Integrative Biology, Pharmacology and Physiology, University of Texas Medical School at Houston, P.O. Box 20707, Houston, TX 77030, USA

ABSTRACT

The role of nitric oxide in cellular signaling in the past 22 years has become one of the most rapidly growing areas in biology with more than 20,000 publications to date. Nitric oxide is a gas and free radical with an unshared electron that can regulate an ever-growing list of biological processes. In many instances nitric oxide mediates its biological effects by activating guanylyl cyclase and increasing cyclic GMP synthesis from GTP. However, the list of effects of nitric oxide that are independent of cyclic GMP is also growing at a rapid rate. For example, nitric oxide can interact with transition metals such as iron, thiol groups, other free radicals, oxygen, superoxide anion, unsaturated fatty acids and other molecules. Some of these reactions result in the oxidation of nitric oxide to nitrite and nitrate to terminate its effect, while other reactions can lead to altered protein structure, function, and/or catalytic capacity. These diverse effects of nitric oxide that are either cyclic GMP dependent or independent can alter and regulate important physiological and biochemical events in cell regulation and function. Nitric oxide can function as an intracellular messenger, an autacoid, a paracrine substance, a neurotransmitter, or as a hormone that can be carried to distant sites for effects. Thus, it is a unique simple molecule with an array of signaling functions. However, as with any messenger molecule, there can be too little or too much of the substance and pathological events result. Some of the methods to regulate either nitric oxide formation, metabolism, or function have been in clinical use for more than a century as with the use of organic nitrates and nitroglycerin in angina pectoris that was initiated in the 1870's. Current and future research with nitric oxide and cyclic GMP will undoubtedly expand the clinicians' therapeutic armamentarium to manage a number of important diseases by perturbing nitric oxide and cyclic GMP formation and metabolism. Such promise and expectations have obviously fueled the interests in these signaling molecules for a growing list of potential therapeutic applications.

INTRODUCTION

As a combined degree student in the laboratories of Earl Sutherland and Theodore Rall shortly after the discovery of cyclic AMP, they taught me to enjoy science and I quickly became convinced of the important role of second messengers in hormonal signaling systems. After additional clinical training, I went to NIH in 1967 to work with Martha Vaughan and cyclic GMP began to emerge as another potential second messenger. Sutherland's laboratory and Aurbach's laboratory independently described the enzyme guanylyl cyclase that catalyzed its formation from GTP (1,2). Also, a new isoform of cyclic nucleotide phosphodiesterase that preferred to hydrolyze and inactivate cyclic GMP (3) and a subsequent novel cyclic GMP-dependent protein kinase, presumably its molecular target (4), were described (Figure 1).

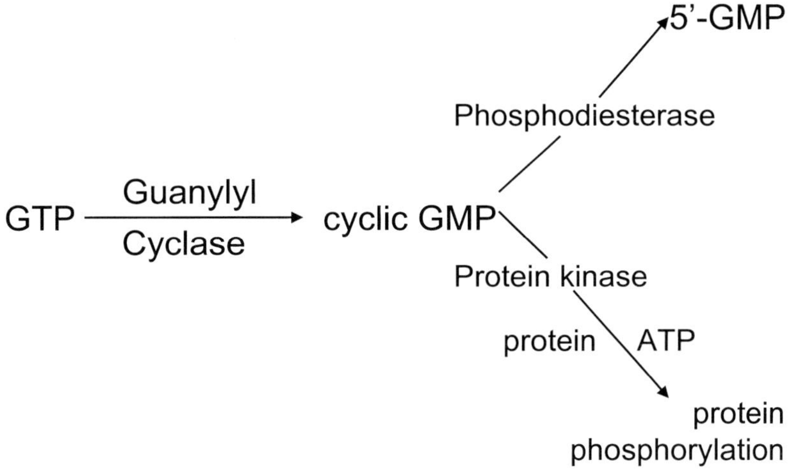

Figure 1. Enzymes involved in cyclic GMP synthesis, metabolism and function.

The cyclic AMP area was becoming very popular and crowded due to new and simpler assay systems and more and more hormones mediating their effects through this intracellular messenger. Thus, cyclic GMP became more appealing to me as I planned to launch an independent academic career when I moved to the University of Virginia in 1970. In our earliest studies, we did numerous descriptive experiments where we added one agent or another to tissues or cells to correlate cyclic GMP accumulation with the function of the preparation. While these early studies seemed necessary at this time and were productive, we gained little insight into the true regulation of cyclic GMP synthesis and its functions. We wanted to address two simple, and obviously, naïve questions: 1) how do hormones, neurotransmitters and various ligands regulate guanylyl cyclase activity and cyclic GMP synthesis? (i.e. what are the molecular coupling events between hormone binding to its appropriate receptor and guanylyl cyclase activation), and 2) what are some biological functions resulting from increased cyclic GMP levels?

POSSIBLE ISOFORMS OF GUANYLYL CYCLASE

After two to three years of what some of my trainees called "dumping experiments" where we added various hormones and agents to preparations and measured cyclic GMP accumulation, we decided that there must be a better fundamental approach to our questions. About 1973, Hiroshi Kimura, a fellow with me, and I turned our attention to an examination of guanylyl cyclase in cell-free systems. We quickly learned that there was enzyme activity in both high-speed supernatant and particulate fractions of most tissue homogenates. Furthermore, the activities were quite different (see Table 1) (5–7). While ATP inhibited both activities, the soluble activity was more sensitive to ATP inhibition. Calcium ion could either activate or inhibit activity dependent upon its concentration, but the soluble and particulate activates had different sensitivities to these effects. The most striking difference was that the soluble activity gave linear kinetics (double reciprocal plots) with regard to the substrate GTP, while the particulate activity gave curvilinear double reciprocal plots indicating cooperativity with respect to GTP with presumably multiple GTP binding sights. The Hill coefficient of the soluble enzyme was 1.0 while the particulate activity had a Hill coefficient of about 1.7 (5–7).

Table 1. Guanylyl cyclase in cytosolic and particulate fractions of homogenates with different properties.

	soluble	particulate
Ca^{++}	stimulates	inhibits
ATP, IC_{50}	0.4mM	>1mM
GTP, Hill coef.	1.0	1.74
detergent	stimulates 50-100%	stimulates >300%

From our subsequent work with the characterization and purification of the enzyme and the cloning of the cDNA's for guanylyl cyclase we were able to demonstrate definitively that several isoforms and separate gene products were present in tissues (8–11). Similar studies have since been done by other laboratories, most notably in the laboratory of David Garbers (12, 13). Today we know that numerous soluble and particulate isoforms can exist. This topic is beyond the scope of this review and interested readers are referred to other reviews (8–13).

EFFECTS OF AZIDE, HYDROXYLAMINE AND NITRITE

Our initial reaction to these early data was that we were dealing with two different isoforms of guanylyl cyclase, one in the cytosolic compartment and one in the membranous compartment or organelles. However, we could not exclude that such differences in the activities were due to artifacts and spurious results with crude high-speed soluble and particulate preparations. We thought that contaminating phosphatases, nucleotidases and perhaps cyclic nucleotide phophodiesterases might also explain our observations. At the time we had no desire or plans to purify the protein(s). A simpler approach was to add azide, pyrophosphate, hydroxylamine, phosphodiesterase inhibitors, fluoride, etc. individually or in combination to preserve the substrate or product and recharacterize the crude soluble and particulate guanylyl cyclase. Quite surprisingly, we found that azide, hydroxylamine and nitrite activated many, but not all of preparations of guanylyl cyclase (14–17). This was an exciting serendipitous observation for us, because any hormones or ligands that increased cyclic GMP accumulation in intact cells failed to increase cyclic GMP synthesis in cell-free preparations. We, therefore, thought that hormone-receptor coupling to guanylyl cyclase activation would be a complicated molecular event that was disrupted in cell-free preparations.

In order to establish the molecular events between hormone binding and guanylyl cyclase activation, it was necessary to have a hormonally responsive cell free system and ultimately purified and reconstituted components of the signaling cascade. We thought that an understanding of the activation of guanylyl cyclase with azide, hydroxylamine and nitrite could assist us in unraveling the complex signaling cascade, perhaps analogous to the stimulatory effects of fluoride that assisted with the characterization of the adenylyl cyclase system. This intuition was correct in ultimately leading us to the mechanism of hormonal activation of guanylyl cyclase as summarized below. Thus, we made a major commitment to understand the mechanisms of guanylyl cyclase activation by these agents.

DISCOVERY OF PROTEINS REQUIRED FOR AZIDE ACTIVATION AND INHIBITION

The activation by azide was tissue specific and had a brief time lag of several minutes before the time course became linear (14–17). Furthermore, the activation was absent in an anaerobic environment requiring air or oxygen and could be enhanced with the addition of thiols (14–17). These results suggested to us that azide might be converted to another compound that served as the activating species. We then did a simple and classical biochemical experiment where we mixed crude soluble cell extracts that were activated by azide with crude soluble extracts that were not activated.

In our first experiment (Table 2) we mixed crude supernatant fractions of rat liver that could be activated with azide with fractions of rat brain and rat heart, tissues without azide effects on guanylyl cyclase (14–17). Mixing extracts of liver with cerebral cortex resulted in an increased azide effect, while mixing liver extracts with heart extracts resulted in a loss of the azide effect. Our interpretation of these data was that azide activation required the presence of other factors present in liver extracts that when added to cerebral cortex permitted azide to activate the enzyme from both tissue extracts. Furthermore, heart extracts possessed a factor(s) that blocked azide activa-

Table 2. Effects of sodium azide on soluble guanylyl cyclase from rat liver, heart and cerebral cortex (with permission).

Enzyme	Cyclic GMP $-NaN_3$ (pmol/min/mg protein)	Cyclic GMP $+NaN_3$ (pmol/min/mg protein)	Ratio $+N_3/-N_3$
Liver	38.8	595.4	15.3
Heart	23.0	23.2	1.0
Cerebral cortex	46.0	42.0	0.9
Liver + heart	27.3	23.1	0.8
Liver + cerebral cortex	23.0	899.0	39.1

tion in the heart extracts and in liver extracts in the mixing experiments. Subsequent experiments proved these hypotheses to be true. The apparent activating factor(s) in liver and the inhibitor(s) in heart were nondialyzable and heat labile. Thus, we began to purify and characterize these activators and inhibitors (14–21).

We thought that if we thoroughly understood the mechanism of azide activation of guanylyl cyclase, perhaps some day we could reconstitute the appropriate components and factors to observe hormonal activation of guanylyl cyclase in cell-free preparations, a major goal of the laboratory. This was a major research commitment that, while risky, turned out to be the correct approach and that led us to the discovery of the first biological effects of nitric oxide. Our University of Virginia colleagues and others thought we were crazy to waste our efforts in pursuing the effects of azide, a known metabolic poison that was often added to stored buffers and columns to inhibit bacterial growth.

CHARACTERIZATION OF THE AZIDE ACTIVATORS AND INHIBITORS

Hiroshi Kimura, Chandra Mittal and I purified rat liver extracts and found that the macromolecular factor required for azide activation was catalase (14–21). Other heme-containing proteins such as horseradish peroxidase and some of the cytochromes (18–21), however, could substitute for the catalase requirement. The inhibitory factors in heart extracts were also purified and were found to be hemoglobin and myoglobin (14–21). It rapidly became apparent that heme-containing proteins either permitted azide to activate guanylyl cyclase or blocked its activation. These stimulatory and inhibitory effects of various heme-containing proteins were important findings in that they led us to some much older literature describing the interactions of azide with catalase to generate nitric oxide.

EFFECTS OF AZIDE AND NITROVASODILATORS ON GUANYLYL CYCLASE ACTIVATION, CYCLIC GMP ACCUMULATION IN INTACT CELLS, AND SMOOTH MUSCLE RELAXATION

Azide, hydroxylanine and nitrite not only activated guanylyl cyclase in cell-free extracts but also increased cyclic GMP accumulation in numerous intact tissues and cell types including brain, liver and some cell culture models (14–21). One of the new fellows in my laboratory, Shoji Katsuki, was working with bovine tracheal smooth muscle preparations that we developed to examine the role of cyclic AMP and cyclic GMP in smooth muscle motility. We developed a rather homogenous smooth muscle preparation (about 90% smooth muscle) with a sufficient amount of tissue that permitted us to monitor motility in an organ bath after adding various agents. After quick freeze-clamping of tissue segments in liquid nitrogen, we could also measure cyclic nucleotide levels or evaluate adenylyl or guanylyl cyclase activities (22–25). It was obvious to us that we should also add these new guanylyl cyclase activators to bovine tracheal smooth muscle segments and evaluate motility and cyclic GMP accumulation. We found that these agents as with other tissues increased cyclic GMP levels as expected and simultaneously caused relaxation in precontracted muscles (22–25). Similar results were observed when we examined clean gastrointestinal smooth muscle segments (22–25). We intentionally avoided vascular segments due to the heterogeneity of the preparations with endothelial cells, blood cells, fibroblasts, etc.. We felt that we could not determine what cell types the cyclic GMP accumulation would occur in due to the marked cellular heterogeneity. In my previous studies with rat fat pads in Martha Vaughan's lab we had great difficulty correlating the effects of some known lipolytic agents with cyclic AMP accumulation due to the cellular heterogeneity of our preparations. The studies only become meaningful and helpful when we examined cyclic AMP accumulation in homogeneous isolated fat cells (26–28).

After finding that these agents relaxed tracheal and gastrointestinal smooth muscle and that the increase in cyclic GMP was coincident with or preceded relaxation, we began to examine other smooth muscle relaxants such as nitroglycerin, nitroprusside, hydrazines, etc. (See Figure 2) (19–25). All of these agents increased cyclic GMP levels in several smooth muscle preparations, as well as other tissues, caused smooth muscle relaxation and activated guanylyl cyclase in most cell-free tissue extracts. We began to call these new guanylyl cyclase activators "nitrovasodilators," a simplified "lab jargon" that permitted us to communicate more readily. While these new nitrovasodilators did not require catalase to activate guanylyl cyclase preparations, these effects were inhibited by hemoglobin and myoglobin. The effects of altered redox conditions, thiols and heme-containing proteins with this growing list of nitrovasodilators were to become very important for us later to determine the effects of endothelial-derived relaxant factor (EDRF) on cyclic GMP accumulation in vascular preparations subsequent to Furchgott's discovery of EDRF in 1980 (29).

DISCOVERY OF THE BIOLOGICAL EFFECTS OF NITRIC OXIDE

The effects of the growing list of "nitrovasodilators" with nitro or nitroso functionalities and the effects of heme-containing macromolecules as factors for either activation or inhibition of these agents suggested to us that perhaps the active intermediate or proximal activator of guanylyl cyclase could be nitric oxide or one of its oxidized or reduced products. This was indeed the case (20–25, 30). The formation of nitric oxide from various nitrovasodilator precursor molecules or "pro-drugs" could be influenced by a variety of conditions as noted in Figure 2. Furthermore, the activity of nitric oxide could also be diminished by a variety of oxidants such as methylene blue, $K_3Fe(CN)_6$, excess oxygen, etc. as well as sinks to trap nitric oxide such as hemoglobin and myoglobin (Figure 2). Delightfully, our first chemical generator to produce nitric oxide and ventilate the gas into guanylyl cyclase incubations was a success (22–25, 30) (Figure 3). The experiment was conducted by Shuji Katsuki and William Arnold, post-doctoral fellows in my laboratory the evening of December 2, 1976, the night before Katsuki was to complete his fellowship and return with his family to Japan. I have often thought that had Katsuki completed his fellowship earlier, our discovery of the first biological effects of nitric oxide could have been delayed considerably. Nitric oxide unlike some other activators and nitrovasodilators increased the activity of most guanylyl cyclase preparations and increased cyclic GMP levels in virtually all tissues tested with few exceptions (Table 3). Furthermore, the stimulatory effects of nitric oxide and various nitrovasodilators on cyclic GMP synthesis were not additive suggesting that they all shared a similar mechanism of activation of guanylyl cyclase (30).

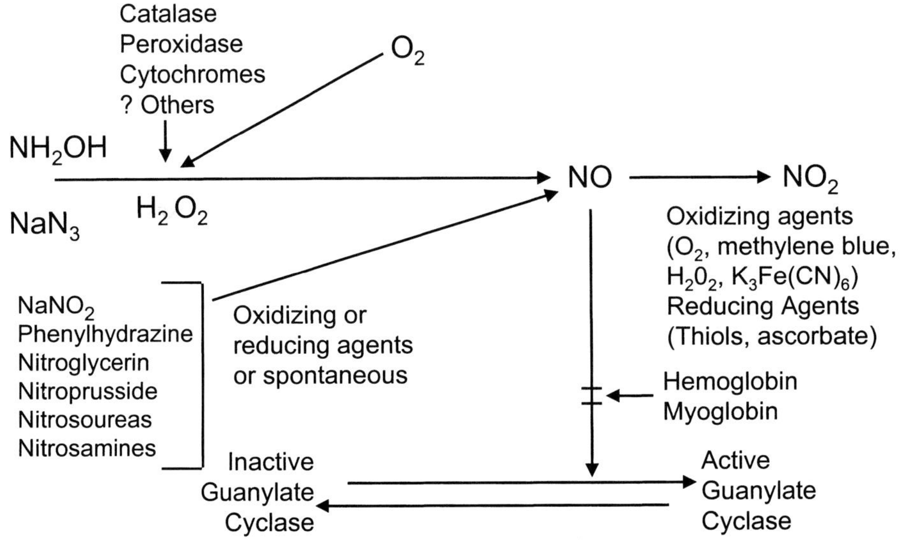

Figure 2. Effects of some nitrovasodilators on cyclic GMP synthesis. (with permission) [19].

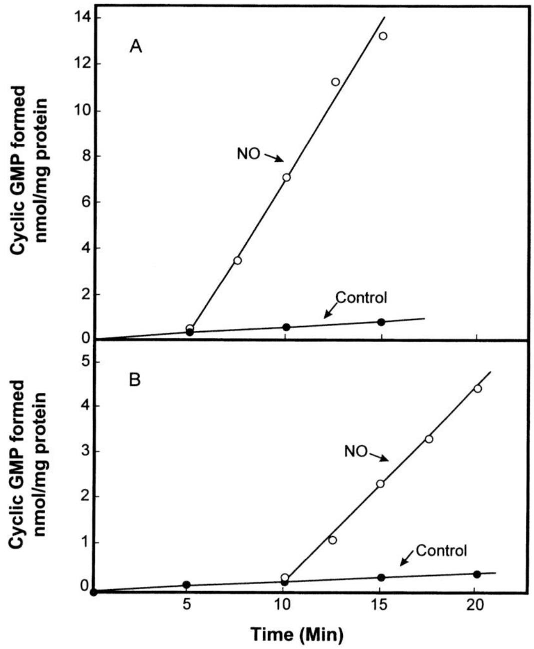

Figure 3. The first demonstration of the biological effect of nitric oxide to activate guanylyl cyclase preparations from rat lung (panel A) and bovine tracheal smooth muscle (panel B). (with permission) [22].

Table 3. Effect of nitric oxide on guanylyl cyclase activity from various tissues.

Tissue	Fraction	Cyclic GMP formed, pmol/mg protein per min		
		-NO	+NO	Ratio +NO/-NO
Rat liver	S	22.1	674.2	30.5
	P	12.4	37.3	3.0
Bovine lung	S	111.7	3625.5	32.5
Bovine tracheal	S	8.9	297.1	33.4
smooth muscle	P	27.0	35.3	1.3
Rat heart	S	10.5	242.8	23.1
Rat kidney	S	51.6	975.9	18.9
Rat cerebral cortex	S	55.7	1122.6	20.2
	P	14.2	209.1	14.7
Rat cerebellum	S	23.7	784.4	33.1
	P	20.6	201.2	9.8
Rat skeletal muscle	S	6.1	84.0	13.8
Rat spleen	S	73.5	381.7	5.2
Rat small intestinal muscle	S	42.9	250.6	5.8
Rat adrenal	S	37.5	394.2	10.5
Rat epididymal fat	S	9.4	108.5	11.5
Rat liver	DEAE-cellulose	296.2	1406.3	4.7
Rat cerebral cortex	DEAE-cellulose	57.2	174.3	3.0
Bovine lung	Sephadex G-100	61.4	1672.1	27.2
Rat heart	Sephadex G-100	30.2	171.9	5.7

Since this was the first and only example of a free-radical activating an enzyme, we committed ourselves to purifying the isoforms of guanylyl cyclase to homogeneity in order to reexamine the effects of nitric oxide. We could not exclude the possibility that nitric oxide was being converted to yet some other activating species or that it was not inhibiting an inhibitor of the enzyme in our crude preparations. The inhibition of a guanylyl cyclase inhibitor could appear in our assays as a possible activation due to disinhibition.

After purifying the soluble guanylyl cyclase to homogeneity, nitric oxide activation was still observed (21,31). In fact, the apparent K_{act} for nitric oxide continued to decrease as we purified the enzyme. Purified soluble guanylyl cyclase has a K_{act} or EC50 for nitric oxide in the order of 1 to 10 nM depending upon the conditions of the assay and the presence of other materials such as thiols, proteins, sugars, lipids, etc. that can form complexes with nitric oxide and alter its effective concentration to activate guanylyl cyclase (31). These later agents can act as traps or scavengers to form nitro or nitroso adducts and complexes to increase the K_{act}/EC50. In some cases these complexes can themselves also become nitrovasodilators or nitric oxide "prodrugs" to re-release nitric oxide under favorable conditions.

IS NITRIC OXIDE A NATURAL ENDOGENOUS GUANYLYL CYCLASE ACTIVATOR?

We proposed in 1978 that nitric oxide formation from some endogenous precursor could perhaps explain hormone, autacoid and neurotransmitter effects on cyclic GMP synthesis in intact cells and tissue (19, 21). We thought that perhaps appropriate hormone treatment could influence the redox state of cells and the formation of nitric oxide from an endogenous precursor or the metabolism of nitric oxide. We also thought that nitric oxide could be an intracellular second messenger to mediate the effects of hormones. Thus, it could perhaps join the short but growing list of second messenger molecules that included cyclic AMP, cyclic GMP, Ca++, eicosanoids, lipids, proteins, etc. Again our colleagues thought we were crazy to believe that a reactive free radical could activate an enzyme and function as a second messenger. After all, nitric oxide was a known pollutant from the combustion of numerous substances that inactivated or destroyed various macromolecules and depleted the ozone layer. It surely could not be a second messenger signaling molecule we were told by the skeptics.

In order to prove this hypothesis it would require a major technical advance. If, indeed, nitric oxide was an endogenous second messenger and regulator of guanylyl cyclase, one would expect that tissue levels would be about 1 to 10 nM (levels near the K_{act} or EC50) in order to be physiologically relevant and permit the fluctuations in nitric oxide concentrations to control enzyme activity. If the concentrations in tissue were much higher, one would expect the enzyme to always be activated and the physiological significance of nitric oxide as a regulator would be less convincing and meaningful. Unfortunately, the assays for nitric oxide and its oxidation products (nitrite and nitrate) were crude colormetric and spectrophotometric assays that were several orders of magnitude less sensitive than needed (about millimolar to micromolar). Thus, our hypothesis remained untestable for several years until we and other developed more specific and sensitive assays for the free radical.

DISCOVERY OF EDRF

In 1980 Furchgott presented a seminar in our Pharmacology Department while I was at the University of Virginia. He was excited about his recent discovery of endothelium-derived relaxing factor (EDRF) and the ability of endothelial cells to generate and release a labile substance that produced relaxation of the underlying smooth muscle in vascular segments (29). The effects of EDRF shared many of the features and properties of nitrovasodilators and I suggested to him that perhaps increased cyclic GMP levels could explain his observations with light-induced and endothelium-induced relaxation. We even planned a collaboration to test this hypothesis which never materialized because of our subsequent move to Stanford University in 1981.

At the time, Robert Rapoport was one of my new fellows working with tracheal and vascular smooth muscle preparations and impatiently waited for the collaboration to begin. After a brief visit to our new laboratory by the late Michael Peach in 1982 and his queries about our interests in EDRF, we then proceeded to demonstrate independently that EDRF did, indeed, increase cyclic GMP synthesis in the smooth muscle compartment of rat aorta segments (32–38) (Figure 4). We observed similar effects of EDRF on cyclic GMP formation with a variety of endothelium-dependent vasodilators such as

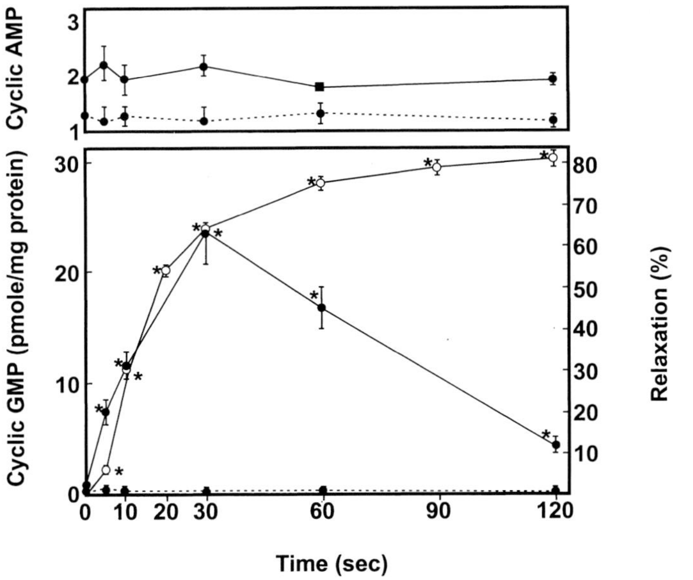

Figure 4. Effects of acetylcholine on cyclic GMP and cyclic AMP accumulation (solid symbols) and relaxation (open symbols) of rat aorta segments with (solid lines) and without (dashed lines) endothelium. (with permission) [32].

acetylcholine, A23187, thrombin, ATP, bradykinin, etc. (32–38). We continued the work to show that EDRF formation as well as nitrovasodilators also increased cyclic GMP-dependent protein kinase activity and altered protein phophorylation of many endogenous smooth muscle proteins including the dephosphorylation of myosin light chain (32–38).

Cyclic GMP accumulation with nitrovasodilators or endothelium-dependent vasodilators produced similar effects on cyclic GMP accumulation, cyclic GMP-dependent protein kinase activation and altered $^{32}PO_4$ incorporation in the same family of proteins isolated with two dimensional polyacrylamide gel electrophoresis (32–38). Subsequently, we and others found that cyclic GMP decreased phosphoinoside metabolism and the formation of inositol phosphates including inositol trisphosphate by decreasing phospholipase C activity (39). These and other effects of cyclic GMP would be expected to decrease cytosolic free calcium and the activity of myosin light chain kinase, a calcium/calmodulin-dependent enzyme. The effect of cyclic GMP elevating agents to decrease cytosolic free calcium levels in smooth muscle and other tissues has been subsequently shown by many laboratories.

Because of the biochemical similarities of the effects of nitrovasodilators and EDRF producing agents, we came to view EDRF as the "endogenous nitrate or endogenous nitrovasodilator" (40). Shortly afterwards, Furchgott and Ignarro independently proposed that EDRF is nitric oxide at a meeting at the Mayo Clinic in the summer of 1986. I previously and continually have argued with them and others that EDRF is probably a family of nitric oxide adducts or complexes as well as nitric oxide. This controversy remains unresolved even today in that many nitro and nitroso complexes can meet the definition of EDRF as originally defined by Furchgott (29). These complexes are readily formed between free nitric oxide in cells and the interstitial space with various cellular constituents such as thiols, lipids, proteins, etc. Presumably, many of these substances can readily dissociate their nitric oxide in the correct environment to account for the free nitric oxide found in some vascular preparations. Unfortunately, this pressing issue may not be resolved in the near future due to the low concentration of EDRF expected. As noted above, nitric oxide at concentrations of 1 to 10 mM are sufficient to activate guanylyl cyclase and permit the signaling event. Clearly endothelial cells and many other cells produce nitric oxide and the activator of the smooth muscle guanylyl cyclase is nitric oxide. The question outstanding is: what are the intervening or intermediate molecules presumably complexed with nitric oxide that Furchgott has called EDRF?

We and a number of laboratories showed that the effects of EDRF could be blocked with methylene blue, hemoglobin and other inhibitors of nitrovasodilator activation of guanylyl cyclase (see Figure 2). The effects of nitric oxide or EDRF can also be mimicked by atriopeptins which selectively activate particulate guanylyl cyclase (41,42) (Figure 5). In fact, we showed that atriopeptin receptors were heterogeneous with two receptor subclasses (ANF-R1 and ANF-R2) (43, 44). One of the atriopeptin receptors (ANF-R1) is the extracellular domain of the transmembrane particulate guanylyl cyclase (45).

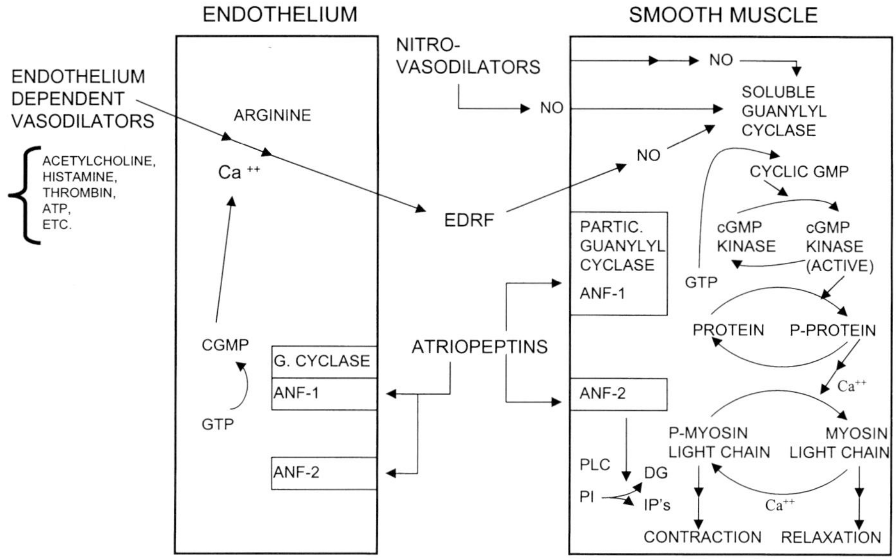

Figure 5. Effects of endothelium-dependent vasodilators, nitrovasodilators and atriopeptins in rat aorta segments. PLC is phopholipase C; PI, phosphoinositides; DG, diacylglycerol; IP's, inositol phosphates. (with permission) [50].

This work has been elegantly confirmed by the cDNA cloning studies from Garbers' laboratory (46).

EFFECTS OF L-ARGININE ON GUANYLYL CYCLASE ACTIVATION AND NITRITE AND NITRATE FORMATION

Degucci's laboratory was also one of the earlier few laboratories interested in guanylyl cyclase. In 1982, Degucci's laboratory reported that brain and/or neuroblastoma extracts possessed an endogenous substance that activated crude or partially purified preparations of soluble guanylyl cyclase (47). Furthermore, this substance was identified as L-arginine. Activation by L-arginine was blocked by hemoglobin and methylhydroxylamine, which blocked the effects of nitrovasodilators. We confirmed their observations with crude or partially purified preparations of guanylyl cyclase but found that L-arginine failed to block our highly purified guanylyl cyclase preparations (our unpublished observations). These observations were the first to report arginine activation of guanylyl cyclase with effects similar to those of nitric oxide and nitrovasodilators. Unfortunately, neither they nor we pursued these arginine effects at the time which could have led to the earlier discovery of nitric oxide synthase.

Subsequently Hibbs' laboratory in 1987 found that the cytotoxic effects of macrophages on tumor cells in culture correlated with the accumulation of nitrite and nitrate in the conditioned media (48). The cytotoxic effects and accumulation of nitrite and nitrate were increased with L-arginine and blocked with L-arginine analogues such as L-N-methyl arginine, an agent that later proved to be an effective competitive inhibitor of nitric oxide synthase.

The findings of Degucci and Hibbs were important, timely, and critical observations that subsequently permitted our laboratory and many other laboratories to move forward rapidly in understanding the role of nitric oxide in cellular signaling (49–52). Unfortunately, their laboratories have not received sufficient recognition for these important observations.

CHARACTERIZATION OF NITRIC OXIDE SYNTHASE AND ITS ISOFORMS

Within several years many laboratories including our own began to characterize, purify and clone this novel synthetic pathway that converted L-arginine to nitric oxide and citrulline. Initially the most active and productive laboratories included Marletta and Stuehr, Palmer and Moncada, Bredt and Snyder, Stuehr and Nathans, and our own laboratory (see 9–11 and references therein).

The first nitric oxide synthase (NOS) isoform to be purified was the neuronal or brain NOS or NOS-1, also called Type 1 NOS or constitutive NOS (Table 4). The first investigators to purify and characterize NOS-1 were Bredt and Snyder (53) and our own laboratory (54–57). This was followed shortly thereafter by NOS-2 or inducible NOS, also Type II NOS by Stuehr and Nathans (58), and then NOS-3 or endothelial NOS, Type III NOS in our lab (59–62). Monoclonal and polyclonal antibodies to the purified isoforms and synthetic peptide fragments were produced as were the cDNA clones for these three gene products (see 9–11 and the references therein). The chromosomal location for each isoform has been identified as have the numerous cosubstrates, cofactors and prosthetic groups (O_2, NADPH, FMN, FAD, tetrahydrobiopterin, and heme) (see 11 and references therein). All three isoforms are regulated by calmodulin but only NOS-1 and NOS-3 have

Table 4. Nitric oxide synthase isoforms.

NOS-1 (155kD)	neuronal, brain, Type I-NOS; central and peripheral neurons, NANC neurons, islets, endometrium, skeletal muscle, etc.
NOS-2 (125kD)	inducible, Type II-NOS; macrophage, liver, smooth muscle, endothelium, heart, etc; effects of LPS, cytokines and glucocorticoids
NOS-3 (135kD)	endothelial, Type III-NOS; endothelium, brain, heart, etc.; acylation, phosphorylation

apparent and obvious regulation with altered cytosolic free calcium concentrations. NOS-2 contains calmodulin as a tightly bound component or subunit after translation and assembly such that dependency upon cytosolic free calcium levels is not demonstrable. The catalytically active isoforms exist as homodimers with a heme prosthetic group serving to facilitate dimer formation. The carboxy terminal domain has considerable homology between the isoforms, and is homologous to cytochrome P450. The amino terminal domain has less homology and may serve to bind to regulatory proteins and/or chaperones for its subcellular compartmentation. The homology of the three isoforms is only about 50 to 60% while the homology of a given isoform between species can be as great as 85 to 92%. Some of the properties of the NOS isoforms are summarized in Table 4.

NOS converts L-arginine to L-hydroxyarginine and subsequently to nitric oxide and citruline as summarized in Figure 6. The transfer of electrons for

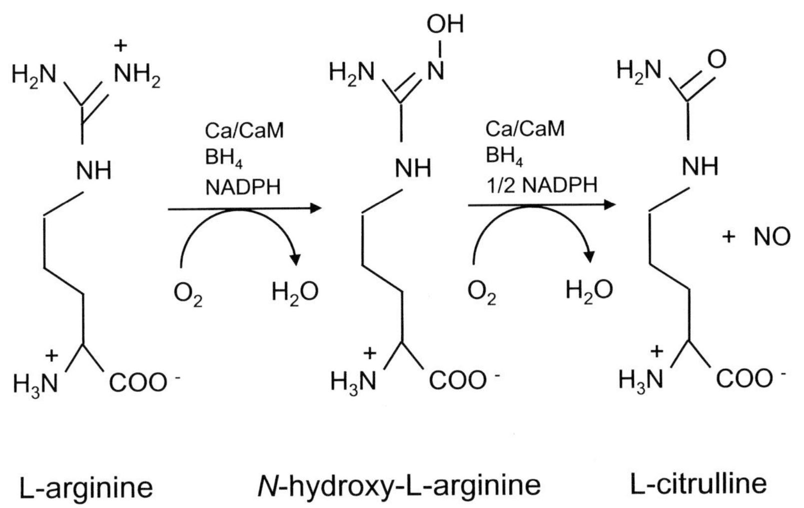

Figure 6. The nitric oxide synthetic pathway.

the oxidation of the guanadino nitrogen of L-arginine and the precise role of each cofactor remains unresolved. This continues as an active area of investigation in many laboratories. Recent x-ray diffraction studies with crystals of some NOS isoform fragments are currently underway in several laboratories. These studies should shed some light upon the precise roles of the cofactors and prosthetic groups. These studies may also permit the development of highly selective and effective inhibitors of each of the NOS isoforms in order to determine other important biological roles of nitric oxide and develop new and important therapeutic agents (see below).

Presumably the *in situ* activity of NOS can be regulated by the availability of substrate, cofactors and prosthetic groups. However, to date most of the regulation of NOS-1 and NOS-3 has focused on the alterations of cytosolic free calcium caused by a myriad of hormones, autacoids, neurotransmitters, etc. (see Figure 7).

NOS-2 does not appear to be present under normal conditions in most cells. However, upon exposure to endotoxin (LPS) interferon γ, IL 1, TNF α and other proinflammatory cytokines, the levels of mRNA, protein and catalytic activity begin to increase within one hour and reach maximal levels in 6 to 18 hours. Various anti-inflammatory cytokines and glucocorticoids can decrease the induction of NOS-2. It is thought that some of the pharmacological and biological effects of these factors may be explained by increased or decreased nitric oxide formation. For example, the hypotension observed in septic shock as well as the ensuing "multiple organ failure syndrome" (MOF) may be due to excessive nitric oxide formation from NOS-2 induction.

TRANSLATIONAL AND POSTTRANSLATIONAL MODIFICATIONS OF NOS

All of the isoforms can probably be phosphorylated by a variety of protein kinases including cyclic AMP-dependent protein kinase, cyclic GMP-dependent protein kinase, protein kinase C and Ca++/calmodulin-dependent protein kinase (63). With phosphorylation, the activity of the enzyme can be increased or decreased. However, the physiological relevance of phosphorylation to regulate enzyme activity or enzyme translocation from one cellular compartment to another is unknown (9–11 and references therein). In some cell-free systems with purified NOS-1, as many as 6 to 7 moles of phosphate

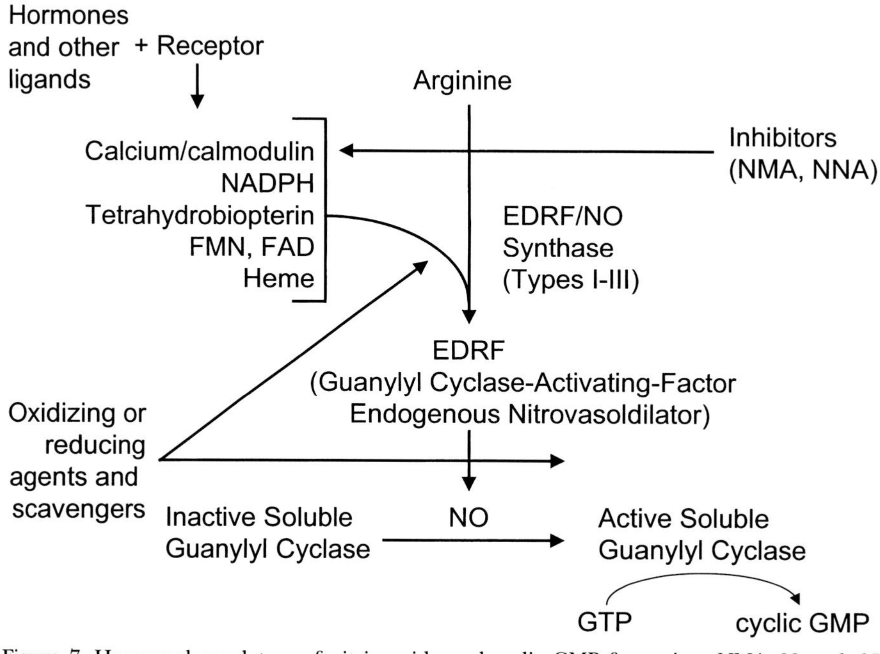

Figure 7. Hormonal regulators of nitric oxide and cyclic GMP formation. NMA, N-methyl-L-arginine; NNA, N-nitro-L-arginine. (with permission) [9].

can be incorporated into one monomer of protein (63). Studies with intact cells and tissues with permeable and specific protein kinase inhibitors are required to determine the physiological relevance of NOS phosphorylation with regard to catalytic activity, enzyme compartmentation, enzyme regulation and participation in various biological processes. Unfortunately, selective and permeable protein kinase inhibitors are not widely available.

NOS-3 is also myristoylated and palmitylated and acylation probably participates in the location of this isoform in the plasma membrane and/or other organelles (61, 64–66). Recently, Michels' laboratory has described the association of NOS-3 with the caveoloae of the plasma membrane (65, 66). These structures also contain many other proteins participating in cell signaling events. Mechanisms to regulate and chaperon NOS-3 from the Golgi to the caveolae and the recycling of NOS-3 to the cytosol and perhaps back to the plasma membrane are areas of active investigation. These processes could perhaps serve as important sites for regulation and drug targeting.

The three gene products of NOS described to date, possible alternate mRNA splicing due to multiple promoters (i.e. NOS-1) (67) and various post-translational modifications should permit a multitude of NOS isoforms and activities for regulation. Undoubtedly this will be a complex problem to resolve with regard to NOS functions, regulation and selective inhibition (see below). Some cells possess only one isoform while other cells have several isoforms with different subcellular locations and perhaps different functions. The precise location of pools of nitric oxide in cells and their proximity to guanylyl cyclase, thiols, fatty acids, heme-containing scavengers or sinks, other free radicals such as superoxide anions, etc. should have marked and profound effects on the biological and/or cytotoxic properties of nitric oxide. Thus, numerous questions regarding nitric oxide formation, location, and function remain to be addressed.

NOS INHIBITORS

Numerous NOS inhibitors have been described. Novel arginine based and guanidine based antagonists are areas of active investigation in both academic and industrial laboratories (see 9–11 and references therein). Most inhibitors to date are competitive antagonists and may be partly selective for one or another isoform. Commercially available compounds may show as much as 100- to 200-fold selectivity for one or another isoform. Unfortunately, to date, there are no highly specific inhibitors of one or another isoform. Such agents could prove invaluable to sort out the NOS isoforms in various cell types and the biological events that they regulate. Furthermore, investigations in this area could result in highly selective and perhaps specific and efficacious therapeutic agents without numerous side effects since one or more isoforms of NOS is found in virtually all cells with few exceptions. Currently, some NOS inhibitors are in clinical trials for patients in septic shock. Other clinical studies are under discussion and consideration.

Unfortunately, the earliest clinical studies with NOS inhibitors were not sufficiently selective as expected from previous *in vitro* studies with the inhibition of NOS isoforms. Thus, the initial enthusiasm for the clinical utility of NOS inhibitors is beginning to meet with some concerns and skepticism as unwanted side effects become apparent, some of which were predictable from earlier basic biological studies. Perhaps this could serve as a lesson regarding the selection of nonspecific agents for clinical development prematurely. None the less, many believe that these molecular targets will prove to be valuable approaches to some diseases with excess nitric oxide production. However, more selective agents are necessary due to the broad distribution of the many NOS isoforms and their important and diverse roles in biological regulation.

NO DONORS OR PRODRUGS IN CLINICAL MEDICINE

A number of NO prodrugs have been used successfully since the use of nitroglycerin and other organic nitrates for angina pectoris more than a century ago. Nitroglycerin was discovered in 1847, and it and other organic nitrates were first used clinically in the 1870s. Ironically, Alfred Nobel, who discovered how to safely formulate dynamite from nitroglycerin, was prescribed nitroglycerin for his angina pectoris late in life, but he refused to take it because of the known vascular headaches of his factory workers. Other therapeutics of the "nitrovasodilator class" include nitroprusside, organic and inorganic nitrites and nitrates, nitrosamines, nitrogen mustards, hydrazines, nitrosoureas, etc. Many academic and industrial laboratories are actively synthesizing additional "NO-donors" that release NO at predictable rates in specific tissues or environments or can be attached to various grafts and supports for vascular surgical procedures (see 11 and references therein). The controlled delivery of nitric oxide with a prodrug or device at a specific target site or vascular bed without the tolerance that is developed to nitroglycerin would be a most significant contribution to therapy of several vascular disorders and other diseases. There is currently considerable promise that such agents and delivery methods can be achieved in that the work is becoming more of a medicinal chemistry and/or engineering effort with obvious targets. An important clinical indication is reperfusion injury after surgery or instrumentation for vascular patency. There are obviously many other indications that can be listed such as vascular remodeling, tissue transplantation, cancer vascularization, etc. with the known effects of nitric oxide on platelet aggregation, vascular smooth muscle proliferation, atherogenesis, angiogenesis, cytotoxicity, etc.

One of the most rapidly advancing area of clinical application is the inhalation of low concentrations of the NO gas (see 11 and references therein). Interestingly, nitric oxide at low concentrations is quite stable and minimally reactive. Its reactivity and toxicity is a second order reaction dependent upon the concentration of nitric oxide and its interaction with other free radicals and reactive oxygen species such as superoxide. At higher concentrations NO can interact with many transition metals, heme-containing proteins, thiol groups and can oxidize functionalities on polynucleotides (RNA and DNA) and proteins and can form strand breaks in polynucleotides (Figure 8). There is a rapid and almost diffusion-limited interaction of nitric oxide with superoxide to form the very reactive peroxynitrite. This is becoming a very interesting area of investigation that undoubtedly relates to the toxic properties of these free radicals.

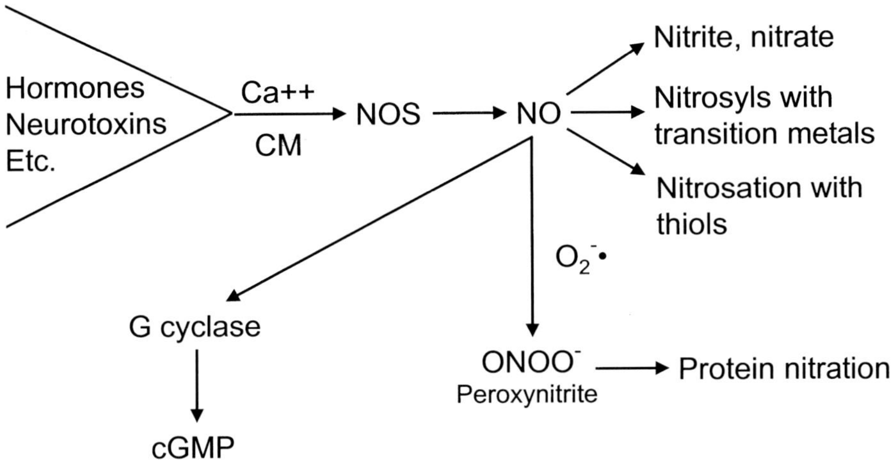

Figure 8. Formation and fate of nitric oxide.

When low concentrations of nitric oxide are administered to premature infants with low birth weights via nasal catheters, nitric oxide has marked beneficial effects on infant mortality. The pulmonary hypertension is improved as is the pulmonary and systemic oxygenation and systemic hypertension resulting from the right to left shunting via the patent ductus arteriosus. There may also be beneficial effects in infants and children with congenital heart defects with right to left shunting and hypoxia. The beneficial effects in adult patients with various pulmonary disorders including pulmonary hypertension and/or acute respiratory distress syndrome (ARDS) are less convincing and more controversial. The controversy and variability of the response to inhaled nitric oxide in adults undoubtedly relates to the etiology of their disease and the presence or absence of some "reversible components" of their pathology. Some of the molecular targets and processes regulated by NO and/ or cyclic GMP are summarized in Tables 5 and 6. These targets can obviously suggest other therapeutic uses for NOS inhibitors and NO prodrugs.

SUMMARY

Since our discovery of the first biological effects of nitric oxide more than two decades ago, more than 20,000 publications have appeared on this topic. There have been more than 4500 to 5000 publications per year in the recent several years and the number of publications continues to grow exponentially. It has become almost impossible for investigators to read all the abstracts of

Table 5. Nitric oxide and cyclic GMP targets.

Guanylyl cyclase
Cyclic nucleotide protein kinases
Cyclic nucleotide phosphodiesterases
Cyclooxygenase (COX II)
Heme proteins, iron centers and thiol groups
DNA modifications
Glutamate (NMDA) receptor regulation
Phospholipase C

Table 6. Some processes regulated by nitric oxide and/or cyclic GMP.

Smooth, cardiac and skeletal muscle relaxation
Retinal phototransduction
Intestinal secretion and ion transport
Renal tubular-glomerular feedback
Endothelial permeability
Smooth muscle proliferation
Platelet adhesion and aggregation
Insulin secretion
Hormone production and secretion
Neurotransmission
Long-term potentiation and memory
Regulation of transcription
Tissue injury and inflammation
Pathogen cytotoxicity
Tumor cytotoxicity
Calcium transport and redistribution

these publications, let alone the full papers. At times it can be difficult to keep up with just the titles, authors and laboratories and remain current. This will undoubtedly test the mentorship of many of us as we train students and fellows who need exciting and novel publications and the priorities required to find positions in our competitive research society and remain funded. However, there seems to be no obvious saturation of the talent required in this exciting and growing field. Numerous important questions remain to be addressed and answered. I have attempted to raise some of these areas and questions here. What excitement it has been for those of us who joined the field early and were able to see such major advances and progress in recent years. There still remain many nuggets available in a vast gold mine of nitric oxide biology. By no means has the field been over-mined with a paucity of gems to be found.

In recent years a number of laboratories have made significant advances in nitric oxide research. Some current areas of active research are listed in Table 7. Undoubtedly, other interesting areas will also emerge and require increasing research and attention.

Table 7. Some areas of active nitric oxide research.

transcriptional regulation
overexpression and gene therapy
gene knockouts
histochemistry, hybridization
acylation, phosphorylation, etc.
cell types with NOS isoforms and functions
cyclic GMP mediated effects
non-cyclic GMP mediated effects
selective/specific NOS inhibitors
novel NO donors

ACKNOWLEDGMENTS

I would like to thank the eighty-two trainees to date who have shared in this work with me. I must acknowledge the major contributions chronologically of Hiroshi Kimura, Shoji Katsuki, William Arnold, Mark Braughler, John Lewicki, Scott Waldman, Robert Rapaport, Yoshinari Kamisaki, Takayashi Kuno, Masaki Nakane, Kunio Ishii, Dale Leitman, Ulrich Fostermann, Harold Schmidt, Jennifer Pollock, and many others who participated in various aspects of this work. Their collaboration, curiosity, hard work, sense of humor and present successes have added even more to my excitement of an academic career. I wish to also thank the John S. Dunn Foundation, the University of Texas, and numerous agencies for their past grant support and in particular the National Institute of Health.

REFERENCES

1. Hardman, J., Beavo, T., Gray, J., Chrisman, T., Patterson, W., and Sutherland, E.: The formation and metabolism of cyclic GMP. *Ann. N.Y. Acad. Sci.* **185**: 27–35, 1971.
2. White, A.: Guanylate cyclase activity in heart and lung. *Adv. Cyclic Nucleotide Res.* **5:** 353–373, 1975.
3. Appleman, M., and Terasaki, W.: Regulation of cyclic nucleotide phosphodiesterase. *Adv. Cyclic Nucleotide Res.* **5**: 153–162, 1975.
4. Greengard, P.: Cyclic nucleotides, protein phosphorylation and neuronal function. *Adv. Cyclic Nucleotide Res.* **5**: 585–602, 1975.
5. Kimura, H., Murad, F.: Evidence for two different forms of guanylate cyclase in rat heart. *J. Biol. Chem.* **249**: 6910–19, 1974.
6. Kimura, H., Murad, F.: Two forms of guanylate cyclase in mammalian tissues and possible mechanisms for their regulation. *Metabolism.* **24**: 439–45, 1975.
7. Kimura, H., Murad, F.: Increased particulate and decreased soluble guanylate cyclase activity in regenerating liver, fetal liver, and hepatoma. *Proc. Natl. Acad. Sci.* **72:** 1965–69, 1975.
8. Waldman, S.A., Murad, F.: Cyclic GMP synthesis and function. *Pharm. Rev.* **39:** 163–96, 1987.
9. Murad, F., editor. Cyclic GMP Synthesis, Metabolism and Function, In: Adv. In Pharmacol., Vol. 26, pp 1–335, Academic Press, 1994
10. Murad, F.: The role of nitric oxide in modulating guanylyl cyclase. *Neurotransmission.* Vol. X, 1–4, 1994.
11. Ignarro, L. and Murad, F. (eds), Nitric Oxide: Biochemistry, Molecular Biology, and Therapeutic Implications. *Advances in Pharmacology,* Volume 34, pp 1–516, Academic Press, 1995.
12. Chinkers, M. and Garbers, D.: Signal Transduction by guanylyl cyclase. *Ann. Rev. Biochemistry.* **60:** 553–575, 1991.
13. Garbers, D., Lowe, D., Dongatt, L., Chinkers, M., Thorpe, D., Bentley, C., Ramareo, Goeddel, and Singh, S.: The membrane form of guanylate cyclase. *Cold Spring Harbor Symp. Quart. Biol.* **53:** 993–1003, 1988.
14. Kimura, H., Mittal, C.K., Murad, F.: Activation of guanylate cyclase from rat liver and other tissues with sodium azide. *J. Biol. Chem.* **250:** 8016–22, 1975.
15. Mittal, C.K., Kimura, H., Murad, F.: Requirement for a macromolecular factor for sodium azide activation of guanylate cyclase. *J. Cyclic Nucl. Res.* **1:** 261–69, 1975.
16. Kimura, H., Mittal, C.K., Murad, F.: Increases in cyclic GMP levels in brain and liver with sodium azide, an activator of guanylate cyclase. *Nature.* **257:** 700–02, 1975.
17. Kimura, H., Mittal, C.K., Murad, F.: Appearance of magnesium guanylate cyclase activity in rat liver with sodium-azide activation. *J. Biol. Chem.* **251:** 7769–73, 1976.
18. Mittal, C.K., Kimura, H., Murad, F.: Purification and properties of a protein required for sodium azide activation of guanylate cyclase. *J. Biol. Chem.* **252:** 4348–90, 1977.
19. Murad, F., Mittal, C.K., Arnold, W.P., Katsuki, S., Kimura, H.: Guanylate cyclase: Activation by azide, nitro compounds, nitric oxide, and hydroxyl radical and inhibition by hemoglobin and myoglobin. *Adv. Cyclic Nucl. Res.* **9:** 145–58, 1978.
20. Murad, F., Mittal, C.K., Arnold, W.P., Ichihara, K., Braughler, E., El-Zayat, M.: Properties and regulation of guanylate cyclase: Activation by azide, nitro compounds and hydroxyl radical and effects of heme containing proteins. Proc. of the NATO Adv. Study Inst. on Cyclic Nucleotides, Italy, 1977, in *Molecular Biology and Pharmacology of Cyclic Nucleotides,* Eds., G. Folca and R. Paoletti, Elsevier, Amsterdam, pp. 33–42, 1978.
21. Murad, F., Mittal, C.K., Arnold, W.P., Braughler, J.M.: Effect of nitro-compound smooth muscle relaxants and other materials on cyclic GMP metabolism. Proc. of the 7th International Congress of Pharmacology, Paris, France, July, 1978, in *Advances in Pharmacology and Therapeutics, Vol. 3 Ions, Cyclic Nucleotides, Cholinergy,* Ed., J.C. Stocklet, Pergamon Press, New York, pp. 123–32, 1978.
22. Katsuki, S., Arnold, W., Mittal, C.K., Murad, F.: Stimulation of guanylate cyclase by so-

dium nitroprusside, nitroglycerin and nitric oxide in various tissue preparations and comparison to the effects of sodium azide and hydroxylamine. *J. Cyclic Nucl. Res.* **3:** 23–35, 1977.

23. Katsuki, S., Arnold, W.P., Mittal, C.K., Murad, F.: Stimulation of formation and accumulation of cyclic GMP by smooth muscle relaxing agents. Proc. of the 2nd Japanese Cyclic Nucleotide Conference, July 7–9, 1977, pp. 44–50.
24. Katsuki, S., Murad, F.: Regulation of cyclic 3′,5′-adnosine monophosphate and cyclic 3′,5′-guanosine monophosphate levels and contractility in bovine tracheal smooth muscle. *Molecular Pharmacology.* **13:** 330–41, 1977.
25. Katsuki, S., Arnold, W.P., Murad, F.: Effect of sodium nitroprusside, nitroglycerin and sodium azide on levels of cyclic nucleotides and mechanical activity of various tissues. *J. Cyclic Nucl. Res.* **3:** 239–47, 1977.
26. Murad, F., Manganiello, V., Vaughan, M.: Effects of guanosine 3′,5′-monophosphate on glycerol production and accumulation of adenosine 3′,5′-monophosphate during incubation of fat cells. *J. Biol. Chem.* **245:** 3352–60, 1970.
27. Manganiello, V., Murad, F., Vaughan, M.: Effects of lipolytic and antilipolytic agents on cyclic 3′,5′-adenosine monophosphate in fat cells. *J. Biol. Chem.* **246:** 2195–02, 1971.
28. Vaughan, M., Murad, F.: Adenyl cyclase activity in particles from fat cells. *Biochemistry.* **8:** 3092–99, 1969.
29. Furchgott, R. and Zawadski, J.: The obligatory role of endothelial cells in the relaxation of arterial smooth muscle to acetylcholine. *Nature.* **288:** 373–376, 1980.
30. Arnold, W.P., Mittal, C.K., Katsuki, S., Murad, F.: Nitric oxide activates guanylate cyclase and increases guanosine 3′,5′-monophosphate levels in various tissue preparations. *Proc. Nat. Acad. Sci., USA.* **74:** 3203–07, 1977.
31. Braughler, J.M., Mittal, C.K., Murad, F.: Purification of soluble guanylate cyclase from rat liver. *Proc. Natl. Acad. Sci., USA.* **76:** 219–22, 1979.
32. Rapoport, R.M., Murad, F.: Agonist-induced endothelial-dependent relaxation in rat thoracic aorta may be mediated through cyclic GMP. *Circ. Res.* **52:** 352–57, 1983.
33. Rapoport, R.M., Murad, F.: Endothelium-dependent and nitrovasodilator-induced relaxation of vascular smooth muscle: Role for cyclic GMP. *J. Cyclic Nucl. and Protein Phosphor. Res.* **9:** 281–96, 1983.
34. Rapoport, R.M., Draznin, M.B., Murad, F.: Endothelium-dependent vasodilator- and nitrovasodilator-induced relaxation may be mediated through cyclic GMP formation and cyclic GMP-dependent protein phosphorylation. *Trans. Assoc. Amer. Phys.* **96:** 19–30, 1983.
35. Rapoport, R.M., Draznin, M.B., Murad, F.: Endothelium dependent relaxation in rat aorta may be mediated through cyclic GMP-dependent protein phosphorylation. *Nature.* **306:** 274–76, 1983.
36. Fiscus, R.R., Rapoport, R.M., Murad, F.: Endothelium-dependent and nitrovasodilator-induced activation of cyclic GMP-dependent protein kinase in rat aorta. *J. Cyclic Nucl. and Protein Phosphor. Res.* **9:** 415–25, 1983.
37. Rapoport, R.M., Murad, F.: Effect of cyanide on nitrovasodilator-induced relaxation, cyclic GMP acumulation and guanylate cyclase activation in rat aorta. *Europ. J. Pharm.* **104**: 61–70, 1984.
38. Rapoport, R.M., Draznin, M., Murad, F.: Sodium nitroprusside-induced protein phosphorylation in intact rat aorta is mimicked by 8-bromo-cyclic GMP. *Proc. Nat. Acad. Sci., USA.* **79:** 6470–74, 1982.
39. Hirata, M., Kohse, K., Chang, C.H., Ikebe, T., Murad, F.: Mechanism of cyclic GMP inhibition of inositol phosphate formation in rat aorta segments and cultured bovine aortic smooth muscle cells. *J. Biol. Chem.* **265:** 1268–73, 1990.
40. Murad, F.: Cyclic guanosine monophosphate as a mediator of vasodilation. *J. Clin. Invest.* **78:** 1–5, 1986.
41. Winquist, R.M., Faison, E.P., Waldman, S.A., Schwartz, K., Murad, F., Rapoport, R.M.: Atrial natriuretic factor elicits an endothelium independent relaxation and activates particulate guanylate cyclase in vascular smooth muscle. *Proc. Nat. Acad. Sci.* **81:**

7661–64, 1984.
42. Waldman, S.A., Rapoport, R.M., Murad, F.: Atrial natriuretic factor selectively activates particulate guanylate cyclase and elevates cyclic GMP in rat tissues. *J. Biol. Chem.* **259:** 14332–334, 1984.
43. Leitman, D.C., Molina, C.R., Waldman, S.A., Murad, F.: Atrial natriuretic peptide receptors and the guanylate cyclase-cyclic GMP system. *Proc. of the UCLA Symposium on Atrial Natriuretic Factor in Biological and Molecular Aspects of Atrial Factors,* UCLA Symposia on Molecular and Cellular Biology, **81:** 39–56, 1988.
44. Waldman, S.A., Murad, F.: Cyclic GMP synthesis and function. *Pharm. Rev.* **39:** 163–96, 1987.
45. Kuno, T., Andresen, J.W., Kamisaki, Y., Waldman, S.A., Chang, L.Y., Saheki, S., Leitman, D.C., Nakane, M., Murad, F.: Co-purification of an atrial natriuretic factor receptor and particulate guanylate cyclase from rat lung. *J. Biol. Chem.* **261:** 5817–23, 1986.
46. Garbers, D.L.: Guanyl cyclase linked receptors. *Pharmacol. Ther.* **50:** 337–345, 1991.
47. Degucci, T. and Yoshiako, M.: L-arginine identified as an endogenous activator for soluble guanylate cyclase from neuroblastoma cells. *J. Biol. Chem.* **257:** 10147, 1982.
48. Hibbs, J., Traintor, R. and Vanin, Z.: Macrophage cytotoxicity. Role for l-arginine deaminase and imino nitrogen activation to nitrate. *Science.* **235:** 473, 1987.
49. Murad, F.: The role of cyclic GMP in the mechanism of action of nitrovasodilators, endothelium-dependent agents and atrial natriuretic peptide. *Biochemical Society Transactions.* **16:** 490–92, 1988.
50. Murad, F., Leitman, D., Waldman, S., Chang, C.H., Hirata, M., Kohse, K.: Effects of nitrovasodilators, endothelium-dependent vasodilators and atrial peptides on cGMP. Proc. *Cold Spring Harbor Symposium on Quantitative Biology, Signal Transduction.* **53:** 1005–1009, 1988.
51. Murad, F.: Modulation of the guanylate cyclase-cGMP system by vasodilators and the role of free radicals as second messengers. *Proc. of the NATO Advanced Studies Institute on Vascular Endothelium: Receptors and Transduction Mechanisms.* Porto Carros, Greece, June, 1988, in *Vascular Endothelium,* eds. J.D. Catravas, C.N. Gillis and U.S. Ryan, pp 157–164, Plenum Pub., 1989.
52. Murad, F.: Mechanisms for hormonal regulation of the different isoforms of guanylate cyclase. *Proc. of the 40th Mosbach Colloquium on Molecular Mechanisms of Hormone Action,* eds. Y. Gehring, E. Helmreich and G. Schultz, Springer, Heidelberg, April 1989. pp 186–194, 1989.
53. Bredt, D. and Snyder, S.: Isolation of nitric oxide synthase, a calmodulin-requiring enzyme. *Proc. Nat. Acad. Sci.,* USA. **87:** 682–685, 1990.
54. Gorsky, L., Förstermann, U., Ishii, K., Murad, F.: Production of an EDRF-like activity in the cytosol of N1E-115 neuroblastoma cells. *FASEB Journal,* **4:** 1494–1500, 1990.
55. Förstermann, U., Ishii, K., Gorsky, L.D., Murad, F.: The cytosol of N1E-115 neuroblastoma cells synthesizes and EDRF-like substance that relaxes rabbit aorta. Naunyn Schmiedbergs *Arch. Pharmacol.* **340:** 771–774, 1989.
56. Förstermann, U., Gorsky, L., Pollock, J., Ishii, K., Schmidt, H.H.H.W., Heller, M., Murad, F.: Hormone induced biosynthesis of Endothelium-derived relaxing factor-Nitric oxide-like material in N1E-115 neuroblastoma cells requires calcium and calmodulin. *Molecular Pharmacology.* **38:** 7–13, 1990.
57. Schmidt, H.H.H.W., Pollock, J., Nakane, M., Gorsky, L., Förstermann, U., Heller, M., Murad, F.: Purification of a soluble isoform of guanylyl cyclase-activating-factor synthase. *Proc. Nat. Acad. Sci.* **88:** 365–69, 1991.
58. Stuehr, D., Cho, H., Kwon, N., Weise, M. and Nathans, C.: Purification and characterization of the cytokine-induced macrophage nitric oxide synthase. *Proc. Nat. Acad. Sci., USA.* **88:** 7773–77, 1991.
59. Förstermann, U., Schmidt, H.H.H.W., Pollock, J.S., Heller, M., Murad, F.: Enzymes synthesizing guanylyl cyclase activating factor (GAF) in endothelial cells, neuroblastoma cells and rat brain. *J. Cardiovasc. Pharmacol.* **17** (Sup3), 557–564, 1991.
60. Förstermann, U., Gorsky, L., Pollock, J.S., Schmidt, H.H.H.W., Ishii, K., Heller, M.,

Murad, F.: Subcellular localization and regulation of the enzymes responsible for EDRF synthesis in endothelial cells and N1E-155 neuroblastoma cells. *Eur. J. Pharmacol.* **183:** 1625–26, 1990.

61. Pollock, J.S., Förstermann, U., Mitchell, J.A., Warner, T.D., Schmidt, H.H.H.W., Nakane, M., Murad, F.: Purification and characterization of particulate EDRF synthase from cultured and native bovine aortic endothelial cells. *Proc. Nat. Acad. Sci., USA.* **88:** 10480–84, 1991.
62. Förstermann, U., Schmidt, H.H.H.W., Pollock, J.S., Sheng, H., Mitchell, J.A., Warner, T.D., Nakane, M., Murad, F.: Isoforms of EDRF/NO synthase: Characterization and purification from different cell types. *Biochem. Pharmacol.* **42:** 1849–57, 1991.
63. Nakane, M., Mitchell, J.A., Förstermann, U., Murad, F.: Phosphorylation by calcium/calmodulin-dependent protein kinase II and protein kinase C modulates the activity of nitric oxide synthase. *Biochem. Biophys. Res. Com.* **180:** 1396–1402, 1991.
64. Pollock, J., Klinghofer, V., Förstermann, U., and Murad, F.: Endothelial nitric oxide synthase is myristylated. *FEBS. Lett.* **309:** 402–404, 1992.
65. Robinson, L.J., Busconi, L. and Michel, T.: Agonist-modulated palmitoylation of endothelial nitric oxide synthase. *J. Biol. Chem.* **270:** 995–998, 1995.
66. Shaul, P.W., Smart, E.J., Robinson, L.J., German, Z., Yuhanna, I.S., Ying, Y., Anderson, R. and Michel, T.: Acylation targets endothelial nitric-oxide synthase to plasmalemmal caveolae. *J. Biol. Chem.* **271:** 6518–22, 1996.
67. Xie, J., Roddy, P., Rife, T., Murad, F. and Young, A.: Two closely linked but separate promoters for human neuronal nitric oxide synthase gene transcription. *Proc. Nat. Acad. Sci.* **92:** 1242–1246, 1995.

Physiology or Medicine 1999

GÜNTER BLOBEL

"for the discovery that proteins have intrinsic signals that govern their transport and localization in the cell"

THE NOBEL PRIZE IN PHYSIOLOGY OR MEDICINE

Speech by Professor Ralf Pettersson of the Nobel Committee at Karolinska Institutet.
Translation of the Swedish text.

Your Majesties, Your Royal Highness, Ladies and Gentlemen,

Imagine a large factory that manufactures thousands of different items in millions of copies every hour, and that promptly packages and ships each of them to waiting customers. Naturally, to avoid chaos, each product requires a clearly labeled address tag. Günter Blobel is being awarded this year's Nobel Prize in Physiology or Medicine for having shown that newly synthesized proteins, analogous to the products manufactured in the factory, contain built-in signals, or address tags, that direct them to their proper cellular destination.

An adult human is comprised of approximately 100,000 billion cells, all of which are structurally similar. A striking feature is that each cell contains small compartments or organelles. Organelles are bounded by impermeant, lipid-rich membranes that ensure the physical and functional separation of vital biochemical processes. This compartmentalization enables cells to be compared to a large city in which each public function is housed in a separate building. The blueprint for all cellular processes is maintained in the genome located in the cell nucleus, the City Hall of the cell. Energy production takes place within the mitochondria, the power plant of the cell; the breakdown and recycling of waste takes place in the lysosome etc. The production of new products, proteins in the case of the cell, is carried out by ribosomes in a process resembling an assembly line. There is indeed a feverish amount of activity within cells. Every second, thousands of protein molecules are degraded and replaced by new ones. How does a newly made protein get to its correct intracellular location, and how do proteins enter into and move across the membranes surrounding individual organelles? These two central questions occupied the minds of scientists during the 1960s.

Günter Blobel has provided the answer to both of these questions. In 1967 he joined the renowned cell biology laboratory headed by George Palade at the Rockefeller University in New York. Palade, who received the Nobel Prize in 1974, had defined and charted the route that secretory proteins take from their site of synthesis within the cell to the cell surface. Secreted proteins are made in the cell in association with a membrane system called the endoplasmic reticulum.

Blobel began by examining how a newly synthesized secretory protein is targeted to and then translocated across the endoplasmic reticulum membrane. Based on the results from a series of elegant experiments, Blobel put forth the so-called "signal hypothesis," in a preliminary form in 1971 and a mature final form in 1975, to explain how this process takes place. The signal hypothesis postulated that newly made proteins contain built-in signals, address tags or zip codes, that target proteins to the endoplasmic reticulum and that subsequently lead them across the reticulum membrane through a specialized channel. Proteins that are translocated across to the other side are packaged for subsequent transport to the cell surface.

To test this hypothesis, Blobel developed an ingenious experimental test tube system, which enabled him to individually study each step of the process. The system which relied on components obtained from mouse, rabbit, and dog cells, laid the foundation for the development of the field of molecular cell biological research. In the following 20 years, Blobel and his coworkers characterized this complex process in great detail. The original signal hypothesis, in all its essential parts, has stood the test of time and proven to be correct.

Blobel extended his studies and was able to demonstrate that proteins destined to become transported to other organelles, or that become integrated into different cellular membranes, also contain specific address tags and so-called "topogenic" signals. The guiding principles that Blobel has helped to elucidate are universally applicable and highly conserved. They have remained almost unchanged during the course of evolution, functioning in yeast, plant, and animal cells.

Perhaps the most important consequence of Günter Blobel's discoveries is that we now understand the fundamental principles guiding the formation and maintenance of cell and organelle structure. The signal hypothesis provides a framework to understand the mechanisms underlying many hereditary diseases and other disease processes in which specific proteins become mislocalized. In addition, these discoveries have enabled the pharmaceutical industry to turn cultured cells into efficient mini-factories for the production of protein-based drugs, such as insulin, growth hormone, coagulation factors etc.

Günter Blobel,

Your discovery that proteins contain built-in signals that direct them to their correct destination within cells and across membranes has had a profound impact on our understanding of how a cell and its organelles are assembled and maintained. Your work has also laid the foundation for modern molecular cell biology. On behalf of the Nobel Assembly at Karolinska Institutet I wish to convey to you my warmest congratulations and I now ask you to step forward to receive your Nobel Prize from the hands of His Majesty the King.

Günter Blobel

GÜNTER BLOBEL

In 1936, when I was born in the small Silesian village of Waltersdorf in the county of Sprottau in the then eastern part of Germany, now part of Poland, the fine structure of the cell was still an enigma. After 300 years of staring through light microscopes, essentially all that biologists had learned was that the cell was delimited by a plasma membrane and contained a nucleus. Staining procedures had revealed other distinct territories in the cytoplasm and in the nucleus, but their fine structure remained unknown. A dramatic revolution occurred in 1945, when Keith Porter, Ernest Fullam and Albert Claude at the then Rockefeller Institute for Medical Research in New York City introduced the electron microscope to look at cells. The first structure they saw was a lace-like network in the cytoplasm that they termed the endoplasmic reticulum. This discovery formed the foundation for my future scientific career.

1945 was also a turning point in my life. Until then my childhood was a perfect 19th century idyll. In the cold and snow-rich Silesian winters there were hour-long rides on Sundays in horse-drawn slays to my maternal grandparent's farm to have lunch and to spend the afternoon. The house was a magnificent 18th century manor house in the nearby Altgabel with a great hall that was decorated with hunting trophies. In the summer, of course, horse-drawn landauers were used as means of transportation. The way to school was a long one. We went there on foot and as a pack, usually consisting of one or two of my seven brothers and sisters and of children from neighboring houses.

At the end of January 1945, we had to flee from the advancing Russian Red Army. My father, a veterinarian stayed behind for a few more days and left only hours before the Red Army moved in. My fourteen year-old brother, Reiner, drove my mother, my youngest brother, an older brother, the two younger sisters and me in a small automobile to relatives west of Dresden in Saxony. On the way there we drove through Dresden. We entered the city from the eastern hills. Its many spires and the magnificent cupola of the Frauenkirche (die Steinerne Glocke, the Stone Bell) were a magnificent sight even for the untrained eye of a child. Driving through Dresden, I still remember the many palaces, happily decorated with cherubs and other symbols of the baroque era. The city made an indelible impression on me. Only a few days later, on February 13, 1945, we saw from a distance of about 30 kilometers a fire-lit, red night sky reflecting the raging firestorm that destroyed this great jewel of a city in one of the most catastrophic bombing attacks of World War II. It was a very sad and unforgettable day for me.

The months before and after the end of World War II were chaotic and miserable. None of my relatives had enough space to accommodate our large family leaving us divided among several relatives in different villages. There was no communication and little food. On September 9, 1945, we learned of the death of my beautiful oldest sister Ruth who, at age 19, was killed in an air raid on a train she was travelling in on April 10, 1945. She was buried in a mass grave near the site of the attack in Schwandorf, Bavaria. Ruth was born when my mother was just 20. The two had a sisterly relationship. My mother grieved over Ruth's death until the end of her own life.

Fortunately, things took a turn for the better, when my father was able to continue his veterinarian practice in the charming medieval Saxon town of Freiberg. Most members of our family were reunited there by 1947. We lived in a nice villa surrounded by a large garden on the edge of town. My way to school was along the old medieval city wall. For only 40,000 inhabitants, Freiberg had a rich cultural life with a 175 year old theater. Most impressive were the musical performances in the magnificent gothic cathedral, the Dom, with the splendid great Silbermann organ. Each week Bach cantatas were performed. The great choral works of Bach, Mozart and Haydn were regularly performed and at the highest artistic level at the major religious holidays. I even participated in singing in the cantus firmus of Bach's Matthäus Passion. So, it was almost like a 19th century idyll again, this time in a small medieval town instead of a country village.

However, there was now the ever more oppressive regime of East Germany to deal with on a daily basis. When I graduated from high school in 1954, I was not allowed to continue my education at a university because I was considered a member of the "capitalist" classes. Fortunately, at that time, i.e., before the Berlin Wall, it was possible to escape and to travel freely to West Germany. So, on August 28, Goethe's birthday, I left Freiberg for Frankfurt on the Main in West Germany. The train left in the morning and in the afternoon it passed Weimar, where Goethe spent most of his life, and then Eisenach, where Bach was born and in the evening it arrived in Frankfurt, Goethe's birthplace.

I studied medicine, beginning in Frankfurt and then in Kiel, München and Tübingen, graduating in 1960 from the University of Tübingen. Although I completed two years of internship in various small hospitals, I decided against continuing my medical training. I was much more fascinated by the unsolved problems of medicine than by practicing it. Fortunately, my oldest brother Hans had a similar experience in his field of study, veterinary medicine. He had obtained the prestigious Fulbright Fellowship to study in the U.S., continued his training there in microbiology and rapidly achieved the rank of full professor at the University of Wisconsin in Madison. He was extremely sympathetic to my dilemma and helped me to secure a graduate fellowship to study either with Khorana or with Van R. Potter. So, in 1962, I sailed to Montreal on a German steel freighter, and from there drove to Madison to arrive on a beautiful late day in May. Potter was a marvellous mentor, witty, energetic and stimulating. I graduated in November 1966, and decided to join George Palade's Laboratory of Cell Biology at the Rockefeller University (formerly the Rockefeller Institute). The revolution that began there in 1945, and that led to the discovery of all the major structures of the cell continued in the realm of relating cellular structures to specific cellular functions. My arrival there coincided with the end of this second phase and the exciting beginnings of a third phase, the molecular analysis of cellular functions (see below). I was fortunate enough in helping to initiate this third phase of analysis which is still in full swing.

George Palade has been my most influential mentor, a good friend and a wonderful colleague. He taught me how to conceptualize a collection of disparate facts, to formulate working hypotheses and to design experiments to test these hypotheses. I am greatly indebted to him.

In New York I married Laura Maioglio. Laura studied art history and, at her father's death, took over Barbetta Restaurant founded by her father in 1906. Laura has introduced me to many artistic pleasures that I had not experienced before. She greatly encouraged me in my work and never complained about the many hours I spent in the laboratory.

In 1994, I founded Friends of Dresden, Inc., a charitable organization, with the goal to raise funds in the U.S. to help rebuild the Frauenkirche in Dresden. The rebuilding of many of the historic monuments of Dresden is one of the most exciting consequences of German reunification and the liberation from communism. It is a childhood dream come true.

It was one of the great pleasures of my life to donate the entire sum of the Nobel Prize, in memory of my sister Ruth Blobel, to the restoration of Dresden, to the rebuilding of the Frauenkirche and the building of a new synagogue. This donation also serves to express my gratitude to my fellow Saxons. They received us with open arms when we had to flee Silesia. I spent a wonderful period of my life there and they gave me a thorough and valuable education. A few thousand dollars will also be donated for the restoration of an old baroque church in Fubine/Piemonte/Italy, the home town of my wife's father, Sebastanio Maioglio. We have spent many happy summers there in the parental home of my wife.

PROTEIN TARGETING

Nobel Lecture, December 8, 1999

by

GÜNTER BLOBEL

Laboratory of Cell Biology, Howard Hughes Medical Institute, The Rockefeller University, 1230 York Avenue, New York, NY 10021, USA.

PROLOGUE

I began research in the 60ties, first as a graduate student in the laboratory of Van R. Potter (Fig.1) at the McArdle Institute for Cancer Research of the University of Wisconsin in Madison. I continued as a postdoctoral fellow in the laboratory of George E. Palade (Fig. 2) at The Rockefeller University in New York City. At that time, the intracellular pathway of secretory proteins, from their synthesis to their extrusion from the cell, the so-called 'secretory pathway', had already been worked out by George Palade and his coworkers, primarily Philip Siekevitz, Jim Jamieson and Lucien Caro (for review see 1974 Nobel lecture by George Palade [1]). Using pulse-chase labeling with radioactive amino acids in tissue slices in conjunction with cell fractionation and autoradiography, Palade and coworkers established that, during or shortly

Figure 1. Van R. Potter, 1964. Mc Ardle Laboratory, University of Wisconsin, Madison, Wisconsin.

Figure 2. George E. Palade, 1970. The Rockefeller University, New York.

after their synthesis, secretory proteins cross the rough (ribosome-studded) endoplasmic reticulum (ER) membrane. It was proposed that translocation across the ER yields segregation of secretory proteins from cytosolic proteins. From the ER, secretory proteins are transported via vesicular carriers through the cisternae of the Golgi apparatus. Finally, vesicular carriers budding from the trans-cisternae of the Golgi complex were observed to fuse with the plasma membrane, yielding externalization or exocytosis of the secretory proteins (Fig. 3). The biochemical mechanisms underlying this pathway were unknown at that time.

Several reports in the mid-sixties suggested that mRNAs for secretory proteins are sequestered in rough microsomes that represent the vesicular remnants of the fractured rough ER. However, it was not clear how the sequestration of these mRNAs was accomplished. One possibility was that an *untranslated* region that might be common to all mRNAs coding for secretory proteins mediated their attachment to the ER membrane. Other such distinct *untranslated* regions of other mRNAs might mediate attachment to other organelles. Once attached to their 'cognate' organelles, the translation products might then be vectorially transported into these organelles (2). Another idea was that free and organelle-bound ribosomes might differ in their composition. The distinct organelle-bound ribosomes might function to select 'cognate' mRNAs to the organelles, again via cognate *untranslated* regions. These and other ideas were frequently discussed in the Palade laboratory in the late 60ties.

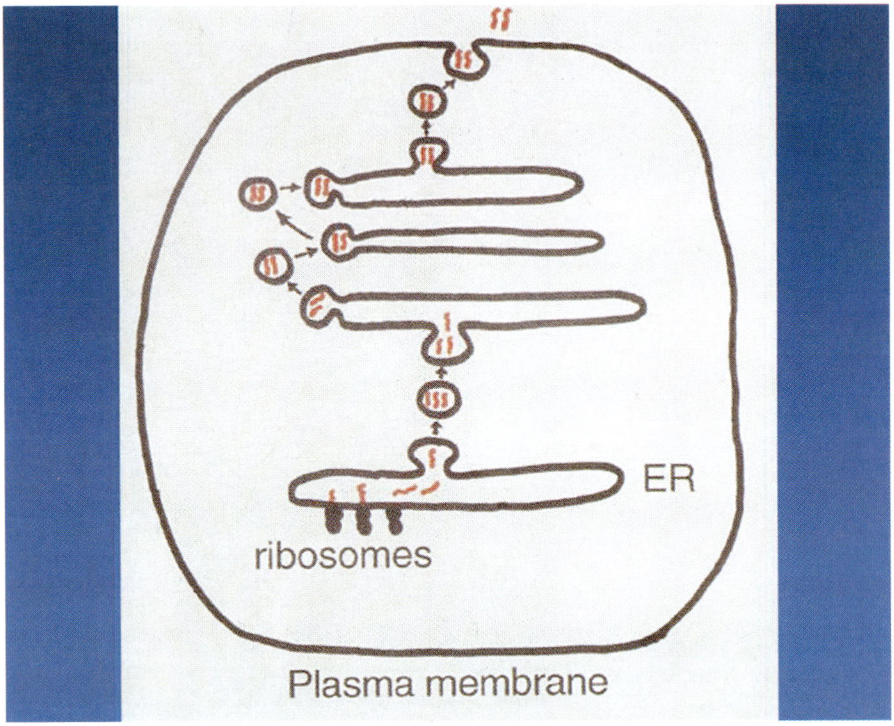

Figure 3. The secretory pathway. Secretory proteins (indicated in red) are synthesized on ribosomes bound to the endoplasmic reticulum (ER). They are then transported via vesicular carriers through the Golgi complex and are finally exocytosed.

This simple cartoon does not show important branches:

1. The branch for lysosomal proteins from the distal Golgi cisternae and the branch for peroxisomal membrane proteins from the ER.
2. The retrograde recycling branches of this pathway, including endocytosis.
3. Most importantly, essentially all integral membrane proteins (except for those of mitochondria and chloroplasts share this pathway.

THE FIRST VERSION OF THE SIGNAL HYPOTHESIS

David Sabatini (Fig. 4) and I tested one of these ideas, namely whether free and ER-bound ribosomes differ in their protein composition. We used one-dimensional sodium dodecyl sulfate (SDS) polyacrylamide gel electrophoresis (PAGE). Although this method did not resolve all ribosomal proteins, we could not detect significant differences in the protein pattern between these two ribosome populations. Therefore, we disfavored the idea that sequestration to the ER of mRNAs coding for secretory proteins occurs via distinct features of ER-bound ribosomes.

Instead, we hypothesized in 1971, that selection of mRNA to the ER membrane is not via direct binding of the mRNA itself, but rather via binding of its nascent translation product (3). We postulated that all mRNAs for secretory proteins code for a signatory amino-terminal sequence element that is recognized by a soluble factor that, in turn, binds the nascent chain-ribosome complex to the ER membrane (Fig. 5). We also postulated that ribosomes,

Figure 4. David D. Sabatini, 1970. The Rockefeller University, New York.

free or ER-bound, are indistinguishable and that they cycle between an ER-bound and a cytosolic pool. At this point, there were only a few sequences of secretory proteins established and there was no evidence that these proteins share a common amino-terminal sequence element. However, it occurred to us that such an amino-terminal sequence tag might be a transient and not a permanent feature of nascent secretory proteins. Therefore, we had no hesitation to publish this idea (3). Nevertheless, our proposals at this point were pure speculation without any supporting evidence.

To experimentally test the predictions of this hypothesis, we sought to develop a cell-free system, in which protein translation and protein translocation across microsomal membrane vesicles was faithfully recapitulated. We hoped to reconstitute such an *in vitro* system from defined components. To accomplish this, we began with a series of deconstruction experiments. We started with polysomes. Using puromycin and high concentrations of salt, we isolated functional ribosomal subunits (4) and mRNA still attached to its binding proteins (5, 6, 7). Using the same methods, we succeeded in disassembling rough microsomes. This yielded a virtually mRNA- and ribosome-free membrane fraction (8). We also isolated 'native' small ribosomal subunits that contained translation initiation factors. These initiation factors could be dissociated as multi-subunit complexes (9, 10). With these components in hand, we then attempted to reconstruct a functional protein translation/translocation system from defined components.

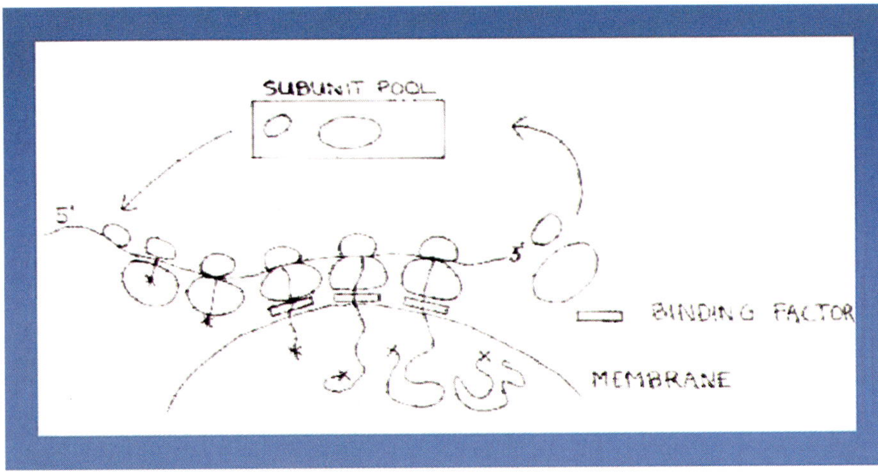

Figure 5. The signal hypothesis in its first version (1971). We announced this idea in a symposium on Biomembranes in 1971 (3) and published it in the proceedings of this symposium. To us the idea was very appealing, even though there was essentially no evidence for it. The signal sequence is indicated by an x and was predicted to be recognized by a "binding factor" that mediates binding to the ER membrane. The signal sequence was not indicated to be a transient feature of the nascent protein, although it occurred to us that it may no longer be present in the mature secreted protein. After completion of translocation the ribosome was predicted to be dissociated into subunits that would join a cytoplasmic pool. No specific proposals were yet made as to how the chain crosses the membrane. From Blobel and Sabatini (3), with permission.

While these deconstruction experiments were going on, two important papers appeared in 1972 from the laboratories of Philip Leder (11) and of Cesar Milstein (12). These investigators studied translation of poly(A) containing mRNAs from myeloma cells (containing primarily mRNA for the light chain of IgG) in a cell-free system that lacked microsomal vesicles. The translation yielded one major product that was larger by about 2–3 kDalton than the mature light chain of IgG which was secreted from the myeloma cells. By peptide mapping of the radioactively labeled translation product they could show that it contained an extension located at the amino terminus. A partial sequence of this amino-terminal extension was subsequently established by Schechter and colleagues (13) and revealed a preponderance of hydrophobic residues.

Could this additional peptide function as the determinant for attachment to the ER and subsequent translocation across the ER, as we had postulated in 1971? Or was the larger form the result of an *in vitro* translation artifact, e.g., perhaps the consequence of an erroneous upstream initiation? Interestingly, *in vitro* completion of nascent chains in isolated rough microsomes of myeloma cells yielded the mature light chain and not the larger form (12). Therefore, it was suggested that the microsomal membranes might contain a protease that converts the larger form into the mature form of the light chain of IgG by removing its amino-terminal extension (12).

Figure 6. Bernhard Dobberstein, 1975. The Rockefeller University, New York.

SIGNAL PEPTIDASE

Together with Bernhard Dobberstein (Fig. 6), we succeeded in reproducing (14) the results of Leder's and Milstein's laboratories. We translated poly(A) containing mRNA from myeloma cells in a cell-free translation system that was reconstructed from heterologous components. We obtained a major polypeptide (Fig. 7, lane 1) that was larger by about three kDalton than the mature form of the light chain that was secreted from these myeloma cells (Fig. 7, lane 2).

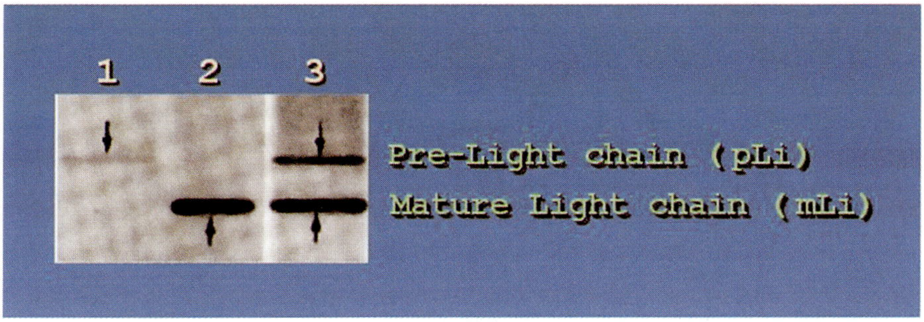

Figure 7. Synthesis of the large (pLi) and mature (mLi) form of the light chain of IgG. ^{35}S–methionine-labeled proteins were separated by SDS-PAGE and visualized by autoradiography. Lane 1, translation products of poly(A) containing mRNAs isolated from myeloma cells; lane 2, mature light chains secreted from myeloma cells; lane 3, in vitro readout of polysomes ("detached polysomes") prepared by detergent from isolated rough microsomes of myeloma cells. From Blobel and Dobberstein (14) with permission.

To address the question of whether the larger form of the IgG light chain was a physiological precursor of the mature chain or an *in vitro* artifact, we were guided by three conjectures. First, the processing enzyme (subsequently termed signal peptidase) is a membrane-associated protease, as had been postulated by Milstein and co-workers (12). Therefore, this enzyme might be solubilized when isolated rough microsomes are treated with detergent; the 'detached' polysomes that could subsequently be sedimented by centrifugation might therefore be free of signal peptidase. Second, as the signal sequence was presumed to mediate the attachment of the ribosome to the cis-side of the membrane (see Fig. 5), the signal peptidase activity is likely associated with the trans-side of the microsomal membrane. This would assure that signal sequence removal occurs only after signal sequence-mediated attachment on the cis-side of the membrane. Third, after signal sequence-mediated attachment to the membrane the signal sequence would be translocated across the membrane and thereby gain access to its trans-side. There it would be cleaved 'co-translationally' and 'co-translocationally', i.e., during but not after translation and translocation were completed. Hence, in 'detached polysomes', the ribosomes near the 5' end of the mRNA might contain nascent chains with a signal sequence. In contrast, ribosomes located near the 3' end of the mRNA should contain nascent chains that already had their signal peptide removed (see model in Fig. 8).

In agreement with these assumptions, we observed (14) that nascent chains present in detached polysomes can be completed in a cell-free translation system (in the presence of an initiation inhibitor to prevent re-initiation) to yield both the larger and the mature forms of the light chain (Fig. 7, lane 3). Moreover, in a time course experiment, we observed that nascent chains lack-

Figure 8. The 1975 version of the signal hypothesis. Signal peptidase, not indicated here was postulated to be associated with the trans side of the microsomal membrane. When the membrane is solubilized by detergent, signal peptidase would be absent in isolated detached polysomes. The detached polysomes should contain two kinds of chains: those still containing their signal peptide at the N-terminus and those with their signal peptide cleaved off by signal peptidase. If signal peptide cleavage occurs during translocation then those ribosomes near the 3' end of mRNA should contain chains without signal peptide and those near the 5' end of mRNA should still contain their signal peptide. In a time course of translation one should therefore first see the completion of chains that have lost their signal peptide and then the completion of chains that still contain their signal peptide.

ing a signal peptide were completed first, followed by the completion of chains still containing their signal peptide (Fig. 9). These results supported the notion that microsomal ribosomes near the 3'end of mRNA contained nascent chains from which the signal sequence had already been removed, co-translationally and co-translocationally. In contrast, ribosomes near the 5'end still contained their signal sequence. This signal sequence could no longer be cleaved as the signal peptidase had been removed during the preparation of the 'detached polysomes'. These 'readout' data with 'detached polysomes' suggested that a *completed* larger form of the light chain of IgG is indeed an *in vitro* artifact. *In vivo,* the signal peptide is cleaved off the *nascent* chain but cannot be cleaved off the c*ompleted* chain. Hence, the *completed* larger form is not a physiological precursor of the mature form of the light chain.

RECONSTITUTION OF CO-TRANSLATIONAL TRANSLOCATION

The data on signal sequence removal provided us with important clues for our subsequent attempts to reconstitute translation and translocation of the light chain of IgG. In view of the signal hypothesis, it appeared unlikely that the larger form would be translocation-competent. That is, cell free synthesis of the completed larger form in the absence of membranes followed by post-translational incubation of the reaction with microsomal membrane vesicles was unlikely to yield translocation and conversion of the larger form to the mature form. Indeed, we found that the post-translational addition of microsomal vesicles to the *in vitro* translation reaction containing the larger form of the light chain of IgG did not yield conversion to the mature form.

At this point, we experienced an extremely frustrating period in our efforts to reconstitute co-translational translocation. All microsomal vesicle preparations that we prepared from several tissues of the usual laboratory animals (mice, rats, chicken, guinea pigs, pigeons and rabbits) severely inhibited the synthesis of the light chain of IgG in the translation reaction. As translation

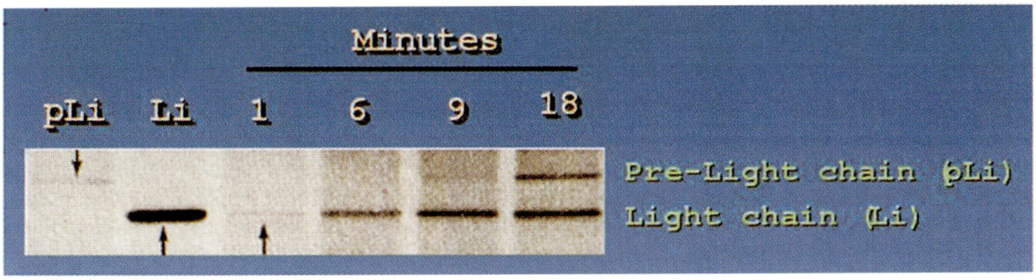

Figure 9. Time course of readout of "detached polysomes" (see Figs 7 and 8) from myeloma cells in a cell-free translation system. ^{35}S-methionine-labeled proteins were separated by SDS-PAGE and visualized by radioautography. Lane 1, major in vitro translation products of poly(A) containing mRNAs from myeloma cells; lane 2, light chains secreted from myeloma cells; lanes 1, 6,9,18, readout products from detached polysomes after 1, 6, 9 and 18 min of incubation in a translation system in the presence of an inhibitor of initiation. From Blobel and Dobberstein (14), with permission.

was completely inhibited by the added microsomal membranes, we were not able to detect translation-coupled translocation across these membranes.

A way out of this dilemma came in December of 1974. The chance arose to prepare and to test yet another rough microsome preparation from yet another source: canine pancreas. Surprisingly, addition of canine pancreas microsomal membranes to the translation reaction did not inhibit translation!! And, indeed, analysis of the translation products by SDS-PAGE showed that in the co-translational presence of microsomal vesicles most of the light chain was synthesized in its mature form (Fig.10, lane 3) (15). This strongly suggested that the signal sequence of the nascent chain engaged the translocation machinery of the membrane and gained access to signal peptidase on the trans-side of the membrane. The mature chain co-sedimented with the vesicles, suggesting that it was segregated within the vesicle lumen.

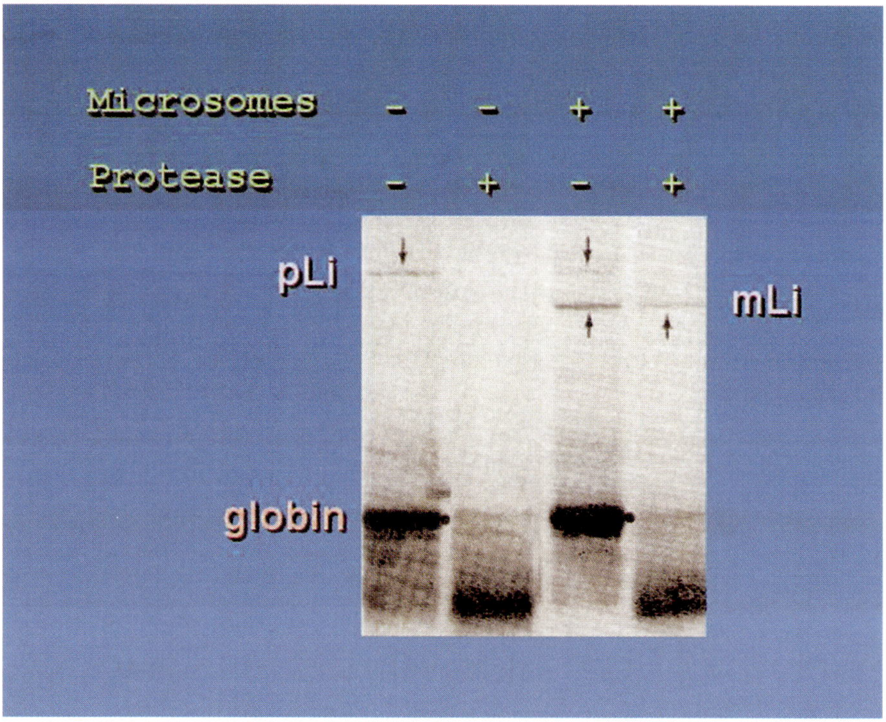

Figure 10. Reconstitution of translocation. Poly(A) containing mRNA form rabbit reticulocytes (coding primarily for the two globin chains) and from myeloma cells (coding primarily for the pre-light chain if IgG) were translated in the absence or presence of canine pancreas rough microsomes. Aliquots were post-translationally incubated in the absence or presence of protease. ^{35}S-methionine-labeled translation products were separated by SDS-PAGE and visualized by radioautography. Note that the mature light chain is synthesized only in the co-translational presence of microsomes and is protected from protease. From Blobel and Dobberstein (15), with permission.

As a control, we also translated rabbit reticulocyte mRNA (15). Translation of this mRNA yielded primarily the two globin chains (Fig. 10, lane 1). The two globin chains are cytosolic proteins and therefore should not contain a signal sequence for translocation into the ER lumen. Moreover, there should not be any shortening of the two globin chains as a result of signal peptidase cleavage in the co-translational presence of pancreas microsomal vesicles. Indeed, we did not observe conversion to smaller forms of the two globin chains in the co-translational presence of microsomal membranes (Fig 10, lane 3). Unexpectedly, however, the two globin chains co-sedimented with the microsomal vesicles. Could co-sedimentation result from co-translational segregation of the two globin chains into the lumen of the ER? Or was co-sedimentation of the two globin chains with the microsomal vesicles the consequence of non-specific binding of the newly synthesized globin chains to the cis-side of the vesicle membrane rather than of translocation to the trans-side of the vesicle membrane, i.e., into the vesicle lumen. To distinguish between these possibilities, we carried out post-translational incubation with proteolytic enzymes. We had shown earlier (16) that membranes of microsomal vesicles are sealed in such a way that they protect the content proteins from attack by externally added proteolytic enzymes while allowing proteolysis of surface-bound proteins and even detachment of ribosomes. Hence, post-translational incubation with proteolytic enzymes should distinguish between proteins bound to the surface of microsomal vesicles and those segregated in the vesicle lumen. We found that the two globin chains were largely degraded by post-translationally added proteases, regardless of whether translation was carried out in the absence of vesicles (Fig. 10, lane 2) or in the presence of vesicles (Fig. 10, lane 4).

Most importantly, post-translational proteolysis of the translation reaction that was carried out in the presence of microsomal vesicles yielded degradation of the larger form of the light chain but not the mature form of the light chain (Fig. 10, lane 4). These results suggested that even in the presence of microsomal membranes some of the nascent chains failed to engage the translocation machinery of the membrane, never gained access to signal peptidase on the trans-side of the vesicles and therefore were completed in the translation reaction as larger forms. Having not been translocated, the larger forms were degraded by added proteolytic enzymes.

Together, these data strongly suggested that we had succeeded in recapitulating the initial events in the secretory pathway in an *in vitro* system. This opened the way for biochemical analysis of protein translocation across the ER. These experiments marked the beginning of molecular cell biology. Successful *in vitro* reconstitution was subsequently achieved for many other membrane-coupled reactions.

AN EXPANDED SIGNAL HYPOTHESIS

The results of the reconstitution experiments encouraged us to further elaborate the simple model of 1971 (Fig. 5). The most significant additional postulate was that protein transport across the microsomal membrane proceeds through a protein-conducting channel (Fig. 11) (14). The protein-conducting channel was envisaged to consist of intergral membrane protein sub-

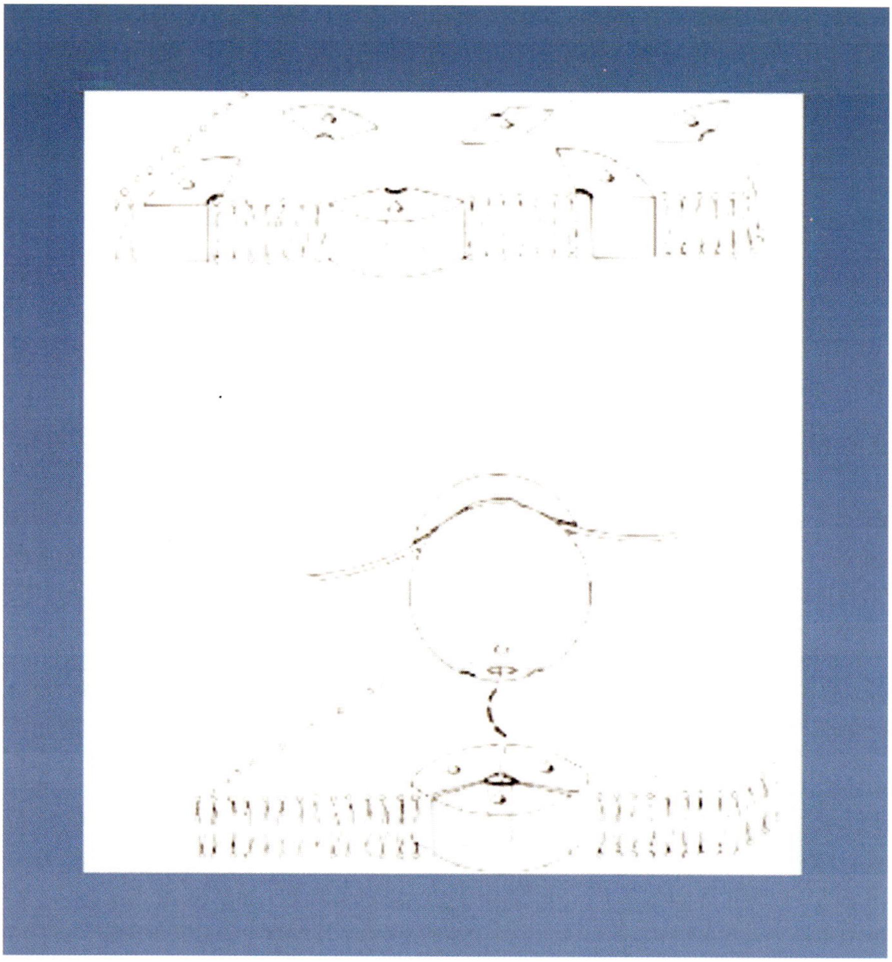

Figure 11. Signal hypothesis 1975. Proposal of a protein-conducting channel (PCC). The PCC was envisaged to be assembled from integral membrane proteins. The signal sequence, cooperatively with the ribosome, was proposed to gate open the channel, with the ribosome attaching to the channel. This arrangement provided for an alignment of a protein-conducting tunnel in the large ribosomal subunit with the protein-conducting channel in the membrane, allowing translation-coupled (co-translational) translocation of the chain across the membrane (see below [Fig. 46] showing relevant data obtained twenty-two years later). This proposal became one of the most controversial aspects of the signal hypothesis. Other proposals suggested a direct transfer of the nascent protein across the lipid bilayer. From Blobel and Dobberstein (14), with permission.

units. The signal sequence and the ribosome were proposed to bind to these subunits and to assemble them into an aqueous channel (Fig. 11) allowing transport of the chain in an aqueous conduit across the membrane. The concept of a protein-conducting channel remained controversial for many years until definitive evidence for it was provided by electrophysiological experiments in 1991 (17). In the intervening 15 years, alternative hypotheses that proposed transfer of secretory proteins across the membrane directly through the lipid bilayer, unassisted by proteins, flourished and became widely accepted.

We also suggested that integral membrane proteins that experience a partial translocation of a segment of the nascent chain to the trans-side of the membrane might use a signal sequence to initiate translocation.

Finally, we predicted that the translocation of proteins across other cellular membranes, e.g., import of cytoplasmically synthesized proteins into mitochondria, is mediated by a signal sequence that would be distinct from that of a secretory protein.

MEMBRANE PROTEIN ASSEMBLY

To test whether an integral membrane protein does, in fact, use a signal sequence to translocate a domain to the trans-side of the ER, we studied the glycoprotein of the vesicular stomatitis virus (G protein of VSV) (18, 19). This protein is part of the viral membrane and contains an amino-terminal domain exposed on the surface of the viral membrane, a single transmembrane segment and a carboxy-terminal domain that interacts with the viral capsid. To achieve this topology in the viral membrane, the amino – terminal domain of the G protein must be translocated to the trans-side of the ER, whereas the carboxy-terminal domain needs to remain untranslocated on the cis-side of the ER.

To accomplish this asymmetric integration into the ER membrane, we reasoned that a signal sequence, indistinguishable from those of secretory proteins, might initiate the translocation process in the ER. An additional sequence, that we termed 'stop-transfer' sequence, would then terminate the translocation process. We speculated that the stop-transfer sequence would open the channel laterally, thereby allowing displacement of the stop-transfer sequence from the aqueous protein-conducting channel to the lipid bilayer. The stop-transfer sequence would then become the trans-membrane segment that is embedded in the lipid bilayer. The carboxy-terminal remainder of the protein would not be translocated, and would remain on the cis-side of the membrane (see Fig. 12).

It should be remembered that in those days there were still no pure mRNAs available that could be readily obtained by *in vitro* transcription from recombinant DNA. All we could do was to isolate total mRNA from cells or tissues that contained major species of mRNA whose translation would give rise to major and readily detectable translation products. The VSV-infected cell was a good model system as most of the total mRNA from these infected cells coded for VSV proteins. Of these, the G protein is the largest and is readily identifiable on the basis of its migration in SDS-PAGE. These experiments were done in collaboration with Harvey Lodish's laboratory at MIT, particularly his graduate student Flora Katz, and by a graduate student in our laboratory, Vishwanath ('Vishu') Lingappa (Fig. 13).

As expected, translation of mRNA from VSV-infected cells yielded the G protein and other VSV proteins (Fig. 14, lane 1, arrow pointing to G_0). However, contrary to our expectation, in the co-translational presence of the

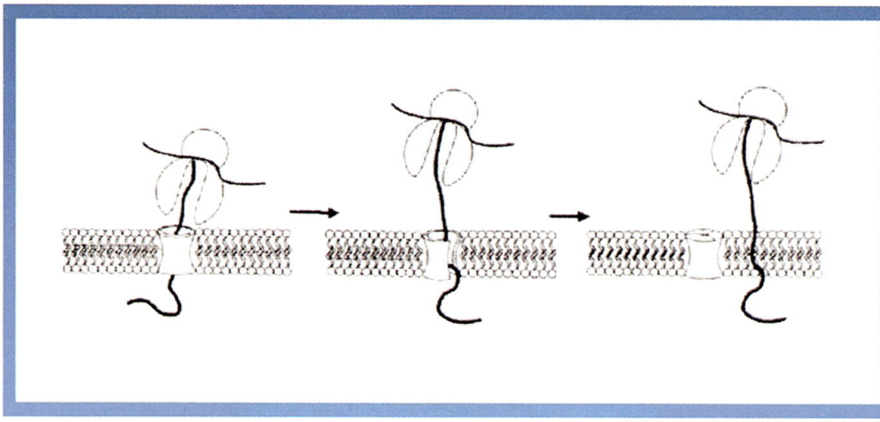

Figure 12. Model for catalyzed integration into the lipid bilayer of integral membrane proteins. The proposal that the initial steps in the integration of membrane proteins are the same as those for the translocation of secretory proteins was made already in the 1975 version of the signal hypothesis. Translocation across the protein-conducting channel (PCC) was envisaged to be initiated by a signal sequence, as in the case of secretory proteins. A stop-transfer sequence, corresponding to the trans-membrane segment of the translocating chain would be recognized by the PCC and open it laterally. The trans-membrane segment would then be displaced from the aqueous PCC, to the lipid bilayer, resulting in the closure of the PCC and the integration of the nascent chain into the lipid bilayer.

microsomal vesicles, at least half of the G protein molecules appeared to be synthesized as a *larger* form (Fig. 14, lane 3, arrow pointing to G_1). Rather than having cleaved off a putative signal sequence, the membranes appeared to have added something to the G protein. We were very puzzled by this result until it occurred to us that the canine microsomal vesicles might contain ac-

Figure 13. Vishwanath (Vishu) Lingappa, 1979. The Rockefeller University, New York.

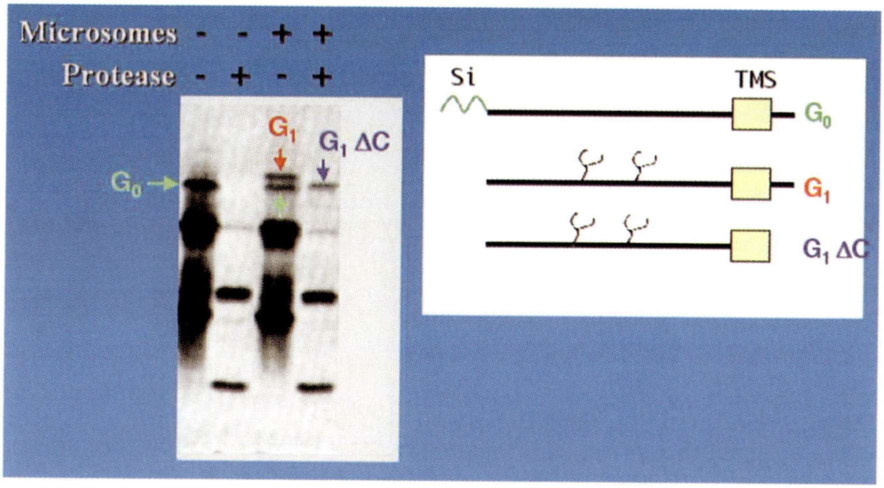

Figure 14. Membrane assembly *in vitro*. Poly(A) containing mRNA from VSV-infected cells was translated in the absence or presence of canine pancreas microsomal membranes. Equal aliquots were post-translationally incubated in the absence or presence of protease. The ^{35}S-methionine-labeled translation products were separated by SDS-PAGE and visualized by radioautography. G_0, translation product in the absence of membranes; it still contains the amino-terminal signal sequence; G_1, translation product in the presence of microsomes; it is core-glycosylated and has lost its signal sequence; $G_1\Delta C$, a form of G_1 that has lost its vesicle-exposed carboxy-terminal tail during post-translational proteolysis. From Katz, *et al.* (18) and Lingappa, *et al.* (19), with permission.

tivated oligosaccharides and oligosaccharidyl transferase to carry out co-translocational core-glycosylation of the G protein. The added core sugars might more than compensate in mass for the loss of a signal sequence and therefore result in a slower migration in SDS-PAGE. Biochemical characterization of both the G_0 and the G_1 species indeed confirmed that G_0 contained an amino-terminal signal sequence that was structurally and functionally equivalent to those of secretory proteins (18, 19). The G_1 species had lost its signal sequence and had acquired core sugars. Moreover, post-translational proteolysis experiments showed complete digestion of G_0 whereas G_1 was converted to a slightly smaller form (Fig. 14, lane 4). This smaller form represented G_1 that had lost its carboxy-terminal tail. These post-translational proteolysis data suggested that the G protein had been correctly integrated into the microsomal vesicles!! As *in vivo*, the bulky amino-terminal domain was translocated to the trans-side of the membrane, whereas the smaller carboxy-terminal domain of about 40 amino acid residues remained untranslocated on the cis-side of the membrane.

Microsomal vesicles were not only able to translocate secretory proteins. They were also competent to asymmetrically integrate membrane proteins and to core glycosylate them! These results demonstrated that the asymmetric integration of membrane proteins is not a spontaneous event but is catalyzed in the ER, presumably by the same machinery that mediates the translocation of secretory proteins.

THE SIGNAL RECOGNITION PARTICLE

Our next goal now was to deconstruct the translocation machinery of the microsomes into functional components. It had been shown that a salt wash of microsomal vesicles removes an activity that is required for translocation of secretory proteins. Such salt washed microsomal vesicles were incapable of co-translational translocation of secretory proteins that were synthesized in a wheat germ cell-free translation system. Addition of the salt wash to the *in vitro* translocation system containing salt-washed microsomal vesicles restored their translocation activity (20). These data suggested that salt extracted a membrane-associated activity that is required for protein translocation. Peter Walter (shown in an Einsteinian pose in Fig. 15), a graduate student in the laboratory, discovered that the activity of the salt wash vanished after only a few hours of storage at 0°C. This frustrated all his attempts to purify this activity until he made the important discovery that minute concentrations of the non-ionic detergent Nikkol stabilized the activity. We reasoned that Nikkol protected a hydrophobic site of the activity, perhaps the binding site for the hydrophobic signal sequence. Incorporating the hydrophobic-site-idea into his purification protocol, he was able to rapidly purify the activity. The activity sedimented at about 11S and by SDS-PAGE consisted of six proteins (21). Based on an extensive analysis of its function (22, 23, 24) we termed the protein complex "signal recognition protein" (SRP).

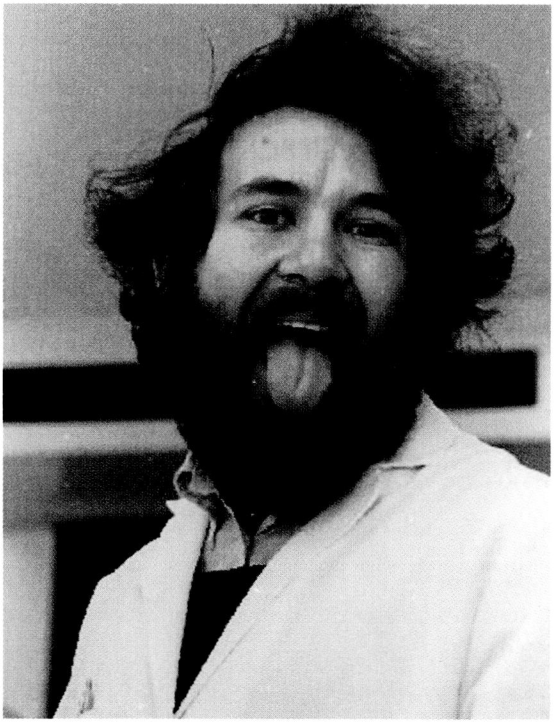

Figure 15. Peter Walter, 1980. The Rockefeller University, New York. Obviously inspired here by the famous photo of Einstein.

It took us another two years before we realized that the purified activity also contained a 7S RNA (25)! We then changed the name to signal recognition particle, leaving the acronym SRP intact. A considerable amount of work in both Peter Walter's and Bernhard Dobberstein's laboratory has led to a functional and structural model for the SRP and to the discovery of an interesting homolog in bacteria (Fig. 16) (26).

Functional analyses (22, 23, 24) revealed that canine pancreas SRP bound with low affinity to wheat germ ribosomes. However, SRP bound with high affinity, when wheat germ ribosomes were programmed with a mRNA coding for a secretory protein (bovine preprolactin), not when programmed with a mRNA coding for cytosolic proteins, such as the globin chains. Also, if translation was carried out in the presence of hydroxy-leucine instead of leucine, there was greatly reduced binding of SRP. As the signal sequence for preprolactin is rich in leucine, this result suggested that SRP interacted directly with the signal sequence. Moreover, the interaction of SRP caused a translation ar-

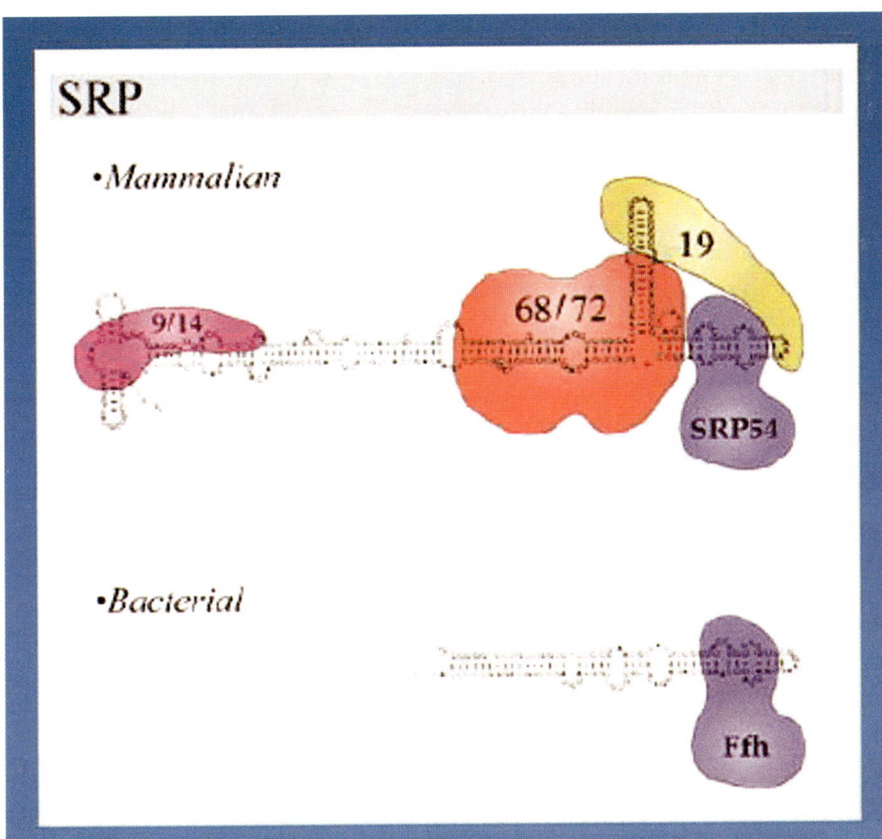

Figure 16. Signal Recognition Particle (SRP) in mammalian cells and in bacteria. Both SRPs contain an RNA and proteins. The mammalian SRP contains six proteins of indicated mol.wt. in kDaltons. The bacterial SRP contain a shorter RNA and a single protein (ffh, fifty-four-homolog). From Walter and Johnson (26), with permission.

rest when ribosomes were programmed with a mRNA for a secretory protein, but not when programmed with mRNA for cytosolic proteins. The size of the arrested nascent secretory protein corresponded to about 70 amino acid residues, suggesting that binding of the signal sequence occurred after the signal sequence was fully exposed following exit from the tunnel in the large ribosomal subunit (Fig. 17). When microsomal membranes were added, there was release of the translation arrest, suggesting that the ER membrane contains an SRP receptor that is capable of releasing SRP-mediated translation arrest.

We found that the *in vitro* assembled polysomes synthesizing secretory proteins, but not those synthesizing cytosolic proteins, bound to SRP-depleted (salt-washed) microsomes, but only in the presence of SRP. Binding was abolished when SRP was pretreated with N-ethyl maleimide.

These data strongly supported the notion that protein translocation across the ER is a receptor-mediated event and ruled out proposals that chain translocation occurs spontaneously and unassisted by proteins.

The discovery of the signal recognition particle was an important milestone in the molecular analysis of protein translocation across the ER membrane. Its function corresponded to that of the "binding factor' whose existence had been predicted in 1971 (see Fig. 5). It was the first component of any of the cellular membrane translocation systems that was isolated and characterized. It provided strong support for the predictions made in the signal hypothesis.

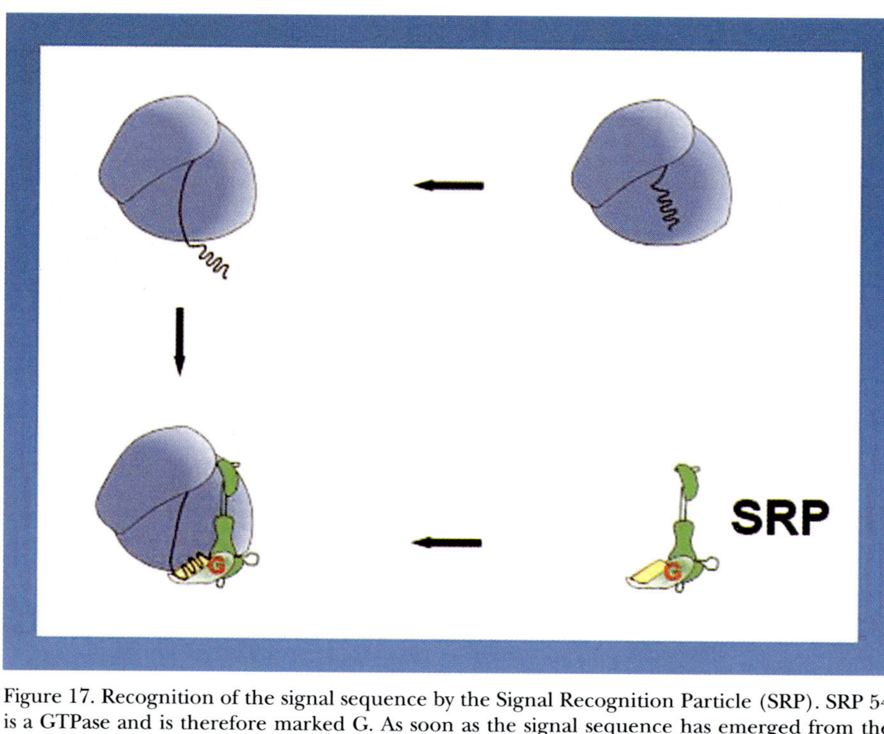

Figure 17. Recognition of the signal sequence by the Signal Recognition Particle (SRP). SRP 54 is a GTPase and is therefore marked G. As soon as the signal sequence has emerged from the large ribosomal subunit, SRP can bind to it.

Figure 18. Reid Gilmore, 1985. The Rockefeller University, New York.

SRP RECEPTOR

Reid Gilmore, a postdoctoral fellow in our laboratory (Fig. 18) used the ability of salt-washed microsomal membranes to release the SRP-mediated elongation arrest of the synthesis of secretory proteins as an assay to follow

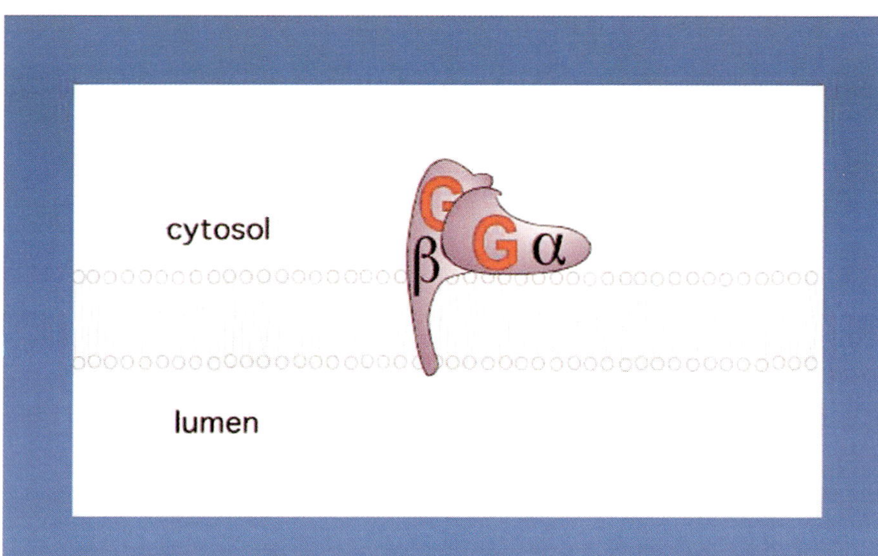

Figure 19. The SRP receptor is a heterodimer of two GTPases; the beta subunit is an integral membrane protein and the alpha subunit a peripheral membrane protein.

the purification of this activity. Together with the membrane's signal peptidase (see below), the arrest-releasing activity could be solubilized by treatment of the salt-washed microsomal membranes with non-ionic detergent and moderate salt concentrations. The purified SRP receptor consisted of two subunits, a larger subunit and a smaller subunit (27–31). The larger subunit is a peripheral membrane protein whereas the smaller subunit is an integral membrane protein with one transmembrane segment. Both proteins were shown to be GTP binding proteins (30, 31) (Fig. 19). SRP and SRP receptor interact with each other (32) (Fig. 20). The SRP receptor is located exclusively to the ER (including the outer nuclear envelope) and is present in the ER in sub-stoichiometric quantities relative to membrane-bound ribosomes (27). This suggested that the SRP-SRP receptor interaction is transient in nature. It serves in the targeting but not in the subsequent translocation of the chain.

With the discovery of the SRP and its cognate SRP receptor, the components involved in signal sequence recognition and targeting to the ER had been isolated.

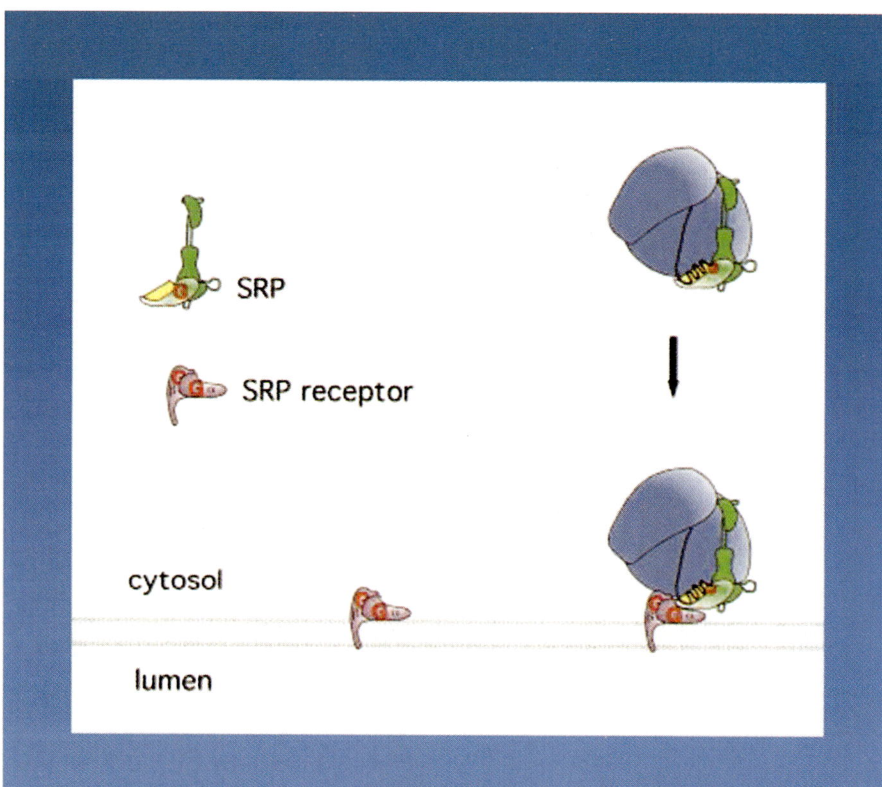

Figure 20. The ribosome/nascent chain/SRP complex binds to the SRP receptor in the ER.

SIGNAL PEPTIDASE

The signal peptidase cleavage site of nascent secretory proteins is accessible to signal peptidase only on the trans-side of the membrane (see above). Therefore, signal sequence cleavage is dependent on translocation. Robert Jackson in our laboratory developed a translocation-independent assay for signal peptide cleavage. He solubilized the microsomal signal peptidase by detergent. Using an *in vitro* synthesized presecretory protein as a substrate he demonstrated correct endoproteolytic removal of the signal sequence by the detergent form of the enzyme (33).

The solubilized microsomal signal peptidase was purified by Emily Evans, a graduate student in our laboratory (Fig. 21). The purification of this enzyme was a major challenge as its activity required the presence of lipids and detergents throughout all purification procedures. Omission of lipids led to rapid inactivation of the enzyme. Surprisingly, the purified microsomal signal peptidase turned out to be a complex of five non-identical subunits (34) (Fig. 22), all of which are integral membrane proteins (35, 36, 37).

With the SRP receptor and signal peptidase, two components of the microsomal membrane that could be solubilized by detergents and assayed independently of intact vesicles and of translocation, had been purified. Our next goal was to purify the putative protein-conducting channel. The proteins making up such a channel were likely to be integral membrane proteins. To purify them from other membrane proteins required detergent solubilization. However, unlike SRP receptor and signal peptidase, channel proteins require reconstitution into proteoliposomes to unambiguously assay their activity.

Figure 21. Emily Evans, 1986. The Rockefeller University, New York.

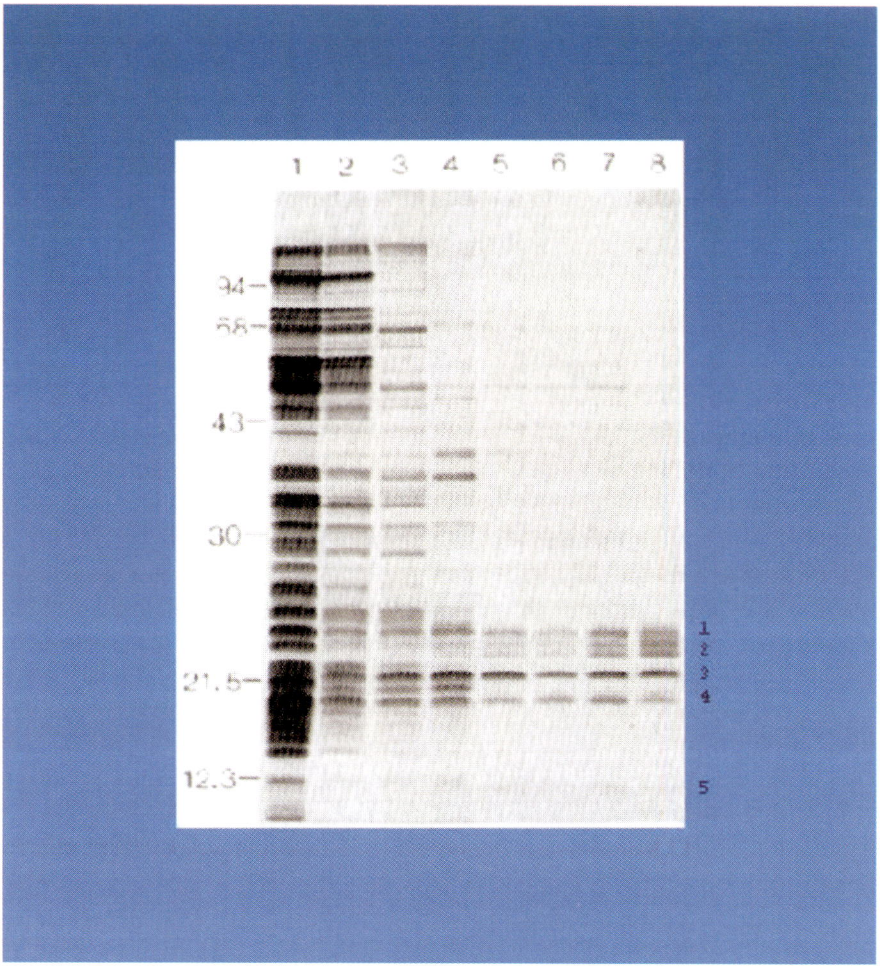

Figure 22. Purification of mammalian signal peptidase, a pentameric complex of integral membrane proteins. Band 2 represents the glycosylated and non-glycosylated form of the same polypeptide. From Evans, *et al.* (34), with permission.

RECONSTITUTION OF TRANSLOCATION-ACTIVE PROTEOLIPOSOMES

The reconstitution of translocation-active proteoliposomes had been a high priority on our agenda for quite some time before Christopher Nicchitta (Fig. 23) joined our laboratory as a postdoctoral fellow. However, many attempts by the most talented people in our laboratory had failed to achieve this goal. We had arrived at the conclusion that reconstitution, if at all possible, would probably occur with very low efficiency. Presumably, after detergent solubilization and removal of the detergent to form proteoliposomes, the components that are required for co-translational translocation are not necessarily reconstituted in the asymmetric orientation and in the stoichiometry in which they exist in the microsomal vesicles. Instead, reconstituted proteoli-

Figure 23. Christopher Nicchitta, 1989. The Rockefeller University, New York.

posomes may contain these proteins in a scrambled fashion and in non-stoichiometric amounts rendering them inefficient or inactive in translocation. The SRP receptor, the signal peptidase complex, or the constituents of the protein-conducting channel and perhaps other components that participate in protein translocation may not regain their asymmetric orientation in the reconstituted proteoliposome vesicles. Signal peptidase and the SRP receptor, for example, exhibit their active sites in opposite orientations, on the trans-side and on the cis-side of the vesicles, respectively.

Through careful choice of detergents and of other conditions (pH as well as salt- and detergent concentrations) we were able to obtain translocation-competent proteoliposomes (38) (Fig. 24).

These experiments were seminal. They showed that after complete detergent solubilization of microsomal membranes, the detergent forms of integral membrane proteins required for protein translocation, including a putative protein-conducting channel, could be reconstituted in an active form. This paved the way for the isolation of the protein-conducting channel. It also allowed the subsequent reconstitution of translocation-active proteoliposomes from purified components.

ELECTROPHYSIOLOGICAL DETECTION OF THE PROTEIN-CONDUCTING CHANNEL

We first proposed the concept of a protein-conducting channel (PCC) made up of integral membrane proteins in 1975 (14). However, it remained a highly controversial idea, particularly as there was no direct evidence for it. Alternative models of protein translocation directly through the lipid bilayer

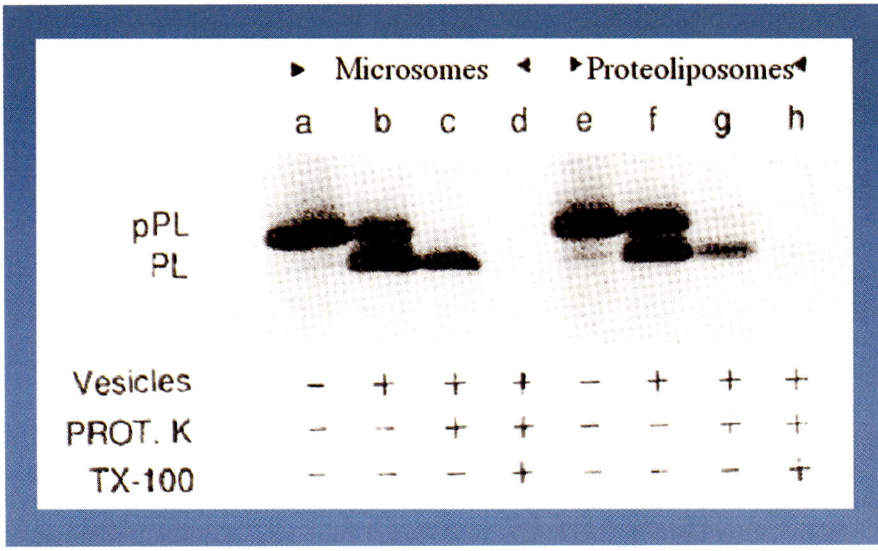

Figure 24. Reconstitution of protein translocation activity into proteoliposomes. Comparison of protein translocation activity of microsomes and reconstituted proteoliposomes. pPL, preprolactin; PL, prolactin; Prot. K, proteinase K. TX-100, the non-ionic detergent Triton X-100. From Nicchitta and Blobel (38), with permission.

flourished (39, 40). In 1985, we showed (41) that nascent chains in the process of translocation are accessible to aqueous perturbants. Although these data were consistent with an aqueous PCC, they could not rule out that an aqueous environment was created transiently by the hydrophilic head groups of the lipids rather than proteins forming such an aqueous channel.

Although electrophysiology was a standard approach for identifying and characterizing ion-conducting channels, it had never been applied to the detection and characterization of any protein-conducting channels. It was difficult to predict how such channels would behave electrophysiologically. We had postulated that they would be opened and closed for each protein translocation event (14). Moreover, we reasoned that the conductance of a protein across the channel is unlikely to be accompanied by a significant co-conductance of ions or small molecules. If this were the case, it would be difficult to maintain the distinct composition of ions (particularly calcium) and small molecules in the ER lumen, the 'reticuloplasm', as compared to that in the cytoplasm. PCCs are likely to be closed when they are not conducting a polypeptide chain. Moreover, they are most likely designed to prevent the co-conduction of ions when they are opened for and during translocation of a protein. Therefore, it was not clear whether PCCs could be revealed at all by electrophysiological measurements. With these caveats in mind, Sanford Simon (Fig. 25), a post-doctoral fellow in our laboratory, set out on a most remarkble and exciting journey into completely uncharted territory: to explore PCCs by electrophysiological approaches. These experiments were begun in collaboration with Joshua Zimmerberg at NIH (42).

Because rough microsomes were too small for patch-clamp experiments,

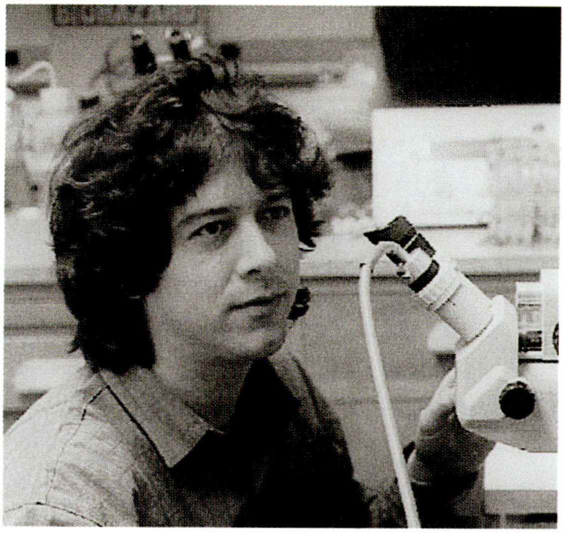

Figure 25. Sanford M. Simon, 1991. The Rockefeller University, New York.

we decided to use the planar lipid bilayer system developed by Mueller *et al.* in 1962 (43). In this system, vesicles are fused to a planar bilayer that is formed in a hole of a partition separating two chambers (Fig. 26). Addition of microsomes to the cis-chamber yielded occasionally fusion of a single rough microsomal vesicle as evidenced by an increase in conductance. Some preparations yielded unitary conductances of 20, 55, 80 and 115 pico Siemens (pS) in 45 mM potassium glutamate (42) (Fig. 27). By electron microscopy it can be estimated that each of the rough microsomal vesicles contains about a hundred membrane-bound ribosomes (Fig. 28). Many of these membrane-bound ribosomes are potentially in the process of conducting a chain across the membrane. Therefore, it was unlikely that the few unitary conductances that we observed were related to PCCs. Nevertheless, the fact that we could

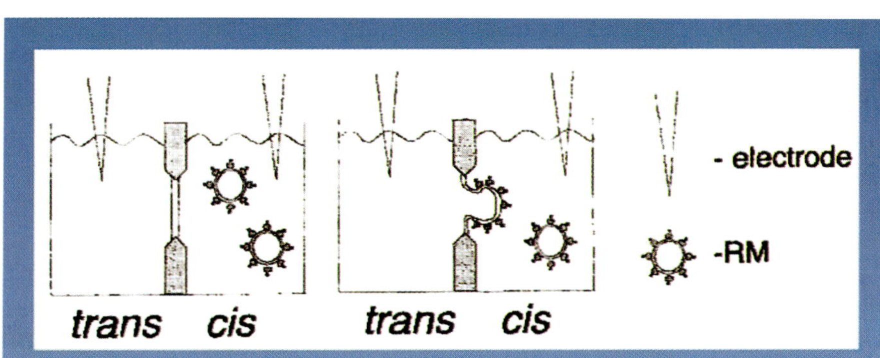

Figure 26. Schematic representation of the bilayer system. The cis chamber is separated from the trans chamber by a plastic division (thick line) with a small circular hole that contains a planar lipid bilayer. Left, rough microsomes (RM) are added to the 'cis' chamber. Right, a single rough microsome has fused with the planar bilayer. From Simon and Blobel (17), with permission.

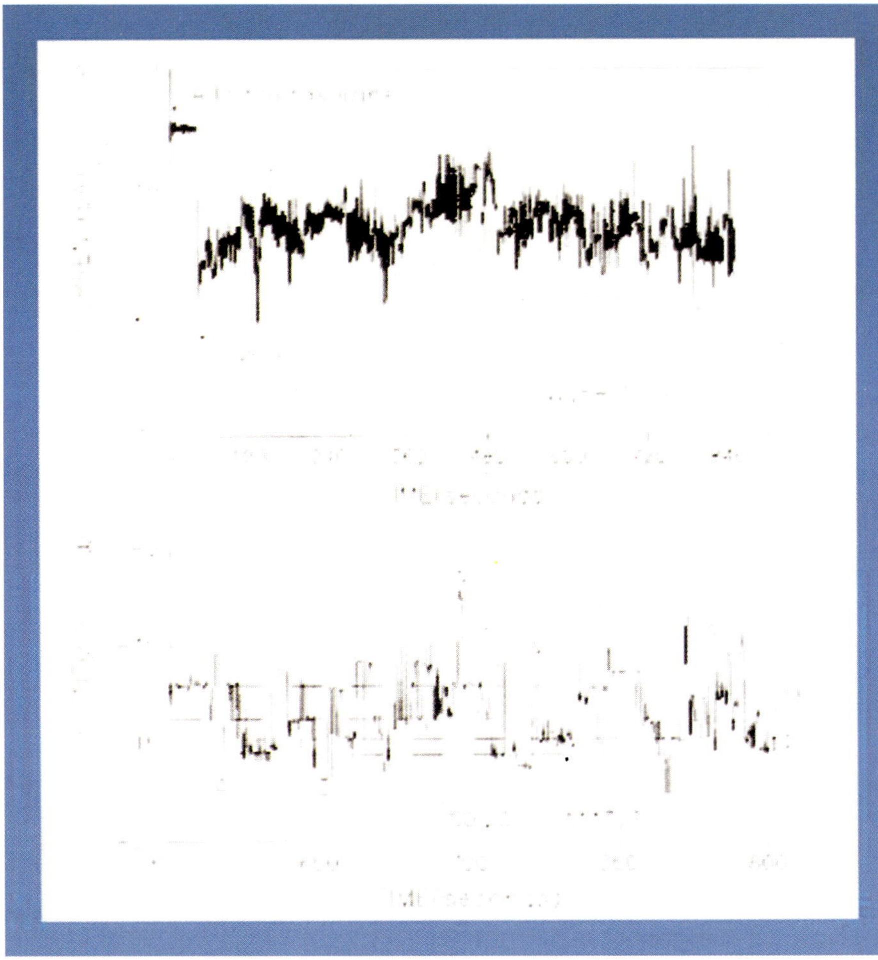

Figure 27. Channels in rough microsomes. (A) Fusion of a single rough microsome vesicle to the planar bilayer. After 45 sec there was a substantial increase in the fluctuations of the current (B). A segment of the record of A from 600 to 800 sec is shown in an expanded time scale. Four distinct channel sizes can be distinguished: 20 pS, 50 pS, 80 pS and 115 pS. The function of these channels remains to be determined. From Simon, *et al.* (42), with permission.

detect conductances at all suggested that fusion of a rough microsomal vesicle with the planar bilayer had taken place.

If the PCCs are indeed either closed or electrically silent when occupied by nascent chains, would it be possible to reveal these channels by releasing the nascent chain without closing them? Could puromycin, an analog of amino acyl tRNA (Fig. 29) achieve this? The peptidyl transferase activity of ribosomes couples puromycin to the carboxy-terminal end of the nascent chain. This causes chain release from the ribosome. In the case of rough microsomes, it had been established some time ago that the chain is vectorially discharged into the lumen of the vesicle (2). When the puromycin reaction is carried out at low salt concentrations the chains are vectorially discharged,

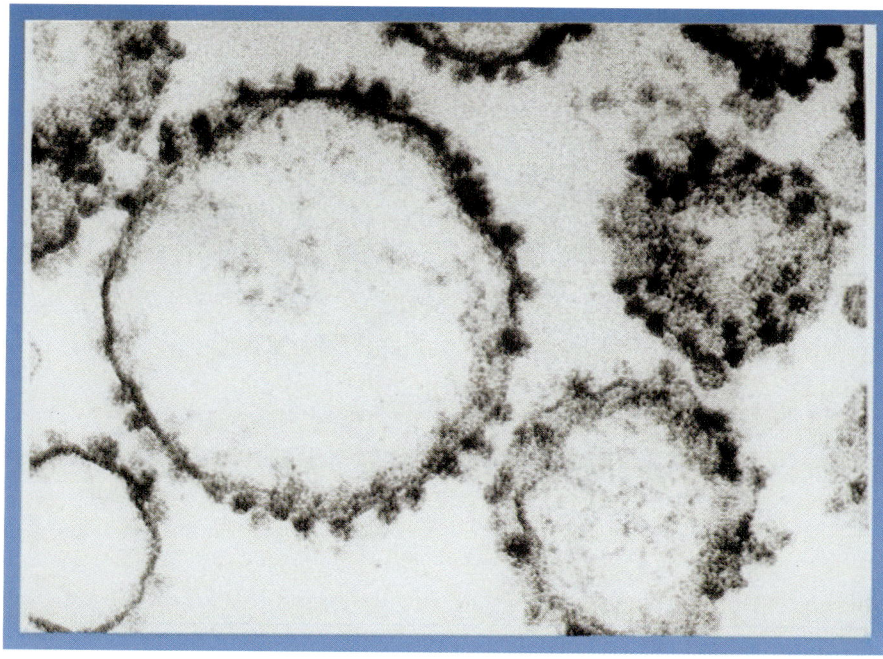

Figure 28. Electron micrograph of isolated canine pancreas rough microsomes.

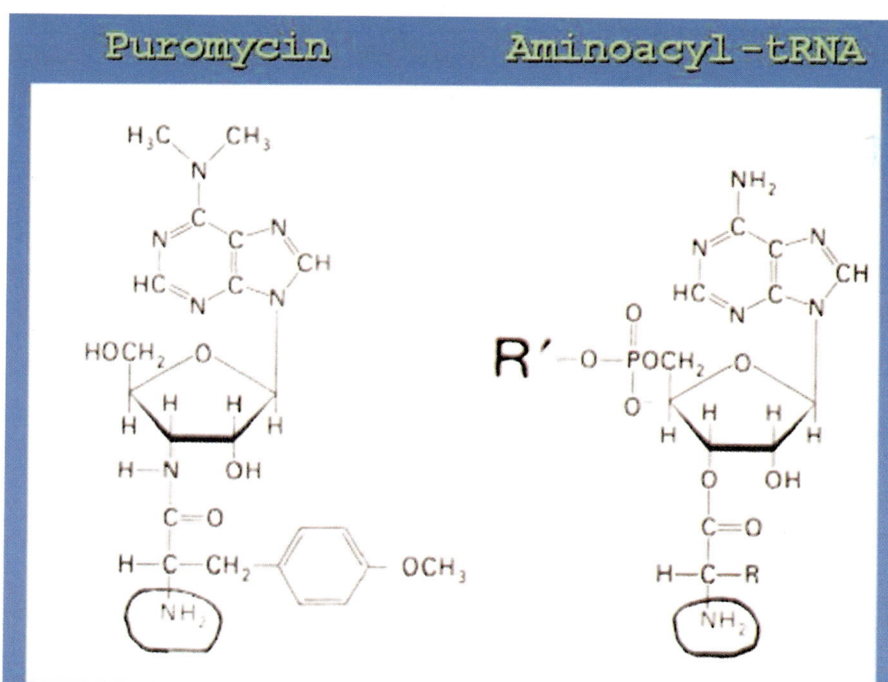

Figure 29. Puromycin is an analog of aminoacyl-tRNA.

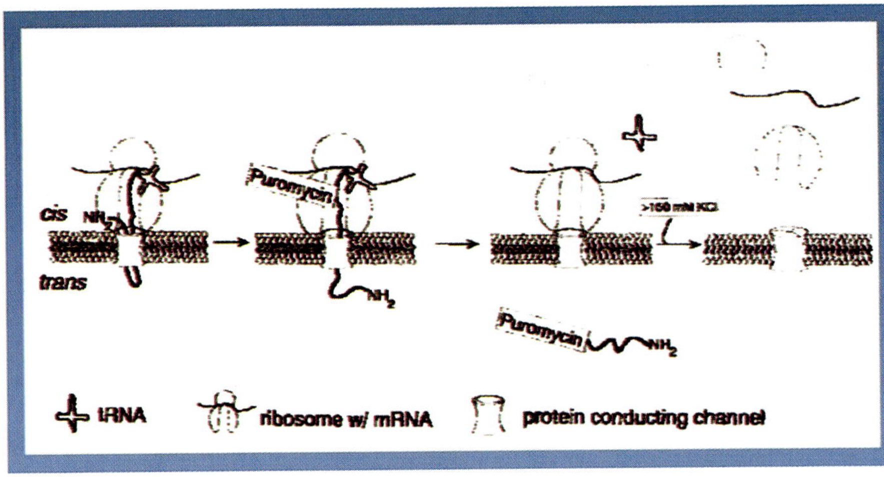

Figure 30. Schematic representation of puromycin-induced vectorial release of the nascent polypeptide chain and subsequent disassembly of rough microsomes by high concentrations of salt. From Simon and Blobel (17), with permission.

but the ribosomes remain attached to the membrane (8). Could the attached ribosome keep a cleared PCC in an open configuration (Fig. 30)? And could subsequent treatment with high salt release the ribosome and close the PCC (Fig. 30) (8)?

Based on these considerations we added puromycin to the cis-chamber of the bilayer that contained a fused rough microsome vesicle (Fig. 26). We observed a huge increase in conductance (17) (Fig. 31). This effect was specific

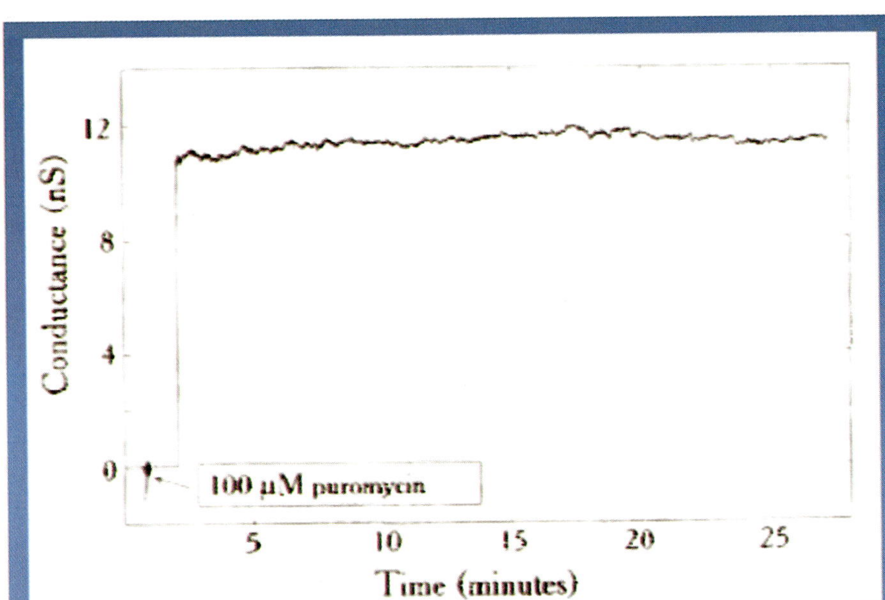

Figure 31. Puromycin-induced clearance of protein-conducting channels in rough microsomes. A sharp increase in conductance occurred within 45 sec of addition of puromycin (100 μM) to the cis side of the microsomal membrane. From Simon and Blobel (17), with permission.

as it was observed only when puromycin was added to the cis-chamber, not when it was added to the trans-chamber (17). The cis-chamber, but not the trans chamber, contained the exposed ribosomes of the fused rough microsomal membrane (see Fig. 26). Puromycin, when added to the trans-chamber, could not access the ribosomes and be coupled to the nascent chain. But when subsequently added to the cis-chamber, it again caused a huge increase in conductance (Fig. 32). These side-specific data strongly suggested that puromycin indeed cleared the nascent chain and that at low salt the PCCs stayed in an open configuration and were able to conduct ions.

How many PCCs were cleared by the addition of puromycin? Could very low concentrations of puromycin make it possible to demonstrate a time-resolved clearance of individual PCCs? Indeed, when much lower concentrations of puromycin were added to the cis-chamber, we observed clearance, one at a time, of individual PCCs (17) of 220 pS (Fig. 33). Hence, the PCC contains a very large aqueous pore that conducts about ten times more ions than an ordinary ion-conducting channel (17).

Could one close the puromycin-revealed PCC by dissociating the ribosome by high salt concentrations? Indeed, when a single PCC (Fig. 34) was revealed at low concentrations of puromycin and at low salt, it was closed by a subsequent increase in salt concentration (Fig. 35). This result was consistent with

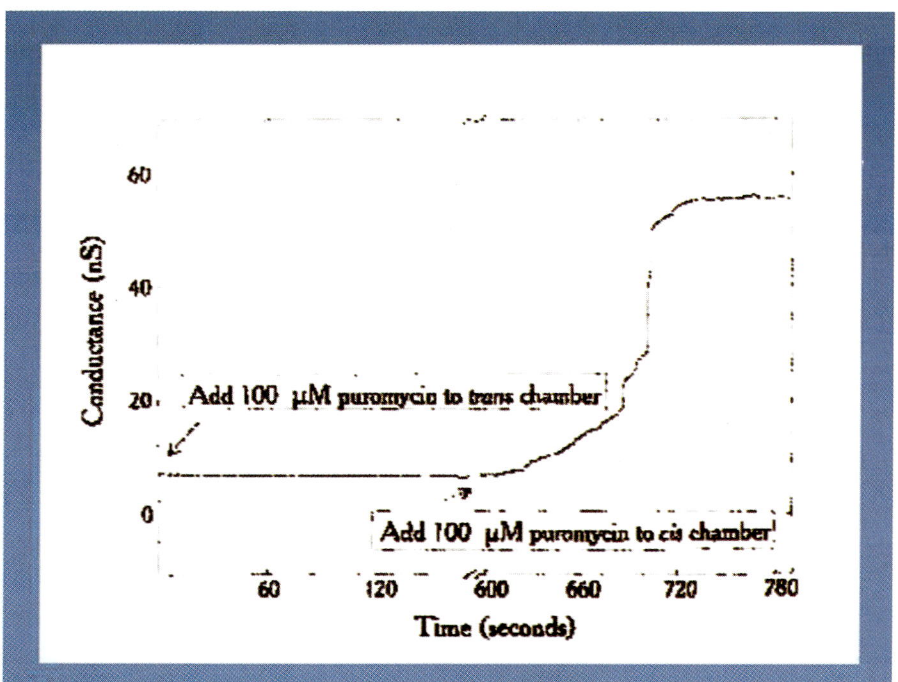

Figure 32. Specificity of puromycin action. At the first arrow puromycin was added to the trans chamber (lumenal side of the microsomal membrane). After 10 min, puromycin was added to the cis chamber (ribosomal side of the microsomal membrane). A substantial increase in conductance was observed. From Simon and Blobel (17), with permission.

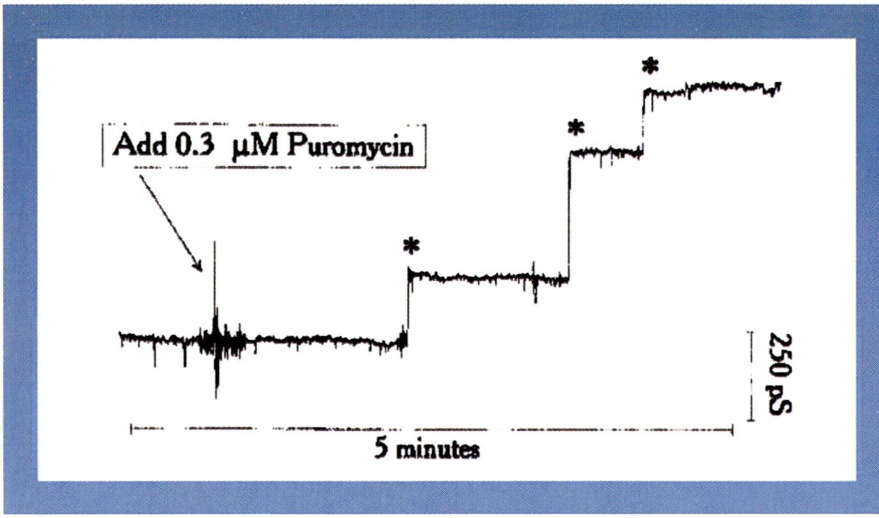

Figure 33. Single puromycin-revealed channels. Puromycin (0.3 µM) was added to the ribosomal side of the microsomal membrane. Discrete consecutive jumps in the conductance of 220 pS, 440 pS and 220 pS were observed (at asterisks). At faster time resolution, the 440 pS jump resolved into two discrete 220 pS steps. From Simon and Blobel (17), with permission.

the idea that the puromycin-revealed PCCs are kept open by attached ribosomes but are closed when ribosomes are dissociated (see model Fig. 30).

Can PCCs be demonstrated also in other membranes and by methods other than puromycin-induced clearance of translocating chains? Does the signal sequence serve as a ligand to open (or assemble) the PCC as was postulated in the signal hypothesis (14)? To answer these questions we turned to the prokaryotic plasma membrane. It had been suggested that in evolution

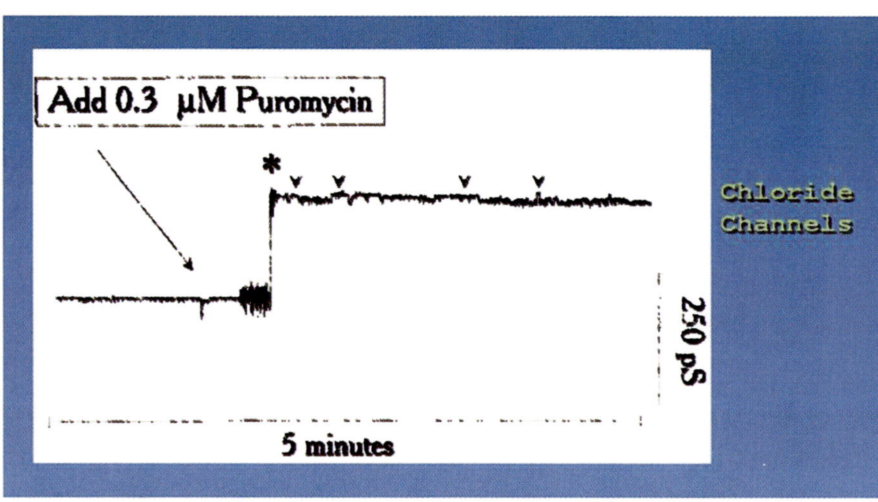

Figure 34. A single puromycin-revealed channel. Puromycin was added to the ribosomal side of the microsomal membrane and then removed. Small chloride channels are marked by asterisks. From Simon and Blobel (17), with permission.

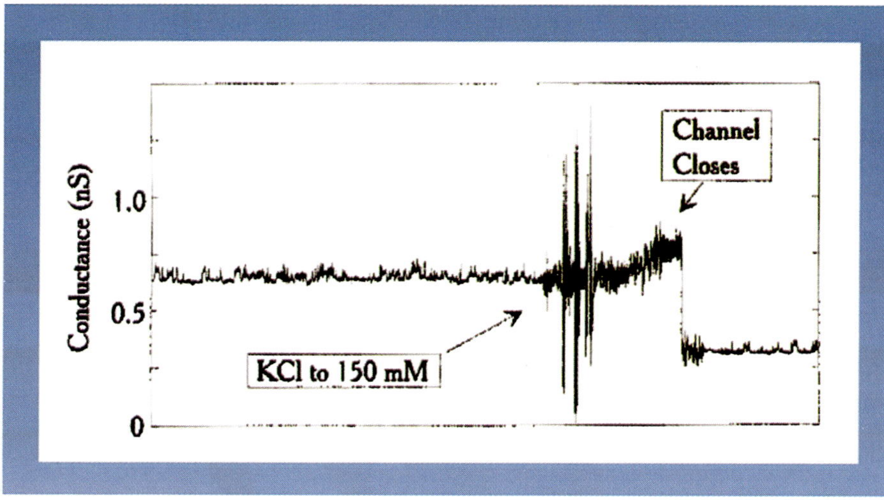

Figure 35. Closure of puromycin-revealed channel. Addition of 150 mM KCl closed a single puromycin-revealed channel. From Simon and Blobel (17), with permission.

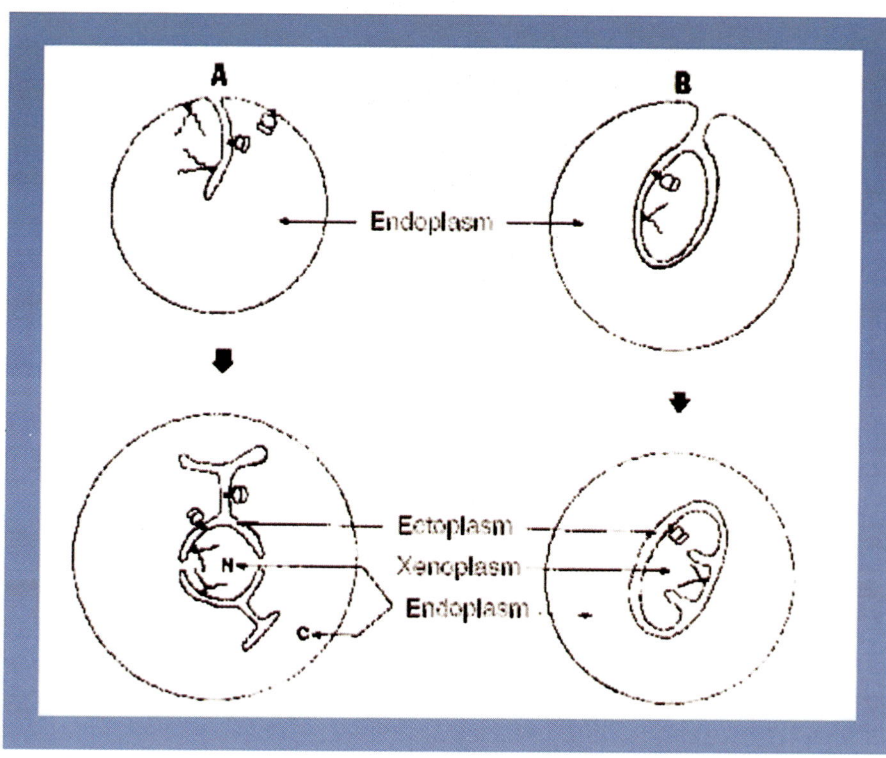

Figure 36. Evolution of eukaryotic intracellular membranes. Left, invagination of the prokaryotic plasma membrane with indicated ribosome and DNA binding sites yields the endoplasmic reticulum and the nuclear envelope. Right, uptake of a prokaryotic progenitor cell yields mitochondria and chloroplasts. From Blobel (44), with permission.

the eukaryotic ER arose by invagination of the prokaryotic plasma membrane (Fig. 36) (44). In fact, signal sequences for bacterial secretory proteins function in translocation across the ER of eukaryotic cells and, vice versa, signal sequences addressed to the eukaryotic ER function in translocation across the prokaryotic plasma membrane. It was therefore conceivable that the PCC of the prokaryotic plasma membrane resembles that of the eukaryotic ER.

When protoplasts of *Escherichia coli* (Fig. 37) are fused from the cis-chamber to a planar bilayer, the putative signal sequence-binding site of the PCC will be exposed on the trans-chamber (Fig. 38). The addition of low concentrations of synthetic signal peptide of the bacterial LamB protein (Fig. 39) to the trans-chamber, but not to the cis-chamber, caused a stepwise increase in conductance (Fig. 40). The stepwise increase in conductance was observed only in the presence of high salt. Each step is likely to represent the binding of one signal peptide to one signal peptide binding site of the PCC. High salt would reinforce the largely hydrophobic interactions between the hydrophobic signal peptide and the PCC and thus keep the PCC in an open configuration for long periods of time (45). The conductance properties of the prokaryotic plama membrane PCC were found to be similar to those of the PCC

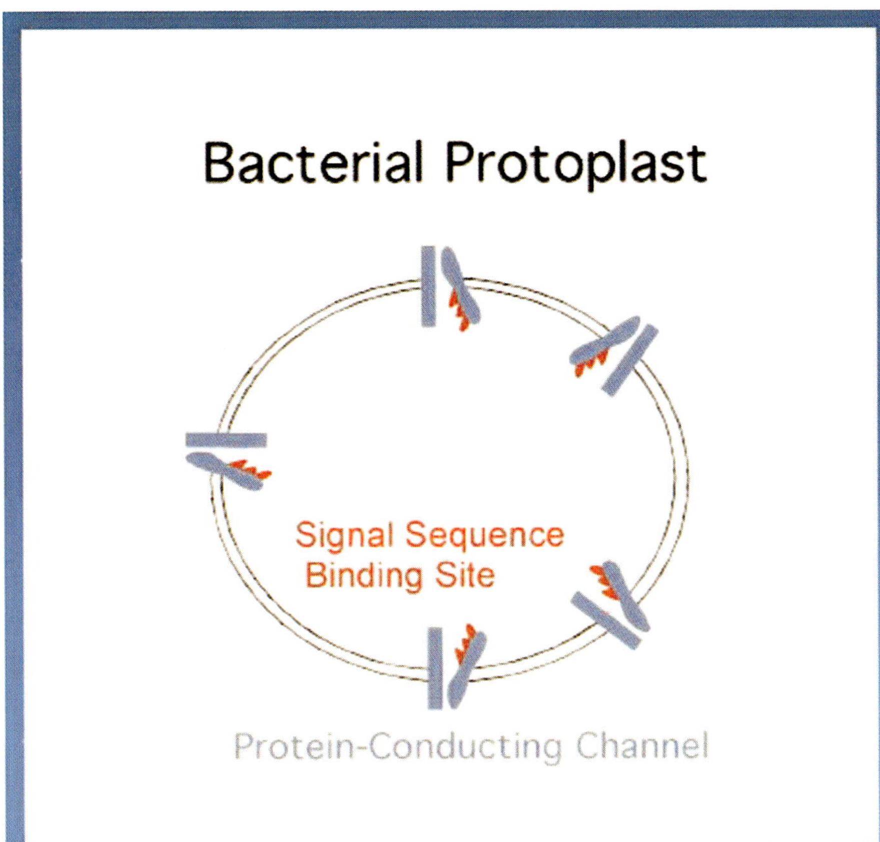

Figure 37. Schematic drawing of a bacterial protoplast with protein-conducting channels and their signal sequence binding sites.

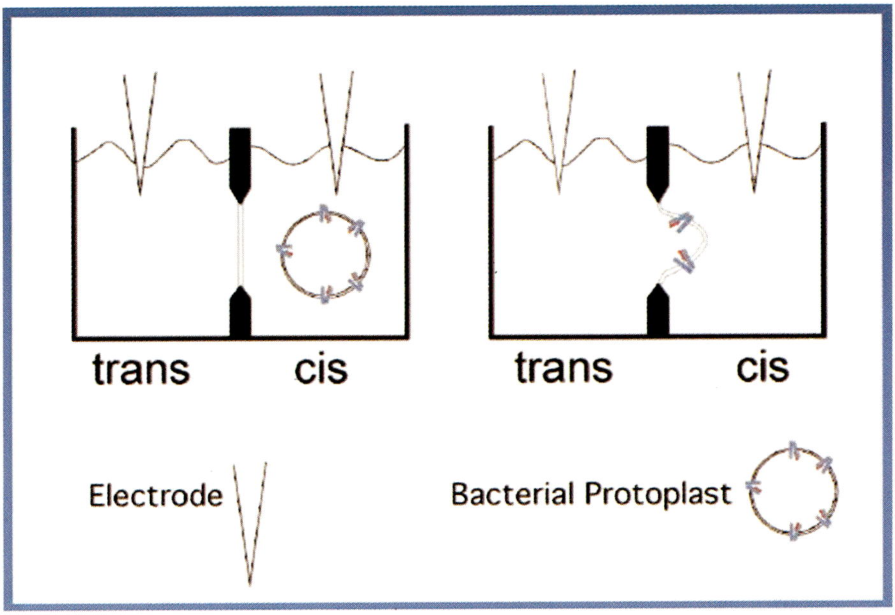

Figure 38. Schematic drawing of a bacterial protoplast added to the cis chamber and fused to the planar bilayer. Note that the cis side of the bacterial plasma membrane with the signal sequence binding sites of the protein-conducting channels is exposed to the trans chamber.

in the ER, in support of the notion that the two channels share a common history and are conserved.

Taken together, these data strongly supported the existence of aqueous protein-conducting channels that, like ion-conducting channels, were composed of proteins and that the signal sequence indeed served as one of the ligands to open the channel. These results led to a paradigm shift in cell biology and strongly argued against models that had reigned supreme for many years and that had proposed transfer of proteins directly through the hydrophobic core of the lipid bilayer.

VISUALIZATION OF THE PROTEIN-CONDUCTING CHANNEL

By genetic approaches, Schekman and co-workers (46) had identified Sec61 as a candidate for the protein-conducting channel in yeast. Rapoport and co-workers, used subfractionation and reconstitution into proteoliposomes (see

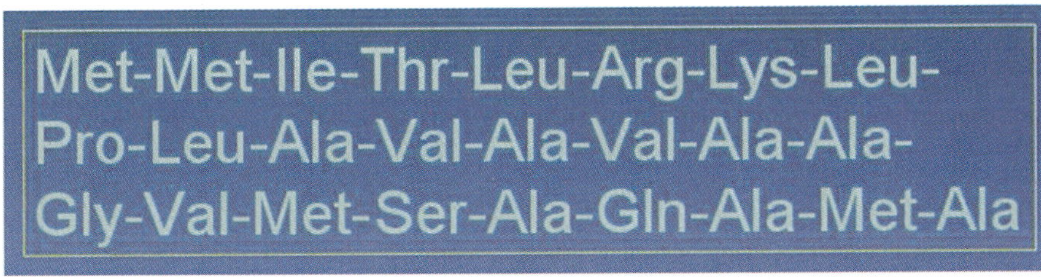

Figure 39. Synthetic signal peptide of the preLamB protein of *E. coli*.

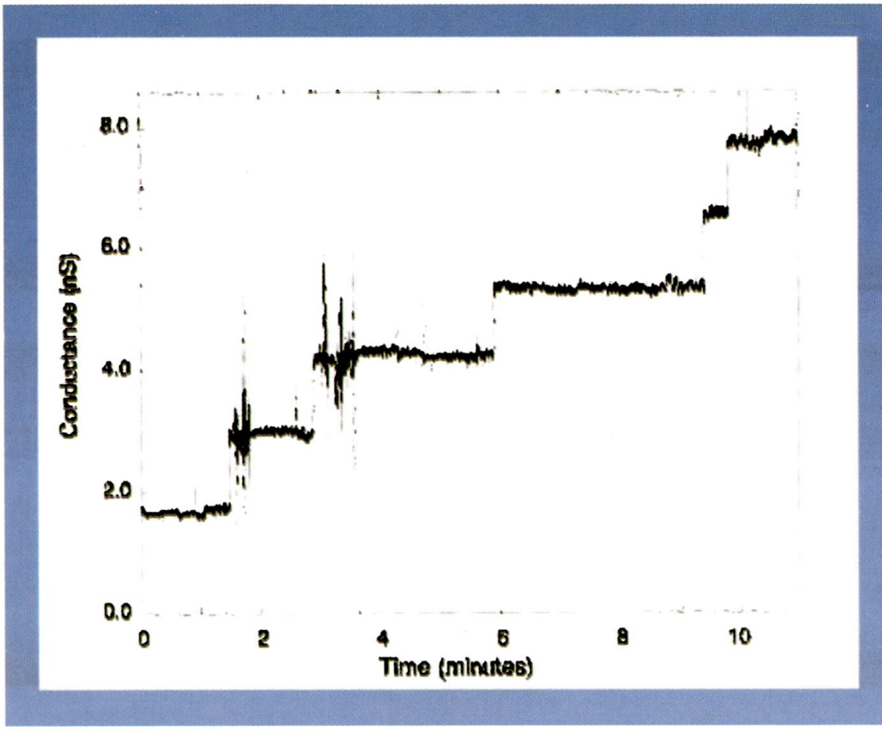

Figure 40. Signal peptide-gated channel remains open in high salt. *E. coli* protoplasts were fused to the planar bilayer. Synthetic preLamB signal peptide was added to the trans chamber (see Figure 38) in a final concentration of 0.2 nM and in the presence of 700 mM KCl. Single channels of 220 pS remained open in high salt. Presumably, the high concentration of salt reinforced the hydrophobic interactions of the signal peptide with the signal peptide binding site of the protein-conducting channels, leaving them in an open configuration. From Simon and Blobel (45), with permission.

above) to isolate the mammalian counterparts of the yeast PCC (47, 48). Their work established that the PCC consisted of a heterotrimer, termed the alpha, beta and gamma subunits of Sec61 (Fig. 41).

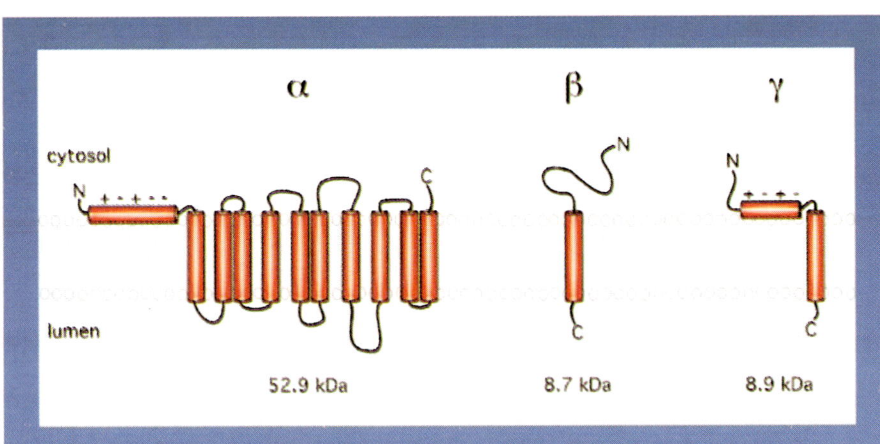

Figure 41. Topology of the Sec61 heterotrimer in the endoplasmic reticulum membrane.

Figure 42. Roland Beckmann, 1999. The Rockefeller University, New York.

To visualize the PCC attached to the ribosome, Roland Beckmann, a post-doctoral fellow in our laboratory (Fig. 42), was able to isolate the Sec61 complex by detergent solubilization of yeast microsomes (Fig. 43) and to bind it to isolated yeast monomeric ribosomes (48). Binding was saturable and sug-

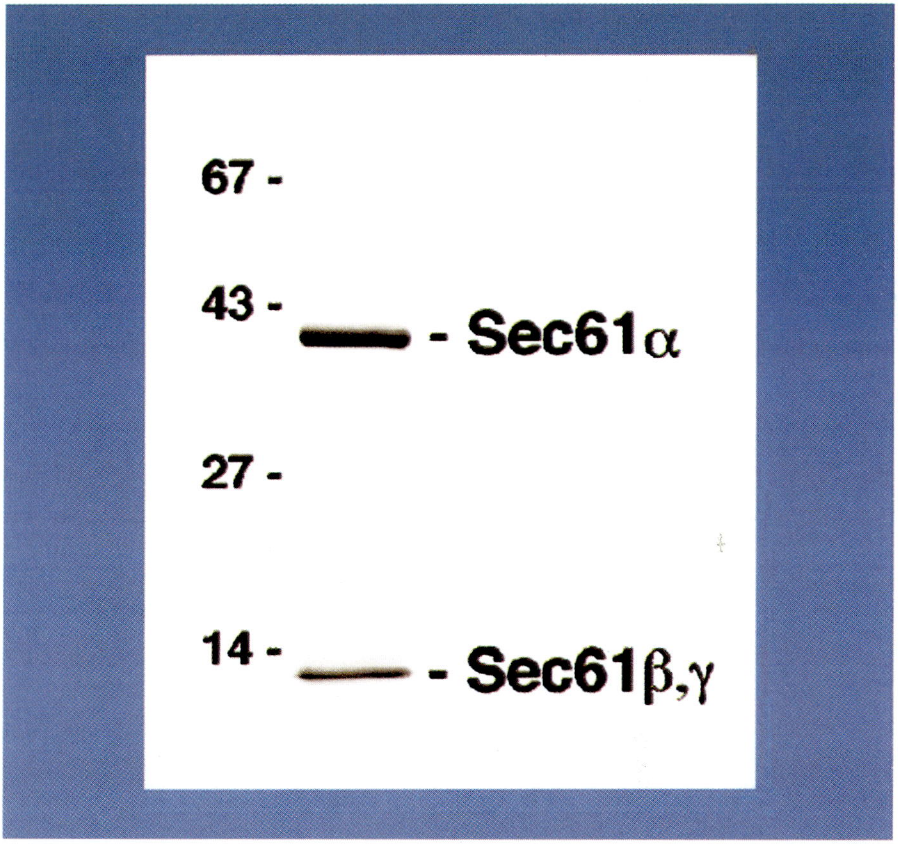

Figure 43. The purified Sec61 heterotrimer. The proteins of the purified yeast Sec61 heterotrimer were separated by SDS-PAGE and visualized by Coomassie Blue staining. From Beckmann, *et al.* (49), with permission.

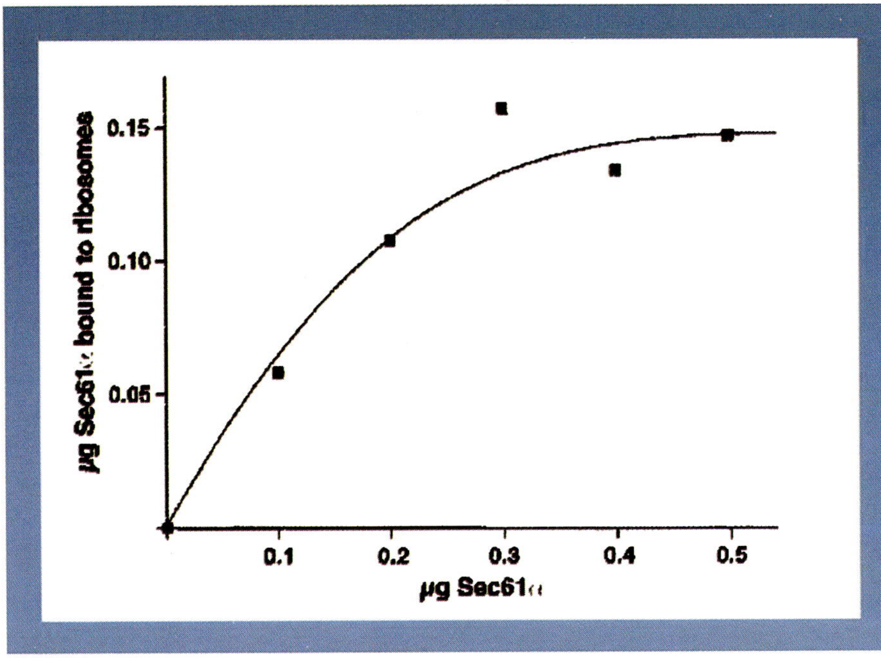

Figure 44. Binding of purified Sec61 to purified yeast ribosomes. Binding is saturable at an estimated binding of two Sec61 heterotrimers per ribosome. From Beckmann, *et al.* (49), with permission.

gested that two Sec61 trimers bound per ribosome (Fig. 44). Cryo-electron-microscopy of these complexes (Fig. 45) and three-dimensional image reconstruction (Fig. 46, left panel), done in collaboration with Joachim Frank's laboratory, revealed that the Sec61 trimer bound to a region of the large

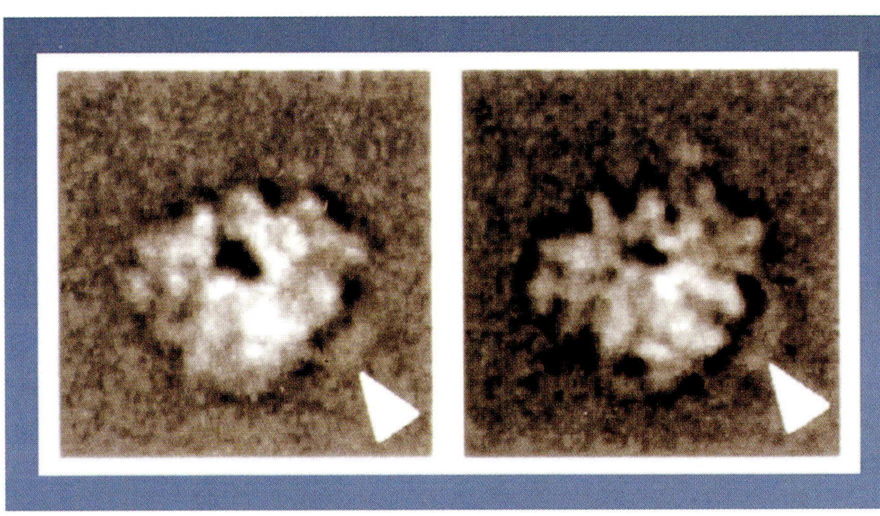

Figure 45. Cryo-electron micrograph of ribosome/Sec61 complexes. Arrowhead points to the Sec61 complex visible as a mass lying parallel to the ribosome surface. From Beckmann, *et al.* (49), with permission.

Figure 46. Three dimensional reconstruction of the ribosome/Sec61 complex. Small ribosomal subunit in yellow; large ribosomal subunit in blue; and Sec 61 complex in red. Left, note single attachment site of Sec61 to the large ribosomal subunit. Right, same orientation as left, but cut along a plane that sections the PCC of the Sec61 and the ribosome tunnel. The space between the two ribosomal subunits is indicated by an asterisk. The ribosomal tunnel and its alignment with the Sec61 PCC is indicated by a broken red line. Hence the conduit of the nascent polypeptide chain in the large ribosomal subunit tunnel and the Sec61 PCC is aligned. From Beckmann, *et al.* (49), with permission.

ribosomal subunit that was near to the previously identified exit site of the nascent chain. The Sec61 complex formed a funnel-shaped structure with a diameter of about 35 Ångström at the side facing the ribosome and a diameter of 15-20 Ångström at the side facing the ER lumen. The Sec61 complex was attached to the large ribosomal subunit at one site. Most interestingly, a section across the center of the ribosome (Fig. 46, right panel) revealed that a tunnel in the large ribosomal subunit aligned with the Sec61 channel. This arrangement suggests a likely conduit of the nascent chain within the large ribosomal subunit and across the Sec61 channel into the lumen of the ER.

The visualization of the PCC attached to the large ribosome and the alignment of the large ribosomal subunit tunnel with the PCC was an extremely gratifying confirmation of the predictions that had been made in 1975 (see Fig. 11) (14).

We are presently repeating these experiments with ribosomes that contain nascent secretory proteins in order to visualize the PCC in its active (open) form. For these experiments the ribosomes are programmed with truncated mRNA whose translation yields nascent chains long enough for the signal sequence to be exposed on the surface of the large ribosomal subunit. These nascent chain-ribosome complexes can then be used to engage SRP, SRP receptor and Sec61. These experiments should sublocalize SRP and SRP receptor on the active ribosome and should reveal the PCC in an active (open) configuration.

Likewise, truncated mRNAs coding for integral membrane proteins can be translated on ribosomes to investigate the PCC morphology that is accompanied by the proposed lateral opening of the PCC to the lipid bilayer. Attempts will also be made to crystallize the Sec61 complex. Moreover, the signal peptidase complex and the oligosaccharidyl transferase complex associate with the PCC to allow co-translational cleavage of the signal sequence and co-translational addition of core sugars to the translocating chain (Fig. 47). These enzymes will be isolated and bound to the nascent chain/ribosome/Sec61 complex and will then be visualized by cryo-electronmicroscopy and three-dimensional image reconstruction.

The PCC is clearly one of the marvels of nature. Unlike an ion-conducting channel that opens and closes in only one dimension, the PCC opens and closes in two dimensions, across the lipid bilayer and in the plane of the bilayer. The PCC is not merely a passive conduit but it scans the unfolded nascent chain as it passes across it, responding to a passing stop-transfer sequence by lateral opening. Moreover, the PCC is constructed in such a way that it does not leak significant amounts of small molecules. More recently, evidence has accumulated for reverse translocation and disintegration of membrane proteins. In order to allow degradation of secretory and integral membrane proteins by the cytoplasmic proteasomes, proteins can be reverse-translocated from the lumen of the ER or disintegrated (membrane proteins) from the ER membrane. Reverse translocation and disintegration are likely to occur by opening the channel from the trans-side of the membrane (for translocated proteins) and by additional lateral opening of the PCC from the bilayer (for integral membrane proteins).

The demonstration that the PCC consists of integral membrane proteins confirmed a prediction made in 1980 (44). Virchow's dictum on cells: 'omnis cellula e cellula' (each cell comes from a pre-existing cell) was extended to membranes: 'omnis membrana e membrana' (each membrane comes from a pre-existing membrane). Membranes and compartments are not created *de*

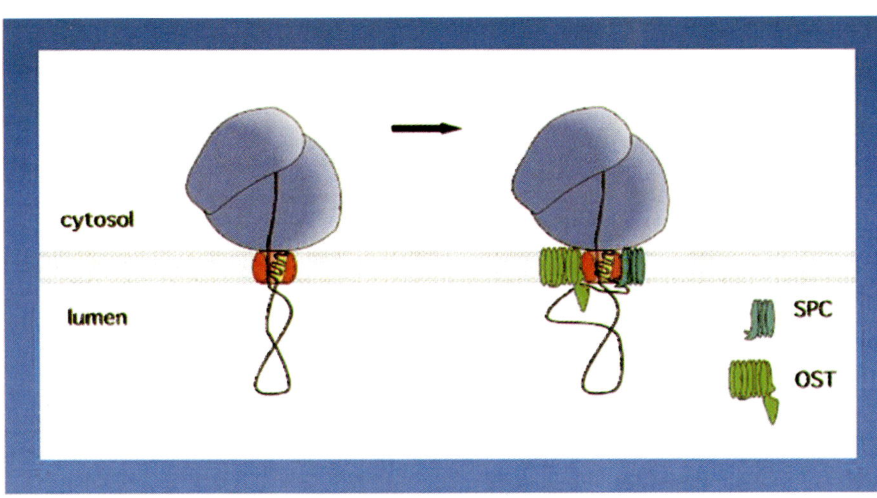

Figure 47. Recruitment of the signal peptidase complex (SPC) and the oligosaccharidyl transferase complex (OST) to the ribosome/nascent chain/Sec61 complex in the endoplasmic reticulum membrane. SPC endoproteolytically removes the signal peptide. OST attaches core oligosaccharides to the translocating nascent chain.

novo, but recreate themselves. The asymmetric integration of the integral membrane proteins of the Sec61 complex requires the pre-existence of asymmetrically integrated Sec61 complex. The asymmetrically oriented Sec61 complex catalyzes asymmetric integration of all other integral membrane proteins of the exocytotic and endocytotic membranes as specified by their intrinsic sequence elements (*signal*- and *stop transfer* sequences) (50). This assures that the asymmetric integration of a membrane protein proceeds with high fidelity which is absolutely crucial for the membrane protein's function. In contrast, reconstitution of proteoliposomes by mixing lipids, detergents and proteins usually does not result in a high fidelity asymmetric topology of membrane proteins. It should be noted, however, that some proteins are designed to insert "spontaneously" into lipid bilayers (e.g., certain toxins). The asymmetric integration of most membrane proteins, however, is catalyzed by PCCs of the ER or by PCCs of mitochondrial or chloroplast membranes (see below).

POST-TRANSLATIONAL TRANSLOCATION ACROSS THE ER

Protein translocation across the ER can also occur post-translationally. This was first discovered in yeast (51, 52). Instead of SRP and SRP receptor this system uses other factors, primarily heat shock proteins, to keep the protein in an unfolded and translocation-competent configuration (53, 54). Moreover, additional membrane proteins interacting with the Sec61 complex are required to facilitate presentation of the signal sequence for opening the Sec61 PCC.

OTHER PROTEIN TRANSLOCATION SYSTEMS

Work on the protein translocation system of the ER established the basic principles. The other systems for protein translocation across distinct cellular membranes (Fig. 48) were subsequently shown to work on similar principles. Entry of a protein into any of these other "public" translocation systems requires a cognate signal sequence. As in the case of a signal sequence addressed to the ER, each signal sequence is recognized by a cognate signal recognition factor that, in turn, targets the complex to a cognate receptor. These receptors are restricted in their localization to distinct cellular membranes (Fig. 48). Translocation then occurs through channels to the other side of the membrane. There are fascinating similarities and differences in the construction of these various translocation systems.

The translocation system of the ER has evolved from the major protein translocation system in the plasma membrane of prokaryotes. Most of the eukaryotic cellular membranes arose by invagination of the prokaryotic plasma membrane and by relegation of specific membrane protein functions of the prokaryotic plasma membrane to these intracellular membranes (Fig. 36) (44). The signal sequence of the ER is related to the signal sequence addressed to the major prokaryotic plasma membrane translocation system.

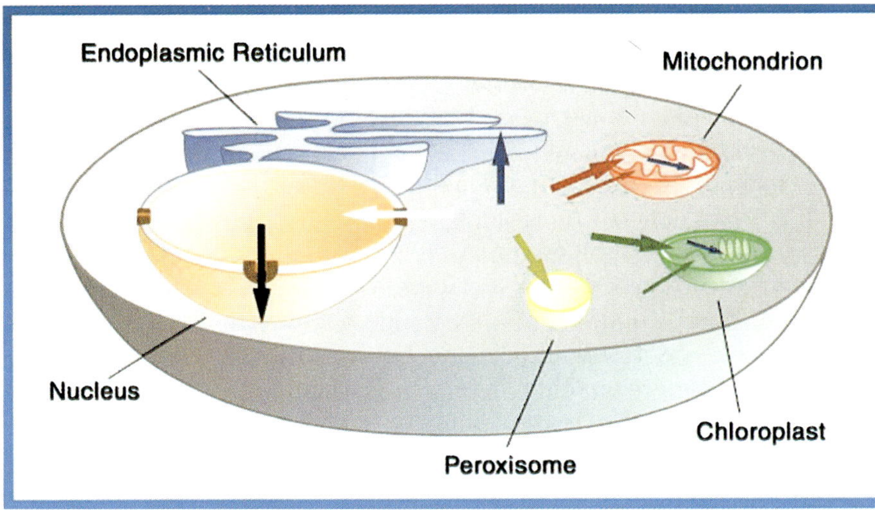

Figure 48. "Public" protein translocation systems in the cell. Bold blue arrow indicates protein translocation system in the ER; slim blue arrows indicate the related systems within the mitochondrial matrix and the chloroplast stroma. Bold and slim red and green arrows indicate protein translocation systems for import across the outer and inner mitochondrial or chloroplast membrane, respectively. White and black arrows indicate transport into and out of the nucleus across the nuclear pore complex.

Moreover, homologs of SRP (see above) and SRP receptor exist in prokaryotes and the SecY complex of bacteria is the homolog of the Sec61 complex of the ER. Hence, protein translocation across membranes and protein integration into membranes are very ancient and highly conserved membrane functions. Related systems exist for protein translocation from the mitochondrial matrix or the chloroplast stroma across the inner mitochondrial membrane or the thylakoid membranes of chloroplasts, respectively (Fig. 48).

Although the protein import system into mitochondria and chloroplasts are not related to those of the ER, the principles by which they operate are similar: distinct signal sequences, cognate signal recognition factors, cognate receptors and protein conducting channels. Like the PCCs in the ER, the PCCs in the outer and inner membrane of these organelles can also open laterally to achieve integration of membrane proteins. Proteins that are targeted to the matrix contain two signal sequences, one for translocation across the outer membrane and one for translocation across the inner membrane. The two PCCs in the outer and inner membrane can transiently interact with each other forming a continuous conduit for the unfolded nascent chain from the cytosol into the matrix. Protein translocation can occur post-translationally. Hence, in addition to signal recognition factors there are additional proteins required that keep the completed chain in an unfolded and translocation-competent configuration.

The protein import system into peroxisomes operates with at least two distinct signal sequences and cognate signal recognition factors and receptors. However, it appears that *integral membrane proteins* of the peroxisomal membrane are integrated using the ER translocation system and are then sorted to the peroxisomal membrane, as was first suggested in 1978 (55). It appears that peroxisomal *content proteins* can be transported across the peroxisomal membrane post-translationally and in a folded and even oligomeric form. A putative peroxisomal membrane pore may not be able to open laterally to the lipid bilayer and, therefore, may not be competent for the integration of membrane proteins. This would explain why integral membrane proteins of the peroxisomal membrane need to be integrated by another translocation system, namely, that of the ER.

Import and export of macromolecules into and from the nucleus is not restricted to proteins but includes ribonucleoproteins and deoxyribonucleoproteins (viruses). The nuclear pore complex (NPC) is the common conduit for both import and export of all these molecules and molecular complexes. It is a huge organelle, at least 20 times the mass of the ribosome with a central pore of about 25nm diameter. Again, protein import and export are mediated by distinct signal sequences. These signal sequences are recognized by cognate signal recognition factors (karyopherins or kaps) that ferry between the nucleus and the cytoplasm. The kaps target the import and export substrates to receptors that are represented by a subset of nuclear pore complex proteins (collectively termed nucleoporins or nups). The NPC is connected on the nuclear side to tubes that extend all the way to the nucleolus. These tubes form a chromatin-free zone and are likely to facilitate intranuclear diffusion of macromolecules. Nuclear import and export, the nuclear pore complex and intranuclear traffic in these tubes are at present intensely investigated by many laboratories. The next years will bring considerable progress in the detailed understanding of these important cellular activities.

OTHER TOPOGENIC SEQUENCES

It is clear that *signal* sequences for protein translocation across distinct cellular membranes and a combination of *signal-* and *stop-transfer* sequences for the asymmetric integration of proteins into membranes are not the only "topogenic" sequences that serve as determinants for protein localization (44). Following translocation across or asymmetric integration into membranes, many proteins undergo further traffic. For example, lysosomal proteins contain signal sequences indistinguishable from those of secretory proteins) (56, 57) and like secretory proteins, they are first segregated within the lumen of the ER. Subsequently, they have to be sorted to the lysosomes. *Sorting* sequence elements are required to achieve sorting to the lysosome.

Another example is the polyIg receptor. It is asymmetrically integrated into the ER using a signal sequence and a stop-transfer sequence (58, 59). It is then directed to the basolateral plasma membrane where it binds IgA or IgM. The ligand receptor complex is then transcytosed to the apical plasma membrane where the ligand-binding domain of the polyIg receptor is cleaved and shed as the so called 'secretory component' of secreted IgA or IgM. *Sorting* sequences direct this transcytotic pathway. The existence of *sorting* sequences has been postulated in 1980 (44), long before such sequences were actually identified. Like signal and stop-transfer sequences, sorting sequences need to be recognized and various effectors are required to decode them and to direct them into the various pathways they specify.

EPILOGUE

What began as an inquiry about how secretory proteins are translocated across the ER proceeded into an exciting voyage that revealed the principles by which cells organize themselves into distinct membranes and compartments. Most cellular proteins, including integral membrane proteins, contain intrinsic sequence elements. These sequence elements are decoded by cognate recognition factors. Cognate receptors and effectors localize the protein to their proper location.

ACKNOWLEDGEMENTS

Foremost, I want to thank my two mentors, Van R. Potter and George E. Palade. David Sabatini was a particularly inspiring and forming influence, especially in the early stages of my career. David Luck's critical input was extremely valuable, steering me away from the ancillary and keeping me focused on the major questions.

I was fortunate to have extremely talented graduate students and postdoctoral fellows working with me. Those who made seminal contributions on the ER translocation system were mentioned in the text. Other major contributors in this area were: Mark Adelman, Nica Borgese, Christoph Freienstein, Anne Devillers-Thiery, Dennis Shields, Stefano Bonatti, Ann Erickson, Mike McCune, Keith Mostov, David Anderson, Ibrahim Ibrahimi, Gregory Conner, Martin Friedlander, Gerald Waters, Yves Audigier, William Chirico, Sanjay Nigam, Gregory Shelness, Yashpal Kanwar, Jacques YaDeau, Gary Greenburg, and Giovanni Migliaccio. Our work on the bacterial translocation system was carried out by: Chung Nan Chang, Matthias Müller, Irene Fecycz and Makoto Watanabe. Major contributors to the work on mitochondrial protein import were: Maria-Luisa Macchecchini, Katsuyoshi Mihara, Sherry Goltz, Jim Kaput, Debkumar Pain, Norbert Schuelke, and Hiroshi Murakami. Our work on protein import into chloroplasts was carried out by: Debkumar Pain, Danny Schnell and Felix Kessler. Work on peroxisomal protein import was done by: Barbara Goldman and Ralf Erdmann. The work on nuclear structure and nuclear import was carried out by: Ariane Monneron, Robert Aaronson, Nancy Dwyer, Larry Gerace, Paul Fisher, Miguel Berrios, David Fisher, Daniel Fisher, Nilabh Chaudhary, Laura Davis, Spyros Georgatos, Richard Wozniak, John Aris, Howard Worman, Jeffrey Yuan, Eckart Bartnik, Ruben Henriquez, Jerome Appelbaum, Jean-Claude Courvalin, Thomas Meier, Karima Djabali,

Mary Shannon Moore, Susan Wente, Michael Rout, Jun Sukegawa, Susan Smith, Aurelian Radu, Einar Hallberg, Doris Kraemer, Lucy Pemberton, Junona Moroianu, Jian Wu, Michael Matunis, Elias Coutavas, Makoto Hijikata, Cordula Enenkel, Michael Rexach, Caterina Strambio-de Castillia, Michael Hurwitz, John Aitchison, Monique Floer, Ulf Nehrbass, Nabeel Yaseen, Neris Bonifaci, Erica Johnson, Jonathan Rosenblum, Markus Albertini, Beatriz Fontoura, Yuh Min Chook, Anton Titov and Kimihasa Yoshida.

Absolutely essential were the office and technical support. I would like to mention Peggy Beaudet-Fisher, Gisele Nimic, Nancy Dwyer, Parkash Cashmore, Evette Ellison and Sean Campbell. Special thanks are due to Lourdes R. Quirolgico, who has managed the laboratory and office for over 14 years in a flawless and highly efficient manner.

Our research was supported by fellowships and grants from the National Institutes of Health, the American Cancer Society, The Damon Runyon-Walter Winchell Cancer Fund, The Helen Hay Whitney Foundation, The Jane Coffins Child Foundation, The Leukemia Society of America, The Life Sciences Research Foundation, Boehringer Fond, Deutsche Forschungsgemeinschaft, EMBO Fund, Humboldt Stiftung, Thyssen Stiftung, Human Frontier Science Program, NATO, and Canadian, French, Japanese, Swedish and Swiss research organizations.

Most importantly, however, was the **steady** support from the Howard Hughes Medical Institute. Without the encouragement and support of Purnell Choppin, Max Cowan, Claire Winestock, Dinah Singer and Carolyn McGinnis and their staff we would not have been able to initiate or execute many of the risky and exciting experiments of the past 14 years.

I thank the Geschwister Scholl Gymnasium in Freiberg/Saxony where I received a very gratifying education; the Universities of Frankfurt/Main, München, Kiel and Tübingen, where I trained for my M.D.; the hospitals in Neuenbürg, Obergünzburg and Wolfratshausen, where I did my internship; and the University of Wisconsin in Madison, where I was generously supported by a fellowship and where I received a Ph.D.

Special thanks go to Rockefeller University where I spent the past 33 years of my life. It is an extraordinarily beautiful, congenial and inspiring place to work. The administrations of Detlev Bronk, Frederick Seitz, Joshua Lederberg, David Baltimore, Torsten Wiesel and Arnold Levine have been supportive and encouraging at every point in my career. David Rockefeller, in particular, has been an inspiring presence throughout my entire career here at Rockefeller University.

My parents, Margarete and Dr. Bruno Blobel; my sisters Ruth, Siegrid and Ingeborg; my brothers Hans, Reiner, Karl, and Albrecht and their families; and especially my wife, Laura Maioglio, were the pillars of my life. I thank them from the bottom of my heart.

This lecture is dedicated to the memory of my beloved oldest sister, Ruth Blobel, who died at age 19, on April 10, 1945, during a bomb attack on a train near Schwandorf, Bavaria.

ABBREVIATIONS

EMBO	European Molecular Biology Organization
ER	endoplasmic reticulum
IgA	immunoglobulin A
IgG	immunoglobulin G
IgM	immunoglobulin M
Kap	karyopherin
mRNA	messenger ribonucleic acid
NATO	North Atlantic Treaty Organization
NIH	National Institutes of Health
NPC	nuclear pore complex
Nup	nucleoporin
PCC	Protein conducting channel
pS	pico Siemens
RM	rough microsome
SDS-PAGE	sodium dodecylsulfate polyacrylamide gel electrophoresis
SRP	signal recognition particle
VSV	vesicular stomatitis virus

REFERENCES

1. Palade, G.E. *Science* **189**:347–358, 1975.
2. Redman, C.M. and Sabatini, D.D. *Proc. Natl. Acad. Sci. USA.* **56**:608–615, 1966.
3. Blobel, G. and Sabatini. D.D. *In* Biomembranes, ed. by L.A. Manson, pp. 193–195, Plenum Publishing Corporation, New York, USA, 1971.
4. Blobel, G. and Sabatini, D.D. *Proc. Natl. Acad. Sci. USA.* **68**: 390–394, 1971.
5. Blobel, G. *Proc. Natl. Acad. Sci. USA.* **68**:832–835, 1971.
6. Blobel, G. *Biochem. Biophys. Res. Comm.* **47:**88–95, 1972.
7. Blobel, G. *Proc. Natl. Acad. Sci. USA.* **70**:924–928, 1973.
8. Adelman, M., Sabatini, D.D. and Blobel, G. *J. Cell Biol.* **56**:206–229, 1973.
9. Freienstein, C. and Blobel, G. *Proc. Natl. Acad. Sci. USA.* **71**: 3435–3449, 1974.
10. Freienstein, C. and Blobel, G. *Proc. Natl. Acad. Sci. USA.* **72**:3392–3396, 1975.
11. Swan, D., Aviv, H., and Leder, P. *Proc. Natl. Acad. Sci. USA.* **69**:1967–1971, 1972.
12. Milstein, C., Brownlee, G.G., Harrison, T.M., and Mathews, M.B. *Nature New Biology* **239**:117–120, 1972.
13. Schechter, I., McKean, D.J., Guyer, R., and Terry, W. *Science* **188**:160–162, 1975.
14. Blobel, G. and Dobberstein, B. *J. Cell Biol.* **67**:835–851, 1975.
15. Blobel, G. and Dobberstein, B. *J. Cell Biol.* **67**:852–862, 1975.
16. Sabatini, D.D. and Blobel, G. *J. Cell Biol.* **45**: 146–157, 1970.
17. Simon, S.M. and Blobel, G. *Cell* **65**: 371–380, 1991.
18. Katz, F.N., Rothman, J. E., Lingappa, V.R., Blobel, G. and Lodish, H.F. *Proc. Natl. Acad. Sci. USA.* **74**: 3278–3282, 1977.
19. Lingappa, V.R., Katz, F.N., Lodish, H.F. and Blobel, G. *J. Biol. Chem.* **253**: 8667–8670, 1978.
20. Warren, G. and Dobberstein, B. *Nature* **273**:569–571, 1978.
21. Walter, P. and Blobel, G. *Proc. Natl. Acad. Sci. USA.* **77**:7112–7116, 1980.
22. Walter, P., Ibrahimi, I. and Blobel, G. *J. Cell Biol.* **91**:545–550, 1981.
23. Walter, P. and Blobel, G. *J. Cell Biol.* **91**:551–556, 1981.
24. Walter, P. and Blobel, G. *J. Cell Biol.* **91**:557–561, 1981.
25. Walter, P. and Blobel, G. *Nature (London)* **299**: 691–698, 1982.

26. Walter, P. and Johnson, A.E. *Annu. Rev. Cell Biol.* **10**:87–119, 1994.
27. Gilmore, R., Walter, P. and Blobel, G. *J. Cell Biol.* **95**:470–477, 1982.
28. Meyer, D.I., Krause, E., and Dobberstein, B. *Nature* **297**:503–508, 1982.
29. Tajima, S., Lauffer, L., Rath, V.L., and Walter, P. *J. Cell Biol.* **103**:1167–1178, 1986.
30. Connolly, T. and Gilmore, R. *Cell* **57**:599–610, 1989.
31. Miller, J.D., Tajima, S., Lauffer, L., and Walter, P. *J. Cell Biol.* **128**:273–282, 1995.
32. Gilmore, R. and Blobel, G. *Cell* **35**:677–685, 1983.
33. Jackson, R.C. and Blobel, G. *Proc. Natl. Acad. Sci. USA.* **74**:5598–5602, 1977.
34. Evans, E., Gilmore, R. and Blobel, G. *Proc. Natl. Acad. Sci. USA.* **83**:581–585, 1986.
35. Shelness, G.S., Kanwar, Y.S. and Blobel, G. *J. Biol. Chem.* **263**:17063–17070, 1988.
36. Greenburg, G., Shelness, G. and Blobel, G. *J. Biol. Chem.* **264**:15762–15765, 1989.
37. Shelness, G.S. and Blobel, G. *J. Biol. Chem.* **265**:9512–9519, 1990.
38. Nicchitta, C. V. and Blobel, G. *Cell* **60**:259–269, 1990.
39. von Heijne, G. and Blomberg, C. *Eur. J. Biochem.* **97**:175–181, 1979.
40. Engelman, D.M., and Steitz, T.A. *Cell* **23**:411–422, 1981.
41. Gilmore, R. and Blobel, G. *Cell* **42**: 497–505, 1985.
42. Simon, S., Blobel, G. and Zimmerberg, J. *Proc. Natl. Acad. Sci.USA.* **86**:6176–6180, 1989.
43. Mueller, P., Rudin, D.O., Ti Tien, H, and Wescott, W.C. *Nature* **194**:979–980, 1962.
44. Blobel, G. *Proc. Natl. Acad. Sci. USA.* **77**:1496–1500, 1980.
45. Simon, S.M. and Blobel, G. *Cell* **69**:677–684, 1992.
46. Stirling, C.J., Rothblatt, J., Hosobuchi, M., Deshaies, R., and Schekman, R. *Mol. Biol. Cell* **3**:129–142, 1992.
47. Goerlich, D., Prehn, S., Hartmann, E., Kalies, K.U., and Rapoport, T.A. *Cell* **71**:489–503, 1992.
48. Goerlich, D., and Rapoport, T.A. *Cell* **75**: 615–630, 1993.
49. Beckmann, R., Bubeck, D., Grassucci, R., Penczek, P., Verschoor, A., Blobel, G. and Frank, J. *Science* **278**:2123–2126, 1997.
50. Friedlander, M. and Blobel, G. *Nature (London)* **318**:338–344, 1985.
51. Waters, M.G. and Blobel, G. *J. Cell Biol.* **102**:1543–1550, 1986.
52. Hansen, W., Garcia, P.D., and Walter, P. *Cell* **45**: 397–406, 1986.
53. Chirico, W.J., Waters, M.G. and Blobel, G. *Nature (London)* **332**:805–810, 1988.
54. Deshaies, R., Koch, B.D., Werner-Washburne, M., Craig, E., and Schekman, R. *Nature* **332**: 800–805, 1988.
55. Goldman, B. and Blobel, G. *Proc. Natl. Acad. Sci. USA.* **75**:5066–5070, 1978.
56. Erickson, A.H. and Blobel, G. *J. Biol. Chem.* **254**:11771–11774, 1979.
57. Erickson, A.H., Conner, G. and Blobel, G. *J. Biol. Chem.* **256**:11224–11231, 1981.
58. Mostov, K.E. and Blobel, G. *J. Biol. Chem.* **257**:11816–11821, 1982.
59. Mostov, K.E., Friedlander, M., and Blobel, G. *Nature (London)* **308**:37–43, 1984

Physiology or Medicine 2000

ARVID CARLSSON, PAUL GREENGARD and ERIC R. KANDEL

"for their discoveries concerning signal transduction in the nervous system"

THE NOBEL PRIZE IN PHYSIOLOGY OR MEDICINE

Speech by Professor Urban Ungerstedt of the Nobel Committee at Karolinska Institutet.
Translation of the Swedish text.

Your Majesties, Your Royal Highnesses, Ladies and Gentlemen,

This year's Nobel Prize in Physiology or Medicine concerns the most complex structure in the universe that we know of – the human brain. It consists of 100 billion nerve cells, which is the same number of cells as the total number of human beings that have ever lived on this earth.

We talk about the "Internet revolution"; 35 million Internet users who communicate now and then – what is that compared to the nerve cells we all carry within ourselves! 100 billion nerve cells that communicate continuously.

It is this communication, "signal transduction in the nervous system," which is the subject of this year's Nobel Prize. A single nerve cell forms thousands of contact points, so-called synapses, with other nerve cells. In these synapses the nerve cells communicate by chemistry; one cell releases a transmitter, which reaches the other cell.

Professor Arvid Carlsson proved that dopamine is such a transmitter. The general belief was that dopamine was a precursor of other transmitters and of little functional importance. However, Professor Carlsson was able to show that dopamine existed in specific parts of the brain and concluded that it was a transmitter in its own right.

He then used a naturally occurring substance, reserpine, which empties the dopamine from the nerves, and found that the animals lost their ability to move. He realized that it must be possible to restore the dopamine levels with L-DOPA, a precursor of dopamine. In a conclusive, dramatic experiment he showed that the animals regained their ability to move when he gave them L-DOPA.

Reserpine had depleted dopamine and had given the animals the symptoms of Parkinson's disease, that is, rigidity and inability to move and react to stimuli in the environment. When the animals were given L-DOPA, dopamine was produced again in their brains. In this way the idea of treating Parkinson patients with L-DOPA was born. This enables millions of patients around the world to live a normal life.

Professor Paul Greengard showed what happens when dopamine and other similar transmitters stimulate a nerve cell. Receptors on the cell surface activate enzymes in the cell wall, which starts the production of second messengers. These messengers travel into the cell and activate a protein kinase, which starts to bind phosphate groups to other proteins, in this way altering their function. This leads, for example, to the opening of ion channels in the cell membrane and a change in the electrical activity of the cell.

Professor Greengard then showed that dopamine and other transmitters affect a central regulatory protein, which has been called DARPP-32. Like the conductor of an orchestra, it tells other proteins when and how to be activated.

This so-called "slow synaptic transmission" controls our movements and also those processes in the brain that elicit emotions or react to addictive drugs such as cocaine, amphetamine and heroin.

Professor Eric Kandel showed that transmitters of the same type as studied by Arvid Carlsson, via the protein kinases characterized by Paul Greengard, are involved in the most advanced functions of the nervous system such as the ability to form memories.

Imagine how difficult or impossible it must be to study how memory is formed in a human brain with 100 billion nerve cells. Eric Kandel, therefore, did something which is classical in all natural science: He chose to study a simpler model system, a sea slug, *Aplysia*, which has 20,000 nerve cells. He did it with the conviction that even primitive animals must learn in order to survive.

The sea slug has a withdrawal reflex protecting its gills. If they are touched repeatedly, they react less and less – just as human beings do when subjected to an unexpected touch. If, on the other hand, the touch is forceful the reflex is amplified and becomes stronger and stronger.

The habituation or amplification effect lasts only for a few minutes. One may say that the sea slug exhibits a short-term memory. If the forceful stimulus is repeated several times, the sensitization may remain for weeks, that is, the sea slug develops a long-term memory.

Professor Kandel was able to show that habituation to touching was due to changes in the synapse, the contact point between the nerve cells. During habituation less and less transmitter was released.

The forceful stimulus that formed the long-term memory worked in a completely different way. Second messengers activated protein kinases that entered the cell nucleus and started the production of new proteins. This, in turn, brought about a change in the form and function of the synapse. What we call memory is, thus, elicited by direct changes in the billion of synapses that form the contact points between the nerve cells.

I am convinced that you and I will remember this Nobel ceremony for many years. This is because of the dopamine which Arvid Carlsson discovered, enabling the brain to react to what we see and hear; the second messengers that Paul Greengard described, carrying the signals into the nerve cell; and the memory functions that Eric Kandel found to be due to changes in the very form and function of the synapses.

Dear Arvid Carlsson, Paul Greengard and Eric Kandel,

Your discoveries concerning "signal transduction in the nervous system" have truly changed our understanding of brain function.

From Arvid Carlsson's research we now know that Parkinson's disease is due to failure in synaptic release of dopamine. We know that we can substitute the lost function by a simple molecule, L-DOPA, which replenishes the emptied stores of dopamine and in this way, give millions of humans a better life.

We know from Paul Greengard's work how this is brought about. How second messengers activate protein kinases leading to changes in cellular reactions. We begin to see how phosphorylation plays a central part in the very orchestration of the different transmitter inputs to the nerve cells.

Finally, Eric Kandel's work has shown us how these transmitters, through second transmitters and protein phosphorylation, create short- and long-term memory, forming the very basis for our ability to exist and interact meaningfully in our world.

On behalf of the Nobel Assembly at Karolinska Institutet, I wish to convey our warmest congratulations and I ask you to step forward to receive the Nobel Prize from His Majesty the King.

ARVID CARLSSON

I grew up in an academic middle-class family. In 1926, when I was 3 years old, my father was appointed professor of history at the University of Lund, Sweden, and moved there with his family from Uppsala, Sweden. My mother had passed a master-of-arts examination and my father a Ph.D. degree at the University of Uppsala. My mother had a keen interest in research throughout her life but gave priority to raising her children and to assisting her husband in his research. However, when her husband died at the age of 76 she, then 71 years old, started to devote herself entirely to her favorite area of research, that is the legal status of women in the Middle Ages in Sweden. She published a couple of books and a number of articles on this subject in Swedish, which rendered her an honorary Ph.D. degree at the University of Uppsala several years later.

We were four children in the family, and we were all to take academic degrees at different levels. The family had a strong orientation toward humanities. However, in contrast to my elder brother and sister, who also chose humanities, I decided to study medicine, and my 7 years younger brother later joined me. The reason for this deviating behavior of mine was partly the kind of opposition often occurring in youngsters, partly some vague idea of science being more "useful" than arts.

My childhood and youth were characterized by a happy life in a stable environment with loving and supportive parents. I was probably just about average in terms of disobedience and escapades. At school, which I found fairly endurable, I had very good marks without too much effort.

At the age of 16, in June 1939, I took a hitch-hiking trip to Germany for two weeks together with a boy of the same age. This was to be my only trip outside the Nordic countries until the age of 32. It occurred only two and a half months before the outbreak of World War II. We had then the opportunity to talk to many Germans of varying social standing. Most of them felt convinced that a war would begin as soon as the harvest was done, and they seemed to accept that, though reluctantly. In Berlin I spent one night in a hospits run by an Evangelic foundation and inhabitated by obviously very poor people. I especially remember several Jews with long beards and sad faces, mumbling while reading a heavy book that could have been the Talmud, apparently searching for an answer and a solution in a desperate situation. Otherwise I was not aware of the terrible actions against Jews that were probably ongoing around me.

My medical studies started in 1941 and went on smoothly, apart from several interruptions due to altogether a couple of years of military service. Lund was, and still is, a fairly typical small university town somewhat reminiscent of

Oxford. The quality of research and teaching, including that of the medical school, was good. Among the professors of that school a couple of prominent names may be mentioned. Torsten Thunberg, professor of physiology, had discovered "the tissue respiration", i.e. the dehydrogenases, and had developed a method to measure their activity. His work was simultaneous but independent of Warburg's research on the same subject in Berlin. Warburg, but not Thunberg, was awarded the Nobel Prize. Ernest Overton, professor of pharmacology, was famous for his revolutionary theories regarding the structure and lipid properties of the cell membrane and on the mechanism underlying narcosis. However, Overton had died before I started my medical studies, and Thunberg was already emeritus at that time. Sadly Overton became mentally ill in his later life; he was said to propose himself repeatedly for the Nobel Prize, which he like Thunberg certainly deserved but did not get. Among the most innovative of my contemporaries at the medical school of the University of Lund was Nils Alwall, although his seminal contributions were not generally recognized at that time. Alwall is one of the pioneers in the development of the "artificial kidney". He successfully ran a department of kidney dialysis in the Medical University Clinic. The internationally renowned Gambro Company developed out of this work. It is still quite successful in kidney dialysis and other biomedical technology. Another major innovation created at the University of Lund at this time was the ultrasonar cardiography, developed by Helmuth Hertz and Inge Edler. Finally Jörgen Lehmann may be mentioned as the most prominent among Thunberg's students. In the 1940s, then in Göteborg, Sweden, he discovered para-aminosalicylic acid (PAS), which together with streptomycin, discovered by Waksman, provided the first effective drug treatment for tuberculosis. Waksman, but not Lehmann, was awarded the Nobel Prize for his discovery.

During the first five years of my medical studies World War II was going on, rendering Sweden almost entirely isolated. During my first year, however, five Danish medical students were allowed to leave Denmark, then occupied by Germany, to do their anatomy studies together with us in Lund. I became especially friendly with one of them: Ib Munkvad. Among other things we played check together. Munkvad later became a psychiatrist with an orientation towards biological psychiatry. He was instrumental in developing biological psychiatry in Denmark and is also known for formulating the hypothesis that the dyskinesias induced by dopaminergic agonists could make a model of psychosis.

Another remarkable contact with the war occurred during a few nights in October, 1943, when thousands of Jews managed to escape in small fishing boats across the sound between Denmark and Sweden. This happened because one of the top members of the German occupation establishment had informed Danish authorities that a major deportation of Jews from Denmark to Germany was forthcoming. Among the Jews who came to Sweden at this time were a number of prominent medical researchers. These people were immediately enrolled and became an important acquisition as teachers and researchers at our medical school.

An even sadder contact with the war occurred in the spring of 1944, when I was in my first year of clinical training. At this time Count Folke Bernadotte, a member of the royal Swedish family, had managed to persuade German authorities to allow him to go from Sweden into Germany with a number of "white busses" and to pick up prisoners at concentration camps. These transportations turned out to be successful, allowing about 30,000 prisoners, among them 11,000 Jews, to escape. Some of the prisoners were taken to Lund, where a big tent was erected in a park to house them. As a medical student I was given the task to examine several of these prisoners. Many of them were children, suffering from undernutrition. Tuberculosis was not uncommon. However, most shocking was their mental status. They behaved like wild animals, obviously suffering from severe anguish and suspiciousness and trusting nobody.

STARTING IN PHARMACOLOGY

Already at the outset of my medical studies I had decided to try to go into research, and in 1944, I was offered a modest position as "amanuens" (initially without salary) at the Department of Pharmacology, after I had passed a successful examination in that discipline. The head of that department, and at that time the only professor, was Gunnar Ahlgren. He had been Thunberg's favorite student and apparently been an asset in the group around Thunberg. However, when on his own as professor of pharmacology he was not scientifically productive. But he managed to attract a number of talented students, who were to make successful careers in various medical disciplines. One of these people was the already mentioned Nils Alwall.

Ahlgren gave me and another student, named Georg Theander, a topic to investigate, i.e. to measure the duration of action of pentylenetetrazol (Cardiazol), a convulsant at that time frequently used as an "analeptic" to wake up people who had taken an overdose of a hypnotic or sedative, and also used as a forerunner of electroconvulsive therapy. Nowadays this drug has a modest use as a diagnostic and experimental tool. To measure the action of pentylenetetrazol Ahlgren proposed to employ its awakening action against central depressants. At first we tried nitrous oxide, but to our surprise the drug was unable to antagonize its anesthetic action. We then added a subhypnotic dose of barbital and found that the combined action of this drug and nitrous oxide could be antagonized, and the increase in partial pressure of nitrous oxide needed to inhibit the righting reflex could be used as a measure of pentylenetetrazol action. The result of this study was that the duration of action of pentyleneterazol was much longer than expected from observations on its effect in animals not treated with central depressants. We published our results (Carlsson and Theander, 1946), and this first publication of ours was awarded a modest prize, intended for young medical scientists at the University of Lund. One might wonder why Ahlgren was interested in the duration of action of this drug. He never explained that to me, but when I told him about our results, he was obviously pleased, but not surprised. A plausible reason for this was that he had actually tried it on himself to antagonize a sedative and thus discovered its long duration of action.

Together with another, five years older pharmacologist, named Folke Serin, I studied the action of another "analeptic", i.e. nikethamide, and discovered a circadian rhythm in its lethal action. The publication of these findings (Carlsson and Serin, 1950) appears to have been the first describing a circadian rhythm of a drug's action, according to a review article on chronobiology, that was published much later when this subject became fashionable in connection with space research.

WORK ON CALCIUM METABOLISM

In 1948 Gunnar Ahlgren persuaded me to go into an entirely different area of research, i.e. calcium metabolism. At that time radioactive tracers, among them Ca^{45}, had become commercially available, and a Swedish drug company wanted us to test a number of calcium salts regarding their oral availability. I accepted to do this research. The testing of different calcium salts did not disclose anything of interest, but I utilized this opportunity to investigate the absorption of calcium as well as its metabolism in the skeleton by means of the new technique. That work resulted in my thesis for the M.D. degree (corresponding to the American Ph.D. degree) and in several subsequent papers, including the doctoral theses of two students of mine, namely Bertil Lindquist and Göran Bauer, who later became professors of pediatrics and orthopedic surgery, respectively. Briefly, what we found was that it is possible, contrary to the current opinion at that time, to use tracer techniques for measurement of the uptake (or "accretion") and the resorption of the bone mineral differentially (for review, see Bauer *et al.*, 1961). Among other things we discovered that vitamin D in physiological doses can stimulate not only the intestinal absorption and "accretion" but also the resorption of bone mineral. Our findings aroused some international interest and resulted in an invitation to a Gordon Conference in the summer of 1955.

The following events in my research career are summarized in my Nobel Prize Lecture.

CURRICULUM VITAE

Born in Uppsala, Sweden, Jan. 25, 1923. Entered the University of Lund, Sweden, in 1941. "M.L." (corresponding to American M.D.), 1951, "M.D." (corresponding to American Ph.D.), 1951, Univ. of Lund. Assistant of the Department of Pharmacology, Univ. of Lund from 1944, Assistant Professor 1951, Associate Professor 1956. 1959 Professor of Pharmacology, Univ. of Gothenburg, Sweden (Chairman 1959–1976). Emeritus 1989. Visiting Scientist at the Laboratory of Chemical Pharmacology, National Heart Institute, Bethesda, Md., USA, 1955–1956.

AWARDS AND DISTINGUISHED LECTURES

1947 Magnus Blix' Prize, Univ. of Lund, Sweden.

1970 "First Annual James Parkinson Lecture and Award", Parkinson's Disease Foundation, New York.
"Pehr Dubb's Gold Medal", for research in neurochemistry, The Medical Society of Gothenburg.

1973 "Second Annual Grass Foundation Lecture", Society for Neuroscience in San Diego, California.

1974 "Anders Jahre's Medical Prize" University of Oslo, Norway.

1975 "Stanley R. Dean Lecture and Award", American College of Psychiatrists, Phoenix, Arizona.
"Anna-Monika Stiftung, 1st Prize", Munich, for research on antidepressants and neuroleptics.

1978 His Majesty, The King's of Sweden Medal.

1979 "Wolf Prize in Medicine", Jerusalem, Israel (together with R.W. Sperry and O. Hornykiewicz).

1981 "Björkénska Priset", Univ. of Uppsala, Sweden.
"The Thomas William Salmon Lectures and Medal", New York.

1982 "Gairdner Foundation Award", Toronto, Canada.

1985 "Hilda and Alfred Eriksson's Prize", The Royal Swedish Academy of Science.
"Senator Dr. Franz Burda Stiftung", Vienna, Austria, for Research on Parkinson's Disease.

1986 The Medal of Helsinki University, Finland.

1988 "Annual Neurobiology Lectureship Award", Univ. of Lund, Sweden.
The "Jubileumspris" of the Swedish Medical Association.

1989 "Distinguished Lecture Series", Irvine Neuroscience Association, Irvine, California.
"The Second Annual Bristol-Myers Award for Distinguished Achievement in Neuroscience Research", New York (together with J. Axelrod and P. Greengard).
"Varro E. Tyler Distinguished Lectureship", Purdue University, West Lafayette, Indiana.

1990 "The Paul Hoch Prize", American Psychopathological Association, New York.
"The Fred Springer Award", American Parkinson Disease Association, Miami, Florida.

1991 "William K. Warren Schizophrenia Research Award", Tucson, Arizona.

1992 The "Open Mind Award in Psychiatry", Janssen Research Foundation, Paris, France, together with H. Hippius and H. van Praag.
The "Vittorio Erspamer Lecture and Award", Fidia Research Foundation, Washington, D.C.
The "Julius Axelrod Medal", The Catecholamine Club, Anaheim, California.

1993 Lilly Lecture, Indianapolis, Indiana.

1994 Albert Barrett Memorial Lecture, Univ. of Michigan.
The Japan Prize, in Psychology and Psychiatry, The Science and Technology Foundation of Japan.
The "Lieber Prize" for research in schizophrenia, given by the Scientific Council of The National Alliance for Research in Schizophrenia and Depression, USA.
1995 The Research Prize of the Lundbeck Foundation, Roskilde, Denmark.
The Robert J and Claire Pasarow Foundation Award for Neuropsychiatric Research. L.A., USA.
Synthelabo Lecture, VIII ECNP Congress, Venice.
1996 Plaque for "Lifetime Achievement". American Society of Biological Psychiatry.
Gold Medal and Honorary Diploma, The Swedish Parkinson Association.
1997 Kraepelin-Medaille in Gold. Max-Planck-Institut für Psychiatrie, München, Germany.
John Cade Lecture, Melbourne, Australia.
1998 Gold Medal, Soc. of Biol. Psychiatry, Toronto, Canada.
1999 Gold Medal, Distinguished Lecture and Award, Univ. of Bari, Italy.
Antonio Feltrinelli International Award, Accademia dei Lincei, Rome.
2000 Nobel Prize in Medicine or Physiology.

HONORARY DEGREES

1976 M.D. Honoris Causa, Univ. of Cagliary, Italy.
1977 Doctor of Pharmacy Honoris Causa, Univ. of Uppsala, Sweden.
1990 Ph.D. Honoris Causa, Univ. of Helsinki, Finland.
1991 Dr. rer. physiol. honoris causa, Univ. of Marburg, Germany.

MEMBERSHIPS

1962 Member of the Collegium Internationale Neuro-Psychopharmacologicum (President 1978–1980).
1968 Member of the Royal Society for Science and Arts in Gothenburg.
1972 Foreign Member of the Japanese Pharmacological Society.
1974 Foreign Member of the American College of Neuropsychopharmacology.
1974 Corresponding Member of the German Pharmacological Society.
1975 Member of the Royal Swedish Academy of Sciences.
1983 Honorary Fellow of The Medical Society of Gothenburg.
1989 Member of the Academia Europaea.
1991 Honorary Fellow of the World Federation of Societies of Biological Psychiatry.
1992 Honorary Fellow of the Collegium Internationale Neuro-Psychopharmacologicum.
1992 Honorary Fellow of Academia Medicinae & Psychiatriae.

1996 Foreign Associate Member of The Institute of Medicine, National Academy of Sciences, U.S.A.
1997 Honorary Fellow of the Swedish Society of Biological Psychiatry.
1998 Honorary Member of the German Society of Biological Psychiatry.
1999 Honorary Member of the Scandinavian Society for Psychopharmacology
2000 Honorary Member of "Parkinsonföreningen i Västra Sverige".

SELECTED REFERENCES

CARLSSON, A. Pharmacology of selective dopamine receptor agonists. In: *Trends in Medicinal Chemistry* (eds. E. Mutschler and E. Winterfeldt). Proceedings of the Ninth International Symposium on Medicinal Chemistry, Berlin 1986. VCH Verlagsgesellschaft mbH, D-6940 Weinheim, 1987 pp. 313–321.

CARLSSON, A. The dopamine hypothesis of schizophrenia 20 years later. In: *Search for the Cause of Schizophrenia* (eds. H. Häfner, W.F. Gattaz, and W. Janzarik). Springer-Verlag, Berlin Heidelberg 1987, pp. 223–235.

CARLSSON, A. Historical perspective of the chemistry and pharmacological treatment of schizophrenia. In: *Handbook of Schizophrenia, Vol. 2: Neurochemistry and Neuropharmacology of Schizophrenia* (eds. F.A. Henn and L.E. DeLisi). Elsevier Science Publishers B.V., 1987, pp. 1–15.

CARLSSON, A. Perspectives on the discovery of central monoaminergic neurotransmission. *Ann. Rev. Neurosci.* 1987. *10.* 19–40.

CARLSSON, A. The current status of the dopamine hypothesis of schizophrenia. *Neuropsychopharmacol.* 1988. *1.* 179–186.

CARLSSON, M., and A. CARLSSON. The NMDA antagonist MK-801 causes marke locomotor stimulation in monoamine-depleted mice. *J. Neural Transm.* 1989. 75. 221–226.

CARLSSON, M., and A. CARLSSON. Dramatic synergism between MK-801 and clonidine with respect to locomotor stimulatory effect in monoamine-depleted mice. *J. Neural Transm.* 1989. 77. 65–71.

CARLSSON, A. Early psychopharmacology and the rise of modern brain research. *J. Psychopharmacol.* 1990. 4(3). 120–126.

CARLSSON, M., and A. CARLSSON. Interactions between glutamatergic and mono- aminergic systems within the basal ganglia – implications for schizophrenia and Parkinson's disease. *TINS.* 1990. 13. 272–276.

CARLSSON, A., and B. FORNSTEDT. Possible mechanisms underlying the special vulnerability of dopaminergic neurons. *Acta Neurol. Scand.* 1991. 84. 16–18.

HANSSON, L. O., N. WATERS, B. WINBLAD, C.-G. GOTTFRIES, and A. CARLSSON. Evidence for biochemical heterogeneity in schizophrenia: a multivariate study of monoaminergic indices in human post-mortem brain tissue. *J. Neural Transm. [GenSect].* 1994. 98. 217–235.

SVENSSON, K., A. CARLSSON, and N. WATERS. Locomotor inhibition by the D_3 ligand R–(+)–7–OH–DPAT is independent of changes in dopamine release. *J. Neural Transm. [GenSect].* 1994. 95. 71–74.

WATERS, N., L. LÖFBERG, S. HAADSMA-SVENSSON, K. SVENSSON, C. SONESSON, and A. CARLSSON. Differential effects of dopamine D2 and D3 receptor antagonists in regard to dopamine release, in vivo receptor displacement and behaviour. *J. Neural Transm. [GenSect].* 1994. 98. 39–55.

BUNNEY, B. G., W. E. BUNNEY, Jr., and A. CARLSSON. Schizophrenia and glutamate. Chapter 101 in: *Psychopharmacology. The Fourth Generation of Progress* (eds. F. E. Bloom and D. J. Kupfer). Raven Press, Ltd., New York 1995, pp. 1205–1214.

CARLSSON, A. Towards a new understanding of dopamine receptors. *Clin. Neuropharmacol.* 1995. 18 (suppl. 1). S6–S13.

CARLSSON, A. The dopamine theory revisited. In: *Schizophrenia* (eds. S. R. Hirsch and D. R. Weiberger). Blackwell Science, Oxford 1995, pp.379–400.

SVENSSON, A., M. L. CARLSSON, and A. CARLSSON. Crucial role of the accumbens nucleus in the neurotransmitter interactions regulating motor control in mice. *J. Neural Transm. [Gen Sect]*. 1995. 101. 127–148.

A HALF-CENTURY OF NEUROTRANSMITTER RESEARCH: IMPACT ON NEUROLOGY AND PSYCHIATRY

Nobel Lecture, December 8, 2000

by

ARVID CARLSSON

Department of Pharmacology, Göteborg University, Göteborg, Sweden.

BEGINNINGS

My encounter with dopamine followed upon an incredible sequence of fortunate events. I had been working on calcium metabolism, using radioactive isotopes, which had then just become commercially available. This work had resulted in my doctoral thesis in 1951 and a series of subsequent papers, including two doctoral theses by students of mine. It had become somewhat visible internationally, resulting for example, in an invitation to a Gordon Conference in New England in 1955. The reason why I left this research field was that in connection with a competition for an associate professorship in pharmacology the expert committee let me know that in their opinion calcium metabolism did not occupy a central position in pharmacology. I therefore turned to Professor Sune Bergström who was at that time head of the Department of Physiological Chemistry of the University of Lund, Sweden. This department was located in the same building as our Pharmacology Department. Professor Bergström had already been very helpful in several instances when I had a professional problem of some kind. Incidentally, Dr. Bengt Samuelsson was at that time working with Professor Bergström in the same department. Thus the three Swedes who were to become Nobel Laureates in the period 1980–2000, happened to be working under the same roof for a few years.

I asked Sune Bergström if he could help me to get in touch with an outstanding American laboratory working in the area of biochemical pharmacology, which I felt had a great future. He wrote to his friend Dr. Bernard Witkop, a highly talented chemist working in the National Institutes of Health in Bethesda, Md. This letter was forwarded via the late Dr. Sidney Udenfriend, to his boss the late Dr. Bernard B. Brodie (Fig. 1), head of the famous Laboratory of Chemical Pharmacology of the National Heart Institute. That is how I came to work under Dr. Brodie for about five months, starting in August 1955. The timing of my arrival there was extremely fortunate. Brodie and his colleagues had just a few months before made a breakthrough discovery, namely that the administration of reserpine, a recently introduced

Figure 1. Bernard B. Brodie (1907–1989)

antipsychotic and antihypertensive drug, caused the virtually complete disappearance of serotonin from the brain and other tissues (Pletscher *et al.* 1955, 1956, Fig. 2).

"APPRENTICE TO GENIUS"

Brodie was a remarkably charismatic and intensive person. He was generally called Steve Brodie. This referred to a saloon keeper named Steve Brodie, who at the beginning of the previous century had jumped off the Brooklyn Bridge in order to win a bet. Bernard Brodie, too, was a sensation seeker who

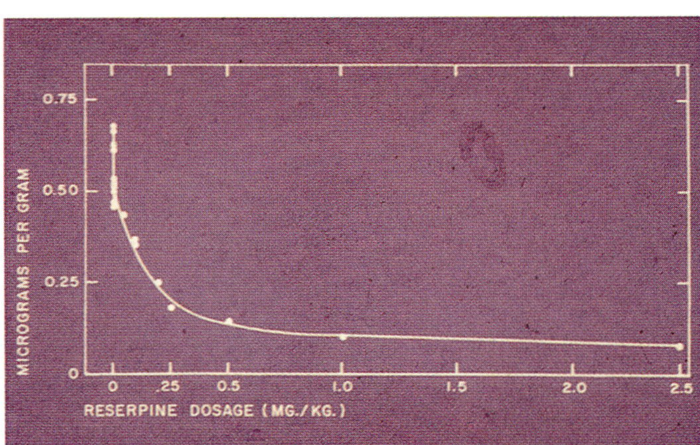

Figure 2. Brain level of serotonin four hours after various intravenous doses of reserpine. From Pletscher *et al.*(1956).

in his youth had started on a career as a boxer but later switched to become an organic chemist. He then confined his sensation seeking to non-physical adventures. He liked to call himself a gambler. He had gained a tremendous reputation as a pioneer in the area of drug metabolism and should perhaps rightly be called the father of modern biochemical pharmacology. A large number of his apprentices, coming from various parts of the world, later became prominent figures in pharmacology (see Kanigel, "Apprentice to Genius"). In the 1950s, after hearing about the sensational clinical actions of the new antipsychotic drugs and the ability of the hallucinogenic LSD to block serotonin effects on various peripheral organs he decided to enter the field of psychopharmacology. While knowing very little about the brain he had a tremendous trump card in being able to determine for the first time serotonin and similar molecules in the brain, using the prototype of a new instrument developed in his own lab together with Sidney Udenfriend and Dr. Robert Bowman. This instrument, the spectrophotofluorimeter, was to replace previous bioassays and to revolutionize drug research and neurotransmitter pharmacology for several decades.

This research soon led to the breakthrough discovery just mentioned, that is, the depletion of serotonin stores by reserpine treatment. For the first time a bridge seemed to have been built between the biochemistry of the brain and some important brain functions, with some obvious neuropsychiatric implications.

Brodie and his colleagues, especially Dr. Parkhurst Shore, generously introduced me into the new analytical methods and the use of the new instrument. I proposed to Brodie that we investigate the effect of reserpine on the catecholamines in view of their chemical similarity to serotonin. But Brodie thought this would be waste of time. He was so sure that serotonin was the target to focus upon.

A "ROSETTA STONE"?

But I felt that a look at the catecholamines might be worth while. To get started quickly I would then need a partner specialized in the catecholamine field. Again I was incredibly lucky. Of all the people working in that field at the time the most clever partner in such a project was located in my home University, the University of Lund: Professor Nils-Åke Hillarp (Fig. 3). I wrote to him from Bethesda and proposed a collaboration, and he agreed. Thus a most fruitful collaboration started, lasting until his untimely death in 1965. Hillarp's personality was different from that of Brodie in many respects, but they were similar in terms of brilliance, charisma and intensity. His background was histology and histochemistry, but his knowledge extended far into physiology and biochemistry.

In the spring of the following year Hillarp and I got the first results. We demonstrated the depletion of catecholamines from the adrenal medulla of rabbits following treatment with reserpine (Carlsson and Hillarp 1956). This was before I had acquired my own miracle instrument, the so-called Aminco-

Figure 3. Hillarp. Photo: Georg Thieme.

Bowman Spetrophotofluorimeter. The only instrument we had for the determination of catecholamines was a colorimeter, using the method of von Euler and Hamberg (1949). But we did not need any instrument because be absence of a colour development in the samples from reserpine-treated rabbits could be seen with the naked eye.

The same results were obtained when we analyzed heart and brain, in the latter case using our new instrument. We also found that sympathetic nerves no longer responded to nerve stimulation following reserpine treatment, apparently due to depletion of transmitter (Carlsson *et al.* 1957a). Thus depletion of catecholamines could be the cause of the behavioral inhibition induced by reserpine. To investigate this we gave DOPA (3,4-dihydroxyphenylalanine) to reserpine-treated animals and thus discovered the dramatic reversal of the reserpine-induced syndrome by this catecholamine precursor (Carlsson *et al.* 1957b, Fig. 4). The reason we used the precursor was that the catecholamines are unable to penetrate from the blood into the brain, because of the blood-brain barrier.

We then analyzed the brains of DOPA-treated animals and much to our disappointment we were unable to detect any restoration of noradrenaline levels. Experiments with monoamine oxidase inhibitors clearly showed that a monoamine rather than DOPA itself was responsible for the behavioral response, and thus we were forced to look for the intermediate in the conversion of DOPA to noradrenaline: dopamine.

At that time dopamine was considered to be without any interest because of its low physiological activity, when tested on various smooth-muscle preparations. We had to develop a method for determining dopamine because no such method was available at the time (Carlsson and Waldeck 1958). We could then show that dopamine occurs normally in the brain in an amount

Figure 4. Rabbits treated with reserpine (5 mg/kg intravenously).before (top) and after DL-DOPA (200 mg/kg intravenously, bottom). From Carlsson (1960). Photo: Tor Magnusson.

somewhat higher than that of noradrenaline, that it is brought to disappear by reserpine treatment, and that the antireserpine action of DOPA is closely correlated to the restoration of dopamine levels in the brain. We also showed that the restoration of serotonin levels by treatment with its precursor 5-hydroxytryptophan did not lead to any reversal of the reserpine syndrome (Carlsson *et al.* 1958).

The classical method in physiology to prove a function of a natural constituent, is to remove the constituent in question and demonstrate a loss of function, and then to reintroduce the constituent, and demonstrate a restoration of the same function. We thought we had done this in the case of dopamine. We could easily exclude possible alternative explanations, such as a role of noradrenaline and serotonin and a direct action of L-DOPA.

In fact, our enthusiasm made us think that now we had found the Rosetta stone that would give us access to the chemical language of the brain.

Later we found the unique distribution of dopamine in the brain, with an accumulation in the basal ganglia, that is structures known to be involved in motor functions. This, taken together with the fact that a characteristic side effect of reserpine is to mimic very faithfully the syndrome of Parkinsonism and to induce a similar symptomatology in animals, led us to conclude that depletion of dopamine will induce the Parkinson syndrome and that treatment with L-DOPA will alleviate that syndrome by restoring the dopamine level. All this I presented at the First International Catecholamine Symposium in October, 1958 (Carlsson 1959, Bertler and Rosengren 1959).

A BATTLE IN LONDON

A year and a half later, in March 1960, a Ciba Foundation Symposium on Adrenergic Mechanisms was held in London (Vane *et al.* 1960). I then presented the same data and some additional support obtained from studies on the action of monoamine oxidase inhibitors. At this meeting practically all of the most eminent experts in this area participated. The central figure was Sir Henry Dale, a Nobel Laureate aged 85 but still remarkably vital. He dominated the scene, and the participants, many of whom were his former students, treated him with enormous respect, like school children their headmaster, although many of them had indeed reached a mature age.

To better understand how our dopamine story was received at this meeting it may be useful to recapitulate briefly the development following Otto Loewi's discovery of chemical transmission in the frog heart (Loewi 1921). During the following decades evidence accumulated, supporting the existence of chemical transmission in various parts of the peripheral nervous system. Dale and his collaborators played an important role here. They had, however, been fiercely attacked by a number of neurophysiologists, who argued in favour of an electrical transmission across the synapses. The most eminent proponent of this view was Sir John Eccles. The debates between Dale and Eccles had been quite vivid, as witnessed by several attendants of these debates between what was called "the sparks" and "the soup". Despite the sometimes harsh wordings the debates between Dale and Eccles over the years ended in mutual respect and admiration, as clearly indicated in the correspondence of more than twenty years between the two (see Katz 1996, Girolami *et al.* 1994). Doubts about a chemical transmission were particularly strongly expressed concerning the central nervous system. In the mid 1950s, however, Eccles had placed one foot in the "soup" camp, based on his own observation that a recurrent collateral of the motor neurone, impinging on the so-called Renshaw cells, seemed to operate by cholinergic transmission. This was, however, a very special case, given the fact that motor neurons are cholinergic. Apart from this finding, as pointed out by McLennan (1963) in his monograph "Synaptic Transmission", there was no evidence in favour of chemical transmission in the central nervous system.

At this meeting in London the debate that followed upon our paper entitled "On the biochemistry and possible functions of dopamine and noradrenaline in the brain" and a subsequent special discussion session, revealed a profound and nearly unanimous skepticism regarding our points of view. Our data as such were not questioned. Actually some confirmatory animal experiments were reported at the meeting, and I referred to a paper by Degkwitz *et al.* (1960), in which an anti-reserpine action of DOPA in humans was reported. Dale expressed the view that L-DOPA is a poison, which he found remarkable for an amino acid. Marthe Vogt concluded that the views expressed by Brodie and us regarding a function of serotonin and catecholamines, respectively, in the brain would not have a long life. W.D.M. Paton referred to some unpublished experiments indicating that the catecholamines are located in glia. In his concluding remarks John Gaddum stated that at this meeting nobody had ventured to speculate on the relation between catechol amines and the function of the brain. But this was what I had insisted upon throughout the meeting, so the clear message to me was that I was nobody!

In retrospect I believe almost everybody would agree that our story and its implications were straight forward and obvious. How come that these eminent experts rejected the whole thing? I have no definite answer. Clearly the pharmacologists had great difficulty in accepting that dopamine could be an agonist in its own right, given its poor physiological effect on smooth-muscle preparations. The idea of DOPA being a mysterious poison probably came out of some experiments reported at the meeting where large doses of this amino acid, given to experimental animals together with a monoamine oxidase inhibitor, could cause paralysis, convulsions and death. In addition, I believe that the previous "sparks-and-soup" debates still had some impact. In these debates some elaborate criteria for a neurotransmitter had been formulated. Our data were of a different kind and these criteria were not applicable.

In this regard I and my collaborators, like my mentor Steve Brodie, simply had the advantage of being ignorant and not so much burdened by dogma.

A PARADIGM SHIFT

But it wouldn't be long until the scene would change dramatically. Hillarp also attended the London meeting. On our trip back to Sweden we agreed we should increase our efforts to convince the world that chemical transmission does indeed exist in the brain. Our idea was that Hillarp join me to work full time on research in our new and well-equipped department of pharmacology of Göteborg University, where I had been appointed professor and chairman the year before. We managed to obtain a grant from the Swedish Medical Research Council to set Hillarp free from his teaching duties in Lund. He could start full-time research in Göteborg already in the autumn of 1960.

We felt that the ability of catecholamines to yield fluorescent conversion products might be useful for their visualization in the microscope. We first tried a modifcation of the trihydroxyindole method (Carlsson *et al.* 1961). It worked beautifully for the adrenal medulla but not in other tissues. Hillarp then turned to another reaction that had been used for the quantitative assay of indoleamines, using formaldehyde as a reagent. Together with his skillful research assistant, the late George Thieme he worked out a model system, in which they managed to optimize the reaction conditions. (These experiments were reported by Falck *et al.* 1962). Subsequently, together with his former student Bengt Falck, Hillarp used air-dried preparations of iris and mesenterium, and discovered that the reaction worked beautifully, thus permitting the visualization of noradrenaline in adrenergic nerves and serotonin in mast cells in the fluorescence microscope. This led to an intense collaboration between our Department of Pharmacology in Göteborg and Hillarp's original Department of Histlogy in Lund, and finally, after Hillarp's move to

Figure 5. Group picture, taken January 1965, showing the group of young researchers recruited by Hillarp after his move to the Karolinska Institute in 1962. From Dahlström and Carlsson (1986). Photo: Lennart Nilsson.

take over the chair of histology at the Karolinska Institute in 1963, with an enthusiastic group of young students in his new department (see Fig.5). Thus within a few years the neuronal localization of dopamine, noradrenaline and serotonin in the central and peripheral nervous system was clearly established (Fig. 6). Moreover, the major monoaminergic pathways could be mapped (Fig. 7), and the site of action of the major psychotropic drugs clarified (see Dahlström and Carlsson 1986, Carlsson 1966, Fig. 8).

As mentioned, a large number of people were engaged in this effort. Sadly, many of these people have passed away, in many cases prematurely. Among these Georg Thieme (Fig. 9) has already been mentioned. Margit Lindqvist

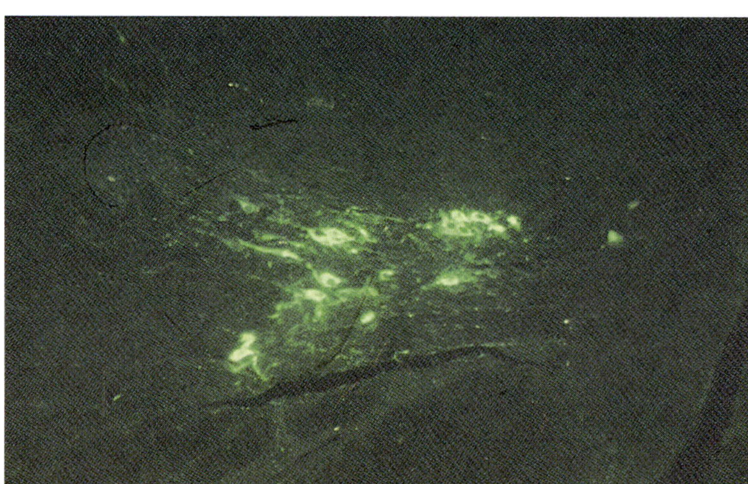

Figure 6. Dopaminergic cell bodies in rat substantia nigra. Green fluorescence developed following treatment with formaldehyde vapour. Courtesy of Annica Dahlström.

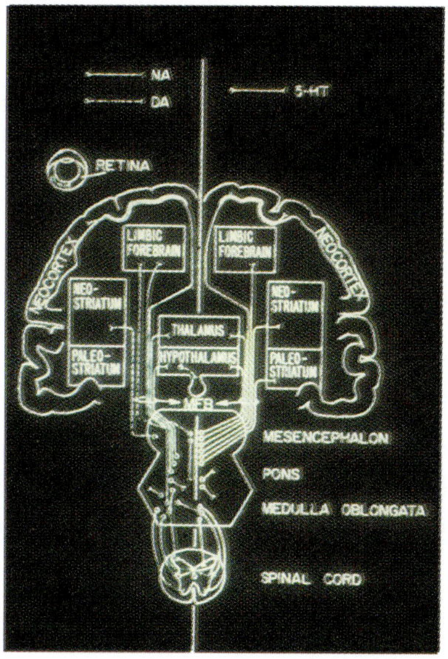

Figure 7. Monoaminergic pathways in brain. From Fuxe and Andén (1966).

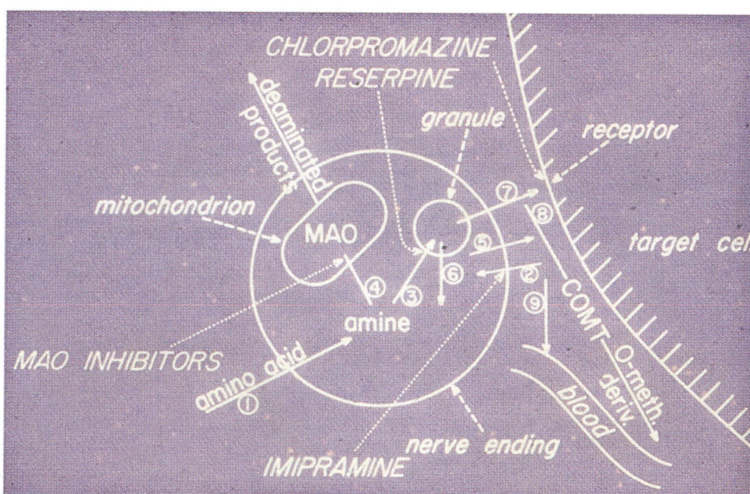

Figure 8. Scheme of monoaminergic synapse, with the sites of action of major classes of psychotropic drugs indicated. From Carlsson (1966).

(Fig. 10), a very skilful laboratrory assistant, who matured to become a qualified research worker, played an enormous role already from the outset of my scientific career. Nils-Erik Andén (Fig. 11) and Jan Häggendal (Fig. 12) were originally students of mine who became outstanding pharmacologists and largely contributed to characterize both central and peripheral monoaminergic transmission (for some of their early work, see Andén, Carlsson and Häggendal 1969). Hans Corrodi (Fig. 13), a very skilful organic chemist, who moved to Sweden because of his love for the mountains in northern Sweden,

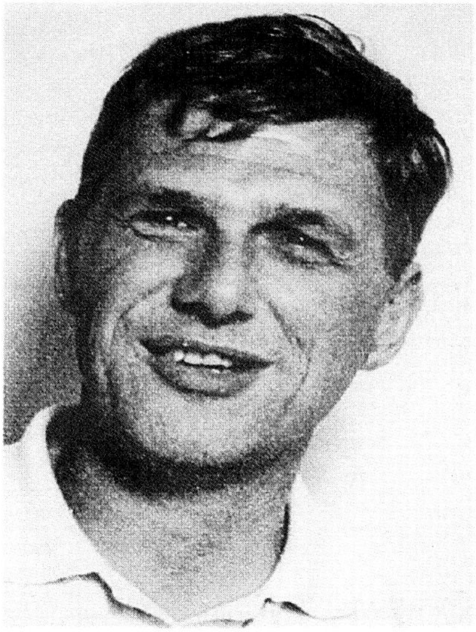

Figure 9. Georg Thieme (1926–1996)

Figure 10. Margit Lindqvist (1924–1978)

contributed much to clarify the chemistry of the formaldehyde histofluorescence method and to many other projects, especially the development of the first selective serotonin reuptake inhibitor SSRI (see below).

In February 1965 an international symposium entitled "Mechanisms of Release of Biogenic Amines" was held in Stockholm (v. Euler *et al.* 1966), with most of the major figures of that research field participating. In his Intro-

Figure 11. Nils-Erik Andén (1937–1990)

Figure 12. Jan Häggendal (1932–1992)

Figure 13. Hans Corrodi (1929–1974)

ductory Remarks Professor Uvnäs mentioned that "...these amines play an important role as chemical mediators in the peripheral and central nervous system...". None of the participants of this symposium expressed any doubt on this point. It looks as though a paradigm shift had taken place between 1960 and 1965.

It goes without saying that the concept of chemical transmission has had a profound impact on practically every aspect of brain research. In so far as neurology and psychiatry are concerned, a couple of examples are summarized below.

"AWAKENINGS"

Following our above-mentioned proposal of a role of dopamine in Parkinsonism some important parallel and apparently independent developments took place in Austria, Canada and Japan. These will now be briefly commented upon, starting out with Austria.

Later in the same year as the Symposium on Adreenergic Mechanisms, there appeared in Klinische Wochenschrift a paper in German, describing a marked reduction of dopamine in the brains of deceased patients who had suffered from Parkinson's disease and postencephalitic Parkinsonism (Ehringer and Hornykiewicz 1960). This was soon followed by a paper by Birkmayer and Hornykiewicz (1961), in which a temporary improvement of akinesia was reported following a single intravenous dose of L-DOPA to Parkinson patients.

As far as I can gather from an autobiography of Hornykiewicz (1992) as well as a personal communication from him, the following had happened. I wish to mention this in some detail, because it illustrates how the interaction of different minds can lead to important progress. In 1958, Hornykiewicz was approached by a mentor Professor Lindner or, according to a different version, by his chief Professor Brücke, who tried to persuade him to analyze the brain of a Parkinson patient, which the neurologist Walter Birkmayer wanted to be analyzed for serotonin. Presumably Birkmayer had been impressed by Brodie's already mentioned discovery in 1955 of the depletion of this compound by reserpine, and in contrast to many neurologists at that time he was aware of its possible implications. Shortly afterwards, in 1959, Hornykiewicz read about our work on dopamine and its role in the Parkinson syndrome. He then decided to include dopamine and noradrenaline in the study. In fact, in the subsequent work serotonin had to be left out initially because of some technical problems.

Hornykiewicz and his postdoctoral fellow Ehringer were now facing a challenge, because they had no adequate equipment to measure dopamine. But they managed to overcome this problem by using the purification of the brain extracts by ion exchange chromatography that our research group had worked out. The subsequent measurement was performed using the colorimetric method of Euler and Hamberg. Although this method by itself is highly unspecific, specificity could be obtained by using our purification step together with our finding that dopamine is by far the dominating catecholamine in the basal ganglia, where it occurred in high concentrations. They had to work up several grams of tissue and to concentrate the extracts by evacuation to dryness. Following this heroic procedure they were richly rewarded, because the samples from the Parkinsonian brains, in contrast to the controls, turned out to be colourless, as revealed by the naked eye!

The corresponding development of Parkinson research in Canada is summarized in a paper by Barbeau *et al.* (1962), presented at a meeting in Geneva in September the previous year. The main findings of the Canadian workers was a reduction of the urinary excretion of dopamine in Parkinson patients and an alleviation of the rigidity of such patients following oral treatment with L-DOPA.

In Japan some remarkable progress was made, which has not been adequately paid attention to in the Western countries (see reviews by Nakajima 1991, and Foley 2000). In a lecture on the 5th of August, 1959, less than a year after my lecture at the International Catecholamine Symposium mentioned above, the basic concept regarding the role of dopamine in the basal ganglia in Parkinson's disease was presented by I. Sano (1959). In this lecture data on the distribution of dopamine in the human brain were presented for the first time. In a lecture in Tokyo on the 6th of February, 1960, Sano reported on reduced amounts of dopamine in the basal ganglia of a Parkinson patient, analyzed post mortem, and in the same year he published a paper describing alleviation of rigidity in a Parkinson patient following intravenous administration of DL-DOPA (Sano 1960).

Thus treatment of Parkinson patients with DOPA was initiated simultaneously in three different countries only a few years after the discovery of the anti-reserpine action of this agent and the subsequent formulation of the concept of a role of dopamine in extrapyramidal functions. While this treatment led to results of great scientific interest, it took several years until it could be implemented as routine treatment of Parkinson patients. The reason was that the treatment regimens used initially were inadequate and led to but marginal improvement of questionable therapeutic value (Hornykiewicz 1966). It remained for George Cotzias (1967) to develop an adequate dose regimen. After that L-DOPA treatment rapidly became the golden standard for the treatment of Parkinson's disease.

When I had seen Cotzias' impressive film demonstrating the effect of escalating oral doses of L-DOPA at a meeting in Canada I hastened back to Göteborg and initiated studies together with Drs. Svanborg, Steg and others, which quickly confirmed Cotzias' observations (Andén *et al.* 1970), like in many other places at the same time. This success story was soon afterwards told to the general public by Oliver Sacks in "Awakenings" (Sacks 1973), which became a bestseller and was also made into a movie.

ROLE OF SEROTONIN IN DEPRESSION: ZIMELIDINE, THE FIRST SSRI

The so-called tricyclic antidepressants, with imipramine as the prototype, were serendipitously discovered in the late 1950s, thanks to Kuhn, a psychiatrist and a keen clinical observer. In the early 1960s these agents were found to block the reuptake of noradrenaline by nerve terminals, thus enhancing the adrenergic transmission mechanism. In 1968, we discovered that many antidepressants also could block the reuptake of serotonin (Carlsson *et al.* 1968), and this prompted us to develop a compound that selectively blocked the reuptake of serotonin without acting on noradrenaline. Such agents are now known as SSRIs. This first agent was called zimelidine, whose preclinical properties we first described by Berntsson *et al.* (1972). Zimelidine turned out to be an active antidepressant agent with a very favourable side effect profile (Carlsson *et al.* 1981), apart from a very rare, but serious side effect, presumably based on an immunological mechanism, that led to its withdrawal from the market. But zimelidine was followed by several other SSRIs, among which Prozac is especially well known, not least because of a bestseller titled "Listening to Prozac", authored by P.D. Kramer (1993). In this book Prozac is stated to be able to treat not only patients with depression and a variety of anxiety disorders, as had previously been amply demonstrated for many SSRIs, but also to be able to change the personality of people with psychological problems. Kramer was especially astonished by the fact that disturbances, which would have taken several months of psychotherapy to control, could be alleviated within a few days of treatment with Prozac. This favourable action, making people feel and function better, even if they were not mentally ill in the conventional sense, is a fascinating but, needless to say, controversial issue. Less controversial is probably the 25 % reduction in suicide rates in Sweden in the 1990s, apparently related to the introduction of the SSRIs (Isacsson 2000). In any event the SSRIs represent a major therapeutic advance as well as a milestone in rational drug development (Carlsson 1999).

The development of zimelidine was based on our discovery that certain antihistamines are serotonin-reuptake blocking agents, albeit non-selective. The most powerful agents among these were the pheniramines and diphenhydramine (Carlsson *et al.* 1969). We started out from the pheniramines and developed zimelidine. The Lilly scientists started out from our diphenhydramine data and developed Prozac, which was found to act very much like zimelidine, though devoid of its serious side effect.

DOPAMINERGIC STABILIZERS – A NOVEL PHARMACOLOGIC PRINCIPLE

In 1963, Margit Lindqvist and I presented the first evidence supporting the view that the most important group of antipsychotic agents, represented by agents such as chlorpromazine and haloperidol, act by blocking receptors for dopamine, and to some extent also receptors for noradrenaline (Carlsson and Lindqvist 1963, Fig.14). This conclusion has later been confirmed and extended in numerous laboratories, and techniques have been developed to screen for such agents in test tube experiments. One might have expected then that this should have led to the development of drugs with stronger efficacy and less side effects. Unfortunately, this has not happened.

We have hypothesized that the cause of this failure is that treatment with dopamine receptor antagonists can hardly avoid the serious and unpleasant side effects induced by dopamine hypofunction. Even though there is evidence of elevated dopaminergic activity in schizophrenia, this may be limited to psychotic episodes. In fact, we may be dealing with an instability of the

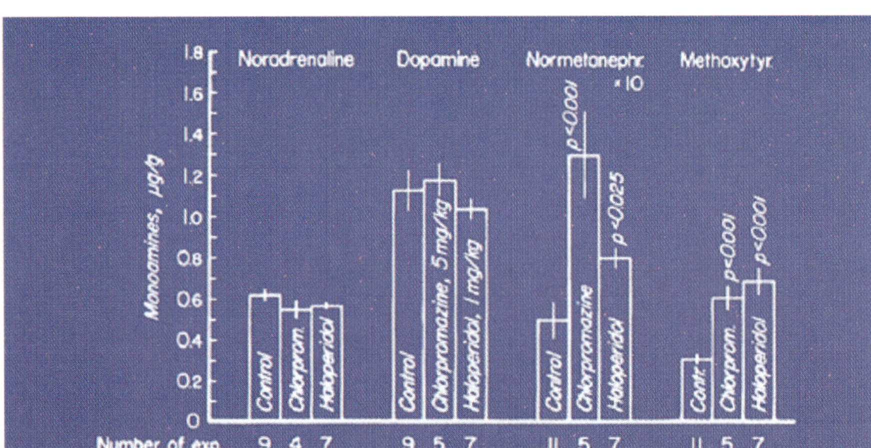

Figure 14. Accumulation of of the basic catecholamine metabolites normetanephrine and 3-methoxytyramine, enhanced by treatment with major neuroleptic agents following monoamine oxidase inhibition. From Carlsson and Lindqvist 1963.

dopamine release rather than a continuously elevated baseline. Thus, between psychotic episodes, the patient would then suffer from a dopaminergic hypofunction, especially during treatment with the currently used antipsychotic agents, showing up as a severe disturbance of the reward system and of cognition, and also as motor disturbances. This may make it impossible to attain an adequate dose level (for discussion and references, see Carlsson *et al.* 2001).

We believe that we can now get around this problem by using a new principle of intervention that we call dopaminergic stabilization. The underlying mechanism is complicated but in principle it rests on the existence of mutually antagonistic subpopulations of dopamine D2 receptors, as regards the final functional outcome. For example, the presynaptically located dopaminergic autoreceptors are inhibitory on the overall dopaminergic activity. Dopaminergic stabilizers are dopamine D2 antagonists or partial D2 agonists capable of occupying mutually opposing receptor subpopulations in such proportions as to leave the normal baseline dopaminergic activity level essentially unchanged. This leads to stabilization by dampening fluctuations of dopamine release, simply because fewer dopamine receptors are unoccupied and thus available for the endogenous neurotransmitter.

Using the dopaminergic stabilizer (-)-OSU6162 (Fig. 15), developed by our research group, partly in collaboration with Upjohn (now merged into Pharmacia Corporation), we have demonstrated the stabilization phenomenon in experimental animals (Fig. 16) and, in preliminary clinical studies, its pharmacotheraputic potential in L-DOPA-induced dyskinesias in Parkinson patients, in Huntington's disease (Fig. 17), and in schizophrenia (Tedroff *et al.* 1999, Ekesbo 1999, Gefvert *et al.* 2000).

The partial dopamine receptor agonist preclamol ((-)-3-PPP) has likewise a dopaminergic stabilizer profile. This agent was discovered by our research group and is in development in collaboration with Dr. Carol Tamminga and her colleagues at the Maryland Psychiatric Research Center (Lahti *et al.* 1988).

Our experience with dopaminergic stabilizers suggests that research into

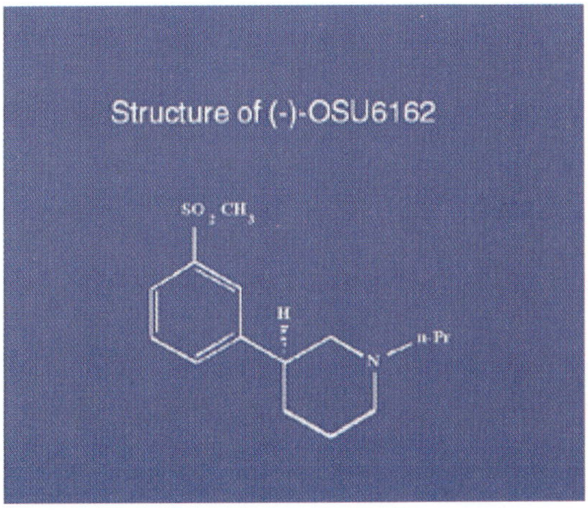

Figure 15. Chemical structure of (-)-OSU6162.

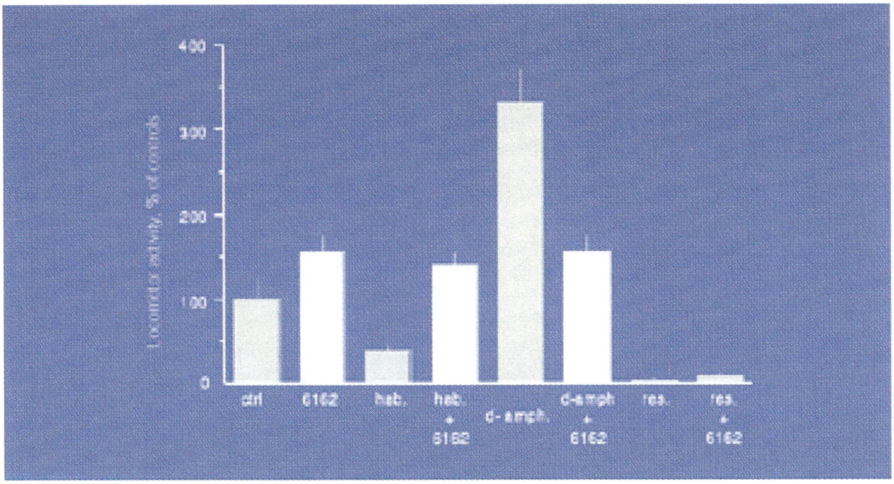

Figure 16. Stabilizing action of (-)-OSU6162 in rats. Filled bars: no treatment with (-)-OSU6162. Open bars: (-)-OSU6162. "Ctr". Actively exploring control rats. "hab": Rats habituated to their environment. "d-amph": rats treated with d-amphetamine. "res": reserpine.

Note. Treatment with one and the same dose of (-)-OSU6162 can induce stimulation of behavior when baseline activity is low (habituated rats) and inhibition when the activity is high (d-amphetamine pretreatment).

neurotransmitter pathophysiology has until now focussed too much on the hyper- versus hypofunction dichotomy. Although the instability concept is by no means new, there has not been much of a goal- directed strategy aiming to stabilize neurocircuits involved in neuropsychiatric disorders. Our preliminary data suggest that such an approach can lead to enormous gains in the treatment of a great variety neurological and psychiatric disorders.

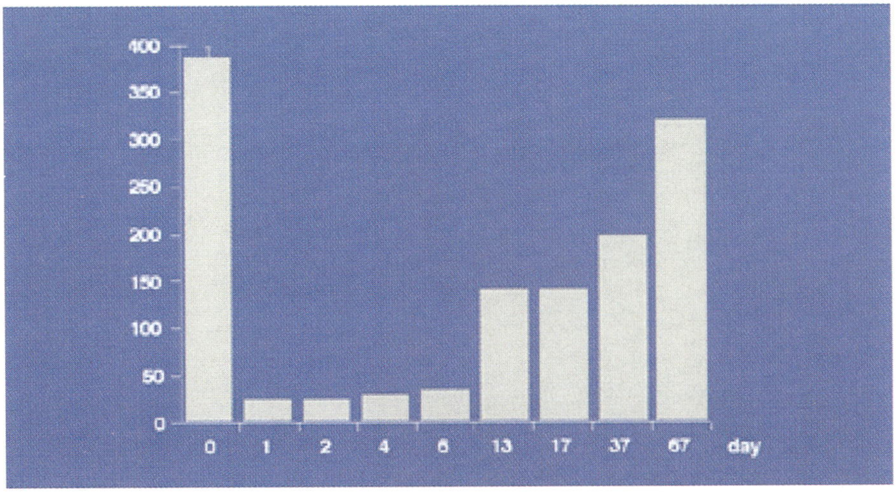

Figure 17. Choreatic events at baseline and following 0.5 mg/kg (-)-OSU6162 as an intravenous infusion during 30 minutes to a patient with Huntington's disease. From Tedroff *et al.* (1999).

OUTLOOK

During the past half-century brain research has been dominated by biochemical approaches, in contrast to the previous half-century, which had a strong electrophysiological emphasis. This switch is understandable in view of the entrance of the neurohumoral transmission concept into brain research in conjunction with the spectacular progress of molecular biology. However, it must be recognized that the brain is not a chemical factory but an extremely complicated survival machine. In order to bring all the forthcoming biochemical observations into a meaningful framework it will prove necessary to emphasize more strongly aspects of neurocircuits and connectivity and to do so both at the microscopic and macroscopic level. For example, the old questions dealing with neurocircuits within a cerebral region such as the cortex and those addressing the interaction between the different regions will in all probability come into focus more strongly in order to make full use of the new knowledge gained from neurotransmitter physiology and molecular biology. Here the new imaging techniques in conjunction with advanced computer-dependent statistics involving pattern recognition derived from a wealth of data with great complexity will probably prove extremely useful and very much help to bridge the gap between animal and human observations. If nothing else, such approaches will help to reveal the enormous width of our present ignorance of the human brain.

ACKNOWLEDGEMENTS

During my scientific career I have had the privilege to work with hundreds of other research workers, highly qualified technicians and other personnel, to whom I owe a lot. Only about forty of these people are mentioned in this text including the reference list. Sadly, a considerable number of these people have passed away, in many cases prematurely. Some of these, to whom I have a special debt of gratitude, have been commemorated with pictures.

Throughout my professional career I have enjoyed excellent working conditions, first for almost two decades at the University of Lund, Sweden, and thereafter, for four decades, at Göteborg University. My 5-month visit to the National Institutes of Health had obviously a decisive and extremely positive impact on my career.

I have received generous support from numerous sources, among which the following need to be mentioned specially: The Swedish Medical Research Council, The Swedish Board of Technical Development, The Knut och Alice Wallenberg Foundation, and during the critical late1950s and early 1960s, U.S. Air Force and National Institutes of Health, U.S.A., and more recently from the Theodore & Vada Stanley Foundation, U.S.A. In addition I have enjoyed a fruitful collaboration with generous financial support from several major pharmaceutical companies, especially Astra/Hässle, Sweden, The Upjohn Company, U.S.A, Organon, The Netherlands, and Aventis (previously Hoechst Marion Roussel), Germany.

To express in a few words my debt of gratitude to my wife Ulla-Lisa and to the rest of my family is not possible. Here I wish to refer to my autobiography (Carlsson 1998), in which I have also had the opportunity to go into further detail in several other respects.

REFERENCES

Andén N-E, Carlsson A, Häggendal J. Adrenergic mechanisms. Ann Rev Pharmacol 1969; 9:119–134.

Andén N-E, Carlsson A, Kerstell J, Magnusson T , Olsson R, Roos B-E, Steen B, Steg G, Svanborg A, Thieme G, Werdinius B.. Oral L-DOPA treatment of Parkinsonism. Acta Med Scand 1970;187:247–255.

Barbeau A, Sourkes TL, Murphy GF. Les catecholamines dans la maladie de Parkinson. In: Monoamines et Système nedrveux central. Génève:Georg et Cie S.A. 1962;247–262.

Berntsson PB, Carlsson PAE, Corrodi HR (1972): *Belg. Pat.* 781105 (72-4-14).

Bertler Å, Rosengren E. Occurrence and distribution of dopamine in brain and other tissues. Experientia 1959; 15:10.

Birkmayer W, Hornykiewicz O. Der L-3,4-Dioxyphenylalanin (=L-DOPA)-Effekt bei der Parkinson-Akinese. Wien Klin Wschr 1961;73:787–788.

Carlsson A. The occurrence, distribution and physiological role of catecholamines in the nervous system. Pharmacol Rev 1959;11:490–493

Carlsson A. Zur Frage der Wirkungsweise einiger Psychopharmaka. Psychiat Neurol 1960;140:220–222.

Carlsson A. Physiological and pharmacological release of monoamines in the central nervous system. In: von Euler US, Rosell S, Uvnäs B, eds. Mechanisms of Release of Biogenic Amines. Oxford: Pergamon Press 1966;331–346.

Carlsson A. Autobiography. In: Squire LR, ed, The History of Neuroscience in Autobiography, Volume 2. San Diego, Academic Press 1998;28–66.

Carlsson A. The discovery of the SSRIs: A milestone in Neuropsychopharmacology and rational drug design. In: Stanford SC, ed. Selective Serotonin Reuptake Inhibitors. Austin: RG Landes Company 1999;1–8.

Carlsson A, Hillarp N-Å. Release of adrenaline from the adrenal medulla of rabbits produced by reserpine. Kungl Fysiogr Sällsk i Lund Förhandl 1956;26:8.

Carlsson A, Lindqvist M. Effect of chlorpromazine or haloperidol on the formation of 3-methoxytyramine and normetanephrine in mouse brain. Acta Pharmacol (Kbh) 1963;20:140–144.

Carlsson A, Lindqvist M. Central and peripheral monoaminergic membrane-pump blockade by some addictive analgesics and antihistamines. J Pharm Pharmac 1969;21:460–464.

Carlsson A, Waldeck B. A fluorimetric method for the determination of dopamine (3-hydroxytyramine). Acta Physiol Scand 1958;44, 293–298.

Carlsson A, Rosengren E, Bertler Å, Nilsson J. Effect of reserpine on the metabolism of catecholamines. In: Garattini S, Ghetti V, eds. Psychotropic Drugs. Amsterdam:Elsevier 1957a; 363–372.

Carlsson A, Lindqvist M, Magnusson T. 3,4-Dihydroxyphenylalanine and 5-hydroxytryptophan as reserpine antagonists. Nature (Lond) 1957b;180:1200 (only).

Carlsson A, Lindqvist M, Magnusson T, Waldeck B. On the presence of 3-hydroxytyramine in brain. Science 1958;127:471 (only).

Carlsson A, Falck B, Hillarp N-Å, Thieme G, Torp A. A new hitstochemical method for visualization of tissue catecholamines. Med Exp 1961;4:123–125.

Carlsson A, Fuxe K, Ungerstedt U. The effect of imipramine on central 5-hydroxytryptamine neurons. J Pharm Pharmacol 1968;20:150–151.

Carlsson A, Gottfries C-G, Holmberg G, Modigh K, Svensson TH, Ögren S-O, eds. Recent Advances in the Treatment of Depression. Acta Physiol Scand 1981, Suppl 290.

Carlsson A, Waters N, Waters S, Carlsson ML. Network interactions in schizophrenia – therapeutic implications. Brain Res Revs 2000; 31: 342–349. Cotzias GC, Van Woert MH, Schiffer LM. Aromatic amino acids and modification of parkinsonism. New Eng J Med 1967;276:374–379.

Dahlström A, Carlsson A. Making visible the invisible. (Recollections of the first experiences with the histochemical fluorescence method for visualization of tissue monoamines). In: Parnham MJ, Bruinvels J eds, Discoveries in Pharmacology. Vol 3, Amsterdam/New York/Oxford: Elsevier, 1986; 97–128.

Degkwitz R, Frowein R, Kulenkampff C, Mohs U. Über die Wirkungen des L-dopa beim Menschen und deren Beeinflussung durch Reserpin, Chlorpromazin, Iproniazid und Vitamin B_6. Klin Wschr 1960;38:120–123.

Ehringer H, Hornykiewicz O. Verteilung von Noradrenalin und Dopamin (3-Hydroxytyramin) im Gehirn des Menschen und ihr Verhalten bei Erkrankungen des extrapyramidalen Sytstems. Klin Wschr 1960;38:1236–1239.

Ekesbo A. Functional Consequences of Dopaminergic degeneration. Thesis, Uppsala University 1999, 1–59.

Euler US von , Hamberg U. Colorimetric determination of noradrenaline and adrenaline. Acta Physiol Scand 1949;19:74–84.

Euler US von , Rosell S, Uvnäs B, eds. *Mechanisms of Release of Biogenic Amines.* Oxford: Pergamon Press 1966;331–346.

Falck B, Hillarp N-Å, Thieme G, Torp A. Fluorescence of catecholamines and related compounds condensed with formaldehyde. J Histochem Cytochem 1962;10:348–354.

Foley P. The L-DOPA story revisited. Further surprises to be expected? The contribution of Isamo Sano to the investigation of Parkinson's disease. In: Riederer P, Calne DB, Horowski R, Mizuno Y,Olanow CV, Poewe W, Youdim MBH, eds, Advances in Research on Neurodegeneration. Wien: Springer Verlag 2000;8:1–20.

Fuxe K, Andén N-E. Studies on central monoamine neurons with special reference to the nigro-neostriatal dopamine neuron system. In: Costa E, Coté LC, Yahr MD, eds. Biochemistry and Pharmacology of the Basal Ganglia. New York; Raven, 1966; 123–130.

Gefvert O, Lindström LH, Dahlbäck O, Sonesson C *et al.* (-)-OSU6162 induces a rapid onset of antipsychotic effect after a single dose. A double-blind placebo-controlled study. Abstract. Nordic J Psychiat 2000;54/2:93–94.

Girolami P, Taborikova H, Nistico G, eds. In Memory of Sir Henry Dale. Romr: Accademia di Scienze Mediche e Biologiche, 1994;1–67.

Hornykiewicz O. Metabolism of brain dopamine in human Parkisonism: Neurochemical and clinical aspects. In: Costa E, Côté LKJ, Yahr MD, eds. Biochemistry and Pharmacology of the Basal Ganglia. New York: Raven Press 1966,171–186.

Hornykiewicz O. From dopamine to Parkinson's disease: A personal research record. In: Samson F, Adelman G, eds, The Neurosciences: Paths of Discovery II. Boston: Birkhäuser 1992;125–148.

Isacsson G. Suicide prevention – a medical breakthrough? Acta Psychiat Scand 2000; 102:113–117.

Kanigel R. Apprentice to Genius, The Making of a Scientific Dynasty. New York: Macmillan 1986;1–271.

Katz B. In: Squire LR, ed. The History of Neuroscience in Autobiography. Volume 1. Washington: Soc. for Neurosci 1996;348–381.

Kramer PD. Listening to Prozac. New York: Penguin Books, 1993.

Lahti AC, Weiler MA,.Corey PK, Lahti RA, Carlsson A, Tamminga CA. Antipsychotic properties of the partial dopamine agonist(-)-3-(3-hydroxyphenyl)-N-n-propylpiperidine (preclamol) in schizophrenia. Biol Psychiat1998;43:2–11.

Loewi O. Über humorale Überträgbarkeit der Herznervenwirkung. (I. Mitteilung). Pflüg Arch ges Physiol 1921;189:239–242.

McLennan H. Synaptic Transmission. Philadelphia: WB Saunders Co. 1963;1–134.

Nakajima T. Discovery of dopamine deficiency and the possibility of dopa therapy in Parkinsonism. In: Nagatsu T, Narabayashi H, Yoshida M, eds. Parkinson's Disease. From Clinical aspects to Molecular Basis. Wien: Springer Verlag 1991;13–18.

Pletscher A, Shore PA , Brodie BB. Serotonin release as a possible mechanism of reserpine action. Science 1955;122: 374–375.

Pletscher A, Shore PA , Brodie BB. Serotonin as a mediator of reserpine action in brain. J Pharmacol. exp Ther 1956;116:84–89.

Sacks O. Awakenings. London: Gerald Duckworth 1973;1–408.

Sano I. Biochemical studies of aromatic monoamines in the brain. In: Japanese Medicine in 1959. The report on scientific meetings in the 15th General Assembly of the Japan Medical Congress, Vol.V. 1959;607–615.

Sano I. Biochemistry of extrapyramidal motor system. Shinkey Kenkyu no Shinpo (Adv Neurol Sci) 1960;5:42–48. English translation in: Parkinsonism and Related Disorders 2000;6:3–6.

Tedroff J, Ekesbo A, Sonesson C, Waters N, Carlsson A. Long-lasting improvement following (-)-OSU6162 in a patient with Huntington's disease. Neurology 1999;53:1605–1606.

Vane JR, Wolstenholme GEW, O'Connor M, eds. Ciba Foundation Symposium on Adrenergic Mechanisms, London: J & A Churchill Ltd 1960; 1–632.

PAUL GREENGARD

My paternal great, great grandfather emigrated from Koenigsberg, now Gdansk, to St. Louis during the 1850s, a time of large-scale movement of Germans, including German Jews, to the Mid-Western United States. My paternal grandfather moved to Binghamton, New York, where my father was born. My father had success in vaudeville as a singer/dancer/comedian. He eventually became a businessman working first in retail and then in wholesale in the cosmetics field.

I was born on December 11, 1925 in New York City under tragic circumstances – my mother, née Pearl Meister, died giving birth to me. My father remarried when I was 13 months old. In contrast to my mother, who had been Jewish, my stepmother, who raised me, was Episcopalian. From that time on I was brought up in the Christian tradition, celebrating Easter, Christmas, etc. I was prevented access to my biological mother's family with whom I became familiar only very recently. I have been delighted to learn that many members of that family are highly creative individuals working in various fields of science, government, etc.

I attended public schools in Brooklyn and Queens. During World War II, I spent three years in the Navy as an electronics technician. After appropriate training, I was assigned to a team at the Massachusetts Institute of Technology that was involved in developing an early-warning system to intercept Japanese kamikaze planes before they could reach the ships of the U.S. fleet.

After the war, I attended Hamilton College, a small liberal arts college located in Clinton, New York, where I majored in mathematics and physics, and from which I graduated in 1948. I had been interested in going to graduate school in theoretical physics, but decided not to do so because at that time the only fellowship support for such graduate studies came from the Atomic Energy Commission. This was only three years after dropping the atomic bombs on Japan, and I didn't want to contribute to research the fruits of which might contribute to creating more powerful weapons of mass destruction. In thinking about various options, I settled on the then nascent field of biophysics. At that time there were two groups of academic biophysicists. One, at the University of California, was engaged in biological and medical applications of radioisotopes. The other, at the University of Pennsylvania, headed by Detlev W. Bronk, used electrophysiological techniques to study nerve function. I chose the latter. Shortly after I arrived in Philadelphia, Bronk announced that he was accepting the Presidency of The Johns Hopkins University and invited a group of us to move there with him and form a new department of biophysics. The most senior member of the group was H. Keffer Hartline, who became Chairman of the Department of Biophysics at Johns Hopkins and who was later to win a Nobel Prize in Physiology or Medicine for his work on vision. I did my first laboratory research under the supervision of Hartline.

Shortly after our move to Hopkins, Allen Hodgkin gave a lecture on the still unpublished work that he and Andrew Huxley had carried out on the ionic basis of the nerve impulse – work that was later recognized by the Nobel Prize in Physiology or Medicine. That work, carried out exclusively with biophysical techniques, filled me with admiration. At the same time, the elegance of that study made me feel that it might be a long time until biophysical techniques would by themselves make further major contributions to our understanding of nerve cell function. Thus it was Hodgkin's lecture that led me to consider combining biophysical and biochemical techniques to understand the molecular and cellular basis of how nerve cells work. Since, at that time, neuroscience as a field had not yet been created, my Ph.D. thesis was carried out under the joint supervision of Frank Brink, a distinguished biophysicist in our Department of Biophysics, and Sidney Colowick, a prominent biochemist who was a Professor in the Department of Biology. I remain to this day very grateful for their nurture and support.

Upon graduation from The Johns Hopkins University in 1953, I went to Europe for postdoctoral studies. My first year was with Henry McIlwain, at the Maudsley Hospital, of the University of London. My second year was with E. C. Slater, first at Cambridge University and then at the University of Amsterdam. At the end of my six-month period in Amsterdam, I returned to London to work in the laboratory of Wilhelm Feldberg, who was Head of the Department of Pharmacology at the National Institute for Medical Research at Mill Hill, London. At that time, the only departments that had both electrophysiological and biochemical facilities were pharmacology departments. Feldberg was a great scientist and a wonderful human being and provided an atmosphere in which I could continue to explore the relationships between biochemistry and electrophysiology in the nervous system. I seriously considered staying in England, which I found particularly compatible with my personality, which at that time was relatively reserved. However, the low level of financial support for scientific research in England, my ignorance of the nuances of the complex British educational system (two sons had been born in England), and the lack of central heating all conspired towards my returning to the United States. Upon my return, I spent one year working in the laboratory of Sidney Udenfriend at the NIH following which I became director of the Department of Biochemistry at the Geigy Research Laboratories. My prime motivation for going to Geigy was the prospect of applying basic scientific principles to the development of new drugs for the treatment of neurological and psychiatric disorders. Unfortunately, at that time, Geigy, like most, if not all, other pharmaceutical companies, was very conservative with regard to the nature of the research programs which they found acceptable. It was extremely difficult to obtain authorization to embark on innovative research approaches. In 1967, I left Geigy. After spending one year as a Visiting Professor, the first semester in Alfred Gilman's Department of Pharmacology at Albert Einstein College of Medicine and the second semester with Sidney Colowick and Earl Sutherland at the Vanderbilt University School of Medicine, I took a position as Professor in the Department of Pharmacology at Yale University. My years at Yale saw the early development of my work on signal transduction in the nervous system. Although I was very happy throughout my 15 years at Yale, the offer to move to The Rockefeller University was irresistible and so I moved to New York in 1983, where I have been located since. It has been at Rockefeller that most of the work described in my Nobel Lecture was performed.

THE NEUROBIOLOGY OF DOPAMINE SIGNALING

Nobel Lecture, December 8, 2000

by

PAUL GREENGARD

Vincent Astor Professor, Laboratory of Molecular and Cellular Neuroscience, The Rockefeller University, 1230 York Avenue, New York, NY 10021, U.S.A.

It is estimated that there are about one hundred billion nerve cells in the brain and that on average each of these nerve cells communicates with one thousand other nerve cells. A vigorous debate went on from the 1930s through the 1960s as to whether intercellular communication across the synapses between nerve cells was electrical or chemical in nature. The electrical school of thought held that the nerve impulse or action potential was propagated along the axon to the nerve ending, changed the electrical field across the postsynaptic plasma membrane, and thereby produced a physiological response. The chemical school believed that when the action potential came down the axon to the nerve terminal, it caused the fusion of neurotransmitter-containing vesicles with the presynaptic plasma membrane, releasing a neurotransmitter, which then diffused across the synaptic cleft and, through activation of a (hypothetical) receptor, produced a physiological response. The chemical school won this debate: over 99% of all synapses in the brain use chemical transmission. Based on those earlier studies, I became interested in the biochemical mechanisms by which neurotransmitters, through activation of their receptors, produce their physiological effects within their postsynaptic target nerve cells.

We know today that there are two categories of chemical transmission between nerve cells, which are referred to as fast and slow synaptic transmission. About half of the fast synapses in the brain are excitatory, and most of these fast excitatory synapses utilize glutamate as their neurotransmitter. The other half of the fast synapses are inhibitory and most of these fast inhibitory synapses use GABA as their neurotransmitter. Synaptic transmission at fast synapses occurs in less than one-thousandth of a second, and is attributable to the ability of the fast-acting neurotransmitters to open what are called ligand-operated ion channels present in the plasma membrane of the post-synaptic cells. In fast excitatory transmission, glutamate binds to its receptor, causing a change in the conformation of the receptor, allowing positively charged sodium ions to rush into the cell, causing a depolarizing, i.e. excitatory, signal to be generated in the target cell. In fast inhibitory transmission, GABA binds to its receptor, causing a change in the conformation of the receptor, allowing negatively-charged chloride ions to permeate the cell, causing a hyperpolarizing, i.e. inhibitory, signal to be generated in the target cell.

Slow synaptic transmission, which occurs over periods of tens of milliseconds to seconds, is enormously more complex than fast synaptic transmission. There are about 150 compounds which are now believed to serve as neurotransmitters in the brain. The vast majority of these putative neurotransmitters appear to work through slow synaptic transmission. Thus, it now seems very likely that all of the biogenic amines, and all of the peptide neurotransmitters, produce their effects on their target cells through slow synaptic transmission. And even the fast acting neurotransmitters, including glutamate and GABA, produce many of their effects through slow synaptic transmission pathways.

Our work in trying to elucidate the molecular basis of slow synaptic transmission was inspired by studies carried out by Earl W. Sutherland[1] and Edwin G. Krebs[2]. Sutherland and Krebs were interested in understanding how the hormones glucagon and adrenaline break down glycogen to glucose in liver and muscle cells. Earl Sutherland and his colleagues found that these hormones stimulated the formation of cyclic AMP from ATP by virtue of activating a class of enzymes that they called hormone-sensitive adenylyl cyclases. They then showed that cyclic AMP could mimic the hormones in causing the breakdown of glycogen to glucose. Edwin Krebs and his colleagues subsequently showed that cyclic AMP caused the breakdown of glycogen to glucose by activating an enzyme which they called cyclic AMP-dependent protein kinase. Protein kinases catalyze the reaction:

Adenosine Triphosphate (ATP) + Substrate	→	Adenosine Diphosphate (ADP) + Phosphosubstrate

They further showed that this substrate, when phosphorylated, was an enzyme which caused the breakdown of glycogen to glucose. The action of protein kinases is reversible, an enzyme called a protein phosphatase catalyzing the reaction:

Phosphosubstrate	→	Substrate + Inorganic Phosphate

When my colleagues and I started our work on the molecular basis of synaptic transmission, my hypothesis was that the same signaling machinery used by the endocrine system, e.g. to break down glycogen to glucose, might be used for communication between nerve cells. One major cause for concern with this concept was that in the case of hormones the distance of communication between the sending cell and the receiving cell can be more than two meters, whereas the distance across a synapse is roughly one millionth of a centimeter. Nevertheless, we decided to test this hypothesis. We searched in the brain for signaling enzymes analogous to those which had been found in liver and muscle. We found a family of neurotransmitter-sensitive adenylyl cyclases, analogous to the hormone-sensitive adenylyl cyclases of Earl Sutherland. These adenylyl cyclases converted ATP to cyclic AMP in the presence of neurotransmitters. The first such neurotransmitter-sensitive adenylyl cyclase, found by John W. Kebabian (photo 1), was a dopamine-sensitive adenylyl cyclase: in the presence of dopamine, this membrane-bound enzyme stimulated

Photo 1. John W. Kebabian, 1977.

Photo 2. Eishichi Miyamoto, 2001.

formation of cyclic AMP. Moreover, the data indicated that this enzyme might play a role in synaptic transmission[3].

At about the same time, Eishichi Miyamoto (photo 2) and J. F. Kuo (photo 3) were able to demonstrate cyclic AMP-dependent protein kinase activity in the brain[4]. In addition, the concentration of this enzyme was enormously higher in brain than in liver. Even more intriguing was the fact that the enzyme was concentrated in the synaptic region of nerve cells. These data were consistent with a possible role for cyclic AMP-dependent protein kinase in synaptic transmission.

Photo 3. J. F. Kuo, (undated).

Photo 4. Howard Schulman, 1968.

Photo 5. S. Ivar Walaas, 1990.

Photo 6. Angus C. Nairn, 1987.

Soon thereafter, we found a second, distinct class of regulated protein kinases, activated selectively by cyclic GMP, rather than cyclic AMP[5]. This cyclic GMP-dependent protein kinase was present both in brain and in non-neural tissues. Subsequently, Howard Schulman (photo 4) discovered a third group of regulated protein kinases, which were stimulated by calcium, in the presence of an unidentified endogenous heat stable protein that was later shown to be the calcium effector protein calmodulin[6]. The discovery of several neurotransmitter-sensitive enzymes that made cyclic AMP and of several second messenger-dependent protein kinases strengthened our belief that second messengers and protein kinases might be involved in signaling in the brain. This idea was supported by the discovery of a large number of brain-specific substrate proteins for these protein kinases. Thus, S. Ivar Walaas (photo 5) and Angus C. Nairn (photo 6) found more than one hundred substrate proteins for protein kinases which were highly enriched in, or exclusively localized to, the brain, some of which were present in very high concentrations[7, 8]. In the ensuing years, we were able to show that injections of various second messengers, protein kinases, protein phosphatases, and activators and inhibitors of these enzymes, were able either to mimic or to antagonize the ability of neurotransmitters to produce physiological responses in nerve cells. Combined, these data have provided overwhelming evidence of a role for those signaling molecules in synaptic transmission[9].

Some of the principal signaling pathways involved in slow synaptic transmission are shown in figure 1. The presynaptic terminal contains synaptic vesicles, which sequester neurotransmitter within them. In response to an action potential, these vesicles fuse with the plasma membrane, releasing the neurotransmitter. In an extensive series of studies, in which Pietro De Camilli (photo 7), Fabio Benfenati (photo 8), Flavia Valtorta (photo 9), and Andrew J. Czernik (photo 10) played leading roles, and which involved a series of col-

Photo 7. Pietro de Camilli, 1981.

Photo 8. Fabio Benfenati, 1999.

laborative studies with Rodolfo R. Llinas and his colleagues, it was found that the efficacy of release of neurotransmitter from the presynaptic terminal was controlled by regulation of the state of phosphorylation of a family of proteins in the presynaptic nerve terminal[10, 11]. We named these proteins synapsins because they are localized to synaptic vesicles in the presynaptic terminals. The synapsins are the most abundant phosphoproteins found in the brain and, possibly for this reason, were the first to be discovered. Another seminal series of studies on the regulation of neurotransmitter release by cyclic AMP-dependent phosphorylation of nerve terminal proteins was car-

Photo 9. Flavia Valtorta, 2000.

Photo 10. Andrew J. Czernik, 1992.

ried out in a collaboration between Eric R. Kandel's laboratory and my own[12, 13]. The various studies of the physiology of the presynaptic terminal demonstrated conclusively that the efficacy of neurotransmitter release in response to the nerve impulse is regulated by protein phosphorylation/dephosphorylation. Those studies lie outside the scope of this presentation and I shall not discuss them further here. Rather, I am going to concentrate on how slow-acting neurotransmitters, through activation of their receptors, produce appropriate physiological responses in their postsynaptic target cells.

Slow-acting neurotransmitters, upon binding to their receptors, change the level of one or another second messenger, e.g. cyclic AMP, cyclic GMP, calcium, or diacyl glycerol. These second messenger molecules in turn activate distinct classes of protein kinases. The activated protein kinases phosphorylate and thereby change the properties of substrate proteins, which serve as downstream physiological effectors. The substrate proteins of the nervous system can be divided into various classes, four of which are diagrammed in figure 1. One major class of substrates are the receptors for neurotransmitters, both fast- and slow-acting neurotransmitters. These protein kinases also phosphorylate various voltage-gated sodium, potassium, and calcium ion channels. Another class of substrate proteins are ion pumps, which restore ionic equilibrium after a burst of neuronal activity. Still another class are transcription factors present in the cell nucleus, which control new protein synthesis. Protein synthesis is involved in long-term changes in the nerve cell in response to activity, and may form an important component of the molecular basis of learning and memory, a field which was pioneered by Eric Kandel, and is described by him in this volume[14].

Figure 1, in highly simplified form, schematizes the basic principles of slow synaptic transmission. Although this scheme is now a component of the scientific dogma, it was greeted initially with enormous skepticism, and at times down-right hostility, by the scientific community. In retrospect, there were at least two major reasons for this. At the time that we started this work, neuroscience was not a clearly defined field. There were two types of people studying the brain. There were biophysicists, working in physiology departments, who believed that everything significant about the brain could be explained in terms of electrical signaling. And there were biochemists working in biochemistry departments who would happily throw a brain into a homogenizer, with as much abandon as they would a liver, and look for enzymes or lipids. But these biochemists were rarely interested in brain function. And so these two groups rarely spoke to each other, which is just as well because when they did they didn't have nice things to say. So, there was almost no one working on the biochemical basis of how nerve cells function. A second reason for the skepticism that biochemical pathways might be involved in synaptic transmission was more substantive: it was not immediately obvious as to how the relatively slow enzymatic reactions that we were studying, involving protein phosphorylation and dephosphorylation, could be involved in fast synaptic transmission, which occurs in less than one-thousandth of a second. The answer to this apparent paradox was that these slow signaling pathways

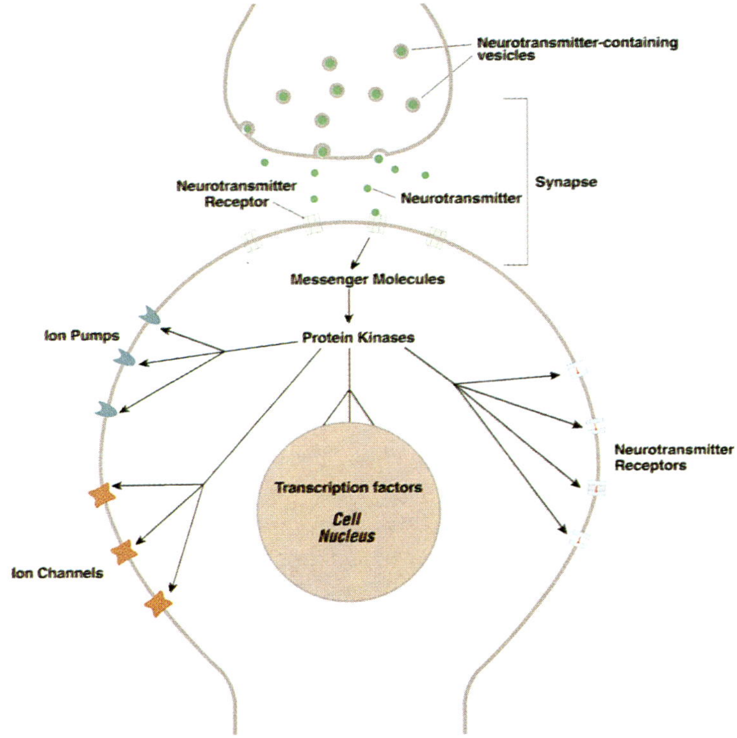

Figure 1. Some of the signaling pathways involved in slow synaptic transmission.

don't *mediate* fast synaptic transmission. Rather they *modulate* fast synaptic transmission, and they do so in two major ways: a) by regulating the state of phosphorylation of synapsins and other key proteins present in the presynaptic terminal, thus modulating the efficacy of neurotransmitter release (the amount of neurotransmitter released from the nerve terminal in response to an action potential), and b) by regulating the state of phosphorylation of neurotransmitter receptors present in the post-synaptic cell, thus modulating the responsivity of these receptors to the released neurotransmitter (responsivity referring to the magnitude of the electrophysiological response to a molecule of neurotransmitter).

The various slow signaling pathways which have been studied to date obey similar principles. The slow-acting neurotransmitter which we have studied most intensively is dopamine. There were several reasons for focusing on this system. First there was the pioneering work of Arvid Carlsson[15], his colleagues, and those who followed in his footsteps, showing that four major and several minor neurological and psychiatric diseases are associated with abnormalities in the dopamine signaling pathway. The four major diseases are Parkinsonism, schizophrenia, Attention Deficit Hyperactivity Disorder (ADHD), and drug abuse. Parkinsonism is associated with the death of dopamine-producing nerve cells, and is treated by giving Levodopa, a precursor of dopamine. Most currently used antipsychotic drugs block a subclass of

Photo 11. Dana W. Aswad, 1979.

Photo 12. Patrick B. Allen, 2001.

dopamine receptors. ADHD is treated by Ritalin, which works in large part by stimulating dopamine release. Virtually all drugs of abuse cause perturbations of dopamine signaling. There were two other reasons for concentrating on dopamine signaling. One was that the neurostriatum, a major target for dopaminergic innervation, is relatively large and homogenous and thus fairly readily permits both electrophysiological and biochemical studies. The other reason is that the simple circuitry of the basal ganglia, compared to the circuitry of the cortex, makes the analysis of signaling systems much more manageable. However, it is worth emphasizing that although much of our research in recent years has concentrated on the dopamine pathways, the basic

Photo 13. James A. Bibb, 2000.

Photo 14. Allen A. Fienberg, 1996.

Photo 15. Gilberto Fisone, 1995.

Photo 16. Jean-Antoine Girault, 1986.

principles elucidated in those pathways appear to be applicable to all slow synaptic pathways in the brain.

In our studies of dopamine signaling, we were very fortunate to discover a molecule that we named DARPP-32, an acronym for *d*opamine and cyclic *A*MP *r*egulated *p*hospho*p*rotein MW = *32* kDa. Dopamine, by activation of a subclass of dopamine receptors, causes an increase in the level of cAMP, the activity of PKA and the phosphorylation of threonine 34 of DARPP-32. DARPP-32 plays an obligatory role in mediating the actions of dopamine and has served as a rosetta stone for understanding the interactions of dopamine with other neurotransmitters, therapeutic drugs, and drugs of abuse. DARPP-

Photo 17. Hugh C. Hemmings, Jr. (right) and Angus C. Nairn, 1986.

Photo 18. Charles C. Quimet, 1985.

Photo 19. Akinori Nishi, 2000.

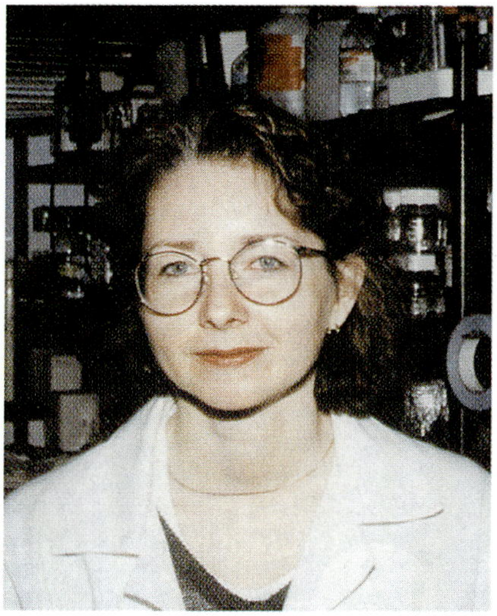

Photo 20. Gretchen L. Snyder, 2000.

32 was discovered by S. Ivar Walaas and Dana W. Aswad (photo 11) while searching for region-specific protein kinase substrates in the brain[16]. DARPP-32 has been characterized by Patrick B. Allen (photo 12), James A. Bibb (photo 13), Allen A. Fienberg (photo 14), Gilberto Fisone (photo 15), Jean-Antoine Girault (photo 16), Hugh C. Hemmings, Jr. (photo 17, together with Angus C. Nairn), Angus C. Nairn, Charles C. Ouimet (photo 18), Akinori Nishi (photo 19), Gretchen L. Snyder (photo 20), Per Svenningsson (photo 21), S. Ivar Walaas, and Zhen Yan (photo 22).

DARPP-32 is highly concentrated in the neostriatum (the caudate and the putamen), and the nucleus accumbens. An enormously simplified version of the neostriatal circuitry in which these DARPP-32-containing nerve cells are located is shown in figure 2. This circuitry has provided a useful model for

Photo 21. Per Svenningsson, 2000.

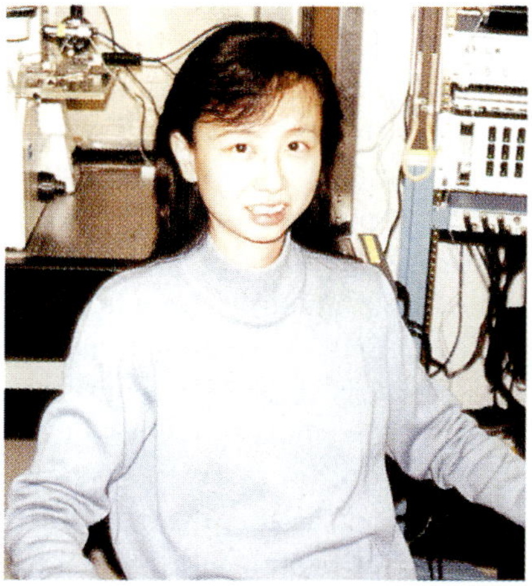

Photo 22. Zhen Yan, 2000.

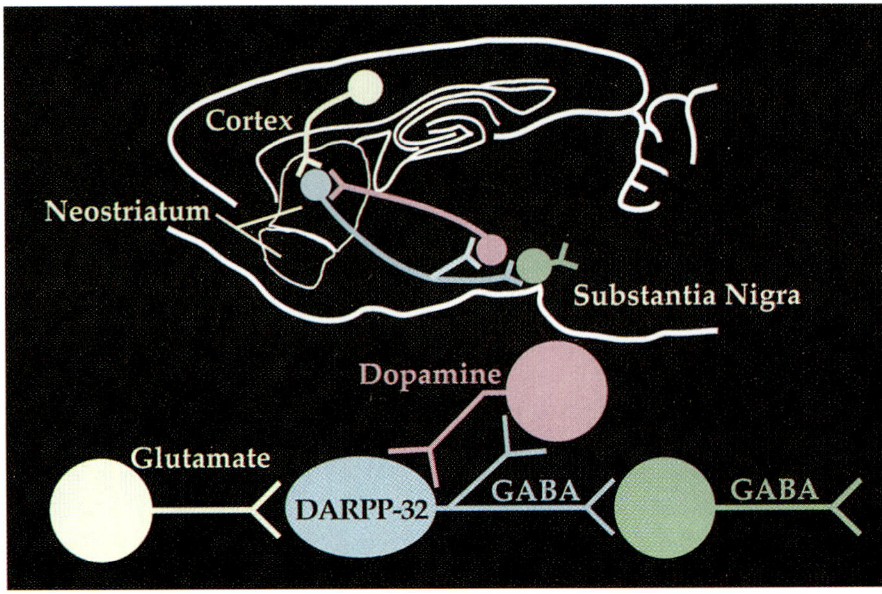

Figure 2. Simplified diagram of some connections between cortex, neostriatum, and substantia nigra.

The upper portion shows the neuroanatomy and the lower portion the corresponding neurochemistry. The cells shown in yellow project from the cortex to the striatum and use glutamate, a fast-acting, excitatory neurotransmitter, to stimulate second-order neurons, shown in blue, which project from the neostriatum to the substantia nigra. These latter neurons contain DARPP-32 and use GABA as their neurotransmitter. In addition to innervating third-order neurons, their axons have branches that innervate dopaminergic neurons, which in turn provide feedback on the DARPP-32-containing cells.

studying the mechanisms by which slow synaptic transmission, as exemplified by dopamine, modulates fast synaptic transmission, as exemplified by glutamate.

There are three major classes of glutamate receptors, designated NMDA, AMPA and Metabotropic, and two major classes of dopamine receptors, designated D1 and D2. The interactions between the dopamine and glutamate signaling pathways are highly complex and in addition are modulated by many other neurotransmitters and their signaling pathways. DARPP-32 plays a central role in the interactions amongst those various complex signaling pathways (figure 3). All four possible classes of mechanisms for regulating the state of phosphorylation of DARPP-32 on threonine 34 have been shown to exist, i.e., increases and decreases in phosphorylation and increases and decreases in dephosphorylation. Why has so much evolutionary machinery gone into regulating the state of phosphorylation of DARPP-32?

The DARPP-32 sequence has been highly conserved within mammals. Rat DARPP-32 is a protein consisting of 205 amino acids. Threonine 34 of DARPP-32 is phosphorylated by PKA or PKG and dephosphorylated by PP2B. Phosphorylation of DARPP-32 on threonine 34 profoundly changes its biological properties, converting it from an inactive molecule into a very potent inhibitor of protein phosphatase 1, with an IC50 (figure 4) and KI of about 10^{-9} M[17]. Since the concentration of DARPP-32 in medium spiny neurons is

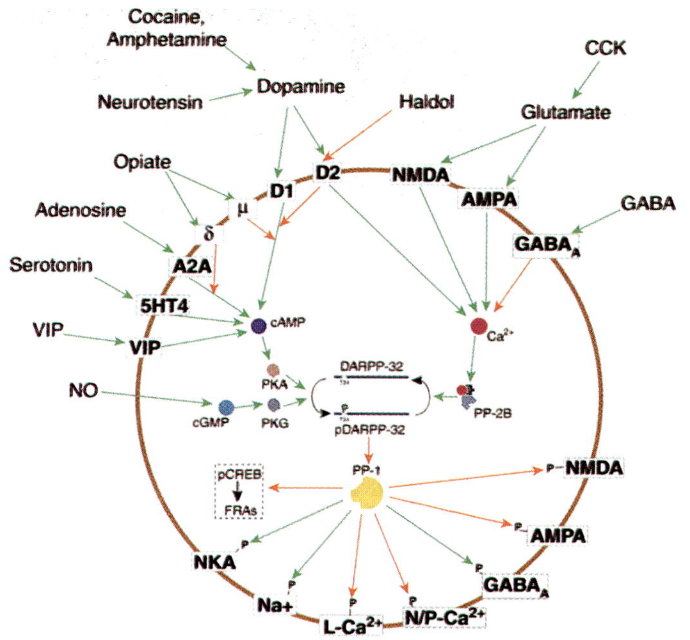

Figure 3. Interactions of signaling pathways in the neostriatum.

Activation by dopamine of the D1 subclass of dopamine receptors stimulates the phosphorylation of DARPP-32 at Thr-34. This is achieved through a pathway involving the activation of adenylyl cyclase, the formation of cyclic AMP, and the activation of cyclic AMP-dependent protein kinase (PKA). Activation by dopamine of the D2 subclass of dopamine receptors causes the dephosphorylation of DARPP-32 through two synergistic mechanisms: D2 receptor activation (a) prevents the D1 receptor-induced increase in cyclic AMP formation, and (b) raises intracellular calcium, which activates a calcium-dependent protein phosphatase, referred to as protein phosphatase 2B (PP2B; calcium/calmodulin-dependent protein phosphatase; calcineurin). Activated PP-2B dephosphorylates DARPP-32 at Thr-34. Glutamate acts as both a fast-acting and slow-acting neurotransmitter. Activation by glutamate of AMPA receptors causes a rapid response through influx of sodium ions, depolarization of the membrane, and firing of an action potential. Slow synaptic transmission, in response to glutamate, results in part from activation of the AMPA and NMDA subclasses of glutamate receptor, which increases intracellular calcium and the activity of PP2B, and causes the dephosphorylation of DARPP-32 on Thr-34. All other neurotransmitters that have been shown to alter the physiology of dopaminoceptive neurons also alter the phosphorylation state of DARPP-32 on Thr-34 in these cells. For example, adenosine acting on the A2A subclass of dopamine receptor, serotonin acting on the 5HT4 subclass of serotonin receptor, and VIP, acting on its receptor, all increase cyclic AMP-dependent phosphorylation of DARPP-32 at Thr-34. Nitric oxide, released from other types of nerve cells, causes an increase in cyclic GMP formation, resulting in activation of cyclic GMP protein kinase (PKG), which like PKA increases DARPP-32 phosphorylation on Thr-34. GABA, acting on the GABA A subclass of GABA receptor, hyperpolarizes the cell membrane, leading to a decrease in intracellular calcium, an inactivation of PP2B, and an increase in the state of phosphorylation of DARPP-32 at Thr-34. Interestingly, dopamine and GABA, both of which are capable of inhibiting activity of these nerve cells, both increase DARPP-32 Thr-34 phosphorylation state, one by increasing phosphorylation, the other by decreasing dephosphorylation. As one would predict, the effects of dopamine and GABA on DARPP-32 phosphorylation are synergistic. Neurotransmitters which act indirectly to affect the physiology of these medium spiny neurons also regulate DARPP-32 phosphorylation: neurotensin, through stimulating the release of dopamine, increases DARPP-32 phosphorylation; conversely cholecystokinin, through stimulating the release of glutamate, decreases DARPP-32 phosphorylation.

Therapeutic drugs and drugs of abuse, all of which affect the physiology of these neurons, also regulate the state of phosphorylation of DARPP-32 on Thr-34. For example antipsychotic drugs, such as Haldol, which block the activation by dopamine of the D2 subclass of dopamine receptor, increase phosphorylation. Agonists for the mu and delta subclasses of opiate receptors block D1 and A2A receptor-mediated increases in cyclic AMP respectively, and the resultant increases in phosphorylation. Cocaine and amphetamine, through increasing extracellular dopamine levels, increase phosphorylation. Marijuana, nicotine, alcohol, and LSD, all of which affect the physiology of the medium spiny neurons, also regulate phosphorylation (unpublished data). Finally, all drugs of abuse have greatly reduced biological effects in animals with targeted deletion of the DARPP-32 gene.

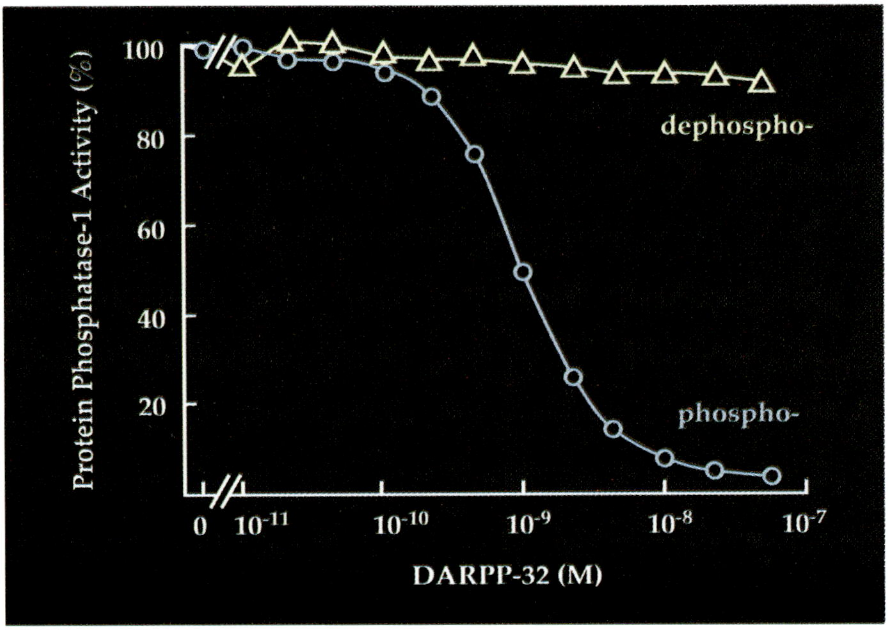

Figure 4. Phospho-DARPP-32, but not dephospho-DARPP-32, inhibits protein phosphatase-1. (Modified from Hemmings, *et al.*, reference 17.)

greater than 10^{-5} M, this means that a small burst of activity in dopaminergic neurons would be expected to result in significant phosphorylation of DARPP-32 and inhibition of protein phosphatase 1. Protein phosphatase 1 has a very broad substrate specificity and controls the state of phosphorylation and activity of a number of physiologically important substrates, including neurotransmitter receptors, voltage-gated ion channels, ion pumps, and transcripton factors. As a result, neurotransmitters which increase or decrease phospho-threonine 34 of DARPP-32, inhibit or activate, respectively, protein phosphatase 1, and thereby increase or decrease the state of phosphorylation and activity of a large array of downstream physiological effectors.

The physiological significance of the DARPP-32/PP1 cascade has been demonstrated in two distinct types of experiments. In one type we injected protein kinases, protein phosphatases, or inhibitors or activators thereof, into medium spiny neurons and obtained physiological responses consistent with the scheme shown in figure 3. The other type of study involved analysis of mice with targeted deletion of the DARPP-32 gene. Allen Fienberg engineered this knockout mouse and, in a multi-institutional collaboration, found that all physiological, biochemical, and pharmacological responses to dopamine, the psychostimulant drugs of abuse, and antischizophrenic drugs, seen in normal mice, were either greatly diminished or abolished in DARPP-32 knockout mice (figure 5)[18].

The scheme shown in figure 3 indicates that the fast excitatory glutamate receptors, AMPA and NMDA, regulate, and are regulated by, the DARPP-32/PP1 cascade. From this scheme one would predict that PP1 would be lo-

Stimulus	Parameter	Examples:
• Dopamine	Ion Pumps	Na^+, K^+-ATPase
• Dopamine	Ion Channels	Ca^{2+} Channels
• Dopamine	Neurotransmitter Release	GABA Release
• Dopamine	Neurotransmitter Receptors	GluR1 Receptor
• Dopamine; Psychostimulants	Immediate Early Genes	Fos Induction
• Psychostimulants	Brain Damage	Reactive Gliosis
• Antischizophrenics	Locomotor Activity	Catalepsy
• Dopamine; Progesterone	Sexual Receptivity	Female Responsiveness

Figure 5. Responses to dopamine, psychostimulants and antischizophrenic drugs are abolished in DARPP-32 knockout mice.

calized in the vicinity of these glutamate receptors. Immunocytochemical experiments, using antibodies prepared against recombinant PP-1, enabled us to localize this phosphatase in neostriatal cells. PP1 was enriched in the spines of the dendrites of medium spiny neurons[19]. To answer the question as to the basis for the high degree of enrichment of PP1 in spines, Patrick Allen used yeast-two hybrid technology to search for a PP1 targeting protein. He found a molecule, called spinophilin, which has the properties of such a targeting protein. Spinophilin is localized almost exclusively in dendritic spine heads at excitatory synapses (figure 6)[20]. This is precisely where the AMPA and NMDA receptors are concentrated.

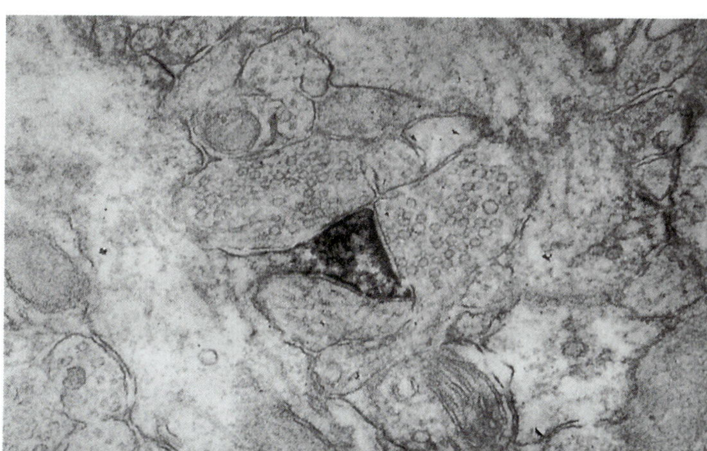

Figure 6. Electron micrograph showing immunoreactivity in a dendritic spine neck and especially in a spine head. Immunoreactivity (black flocculent material) lines the cytoplasmic surface of the spine plasmalemma and coats the postsynaptic density. Immunoreactivity is not present in axon terminals or in the dendritic shaft. (Bar = 100 nm) (Modified from Allen *et al.*, reference 20).

It has recently been demonstrated that spinophilin, through control of PP-1, regulates the conductance properties of AMPA and NMDA receptors[21]. A model to account for the ability of spinophilin and PP1 to regulate the AMPA receptor is shown in figure 7. In its dephosphorylated form, the AMPA receptor is relatively insensitive to activation by glutamate. Its sensitivity to glutamate is greatly enhanced upon phosphorylation by cAMP-dependent protein kinase[23, 24]. In the absence of dopamine, the AMPA receptor is kept in a low conductance state by PP1. Dopamine, through activation of cAMP-dependent protein kinase, causes an increase in the state of phosphorylation of the GluR1 subunit of the AMPA receptor, and this effect is greatly attenuated in the neostriatum of mice lacking the DARPP-32 gene. The activation of PKA in response to dopamine increases the state of phosphorylation of the AMPA receptor by a synergistic mechanism involving direct phosphorylation of the receptor as well as phosphorylation of threonine 34 on DARPP-32, resulting in inhibition of PP-1 catalyzed dephosphorylation of the receptor.

Physiological support for the scheme shown in figure 7 includes evidence that the dephosphorylation, and associated loss of responsivity of AMPA receptors to activation by glutamate, can be prevented by incubation of cells either with a D1 receptor agonist, or with the PP-1 inhibitor okadaic acid, but not with an inactive analog of okadaic acid (figure 7). This scheme is also supported by data demonstrating that injection of a peptide corresponding to residues 6-38 of phospho-DARPP-32, which retains the ability to inhibit PP-1, but not dephospho-DARPP-32 (6-38), prevents loss of AMPA receptor responsivity. Injection of a spinophilin-based peptide, which prevents

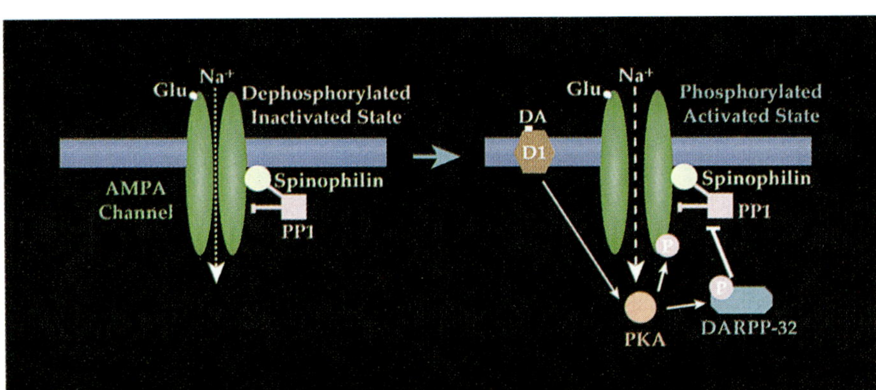

Figure 7. Regulation of AMPA-type glutamate receptors by DARPP-32 and spinophilin.

Model illustrates how regulation of PP-1 could account for the ability of DARPP-32 and spinophilin to control AMPA channels. (Left) Spinophilin, by binding to an unidentified intermediate protein (not shown), localizes PP-1 in the vicinity of the AMPA channel. Under basal conditions, the PP-1/spinophilin complex maintains the AMPA channel in a dephosphorylated, low-activity state. (Right) Following D1 receptor stimulation, AMPA channel phosphorylation is increased due both to direct PKA phosphorylation and to PKA/phospho-DARPP-32-mediated inhibition of PP-1. This synergistic increase in phosphorylation stimulates AMPA channel activity. The kinase anchoring protein (AKAP) believed to localize PKA in the vicinity of the AMPA receptor (reference 22) is not shown. (Modified from Greengard *et al.*, reference 9).

tethering of PP-1 to spinophilin, also prevents AMPA receptor inactivation; a single point mutation abolishes both effects of this peptide. Spinophilin regulates the NMDA receptor in a manner parallel to that by which it regulates the AMPA receptor. Spinophilin does not regulate the fast inhibitory GABA A receptor. Thus spinophilin appears to control fast excitatory, but not fast inhibitory, synaptic transmission. Finally, spinophilin itself is a substrate for PKA[25] and it seems likely that dopamine-induced PKA-mediated phosphorylation of spinophilin plays a role in the complex mechanism by which dopamine controls the efficacy of synaptic transmission at fast excitatory synapses.

MODULATION OF THE BALANCE BETWEEN DOPAMINERGIC AND GLUTAMATERGIC SIGNALING

Our discussion of DARPP-32 up to this point has focused on the ability of various neurotransmitters, by regulating the activity of PKA, PKG, and PP-2B, to regulate DARPP-32 phosphorylation/dephosphorylation at threonine 34 and thereby to control the activity of PP-1, and the state of phosphorylation and activity of a variety of downstream physiological effectors. Clearly, the percentage of DARPP-32 molecules in the phosphorylated state, and the degree of inhibition of PP-1, reflects a balance between the rates of phosphorylation and dephosphorylation of threonine 34 (figure 3). It turns out that the effectiveness of PKA and PP2B in regulating threonine 34 phosphorylation/dephosphorylation is itself regulated by other protein kinases and protein phosphatases. We now know of three sites, in addition to threonine 34, which are phosphorylated in response to activity in other signaling pathways. Each of these three additional pathways modulates the dopamine/D1 receptor/PKA/phospho-threonine 34-DARPP-32/PP1 cascade (figure 8)[26–28]. As is clear from figure 8, DARPP-32 can be either a protein phosphatase inhibitor or a protein kinase inhibitor, depending upon whether threonine 34 or threonine 75 is phosphorylated. Although DARPP-32 is the only bi-functional molecule of this type found to date, it seems likely that other proteins will be found which can serve either as a protein kinase inhibitor or a protein phosphatase inhibitor, depending on the residue phosphorylated.

One major molecular mechanism by which dopamine and glutamate produce opposing physiological effects involves a positive feedback loop that amplifies their mutually antagonistic actions (figure 9). This positive feedback loop involves three components, namely PKA, protein phosphatase 2A (PP2A) and threonine 75 of DARPP-32. In resting animals, threonine 75 is very highly phosphorylated, whereas threonine 34 is only slightly phosphorylated. Tonic activity of the glutamate/CDK5 pathway is probably responsible for keeping threonine 75 phosphorylated, and thereby keeping PKA inhibited. Dopamine, by activating D1 receptors, increases the activity of PKA, leading to phosphorylation of key physiological substrate proteins. Increased PKA activity also increases phosphorylation of threonine 34 on DARPP-32, inhibiting PP1, and thereby decreasing the dephosphorylation of these substrate proteins. PKA also activates PP2A, which dephosphorylates threonine

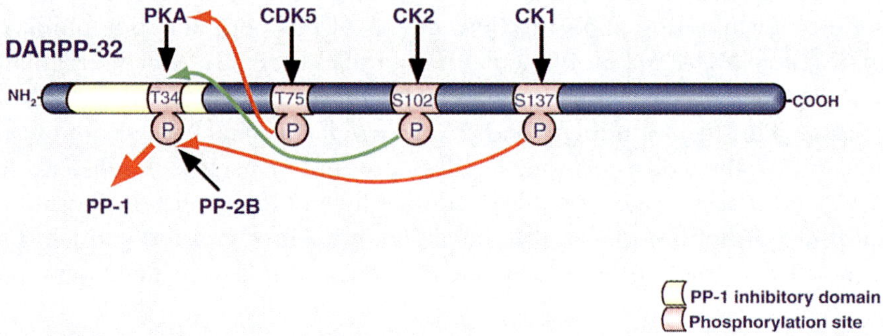

Figure 8. Multiple Phosphorylation sites on DARPP-32.

DARPP-32 is phosphorylated at Thr-34 by PKA (and PKG; not shown) and dephosphorylated by PP-2B (and PP2A; not shown). Phosphorylation at Thr-34 converts DARPP-32 into a potent inhibitor of PP-1. DARPP-32 is phosphorylated on Ser-137 by casein kinase I. Phosphorylation at Ser-137 converts DARPP-32 into a poorer substrate for PP2B-catalyzed dephosphorylation of Thr-34 without affecting phosphorylation of Thr-34 by PKA or PKG and without affecting the ability of PP2B to dephosphorylate other substrates (reference 27). DARPP-32 is phosphorylated on Ser-102 by casein kinase II. Phosphorylation at Ser-102 converts DARPP-32 into a better substrate for phosphorylation by PKA, without affecting its phosphorylation by PKG or its dephosphorylation by PP2B and without affecting the ability of PKA to phosphorylate other substrates (reference 26). Thus, the effects of phospho-Ser-137 DARPP-32 and phospho-Ser-102 DARPP-32 are substrate-directed, not enzyme-directed. DARPP-32 is phosphorylated on Thr-75 by CDK5. Phosphorylation at Thr-75 converts DARPP-32 into an inhibitor of PKA, reducing its ability to phosphorylate any substrate, including DARPP-32 at Thr-34 (reference 28). Thus the effect of phospho-Thr-75 DARPP-32 is enzyme-directed, not substrate-directed. Since dopamine increases and glutamate decreases phosphorylation of DARPP-32 at Thr-34, signaling mediated through phosphorylation/dephosphorylation of Ser-137, Ser-102 and Thr-75 alters the balance between dopaminergic and glutamatergic signaling. Specifically, the casein kinase 1/Ser-137 and casein kinase 2/Ser-102 pathways are prodopaminergic/antiglutamatergic, whereas the CDK5/Thr-75 pathway is antidopaminergic/ proglutamatergic. Phospho-Ser-137 is preferentially dephosphorylated by PP-2C (not shown). NH2-terminal domain of DARPP-32 that binds to PP-1 is shown in yellow. Red arrows indicate inhibition; green arrow indicates stimulation. (Modified from Greengard *et al.*, reference 9).

75, which removes the inhibition of PKA. Thus dopamine causes both an increased activation and a decreased inhibition of PKA. Conversely, glutamate activates CDK5 and, by phosphorylating threonine 75, inhibits PKA, reducing the activity of PP2A, and the resultant dephosphorylation of threonine 75. Through this mechanism, glutamate causes both an increased phosphorylation and a decreased dephosphorylation of threonine 75. Thus the PKA/PP2A/threonine 75 DARPP-32 triad amplifies either the effects of dopamine or the effects of glutamate, whichever is the more dominant neurotransmitter at any given time.

The schemes shown in figures 3, 8 and 9 summarize only a portion of the pathways that we now know to be involved in signal transduction in dopaminoceptive neurons. Both major classes of dopamine receptors and all three major classes of glutamate receptors are involved in the interactions between these neurotransmitters. These signaling pathways are highly complex, and have multiple sites of interaction. For example there are several ways by which the dopamine/D1 receptor/cyclic AMP/PKA/phosphothreonine 34

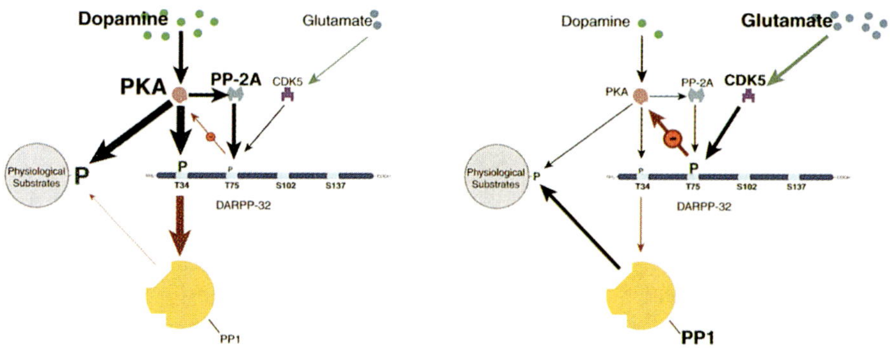

Figure 9. Model illustrating signaling pathways involved in mediating opposing biological actions of the neurotransmitters dopamine and glutamate.

DARPP-32/PP1 pathway can be inactivated. One is by D2 receptor-mediated inhibition of cyclic AMP formation (figure 3), a second is by glutamate-induced phosphothreonine 75 DARPP-32 inhibition of PKA (figure 9), and a third is by AMPA, NMDA or D2-receptor-mediated activation of PP-2B resulting in the dephosphorylation of threonine 34 of DARPP-32 (figure 3).

CONCLUDING REMARKS

The differences between the levels of complexity of fast synaptic transmission, which involves a single ligand-operated ion channel, and the enormously complicated pathways underlying slow synaptic transmission, only part of which we have elucidated, seem amazing. However, when one thinks of fast synaptic transmission as being the hardware of the brain, and slow synaptic transmission as being the software which controls fast transmission, the molecular basis by which nerve cells communicate with each other makes more sense.

The elucidation of the principles underlying slow synaptic transmission, and the discovery of numerous components of the underlying intracellular signaling pathways, have provided a number of novel therapeutic targets for the treatment of neurological and psychiatric illnesses associated with abnormalities of dopamine signaling. Levodopa is effective in most patients with Parkinson's disease. Levodopa is converted to dopamine, which then activates dopamine receptors and alleviates the disease symptoms. Unfortunately, within a short time, many Parkinson's patients become refractory to Levodopa treatment. This refractoriness is probably the result in part of down-regulation of dopamine receptors. It should be possible, through the new-found knowledge of the intracellular signaling pathways by which dopamine produces its physiological effects, to develop various therapeutic substances that activate or inhibit these various intracellular components. Hopefully such agents will be useful for the treatment of Parkinson's disease without the refractoriness associated with Levodopa treatment. Drugs which target intracellular constituents of dopaminoceptive neurons may also be useful for treatment of other neurological and psychiatric disorders associated with dopamine signaling abnormalities. Finally, since the various slow acting neurotransmitters work by similar principles, it should be possible to find drugs working intracellularly for the treatment of diseases affecting parts of the brain where signaling pathways other than dopamine are employed.

ACKNOWLEDGEMENTS

The work summarized here reflects outstanding contributions from many highly gifted associates who have worked in our laboratory, and who have been cited in the text. I would particularly like to mention Angus C. Nairn, who has been a close colleague and friend for more than 20 years. This work has also benefited enormously from collaborations with excellent scientists at several other universities. Our work on regulation of ion pumps was carried out in collaboration with Anita Aperia at the Karolinska Institute. We continue to collaborate with Richard L. Huganir, who was a postdoctoral fellow and assistant professor at The Rockefeller University and is now a professor in the Department of Neuroscience at Johns Hopkins University School of Medicine and with Eric J. Nestler, who was an MD/PhD student in our laboratory and is now Chairman of the Department of Psychiatry at the University of Texas Southwestern Medical Center. Much of our electrophysiological work has been done in collaboration with D. James Surmeier at Northwestern University. The work of our research group has been very generously supported for over 25 years by the National Institutes of Health, including the National Institute of Mental Health, the National Institute on Drug Abuse, and the National Institute on Aging.

REFERENCES

1. Earl W. Sutherland's Nobel Lecture, Les Prix Nobel, 1971.
2. Edwin G. Krebs' Nobel Lecture, Les Prix Nobel, 1992.
3. Kebabian, J. W. & Greengard, P. (1971) Dopamine-sensitive adenyl cyclase: Possible role in synaptic transmission. Science *174*:1346–1349.
4. Miyamoto, E., Kuo, J. F. & Greengard, P. (1969) Adenosine 3':5'-monophosphate -dependent protein kinase from brain. Science *165*:63–65.
5. Kuo, J. F. & Greengard, P. (1970) Cyclic nucleotide-dependent protein kinases. VI. Isolation and partial purification of a protein kinase activated by guanosine 3':5'-monophosphate. J. Biol. Chem. *245*:2493–2498.
6. Schulman, H. & Greengard, P. (1978) Stimulation of brain membrane protein phosphorylation by calcium and an endogenous heat-stable protein. Nature *271*:478–479.
7. Walaas, S. I., Nairn, A. C. & Greengard, P. (1983) Regional distribution of calcium- and cyclic adenosine 3':5'-monophosphate-regulated protein phosphorylation systems in mammalian brain. I. Particulate systems. J. Neurosci. *3*:291–301.
8. Walaas, S. I., Nairn, A. C. & Greengard, P. (1983) Regional distribution of calcium- and cyclic adenosine 3'5'-monophosphate-regulated protein phosphorylation systems in mammalian brain. II. Soluble systems. J. Neurosci. *3*:302–311.
9. Greengard, P., Allen, P. B. & Nairn, A. C. (1999) Beyond the dopamine receptor: The DARPP-32/protein phosphatase-1 cascade. Neuron *23*:435–447.

10. Greengard, P., Valtorta, F., Czernik, A. J. & Benfenati, F. (1993) Synaptic vesicle phosphoproteins and regulation of synaptic function. Science *259*:780–785.
11. DeCamilli, P., Benfenati, F., Valtorta, F. & Greengard, P. (1990) The Synapsins. (Palade, Alberts, and Spudich, editors), Annual Review of Cell Biology *6*:433–460.
12. Castellucci, V. F., Kandel, E. R., Schwartz, J. H., Wilson, F. D., Nairn, A. C. & Greengard, P. (1980) Intracellular injection of the catalytic subunit of cyclic AMP-dependent protein kinase simulates facilitation of transmitter release underlying behavioral sensitization in *Aplysia.* Proc. Natl. Acad. Sci. U.S.A. *77*:7492–7496.
13. Castellucci, V. F., Nairn, A., Greengard, P., Schwartz, J. H. & Kandel, E. R. (1982) Inhibitor of adenosine 3':5'-monophosphate-dependent protein kinase blocks presynaptic facilitation in *Aplysia.* J. Neurosci. *2*:1673–1681.
14. Eric R. Kandel's Nobel Lecture, Les Prix Nobel, 2000.
15. Arvid Carlsson's Nobel Lecture, Les Prix Nobel, 2000.
16. Walaas, S. I., Aswad, D. W. & Greengard, P. (1983) A dopamine- and cyclic AMP-regulated phosphoprotein enriched in dopamine-innervated brain regions. Nature *301*:69–71.
17. Hemmings, H. C., Jr., Greengard, P., Tung, H. Y. L. & Cohen, P. (1984) DARPP-32, a dopamine-regulated neuronal phosphoprotein, is a potent inhibitor of protein phosphatase-1. Nature *310*:503–505.
18. Fienberg, A. A., Hiroi, N., Mermelstein, P. G., Song, W.-J., Snyder, G. L., Nishi, A., Cheramy, A., O'Callaghan, J. P., Miller, D. B., Cole, D. G. , Corbett, R., Haile, C. N., Cooper, D. C., Onn, S. P., Grace, A. A., Ouimet, C. C., White, F. J., Hyman, S. E., Surmeier, D. J., Girault, J.-A., Nestler, E. J. & Greengard, P. (1998) DARPP-32: Regulator of the efficacy of dopaminergic neurotransmission. Science *281*(5378): 838–842.
19. da Cruz e Silva, E. F., Fox, C. A., Ouimet, C. C., Gustafson, E., Watson, S. J. & Greengard, P. (1995) Differential expression of protein phosphatase 1 isoforms in mammalian brain. J. Neurosci. *15*(5):3375–3389.
20. Allen, P. B., Ouimet, C. C. & Greengard, P. (1997) Spinophilin, a novel protein phosphatase 1 binding protein localized to dendritic spines. Proc. Natl. Acad. Sci. U.S.A. *94*:9956–9961.
21. Yan, Z., Hsieh-Wilson, L., Feng, J., Tomizawa, K., Allen, P. B., Fienberg, A. A., Nairn, A. C. & Greengard, P. (1999) Protein phosphatase 1 modulation of neostriatal AMPA channels: regulation by DARPP-32 and spinophilin. Nature Neuroscience *2*(1):13–17.
22. Rosenmund, C., Carr, D. W., Bergeson, S. E., Nilaver, G., Scott, J. D. & Westbrook, G. L. (1994) Anchoring of protein kinase A is required for modulation of AMPA/kainate receptors on hippocampal neurons. Nature *368*:853–856.
23. Greengard, P., Jen, J., Nairn, A. C. & Stevens, C. F. (1991) Enhancement of the glutamate response by cAMP-dependent protein kinase in hippocampal neurons. Science *253*:1135–1138.
24. Wang, L-Y., Salter, M. W. & MacDonald, J. F. (1991) Regulation of kainate receptors by cAMP-dependent protein kinase and phosphatases. Science *253*:1132–1135.
25. Hsieh-Wilson, L. *et al.* (2001) manuscript in preparation.
26. Girault, J.-A., Hemmings, H. C., Jr., Williams, K. R., Nairn, A. C. & Greengard, P. (1989) Phosphorylation of DARPP-32, a dopamine- and cAMP-regulated phosphoprotein, by casein kinase II. J. Biol. Chem. *264*:21748–21759.
27. Desdouits, F., Cohen, D., Nairn, A .C., Greengard, P. & Girault, J. -A. (1995) Phosphorylation of DARPP-32, a dopamine- and cAMP-regulated phosphoprotein, by casein kinase I *in vitro* and *in vivo.* J. Biol. Chem. *270*(15):8772–8778.
28. Bibb, J. A., Snyder, G. L., Nishi, A., Yan, Z., Meijer, L., Fienberg, A. A., Tsai, L.-H., Kwon, Y. T., Girault, J.-A., Czernik, A. J., Huganir, R. L., Hemmings, H. C., Jr., Nairn, A. C. & Greengard, P. (1999) Phosphorylation of DARPP-32 by Cdk5 modulates dopamine signalling in neurons. Nature *402*(6762):669–671.

Eric R. Kandel

ERIC R. KANDEL

PROLOGUE: LIFE IN VIENNA IN THE 1930s

There was little in my early life to indicate that an interest in biology would become the passion of my academic career. In fact, there was little to suggest I would have an academic career. Rather, my early life was importantly shaped by my experiences in Vienna and I spent many years later coming to grips with the circumstances and place of my birth.

I was born in Vienna on November 7, 1929, eleven years after the multi-ethnic Austro-Hungarian Empire fell apart following its defeat in World War I. Although Austria had been radically reduced in size (from 54 million to only 7 million inhabitants) and in political significance, its capital, the Vienna of my youth, was still intellectually vibrant, one of the great cultural centers of the world. A city of one and a half million people, it was home to Sigmund Freud, Karl Kraus, Robert Musil, Arthur Schnitzler, and for a while Arnold Schoenberg. The music of Gustav Mahler and of the earlier 19th Century Vienna school resonated throughout the city, as did the bold expressionist images of Gustav Klimt, Oskar Kokoschka, and Egon Schiele. Even as it thrived culturally, however, Vienna in the 1930s was the capital city of an oppressive, authoritarian political system. I was too young to appreciate its cultural richness, but I sensed later, from the perspective of a more carefree adolescence in the United States, the oppressive conditions in Vienna that affected my early youth.

Even prior to the Anschluss in 1938, anti-Semitism was a chronic feature of Viennese life. Jews, who made up nearly 20% of the city's population, were discriminated against in the Civil Service and in many aspects of social life. Nonetheless, they were fascinated by the city in which they had lived for over a thousand years. My parents genuinely loved Vienna, and in later years I learned from them why the city exerted a powerful hold on them and other Jews. My parents loved the dialect of Vienna, its cultural sophistication, and artistic values. "The greatest grim irony of all was the fierce attachment of so many Jews to a city that through the years demonstrated its deep-rooted hate for them," wrote George Berkley, the American historian of Vienna and its Jews. This fierce attachment was considered by the historian Harvey Zohn to be the most tragically unrequited love in world history.

In spite of the hostile climate, Austrian Jews continued to make remarkable contributions to theater, music, literature, science, and medicine in the period between the two World Wars. The Salzburg Festival was directed by Max Reinhardt; the Vienna Opera was conducted by Bruno Walter. Stefan Zweig and Franz Werfel were two of the most popular writers in the German language, and Elias Canetti, who later won the Nobel Prize in Literature for books describing his youth in Vienna, began writing these in the 1930s. Two of the three Austrians to be awarded the Nobel Prize in Physiology and Medicine in the 1930s were of Jewish origin: Karl Landsteiner was honored in 1930 for his discovery of blood groups and Otto Loewi in 1936 for discovering acetylcholine, a chemical transmitter that slowed the heart. Of the 52 Olympic medals earned by Austrian athletes from the beginning of modern Olympics to 1936, 18 were won by Jewish Austrians. Fully half of the practicing physicians and medical faculty at the University of Vienna were Jewish. This, in fact, was the last period during which Viennese medicine still attracted students and patients from all over the world. They came to study with, or to be treated by, pioneers in diagnostics and therapeutic medicine, such as the pediatrician Béla Schick, the ear specialist Heinrich von Neumann, and the psychoanalyst Sigmund Freud. As this listing makes clear, the period of my early youth has been characterized, appropriately, as "the final flowering of the Austrian Jewish intellectual activity."

My parents were not born in Vienna, but they had spent much of their lives there, having each come to the city at the beginning of World War I when they were still very young. My mother, Charlotte Zimels, was born in 1897 in Kolomea, a town of about 43,000 inhabitants in Galicia, a region of the Austro-Hungarian Empire. (Kolomea now is part of the Ukraine and has been renamed Kolomyya.) Almost half the population of Kolomea was Jewish, and the Jewish community had a very active culture. My mother came from a well-educated middle-class family, and although she had spent only one year at the university in Vienna, she spoke and wrote English as well as German and Polish.

My father Herman was born into a poor family in 1898 in Olesko, a small town of about 3,500 people near Lvov (Lemberg), now also part of the Ukraine. During World War I he was drafted into the Austria-Hungarian Army directly from high school. After the war he worked to support himself and never went back to school.

My parents met in Vienna and married in 1923, shortly after my father had established himself as the owner of a small toy store. My brother Lewis was born on November 14, 1924. I was born five years later. We lived in a small apartment at Severingasse 8 in the 9th district, a middle-class neighborhood near the medical school, and not too far from Freud's apartment, although we had no association with either. Both of my parents worked in the store, and we had a full-time housekeeper to help out at home.

I went to a school near our house. As with most elementary schools in Vienna, it was very traditional and very good, and I followed the well-trodden trail that my exceptionally gifted brother Lewis had blazed five years earlier in the same school with the very same teachers. Throughout my years in Vienna I felt that his was an intellectual virtuosity that I would never match. By the time I began reading and writing, he already was starting to master Greek and to play the piano.

My fondest early memories are of family get-togethers and vacations. On Sunday afternoons my Aunt Minna (my mother's sister) and Uncle Srul would come for tea. This was an occasion for my father and uncle to play cards, games at which my father excelled and which brought out great animation and humor in him. We celebrated Passover in a festive way at the home of my grandparents Hersch and Dora Zimels, and we invariably went on vacation in August to Mönichkirchen, a small farming village in the southeast portion of Lower Austria, not far from Vienna.

It was just as we were about to depart for Mönichkirchen in July of 1934 that the Austrian Chancellor, Engelbert Dollfuss, who had outlawed the Nazi Party, was assassinated by a band of Austrian Nazis disguised as policemen – the first political storm to register on my slowly maturing political awareness. Following the Dollfuss assassination and during the early years of the chancellorship of his successor, Kurt von Schuschnigg, the Austrian Nazi Party went further underground, but it continued nonetheless to gain new adherents, especially among teachers and other civil servants. Paradoxically, the Austrian drive toward authoritarianism was abetted by Dollfuss's own political attitudes and actions. Modeling himself on both Mussolini and Hitler, Dollfuss renamed his Christian Socialist Party the *Fatherland Front*, and took to wearing a modified swastika. To assure his own control of the government he abolished Austria's Constitution and outlawed not only the Nazi Party but *all* opposition parties. Thus, although Dollfus opposed the efforts of the Austrian National Socialist movement to form a Pan-German state with Germany, his abolition of the Constitution and of other political parties helped open the door for Hitler to march in.

And, as I well remember, march in he did. Since his youth, Hitler had dreamed of the union of Austria and Germany. It is therefore not surprising that a key point in the Nazi program, from its beginning in the 1920s, was a merger of all German-speaking people into a Greater Germany. In the fall of 1937 Hitler began to act on this program by raising the level of rhetoric and threatening to move against Austria. Schuschnigg, eager to assert Austria's independence, met with Hitler on February 12, 1938 in Berchtesgaden. Hitler showed up with two of his generals in tow and threatened to invade Austria unless Schuschnigg lifted the legal restrictions on the Austrian Nazi Party and appointed three Austrian Nazis to key ministerial positions in the Austrian Cabinet. Schuschnigg refused, but as Hitler continued to intimidate him, Schuschnigg compromised and agreed to a legalization of the Nazi party and to granting it two cabinet positions. The agreement between Schuschnigg and Hitler so emboldened the Austrian Nazis that they began to challenge the Austrian government in a series of incidents that the police had difficulty controlling. Faced with Hitler's aggression from without and the Austrian Nazi rebellion from within, Schuschnigg took the offensive on March 9th and boldly called for a plebiscite on Austria's autonomy to be held four days later, on March 13th.

This courageous move caught Hitler by complete surprise, an awkward surprise since it seemed almost certain that the vote would favor an independent Austria. Hitler responded by mobilizing troops and threatening to invade Austria unless Schuschnigg postponed the plebiscite, resigned as chancellor, and formed a new government with an Austrian Nazi, Arthur Seyss-Inquart as chancellor. Schuschnigg turned for help to England and Italy, two countries that had formerly supported Austrian independence. But on this occasion both countries failed him and did not respond. Abandoned by his potential allies and concerned about needless bloodshed, Schuschnigg resigned on the evening of March 11th. "Austria is yielding to force," he announced in an emotional farewell radio address to the nation. "God protect Austria." Even though Schuschnigg had resigned and President Miklos of Austria gave in to all the German conditions, Hitler nonetheless invaded Austria.

Hitler's triumphal march into Vienna and his overwhelming reception by the Viennese public made an indelible impression on me. My brother had just finished building his first short-wave radio receiver, and on the evening of March 13th we both were listening with earphones as the broadcaster described the earlier crossing of the Austrian border by German troops on the morning of March 12th. Hitler followed later in the afternoon of that day, crossing the border first at Braunnau am Inn, his native village, and then moving into Linz, the capitol of Upper Austria, where people welcomed him in the marketplace as their native son, screaming "Heil Hitler." Of the 120,000 people of Linz, almost 100,000 came out to greet Hitler. In the background the Horst Wessel song, one of the hypnotic Nazi marching songs that even I found captivating, blared forth. On the afternoon of March 14th Hitler's entourage reached Vienna, where a wildly enthusiastic crowd welcomed him as the hero who had unified the German-speaking people.

The extraordinary reception in Linz and Vienna caused Hitler to change his plan. He now realized the Austrians would not demand the status of a relatively independent protectorate of Germany he had planned for them. The enthusiastic welcome convinced him that Austria would readily accept, indeed would welcome, total annexation. For it seemed as if everyone, from the modest shopkeepers to the most elevated members of the academic community, now embraced Hitler. In a shocking move, even Theodor Cardinal Innitzer, the influential Archbischop of Vienna, welcomed Hitler and ordered all the Catholic churches in the city to fly the Nazi flag and to ring the church bells in honor of Hitler's arrival in Vienna. As the Cardinal personally greeted Hitler, he assured him of his own loyalty and that of all Austrian Catholics – which was most of the population of Austria. The Cardinal promised Hitler that Austria's Catholics would become "the truest sons of the great Reich into whose arms they had been brought back on this momentous day," provided that the liberties of the Church were respected and its role in the education of the young guaranteed.

That night, and for days on end, all hell broke loose. Viennese mobs erupted in nationalistic fervor, expressed by beating up Jews and destroying their property. Foreign commentators, long accustomed to Nazi tactics in Germany, were astonished by the wanton brutality of the Austrian Nazis, and even German Nazis were amazed and emboldened by the viciousness of the attacks in Vienna.

In his autobiography the German playwright Carl Zuckmayer, who had moved to Austria in 1936 to escape Hitler, described Vienna during the days following the annexation of Austria as a city transformed "into a nightmare painting of Hieronymus Bosch." It was as if:

> ... "Hades had opened its gates and vomited forth the basest, most despicable, most horrible demons. In the course of my life I had seen something of untrammeled human insights of horror or panic. I had taken part in a dozen battles in the First World War, had experienced barrages, gassings, going over the top. I had witnessed the turmoil of the post-war era, the crushing uprisings, street battles, meeting hall brawls. I was present among the bystanders during the Hitler Putsch in 1923 in Munich. I saw the early period of Nazi rule in Berlin. But none of this was comparable to those days in Vienna. What was unleashed upon Vienna had nothing to do with seizure of power in Germany ... What was unleashed upon Vienna was a torrent of envy, jealousy, bitterness, blind, malignant craving for revenge. All better instincts were silenced ... only the torpid masses had been unchained ... It was the witch's Sabbath of the mob. All that makes for human dignity was buried."

Having watched the build-up of anti-Jewish laws in Germany following Hitler's rise to power in 1933, my parents did not need much convincing to realize that the violence at the time of the annexation was not likely to fade away. We knew that we had to leave – and to leave as soon as possible. My mother's brother, Berman Zimels, had emigrated a decade earlier to New York and established himself as an accountant. He provided us expeditiously with affidavits that assured he would support us upon our arrival in the United States. Even with these affidavits it took about a year for my parents' Polish quota number to be called. When our number finally was called, we had to emigrate in stages because of United States immigration laws. My mother's parents left first in February 1939, my brother and I next in April 1939, and finally my parents in September 1939, only days before World War II broke out.

During the one year that we lived under Nazi rule, we experienced directly Vienna's humiliating form of anti-Semitism. The day after Hitler marched into Vienna, every one of my non-Jewish classmates – the entire class with the exception of one girl – stopped talking and interacting with me. In the park where I played I was taunted and roughed up. This viciousness toward Jews, of which my treatment was a mild example, culminated in the horrors of Kristallnacht, November 8, 1938. On the morning of November 7, 1938, a 17-year-old Jewish youth, who was distraught over his parent's tragic fate at the hands of the Nazis, shot a third secretary in the German Embassy in Paris, mistaking him for the German Ambassador. In retaliation for this single act, almost every synagogue in Germany and Austria was set on fire. Of all the cities under Nazi control, the destructiveness in Vienna on Kristallnacht was particularly wanton. Jews were taunted and brutally beaten, expelled from their businesses, and temporarily evicted from their homes so that both could be looted by their neighbors. My father was rounded up by the police together with hundreds of other Jewish men. He was released a few days later only because he had fought in the Austria-Hungarian army as a soldier in

World War I. I remember Kristallnacht even today, more than 60 years later, almost as if it were yesterday. It fell two days after my ninth birthday, on which I was showered with toys from my father's shop. When we returned to our apartment a week or so after having been evicted, everything of value was gone, including my toys.

My last year in Vienna was, in a way, a defining year, and it fostered the profound sense of gratitude I came to feel for the life I have led in the United States. It is probably futile, even for someone trained in psychoanalytic thinking as I am, to attempt to trace the complex interests and actions of my later life to a few selected experiences of my youth. Nevertheless I cannot help but think that the experiences of my last year in Vienna helped to determine my later interests in the mind, in how people behave, the unpredictability of motivation, and the persistence of memory. Over the years I have returned to these subjects repeatedly as my professional interests evolved from a youthful interest in European intellectual history at Harvard, where I studied the motivation of German intellectuals during the Nazi era, to an interest in psychoanalysis with its more systematic approach to mental processes, and finally to my interests in the biology of conscious and unconscious memory.

My early experiences in Vienna almost certainly contributed to my curiosity about the contradictions and complexities of human behavior. How are we to understand the sudden release of such great viciousness in so many people? How could a highly educated and cultured society, a society that at one historical moment nourished the music of Haydn, Mozart, and Beethoven, in the next historical moment sink into barbarism?

Clearly the answer to this question is complex, and many scholars of this period have attempted partial answers. One conclusion, troubling to an academic like myself, is that a society's culture is not a reliable indicator of its respect for human life. This rather simplistic conclusion, of course, raises the question: How can values within a society become so radically dissociated? As far as I can tell, the Viennese achieved this dissociation by shifting their frame of reference. By defining Jews in racial rather than religious terms, they were able to exclude Jews from the "more highly evolved European Aryan race," the race they believed to be responsible for the rise of Western civilization.

My last year in Vienna was likely also an important factor in my more specific later interest in the mechanisms of memory. I am struck, as others have been, at how deeply these traumatic events of my childhood became burned into memory – and I would emphasize that my experiences were trivial compared to those of so many who were seriously harmed or killed. For me, the frightening experiences of my last year in Vienna are certainly the most powerful of my "flashbulb memories," the emotionally charged and vivid memory of significant events that came to fascinate me.

RESETTLEMENT IN THE UNITED STATES

Needless to say, arriving in the United States in April of 1939 was like a breath of fresh air. I never actually said "free at last," but I felt it then and have ever since. We settled in Brooklyn and lived at first with my mother's parents. My grandfather Hersch Zimels was a religious and scholarly man who was somewhat unworldly. My brother said that my grandfather was the only man he knew who could speak seven languages but could not make himself understood in any one of them. My grandfather and I liked each other a great deal, and he readily convinced me that he should tutor me in Hebrew during the summer of 1939 so that I might be eligible for a scholarship at the Yeshiva of Flatbush, an excellent Hebrew parochial school that offered both secular and religious studies at a very high level. With his tutelage I entered the Yeshiva in the fall of 1939. By the time I graduated in 1944 I spoke Hebrew almost as well as English, had read through the five books of Moses, the books of Kings, the Prophets and the Judges in Hebrew, and also learned a smattering of the Talmud.

After my parents arrived, my father worked in a toothbrush factory. Even though he was not fond of working in this factory, he threw himself into the work with his usual energy and was soon reprimanded by the union steward for producing toothbrushes too quickly and making other workers appear slow. My father was undeterred. He simply loved America – he often referred to it as the "goldene Medina," the golden state. Even while still in Vienna he had read avidly the novels of Karl May, an author whose books celebrated the conquest of the American West and the bravery of the American Indians.

With time my father managed to save enough money to rent and outfit a modest clothing store at 411 Church Avenue in Brooklyn. We lived in an apartment above the store. My father and mother worked together and sold simple women's dresses and aprons, and men's shirts, ties, underwear, and pajamas. In this way my parents earned enough not only to support us all but also to send me to college and medical school. My father worked in that store until the week before he died at age 78 in 1976. My mother sold the store soon thereafter and died in 1991 at age 94.

ERASMUS HALL HIGH SCHOOL AND HARVARD COLLEGE

In 1944, when I graduated from the Yeshiva of Flatbush elementary school, it did not as yet have a high school. I went instead to Erasmus Hall High School, a local public high school in Brooklyn that was then academically very strong. Here I became interested in history, in writing, and in girls. I worked on the school newspaper and became sports editor. I also played soccer and was co-captain of the track team. At the urging of one of my history teachers, John Campagna, a Harvard alumnus, I applied to Harvard College and was one of two students out of my class of about 1,400 to be admitted, both of us on scholarships! Fair Harvard indeed!

Even though I was thrilled by my good fortune, I was apprehensive about leaving Erasmus, convinced that I would never again feel the sheer joy I had experienced there. It was at Erasmus that I first sensed myself emerging from behind the shadow of my brother Lewis. I now had distinctive interests of my own – jazz music, sports, American constitutional history – things that did not interest Lewis. At Harvard I majored in 19th and 20th century European history and literature and wrote my honors dissertation on *The Attitude Toward National Socialism of Three German Writers: Carl Zuckmayer, Hans Carossa, and Ernst Junger.* Each of these writers was then still alive and represented a different position on the political spectrum of fascism – uncompromising liberal opposition and emigration (Zuckmayer), resigned acceptance and internal (spiritual) emigration (Carossa), and intellectual support (Junger). I came to the rather depressing conclusion that many German artists, intellectuals, and academics succumbed all too eagerly and opportunistically to the nationalistic fervor and racist propaganda of National Socialism. Historical studies have found that Hitler did not have widespread popular support in his first year in office. Had intellectuals mobilized effectively and brought along segments of the general population, Hitler's government might well have been toppled.

I originally thought of doing graduate work in European intellectual history, along the lines of my undergraduate dissertation. However, in the course of my studies at Harvard I befriended a fellow student, Anna Kris, who had also emigrated from Vienna with her parents, Ernst and Marianne Kris, both prominent psychoanalysts from Freud's circle. Anna and her parents were very influential in getting me interested in psychoanalysis. It is difficult to recapture now the extraordinary fascination that psychoanalysis held for young people in 1950. During the first half of the 20th century psychoanalysis provided a remarkable set of insights into the mind – insights about unconscious mental processes, psychic determinism, and perhaps most interesting, the irrationality of human motivation. As a result, in 1950, psychoanalysis outlined by far the most coherent, interesting, and nuanced view of the human mind than did any other school of psychology. In addition, Anna's parents, who represented academic psychoanalysis in its most intellectual and interesting form, were extraordinary people – intelligent, cultured, and filled with enthusiasm. Ernst Kris, a former curator of applied art at the Kunsthistorisches Museum in Vienna, had been a world class art historian before becoming a psychoanalyst. After taking up psychoanalysis, he focused on the psychology of art, an area in which he helped train among others the great historian Ernst Gombrich. Marianne Kris, a wonderful therapist, was the daughter of Oskar Rie, a well-known Viennese pediatrician and Freud's best friend. Marianne in turn was a close friend of Freud's distinguished daughter, Anna Freud.

Both Ernst and Marianne Kris were extremely generous and encouraging to me, as they were to Anna's other friends. As a result of my frequent interactions with them and their colleagues, I was converted to their view that psychoanalysis offered a fascinating new approach, perhaps the only approach, to understanding the mind, including the irrational nature of motivation and unconscious and conscious memory. With time this began to seem much more exciting and interesting to me than European literature and intellectual history.

MEDICAL SCHOOL AT N.Y.U.

To become a practicing psychoanalyst, however, it was best to go to medical school, become a physician, and train as a psychiatrist – a course of study I had not previously considered. So in 1951, almost impulsively, I went to summer school at Harvard and took the required course in introductory chemistry. That summer in Cambridge I shared a house with Robert Goldberger, Henry Nunberg, James Schwartz, and Robert Spitzer, and we all became lifelong friends. A few months later, based on this one chemistry course and my overall college record, I was accepted at N.Y.U. Medical School, with the proviso that I complete the remaining course requirements before I entered medical school in the fall of 1952.

I entered N.Y.U. Medical School dedicated to studying psychiatry and becoming a psychoanalyst. Although I stayed with this career plan through my internship and psychiatric residency, by my senior year in medical school I had become so interested in the biological basis of medical practice (as had everyone else in my class) that I decided I had to learn something about the biology of the mind. In the 1950s most psychoanalysts thought of the mind in nonbiological terms. However, several psychoanalysts – particularly two that I got to know personally and who had a background in neurology, Lawrence Kubie and Mortimer Ostow – had begun to discuss the potential importance of the biology of the brain for the future of psychoanalysis. After considerable discussion with them and with another biologically oriented psychoanalyst, Sydney Margolin, I decided to take an elective period at Columbia University with Harry Grundfest. At that time N.Y.U. had no one on the faculty who was doing basic neural science, and in 1955, Grundfest was the most intellectually interesting neurobiologist in the New York area.

HARRY GRUNDFEST'S LABORATORY AT COLUMBIA UNIVERSITY

Grundfest had obtained his Ph.D. in zoology and physiology at Columbia in 1930 and went on to a post-doctoral fellowship at Columbia, studying with Selig Hecht, an outstanding psychophysicist interested in phototransduction – the transformation of light into neural signals. (Hecht also was the teacher of George Wald, who won the Nobel Prize in 1967 for discovering the chemical structure of the visual pigments.) Grundfest then joined the Rockefeller Institute in 1935, where he remained for a decade collaborating with Herbert Gasser. In 1944, while Grundfest was in his lab, Gasser shared the Nobel Prize in Physiology or Medicine with Joseph Erlanger for introducing the oscilloscope to neurophysiological studies. This methodology allowed accurate temporal resolution of the waveform and conduction velocities of the propagated action potential. In collaboration with Grundfest, Gasser elaborated on his discovery that the conduction velocity of the action potential is a function of the diameter of the axon. Grundfest also carried out reconstructions of the compound action potential from cross-sectional measurements of axonal diameters in mixed nerves, work that formed much of Gasser's Nobel Prize Lecture.

Harry Grundfest

In my decision to work with Grundfest, I was strongly encouraged by a new friend, Denise Bystryn, an extremely attractive and interesting French woman I had just met and would later marry. Denise is also Jewish. Her mother helped her father escape from a French concentration camp, and her parents survived the war by hiding from the Nazis in the southwest of France. During a good part of that time Denise was separated from her parents, hidden in a Catholic convent near Cahors. Denise's experiences, although more difficult, paralleled mine in a number of ways that seemed significant to her but did not seem at all important to me when we first met. However, over the years, our shared experiences in Europe proved to be defining in both our lives.

In 1949, Denise, her brother Jean-Claude, and her parents emigrated to the United States. Denise attended the Lycée Français de New York for one year and was admitted at age 17 to Bryn Mawr College as a junior. On graduating from Bryn Mawr at age 19, she enrolled at Columbia University as a graduate student in sociology. When we met she had just started research for her Ph.D. thesis in medical sociology with Robert Merton. Denise's father, a gifted mechanical engineer who unfortunately died one year before I met Denise, had advised her to marry a poor intellectual because he would likely be sufficiently ambitious to do interesting scholarship. Denise believed she was following that advice (she certainly married someone who was poor) and always encouraged me to make decisions that favored my doing science.

In Grundfest's lab I spent the first several months working on a number of projects with Dominick Purpura, an independent young scientist just starting out on his own career of cortical physiology. To my surprise I found my first experience in a lab really interesting, and very different from the classroom. Of course the research questions we were asking fascinated me and the dis-

Dominick Purpura

cussions were penetrating and enjoyable. Dominick was very bright and very entertaining. (I have referred to him as the Woody Allen of neurobiology.) But the actual performance of the experiments was also pleasurable and, when successful, very satisfying. Nevertheless, I began to worry about the methods we were using to address rather sophisticated questions about the electrical properties of dendrites. We were using evoked responses that were initiated by stimulating small areas of cortex, thereby activating thousands of neurons, and I thought these methods were too indirect to give easily interpretable results. Grundfest and Purpura, of course, were also concerned and talked repeatedly about doing direct intracellular recordings from cortical neurons, but neither thought this was likely to succeed.

AN INTRODUCTION TO STEPHEN KUFFLER

It was in this frame of mind that I was introduced to Stephen Kuffler, a Viennese trained physician turned physiologist, who (together with Bernard Katz and John Eccles) was to become one of my great neurobiological heroes. One evening Grundfest threw into my lap the September 20, 1955 issue of the *Journal of General Physiology*, with three of Kuffler's papers on excitation and inhibition in the dendrites and soma of isolated sensory nerve cells of the lobster and crayfish. Grundfest said something about Kuffler's being very good, and that these papers provided direct evidence for the graded properties of dendrites, evidence that was consistent with what he and Purpura were seeing in cortical neurons. I took the issue home and read the papers as best I could. Although I understood relatively little, one thing stood out immediately. Kuffler was studying the dendrites in a preparation in which he actually saw the dendrites and could record from them directly. For studying dendrites, Kuffler used an invertebrate sensory neuron that sent its dendrites into skeletal muscle much like the muscle spindles of vertebrates. In the introduction to the three papers Kuffler wrote:

> "The greatest advantage of the present preparation lies in its accessibility, since all cellular components can be isolated and visually observed. Further, the state of excitability of the structures could be controlled and graded by utilizing the physiological mechanisms given by the stretch receptor nature of the preparation...
>
> It seems of special interest that the sensory cell of crustacea possessed numerous anatomical features, which bear a striking resemblance to many central nervous system cells of vertebrates."

I learned from Kuffler's papers a new criterion for how good science is done – the importance of having a preparation suitable to testing the questions to be answered. Kuffler taught me to respect the power of invertebrate neurobiology.

On graduating from medical school in June 1956, I married Denise and, after a brief honeymoon in Tanglewood, I started an internship at Montefiore Hospital as she continued her thesis research at Columbia. I returned to Grundfest's lab, spending six weeks with Stanley Crain, who had pioneered the electrophysiological study of nerve cells in tissue culture. Stanley taught me how to make microelectrodes and how to obtain and interpret intracellular recordings from the crayfish giant axon. These experiments confirmed the insights I had gained from reading Kuffler's paper. From Stanley I also received my first insights into the universality of cellular processes.

Based on my two brief periods in his laboratory, Grundfest offered to nominate me for a position at the NIH, an alternative to serving in the physician's draft, which provided medical personnel for the military during the years following the Korean War. On the basis of Grundfest's recommendation, I was accepted by Wade Marshall, Chief of the Laboratory of Neurophysiology at NIMH/NINCDS.

THE LABORATORY OF NEUROPHYSIOLOGY AT THE NATIONAL INSTITUTES OF HEALTH

By the time I arrived in Bethesda, Wade Marshall had passed the peak of what had been a remarkable career. In the 1930s he was arguably the most promising and accomplished young scientist working on the brain in the United States. As a graduate student at the University of Chicago in Ralph Gerard's lab in 1936, he discovered that one could record electrical deflections in the somatosensory processing area of the cerebral cortex by moving the hairs of a cat's limb. He appreciated that one might use this electrical signal (the evoked response) to map the representation of the body surface on the brain.

To study this further, he joined Phillip Bard, Chairman of the Department of Physiology at the Johns Hopkins Medical School, as a post-doctoral fellow.

Wade Marshall

In 1937, Bard had already established himself as a major presence in American neurophysiology. Together with his student Clinton Woolsey, Bard had surgically removed the somatic sensory cortex of the monkey and studied its effect on the "placing reaction," a form of tactile behavior. Marshall joined up with Woolsey and Bard and together they carried out a classic series of studies in which they mapped sensory inputs from the body surface in the somatic sensory cortex and showed that a topographical representation of the entire body is wired into the brain. This provided the first systematic view of the neural representation in the brain of a sensory system. Today, this map is still shown in every textbook of neural science. Marshall next collaborated with John Talbott and mapped the retinal inputs in the striate cortex. Finally, with Harlow W. Ades, he mapped the cochlear inputs in the auditory cortex.

With these classic studies Marshall revolutionized the study of the sensory representations in the brain and showed that the brain had systematic topographical maps of the sensory surface for each of the three major sensations – touch, vision, and hearing. These marvelous scientific achievements came at a price, however, leaving Marshall so psychologically fatigued that he collapsed and, for a number of years, left science altogether. When he returned, in about 1945, he moved on to a completely new problem: the study of spreading cortical depression, a propagating, reversible silencing of cortical electrical activity. Marshall enjoyed doing occasional experiments, but he had lost his scientific drive and now focused much of his energy and interests on administrative matters, which he did well. Although eccentric, moody, and somewhat unpredictable, he was a wonderful lab chief. In particular, he was supportive and generous to young scientists and gave us a great deal of freedom.

Just before I arrived at NIH in 1957, the neurosurgeon William Scoville and the cognitive psychologist Brenda Milner had described the now-famous patient H.M. In order to treat intractable bilateral temporal lobe epilepsy, Scoville had removed on both sides of H.M.'s brain the medial temporal lobes, including a structure deep to them called the hippocampus. As a result of this procedure, H.M.'s seizures were largely eliminated. But, while retaining all cognitive functions, H.M. lost the ability to put new information into long-term memory. These findings pinpointed the medial temporal lobe and the hippocampus as sites specialized for memory storage.

Until the Scoville and Milner paper, the person most identified with attempts to localize memory was Karl Lashley, Professor of Psychology at Harvard and perhaps the dominant figure in American neuropsychology in the first half of the 20th century. Lashley explored the surface of the cerebral cortex in the rat, and systematically removed different cortical areas. In so doing, he failed to identify any particular brain region that was special to or necessary for the storage of memory. Based on these experiments, Lashley formulated the *law of mass action*, according to which memory is not localized to any specific region of the cortex but was a distributed property of the cortex as a whole. The extent of any memory defect, Lashley argued, was correlated with the size of the cortical area removed, not with its specific location.

Since I had already begun to think about problems in psychiatry and psychoanalysis in biological terms, the cell and molecular mechanisms of learning and memory struck me as a wonderful problem to study. I had first become interested in the study of learning at Harvard, where B.F. Skinner, the great behaviorist, was a dominant force in the 1950s. It was clear to me even then that learning and memory were central to behavior, and thus to psychopathology and to psychotherapy. Nothing was known about the cellular mechanisms of learning and memory, and now the cellular techniques for studying them were just becoming available – some beginnings of which I had learned from Stanley Crain in Grundfest's lab.

My initial ideas about how to tackle the biology of memory upon arrival at the NIH were confused and vague. Because intracellular recordings seemed to me such a powerful analytic tool for studying nerve cells, and because the hippocampus seemed particularly important for memory, I wanted to explore the hippocampus in cellular terms. This was made even more attractive for me because, as the great Spanish anatomists Ramón y Cajal and Lorente de Nó had pointed out, the cellular architecture of the hippocampus is remarkably conserved among mammals, and the main cell type, the pyramidal cell, is found in a discrete layer that is easy to target with microelectrodes. In addition, the pyramidal cells send their axons into a large fiber tract (the fornix), which allows the pyramidal cells to be identified electrophysiologically by stimulating the axons in the fornix and backfiring the pyramidal cells. I thought it would be interesting to compare the pyramidal neurons to the only other mammalian neurons that had been well studied at that time – the motor neurons of the spinal cord. I had the idea that the properties of the pyramidal cells themselves might reveal something about memory storage. I was emboldened to try this technically demanding study because Karl Frank was in the laboratory next to ours, pioneering the examination of spinal motor neurons with intracellular recordings in parallel with John Eccles. Although Frank himself thought that studying the hippocampus was chancy, he was not discouraging.

Almost as soon as I began, my research took an extremely fortunate turn in the person of Alden Spencer, who arrived in Marshall's lab having just graduated from the University of Oregon Medical School. Like me, Alden was becoming interested in the biology of learning and memory. It therefore took little effort for me to convince him that we should join forces on the hippocampus. Although Alden had no experience with intracellular recordings, he had done electrophysiological research on the brain at the University of Oregon Medical School, where he worked with John Brookhardt. Among Alden's many remarkable talents, he had good surgical skills and a fine knowledge of the anatomical organization of the mammalian brain.

Being both naïve and brash, we were not reluctant to tackle what appeared to Frank and others to be technically difficult problems, namely obtaining intracellular recordings from cortical neurons in a pulsating brain. Alden and I developed a simple way of reducing pulsations in the hippocampus that allowed us to obtain, on occasion, high-quality recordings for a long enough period – up to one hour – to carry out an initial analysis of the electrical properties of the hippocampal pyramidal cells. By applying to the hippocampus the powerful methodologies we learned from Frank, we easily picked some low-lying intellectual fruit. First, we found that action potentials in hip-

Alden Spencer

pocampal neurons were initiated not only at the axon hillock, as in motor neurons, but also at a second site, which we inferred to be the apical dendrites. These putative dendritic action potentials, which we called fast prepotentials, appeared to trigger the firing at the axon hillock. Second, we found that the hippocampal neurons, unlike motor neurons, were not silent in the absence of synaptic activity, but tended to fire spontaneously, and that this firing often took the form of bursts of spikes that were maintained by summation of depolarizing afterpotentials. Third, we found that the hippocampal neurons engaged a powerful recurrent inhibitory system that gave rise to a prolonged inhibition – several orders of magnitude longer than the inhibition seen in the spinal cord.

The mere technical success of obtaining intracellular recordings from hippocampal neurons, and the few interesting questions we were able to address, caught the enthusiastic attention of, and drew encouragement and help from, our senior colleagues at the NIH – Marshall, Frank, Michael Fuortes, Frank's gifted colleague and the great Japanese-American biophysicist Ichiji Tasaki. When John Eccles visited the NIH, he also was generous in his comments. But even in our brashest moments, we both realized that ours was a typical NIH story. In the Intramural Program at the NIH, young inexperienced people were given the opportunity to try things on their own, knowing that wherever they turned there were experienced people to help out.

Moreover, as Alden and I reviewed our work we realized that the cellular properties of hippocampal neurons were not sufficiently different from those of spinal neurons to explain the ability of the hippocampus to store memory. Thus, it dawned on us what in retrospect is quite obvious: that the neuronal mechanisms of learning and memory probably did not reside in the properties of the neurons themselves. Rather, because the signaling properties of neurons are quite alike, we began to think that what must matter is how neurons are functionally connected. The basis of learning must reside in the modification of interconnections by appropriate sensory signals. This conclusion, so clear in retrospect, emerged only gradually as we learned, mostly through reading and discussions with one another, to think more effectively about the biology of learning and memory.

This realization led us to reappraise our strategy. Since the hippocampus has a large number of neurons and an immense number of interconnections, it was not the place to begin. Even though we were now quite familiar with the hippocampus, it would be extremely difficult to work out how sensory information specific to learning reached the hippocampus or how learned information processed by the hippocampus might influence motor output.

Alden and I therefore became convinced that to make headway with the study of learning at the cellular level required a very different approach. Alden, a committed mammalian neurophysiologist, turned to the study of the spinal cord, particularly the modifiability of spinal reflexes, and went on to make important contributions in collaboration with Richard Thompson. However, even the spinal cord proved too difficult for a detailed cellular analysis, and both Alden and Thompson ended up leaving it.

THE SEARCH FOR A TRACTABLE SYSTEM FOR STUDYING LEARNING

Influenced by Kuffler, Grundfest, and Crain, I yearned for a more radically reductionist approach to the biology of learning and memory. I wanted a system that would serve the cellular study of learning as well as the squid giant axon had served for studies of the action potential, or the nerve-muscle synapse of the frog had served for the study of synaptic transmission. I wanted to examine learning in an experimental animal in which a simple behavior was modifiable by learning. Ideally that behavior should be controlled by only a small number of large and accessible nerve cells, so that the animal's overt behavior could be related to events occurring in the cells that control that behavior.

Such a reductionist approach has been traditional in biology. In neurobiology it is exemplified by the work on the squid giant axon by Hodgkin and Huxley, the nerve-muscle synapse of the frog by Bernard Katz, and the eye of *Limulus* by Keefer Hartline. When it came to the study of behavior, however, most investigators were reluctant to apply a strict reductionist strategy. In the 1950s and 1960s it was often said that behavior was the area in biology in which simple animal models, particularly invertebrate ones, were least likely to produce fruitful results because the brain that really learns, the mammalian brain, especially the human brain, is so complex that inferences from studies of invertebrates would not stand up. It was thought that humans, because of higher-order capabilities not found in simpler animals, must have types of neuronal organization that are qualitatively different from those found in invertebrates. Although these arguments held some truth, they overlooked certain critical issues. Work by students of comparative behavior, such as Konrad Lorenz, Niko Tinbergen, and Karl von Frisch, had already shown that certain behavior patterns, including elementary forms of learning, were common to humans and simple animals. From the outset I therefore believed that the mechanisms of memory storage were likely to be conserved in phylogeny, and that a cellular analysis of learning in a simple animal would reveal universal mechanisms that are also employed in more complex organisms.

Not surprisingly, I was strongly discouraged in the early days from pursuing this strategy by some senior researchers in neurobiology, particularly John Eccles. His concern reflected, in part, the existing hierarchy of acceptable research questions in neurobiology. Few self-respecting neurophysiologists, I was told, would leave the study of learning in mammals to work on an invertebrate. Was I compromising my career? Of an even greater concern to me were the doubts expressed by some very knowledgeable psychologists I knew, who were sincerely skeptical that anything interesting about learning and memory could be found in a simple invertebrate animal. I had made up my mind, however. Since we knew nothing about the cell biology of learning and memory, I believed that any insight into the modification of behavior by experience, no matter how simple the animal or the task, would prove to be highly informative.

After an extensive search that included crayfish, lobster, flies, and the nematode worm *Ascaris*, I settled on *Aplysia*, the giant marine snail. *Aplysia* offered three major technical advantages: (1) its nervous system has a small number of cells, (2) the cells are unusually large, and, as I realized with time, (3) many of the cells are invariant and identifiable as unique individuals. Before leaving the NIH in 1960, I arranged with Ladislav Tauc, one of the two people in the world then working on *Aplysia*, to join him in September 1962, as a postdoctoral fellow, as soon as I had completed my residency training. Here again, Denise's advice was decisive. The only two people working on *Aplysia* were French – Tauc's lab was in Paris, and Angelique Arvanitaki-Chalazonitis worked in Marseilles. So far so good! But, Denise, ever the Parisian chauvinist, thought that living in Marseilles would be like living in Albany (a small town in upstate New York). So Tauc and Paris it was, and that proved an excellent choice.

RESIDENCY TRAINING IN PSYCHIATRY AT THE HARVARD MEDICAL SCHOOL

However, before I would leave for Paris I had already committed to a two-year residency training in psychiatry. I therefore left the NIH in the spring of 1960 to start my psychiatric residency at the Massachusetts Mental Health Center of the Harvard Medical School. When I arrived at Harvard, I found an unanticipated bonus. Steven Kuffler, whose thinking had so influenced my own, had been recruited one year earlier from Johns Hopkins to build up neurophysiology at Harvard. Kuffler brought with him several young post-doctoral fellows – David Hubel, Torsten Wiesel, Ed Furshpan, and David Potter – each of whom was extraordinarily gifted. In this way Kuffler succeeded, in one fell swoop, in setting up at Harvard the premier group of neural scientists in the

Ladislav Tauc

country. I now had my first opportunity to interact with Kuffler and with the remarkable people he had assembled around him. Even though I was in full-time residency training, Kuffler and his group were extremely accessible, and their generosity allowed me to remain intellectually engaged in neurobiology. Moreover, Jack Ewalt, the Professor of Psychiatry at the Massachusetts Mental Health Center, provided me with funds and space so that I even managed to do some research in my spare time. I obtained the first intracellular recording from hypothalamic neuro-endocrine cells and found that these hormone-releasing cells had all the electrical properties of normal nerve cells.

During my psychiatric residency I began to think about simple forms of learning in preparation for work on *Aplysia.* I read Kimble's wonderful revision of Hilgard and Marquis's classical text *Conditioning and Learning,* and I reread Skinner's great book *The Behavior of Organisms.* This reading made me realize that the paradigms of simple learning articulated by Pavlov and Thorndyke, describing changes in behavior in response to controlled stimulation, included precise protocols for stimulating experimental animals. It occurred to me that the paradigms they described – habituation, sensitization, classical conditioning, and operant conditioning – could readily be adapted to experiments with an isolated *Aplysia* ganglion using artificial electrical rather than natural sensory stimuli. While recording the behavior of a single cell in a ganglion, one nerve axon pathway to the ganglion could be stimulated weakly electrically as a conditioned stimulus, while another pathway was stimulated as an unconditioned stimulus, following the exact protocol used for classical conditioning with natural stimuli in intact animals. One could then see whether synapses changed systematically in response to these patterns of stimulation, and, if so, whether the synaptic changes in any way paralleled changes in the overt behavior of intact animals, which classical psychologists had described. It thus dawned on me that in this way one could begin to take an initial step toward the study of learning in the intact animal by analyzing what I soon began to call *analogs of learning* – higher-order stimulus sequences based on patterns of stimulation used in learning experiments in intact behaving organisms, but applied directly to a neuronal system.

PARIS, APLYSIA, AND NEURAL ANALOGS OF LEARNING: CHEMICAL SYNAPSES PROVE TO BE REMARKABLY PLASTIC

Based on this idea, I wrote a successful application for an NINCDS postdoctoral fellowship for work to be done in Tauc's laboratory. And in September 1962, about a year after our son Paul was born, the three of us took off for Paris. Tauc proved an excellent person to work with; both our interests and our areas of competence complemented each other. He was, of course, completely at home with *Aplysia,* but he also had a strong background in physics and biophysics, which I lacked. Born in Czechoslovakia, Tauc had originally studied the electrical properties of plant cells. As a result, he had no experience with behavior and had up to this point thought little about the problems of neuronal integration that dominated thinking about the mammalian brain

– problems that Alden and I had discussed incessantly. Tauc was quite enthusiastic about my approach, which proved even more effective than anticipated. In my cellular studies of analogs of habituation, sensitization, and classical conditioning in *Aplysia,* I found synaptic changes that paralleled the behavioral changes seen in experiments on intact animals. This encouraged us to write in our 1965 paper in the *British Journal of Physiology*:

> "The fact that the EPSPs (excitatory postsynaptic potentials) can be facilitated for over half-an-hour with an input pattern scheme designed to simulate a behavioral conditioning paradigm, also suggests that the concomitant changes in the efficacy of synaptic transmission may underlie certain simple forms of information storage in the intact animal."

A BRIEF RETURN TO THE HARVARD MEDICAL SCHOOL

Upon completing a very productive 16-month stay in Tauc's laboratory, I returned to Harvard in November 1963. More than a year and a half later, in July of 1965, our daughter Minouche was born, completing our family – one boy, one girl – exactly what we had hoped for.

During this period I struggled with three choices that were to have a profound effect on my subsequent career. First, I realized that to do effective science I could not combine basic research and a clinical practice in psychoanalysis, as I had earlier hoped. I therefore decided not to apply to the Boston Psychoanalytic Institute, a decision which meant that I would not attempt to become a psychoanalyst but devote myself full-time to science. It was my strong sense that one of the problems within academic psychiatry, a problem that has become only worse with time, is that young people take on much more than they can handle effectively. I concluded that I could not and would not do that.

The second choice arose a few months later when Dr. Ewalt and Dr. Howard Hiatt, then chairmen of the Department of Medicine at the Beth Israel Hospital at Harvard, suggested that I take on the newly vacated chairmanship of the Department of Psychiatry at the Beth Israel Hospital. For a moment I was forced to rethink my decision to focus full time on science. The person who had just left that position, Grete Bibring, was a leading psychoanalyst who had been a colleague of Marianne and Ernst Kris in Vienna. Earlier in my life achieving this position would have represented my highest aspiration. But by 1965, my thinking had moved in a very different direction, and I decided against it with Denise's strong encouragement. (Denise summarized it simply: "What?" she said, "throw your scientific career away?") Instead, I made my third decision. I decided to leave Harvard and accept an invitation to start a small neurophysiology group focused specifically on the neurobiology of behavior in the Departments of Physiology and Psychiatry at the New York University Medical School.

Harvard was quite wonderful, and it was not easy to leave that intellectually heady neurobiology environment. My interaction with Kuffler had increased after my return from Paris and, until his death in 1980, Kuffler proved a marvelous friend and counselor. Moreover, my interactions during this period with members of Kuffler's group – Hubel, Wiesel, Furshpan, Potter, and Ed Kravitz, a biochemist who joined them later – were extensive and I learned much from them. Many years later, at a small meeting at the Marine Biological Laboratory in Woods Hole in honor of Steve Kuffler, I was surrounded by Steve's Harvard entourage, some of whom were struggling with the decision of whether to leave Harvard for attractive positions elsewhere. I could not resist beginning my lecture with the remark, "I am here as living proof that there is life after Harvard."

NEW YORK UNIVERSITY AND A FOCUS ON THE BEHAVIOR OF *APLYSIA*

The position at N.Y.U. had several great attractions that, in the long run, proved critical. First, it brought us back to New York and closer to my parents and to Denise's mother, all of whom were having medical problems that benefited from our being nearby. Second, N.Y.U. gave me the opportunity to recruit an additional senior neurophysiologist, and Alden Spencer agreed to move to N.Y.U. from the University of Oregon Medical School where he had returned after his stay at the NIH, and to occupy the laboratory next to mine. Although Alden and I never collaborated experimentally again, we talked daily not only about our science – the neurobiology of behavior – but also about almost everything else, until his untimely early death at age 46 from amyotrophic lateral sclerosis in 1977, when we had already moved to Columbia University. During the period he was alive, no one influenced my thinking on matters of science as much as Alden. I still think about him frequently.

Alden and I arrived at N.Y.U. together in the winter of 1965. Within a year we were joined by a biochemist, James H. Schwartz, whom I had first met in the summer of 1951 at Harvard summer school and who was now a member of the Department of Microbiology at N.Y.U. and was becoming interested in behavior. The three of us formed the nucleus of the Division of Neurobiology and Behavior at N.Y.U.

With several important decisions behind me, I made a strong effort to focus on whole-animal behavior. In France I had found that chemical synapses are remarkably plastic; they could readily undergo long-lasting changes in strength. But I had no evidence that these analogs of learning were in fact behaviorally meaningful. I had no reason to believe that these are the sorts of changes that actually occur when an animal learns something. Although during my last few weeks in France I had begun to replicate my results by substituting natural stimuli for electrical stimulation of nerves, I still had not shown that synaptic plasticity actually occurred during behavioral learning. As a first step I thought it essential to show that *Aplysia* was capable of learning. With this in mind, I set about recruiting a postdoctoral fellow with a specific interest in behavioral learning. I was fortunate to recruit, first to Harvard and then to N.Y.U., Irving Kupfermann, an extremely critical and thoughtful student of behavior. We were later joined by another learning psychologist, Harold

Irving Kupfermann

Pinsker, and together we set about delineating a very simple behavior that we could study: the gill-withdrawal reflex. We quickly found that this simple reflex could readily be modified by two forms of learning: habituation and sensitization.

As we explored the two forms of learning, we focused on short-term memory. In 1971, we were joined by another experienced behavioral psychologist, Tom Carew, who brought a new level of energy and insight to our behavioral studies. He arrived as Pinsker was leaving, and soon after we shifted from working with restrained to unrestrained animals, thus opening up the study

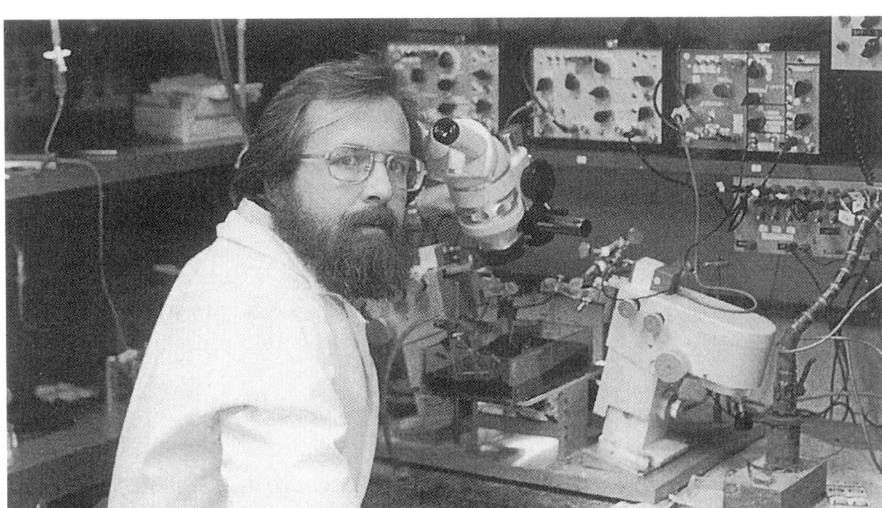

Thomas Carew

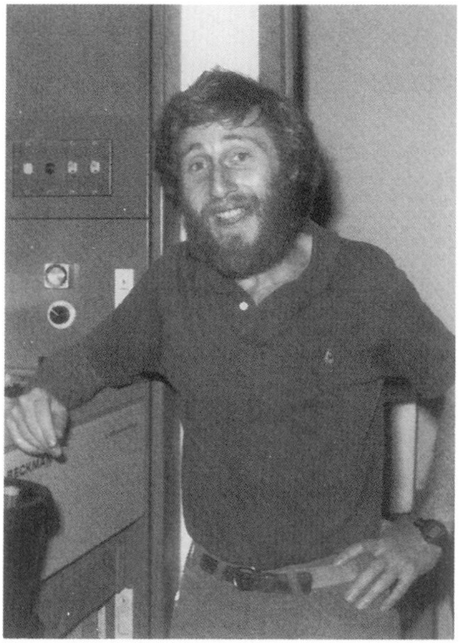

Thomas Abrams

of long-term memory. Tom found that spaced repetition converted the memory for short-term habituation and sensitization to longer-lasting memories. In 1981, after several unsuccessful attempts, Carew, Terry Walters, Tom Abrams, and Robert Hawkins finally were able to define the conditions for reliably producing classical conditioning in *Aplysia.* This was a particularly exciting period; Carew, Walters, Hawkins, and I met regularly to discuss how to explore whether a simple reflex, in a simple invertebrate, could show the higher-order cognitive features of classical conditioning recently demonstrated in mammals by Leo Kamin and somewhat later by Robert Rescorla and Alan Wagner. Soon, Hawkins indeed was able to demonstrate that the gill-withdrawal reflex can undergo second-order conditioning, blocking, overshadowing and other cognitive aspects of associative learning, features that were surprising to uncover in such a simple behavior.

We thus were able to describe a rather rich repertory of learning in *Aplysia.* But long before this inventory of the animal's behavior was complete, we returned to our initial concerns. What happens in the brain of an animal when it actually learns a task? How does it remember? We proceeded, first with Kupfermann and Vincent Castellucci and then with Jack Byrne and Hawkins, to work out most of the neural circuit of the gill-withdrawal reflex. We identified specific sensory neurons and motor cells that produced movements of the gill. Next, we found that the sensory neurons made direct connections to the motor neurons as well as indirect connections through interneurons, both excitatory and inhibitory. The aversive tail stimuli that produced sensitization of the gill-withdrawal reflex activated modulatory interneurons that acted on terminals of the sensory neurons. We now could turn to think about how learning might occur in this reflex.

Robert Hawkins

CELLULAR MECHANISMS OF LEARNING

At the end of the 19th century Ramón y Cajal introduced the principle of *connection specificity*, according to which, during development, a neuron will form connections only with certain neurons and not with others. Kupfermann, Castellucci, and I saw in the circuitry of the gill-withdrawal reflex of *Aplysia* this remarkable regularity of connections that Cajal referred to and we saw, in exquisite detail, that specific identified cells made invariant connections to one another. But this invariant organization of neurons posed deep questions. How could we reconcile hardwired circuits in the nervous system and the

Vincent Castellucci

John Byrne

specificity of connections with the animal's capability for learning? Once acquired, where or how is learned information retained in the nervous system?

One solution was proposed by Ramón y Cajal in his Croonian Lecture to the Royal Society of London in 1894 when he suggested that "... mental exercise facilitates a greater development of the protoplasmic apparatus and of the nervous collaterals in the part of the brain in use. In this way, pre-existing connections between groups of cells could be reinforced by multiplication of the terminal branches of protoplasmic appendices and nervous collaterals."

This remarkably prescient idea was by no means generally accepted. On the contrary, different theories of learning at various times held the attention of neural scientists. Two decades after Ramón y Cajal's proposal, the physiologist Alexander Forbes suggested that memory is sustained not by changes in synaptic strength of the sort suggested by Ramón y Cajal, but by dynamic changes resulting from reverberating activity within a closed loop of self-exciting neurons. This idea was elaborated by Ramón y Cajal's student, Rafael Lorente de Nó, who found in his own and in Ramón y Cajal's analyses of neural circuitry neurons that were interconnected in closed pathways and could thereby sustain reverberatory activity, thus providing a dynamic mechanism for information storage. In his influential book *The Organization of Behavior* (1949), D.O. Hebb proposed that a "coincident activity" initiated the growth of new synaptic connections as part of long-term memory storage. But for short-term memory, Hebb invoked a reverberatory circuit:

> "To account for permanence, some structural change seems necessary, but structural growth presumably would require an appreciable time. If some way can be found of supposing that a reverberatory trace might cooperate with the structural change, and carry the memory until the growth change is made, we should be able to recognize the theoretical value of the trace, which is an activity only without having to ascribe all memory to it. The conception of a transient, unstable reverberatory trace is therefore useful. It is possible to suppose also some more permanent structural change reinforces it."

Similarly, in *The Mammalian Cerebral Cortex*, an influential book of 1958, B. Deslisle Burns challenged the relevance of synaptic plasticity to memory.

> "The mechanisms of synaptic facilitation which have been offered as candidates for an explanation of memory … have proven disappointing. Before any of them can be accepted as the cellular changes accompanying conditioned reflex formation, one would have to extend considerably the scale of time on which they have been observed to operate. The persistent failure of synaptic facilitation to explain memory makes one wonder whether neurophysiologists have not been looking for the wrong kind of mechanisms."

Indeed, some scholars even minimized the importance of specific neuronal connections in the brain, advocating instead mechanisms of learning that were partially or even totally independent of "pre-established" conduction pathways. This view was held by Wolfgang Köhler and the famous Gestalt psychologists, and subsequently by the neurophysiologists Ross Adey and Frank Morrell. Thus, in 1965, Adey wrote:

> "No neuron in natural or artificial isolation from other neurons has been shown capable of storing information in the usual notion of memory. … In particular, the possibility exists that extraneuronal compartments may participate importantly in the modulation of the wave process that characterize the intracellular records, and that these wave processes may rank at least equivalently with neuronal firing in the transaction of information and even more importantly in its deposition and recall."

Finally, there were memory macromolecular notions advocated by Holger Hyden, based upon his finding of changes in the nucleotide composition of RNA. He proposed that learning gave rise to a specific pattern of instructional neural activity that altered the stability of RNA molecules, so that one base can exchange for another. In this way, new RNA molecules are formed with new base sequences that are specific to the instructing pattern of neural activity induced by learning. Hyden's hypothesis thus implied that the patterns of stimulation activated by learning could introduce changes in RNA.

We were now therefore in a position to test experimentally which, if any, of these ideas had merit. Using the gill-withdrawal reflex, we quickly established that memory in the *Aplysia* nervous system is not represented in self-exciting loops of neurons but in changes in synaptic strength. We found that all three simple forms of learning – habituation, sensitization, and classical conditioning – lead to changes in the synaptic strength of specific sensory pathways, and that these changes parallel the time course of the memory process. These findings, which had been fully anticipated by our earlier studies of analogs of learning, gave rise to one of the major themes in our thinking about the molecular mechanisms of memory storage. Even though the anatomical connections between neurons develop according to a definite plan, the strength and effectiveness of those connections is not fully determined developmentally and can be altered by experience.

We therefore concluded the third of our 1970 series of consecutive papers in Science on the cellular mechanisms of learning with the following comments:

> "... the data indicate that habituation and dishabituation (sensitization) both involve a change in the functional effectiveness of previously existing excitatory connections. Thus, at least in the simple cases, it seems unnecessary to explain the behavioral modifications by invoking electrical and chemical fields or a unique statistical distribution in a neural aggregate. The capability for behavioral modification seems to be built directly into the neural architecture of the behavioral reflex.
>
> Finally, these studies strengthen the assumption ... that a prerequisite for studying behavioral modification is the analysis of the wiring diagram underlying the behavior. We have, indeed, found that once the wiring diagram of the behavior is known, the analysis of its modification becomes greatly simplified. Thus, although this analysis pertains to only relatively simple and short-term behavioral modifications, a similar approach may perhaps also be applied to more complex as well as longer lasting learning processes."

A BEGINNING MOLECULAR ANALYSIS OF MEMORY STORAGE

Having defined a critical site of plasticity, the situation became ripe for a molecular analysis. Here again I could not have been more fortunate. As I mentioned earlier, soon after I arrived at N.Y.U. I ran into James Schwartz. Jimmy had attended N.Y.U. Medical School two years behind me, but we had not really talked since I left N.Y.U. in 1956. After medical school Jimmy obtained a Ph.D. with Fritz Lipmann at the Rockefeller University, studying enzyme mechanisms and protein translation in cell-free bacteria extracts. As he and I began to talk again, he mentioned that he was thinking of moving from *E. coli* to the brain. *Aplysia* seemed ideal for biochemical study of individual nerve cells, so in 1966, Schwartz and I joined forces to carry out biochemical studies on individual identified nerve cells of *Aplysia*.

Jimmy soon showed that each nerve cell in *Aplysia* had a specific transmitter biochemistry. Cells that we had presumed on pharmacological grounds to be cholinergic did in fact synthesize and release acetylcholine. With time, Jimmy became interested in the molecular mechanisms of synaptic plasticity, and together we began to examine the role of protein synthesis in memory storage. We knew from the work of Louis Flexner and Bernard Agranoff in the mid 1960s that long-term memory in vertebrates required protein synthesis whereas short-term memory did not. In our first study together in 1971, we found that blocking protein synthesis for 24 hours did not prevent the short-term synaptic changes associated with habituation and sensitization. That finding made us think that short-term changes representing memory

James Schwartz

storage might involve activation of a second-messenger pathway, for example, the cyclic AMP (cAMP) cascade, whose actions might persist for periods longer than the millisecond duration of conventional synaptic actions.

In the discussion of our 1971 paper on the role of protein synthesis and synaptic plasticity, we wrote:

> "Alterations in molecular configuration would not be expected to persist for long periods of time, although molecular changes lasting for several minutes have been observed. ... Most likely, the biochemical mechanisms underlying these short-term plastic changes are composed of a series of sequential reactions which result in a new distribution of transmitter substance. Mechanisms involving cyclic 3',5' AMP might serve as one example of a series of reactions which result in transient enhancement in the activity of a critical enzyme system. A pathway of this kind might trigger the mobilization of transmitter from one component (a long term store) to another (an immediately releasable store).
>
> ... If our conclusion is correct, ... rapidly synthesized RNA cannot immediately play a role in neuronal functions; it might however, be important for long-term neuronal processes."

Sutherland and Rall had already shown in brain slices that several neurotransmitters known to exist in the brain could increase the concentrations of cAMP by activating the enzyme adenylyl cyclase that converted ATP to cAMP. We appreciated that we had a particularly good experimental preparation for examining, on the cellular level, the role of second-messenger pathways in synaptic transmission, synaptic plasticity, and memory storage. In 1972,

Schwartz, Howard Cedar, and I found that stimulation of the pathway involved in sensitization increased the level of cAMP in the entire abdominal ganglion. Schwartz and Cedar next found that the transmitter serotonin could also increase cAMP, providing the initial evidence that serotonin might activate an adenylyl cyclase in *Aplysia.*

COLUMBIA UNIVERSITY AND THE MOLECULAR ANALYSIS OF SHORT-TERM MEMORY

It was at this time that I was invited to move from N.Y.U. to the Columbia University College of Physicians and Surgeons to become the founding director of the Center for Neurobiology and Behavior. I was able to persuade James Schwartz, Alden Spencer, and Irving Kupfermann (who was by then an Associate Professor, having established an independent research program concerned with feeding and motivational state in *Aplysia*) to join me. This move was attractive to me for several reasons. Historically, Columbia has had a strong tradition in neurology and psychiatry, and a friend and former clinical teacher, Lewis Rowland, was about to assume the chairmanship of the Department of Neurology. In addition, I had my first experience in neurobiology at Columbia with Harry Grundfest who was now retiring and I was being recruited to replace him. Finally, Denise was on the Columbia faculty and our house in Riverdale was near Columbia, thereby greatly simplifying our lifes.

In 1974, just after arriving at Columbia, Castellucci and I went back to the elementary circuit of the gill-withdrawal reflex to determine the exact site of the synaptic change produced by short-term sensitization. We wanted to know which component of the synapse changes. Is it, as we suspected, based on indirect evidence, the presynaptic element of the synapse where chemical transmitter is released, or is it the postsynaptic site which contains the receptors which bind and respond to the transmitter? Using a quantal analysis, we found that the synaptic facilitation characteristic of sensitization is presynaptic and that inhibitors of serotonin block this presynaptic facilitation. Later, Hawkins and I found that tail stimuli that initiate sensitization activate a set of modulatory interneurons, the most important of which are serotonergic. The serotonergic and other modulatory interneurons all acted on the sensory neurons and on their presynaptic terminals to enhance transmitter release from their presynaptic terminals. We could now ask for the first time: Was cAMP directly involved in facilitation? In 1976, Marcello Brunelli could take advantage of the size of the *Aplysia* neurons and inject cAMP directly into the presynaptic sensory cell and thereby find a clear enhancement of synaptic transmission. This cAMP-induced enhancement paralleled the enhancement produced by serotonin or tail stimulation.

I now began to interact with Paul Greengard, who was demonstrating that cAMP produced its actions in the brain through the cAMP-dependent protein kinase, or PKA. In 1980, Schwartz, Castellucci, and I collaborated with Greengard. We injected a purified catalytic subunit of bovine PKA into presynaptic sensory neurons and found that it simulated the actions of cAMP or serotonin. Moreover, we could block the actions of serotonin by injecting into the sensory neuron the specific peptide inhibitor of PKA, protein kinase inhibitor PKI. With Steven Siegelbaum we next began to define some of the targets of PKA and focused on one target, a novel K^+ channel. Steve showed that this channel is closed by serotonin and by PKA and that this closure is achieved in a manner consistent with the channel being phosphorylated directly by PKA.

THE HOWARD HUGHES MEDICAL INSTITUTE AND THE MOLECULAR ANALYSIS OF LONG-TERM MEMORY

Just before I arrived at Columbia, Arnold Kriegstein, an M.D.-Ph.D. student, succeeded in culturing embryonic *Aplysia* in the N.Y.U. laboratory, a quest which had intrigued biologists and eluded their efforts for almost a century. Most of us who were there will not readily forget Kriegstein's extraordinary in-house seminar in December, 1973 when he first described his discovery that the red seaweed *Laurencia pacifica* is required to trigger metamorphosis from a free-swimming veliger larva to a small crawling snail, a discovery that allowed him to show the first pictures of the beautiful tiny post-metamorphic juvenile *Aplysia*. I remember saying to myself: "Babies are always so beautiful!" Kriegstein's work opened up the study of development and cell culture in *Aplysia*

Because we now had young animals at all stages of development, we at last had the essential requirements for the generation of dissociated cell culture. This was taken on by Sam Schacher and Eric Proshansky. With the help of Steven Rayport (another M.D.-Ph.D. student at Columbia University), Scha-

Steven Siegelbaum

Arnold Kriegstein and Craig Bailey

cher soon succeeded in culturing the individual sensory neurons, motor neurons, and serotonergic neurons of the gill-withdrawal reflex. The development of the culture system coincided with two other events that allowed me to begin studying the molecular mechanisms of long-term memory storage. The first was my encounter with Richard Axel and my collaboration in 1979, with him and with Richard Scheller, who became a joint post-doctoral fellow. The second was my being recruited to become a senior investigator at the Howard Hughes Medical Institute.

Axel and Scheller's success in 1982 in cloning the gene encoding the egg-laying hormone in *Aplysia* seeded Axel's long-term interest in neurobiology and gave me not only a wonderful friend but also an exposure to the methods of recombinant DNA and modern molecular biology. The very next year, in 1983, Donald Fredrickson, the newly appointed President of the Howard Hughes Medical Research Institute, asked Schwartz, Axel, and me to form the nucleus of a Howard Hughes Medical Research Institute at Columbia devoted to molecular neural science. The Howard Hughes Medical Research Institute gave us the opportunity to recruit from Harvard both Tom Jessell and Gary Struhl, as well as to keep Steven Siegelbaum at Columbia.

My first goal on becoming a Highes Investigator was to examine the molecular mechanisms underlying the synaptic changes that parallel long-term memory storage. In 1885, Herman Ebbinghaus transformed speculation about memory into a laboratory science by having subjects memorize lists of nonsense syllables. In this way Ebbinghaus generated two basic principles about memory storage. First, he found that the transition from short-term memory to long-term memory is graded; practice makes perfect. Second, he anticipated the existence of a fundamental distinction between short- and long-term memory.

What, then, was the molecular basis for this fundamental distinction between short- and long-term memory? As we have seen, in the mid-1960s Flexner and Agranoff examined this distinction biochemically and found that inhibitors of protein synthesis disrupt long-term memory without adversely

Samuel Schacher

affecting learning, or short-term memory. We found that long-term sensitization in *Aplysia* is similarly dependent on protein synthesis, whereas short-term sensitization is not. These findings illustrated the generality of the distinction between short-term and long-term memory processes for both invertebrates and vertebrates. In each case spaced repetition of the learning stimulus acts to transform a transient memory into a more stable (long-term) form by means of a process that depends on new protein synthesis. But how this occurred was a mystery.

We had earlier found in *Aplysia* that long-term sensitization involved a persistent increase in the strength of the same synaptic connection altered by the shortterm process – the connections between the sensory and motor neurons of the gill-withdrawal reflex. To study this process more effectively we turned

Richard Axel

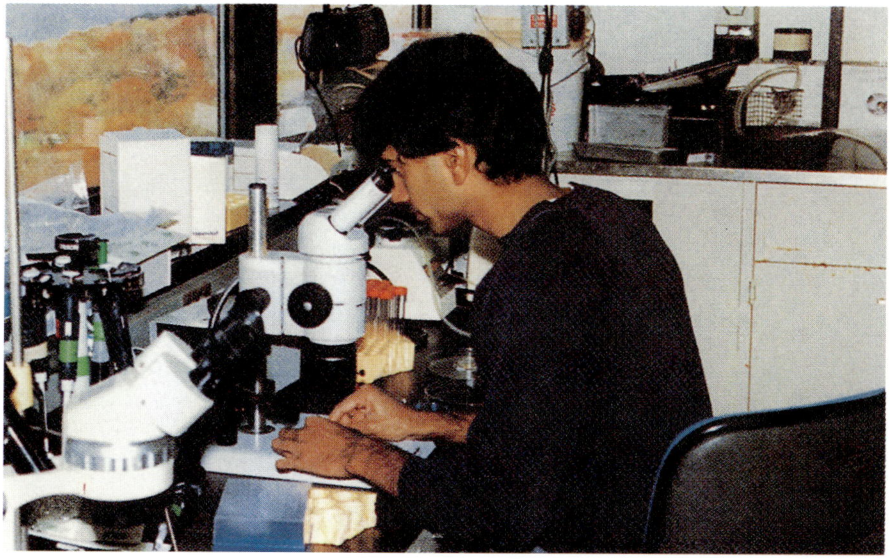

Pramod Dash

to dissociated cell culture and found that we could reconstitute both short- and long-term synaptic facilitation in a culture consisting of only a single sensory neuron and a single motor neuron. We did this together with Sam Schacher, Philip Goelet, and Pier Giorgio Montarolo by applying either one or five brief spaced pulses of serotonin to the sensory neuron and motor neuron in the culture dish. Much like behavioral long-term memory, the long-term synaptic changes required new protein synthesis while the short-term changes did not. Thus, we had trapped the protein synthesis-dependent component of memory storage in the elementary synaptic connection between two identified cells. We now could address directly the question: Why is protein synthesis required for long-term and not short-term facilitation? What are the molecular steps that switch on long-term facilitation and, once switched on, how is it maintained?

We next found that steps for new proteins are activated by a cascade of genes initiated by the cAMP-dependent protein kinase. With repeated application of serotonin, PKA translocates to the nucleus and in so doing activates the MAP kinase (mitogen activated protein kinase), another kinase often recruited for growth. Thus, one of the functions of repeated stimulation was to cause both kinases to move into the nucleus. Pramod Dash and Binyamin Hochner and later Cristina Alberini, Mirella Ghirardi, and Dusan Bartsch provided the first evidence that in the nucleus, these kinases act on a gene regulator called CREB-1 (the cAMP response element binding proteins) to initiate a cascade of gene actions. With David Glanzman and Craig Bailey, we found that the CREB-mediated gene cascade which triggers the synthesis of new protein is required for the growth of new synaptic connections and it is the formation of these new synapses that sustains the long-term change.

The requirement for transcription in long-term facilitation explained why

Dusan Bartsch

long-term memory requires the synthesis of new proteins. However, this requirement now posed a cell-biological puzzle: if long-term synaptic change relies on the activation of genes in the nucleus, that means there is ready communication between the nucleus and the synapse. If that is so, must all such long-lasting changes in the signaling ability of the neuron be cell-wide? Or can long-term synaptic changes be restricted to individual synapses. Experiments by Kelsey Martin, based on a beautiful new cell culture system she developed, revealed that individual synapses or groups of synapses within a cell can be modified independently.

Cristina Alberini and Mirella Ghirardi

David Glanzman

A RETURN TO THE HIPPOCAMPUS: GENETICALLY MODIFIED MICE AND THE STUDY OF COMPLEX SPATIAL MEMORY STORAGE

In our studies in *Aplysia* we focused on the simplest forms of memory, called implicit (or procedural) memory. These memories are concerned with the unconscious recall of perceptual and motor skills and do not require a hippocampus. The hippocampus is involved in explicit (or declarative) memory, memory for people, objects, or places, memories that require conscious participation for recall. For years I tried to encourage people who left my lab to turn their attention to the hippocampus, but to no avail. Finally in 1990, when I reached my 60th birthday, I returned to the study of the hippocampus myself. I was emboldened to do so in great part because of the development of methods for inserting and for knocking out individual genes in mice. This work made it clear to me that mice offered a superb genetic system for examining the role of individual genes in synaptic modification on the one hand, and intact behavior – explicit memory storage – on the other. Mice have a well developed medial temporal lobe and hippocampus, and these are important for explicit memory of objects and space. Moreover, in 1972, Tim Bliss and Terje Lømo in Per Andersen's laboratory in Oslo, had discovered that electrically stimulating any one of the three major pathways in the hippocampus gives rise to a synaptic facilitation, called long-term potentiation or LTP. We were interested in two questions: (1) What are the molecular signaling pathways that are important for LTP? (2) Is LTP important for explicit memory storage? In the move to genetically modified mice, the contributions of Seth Grant and Mark Mayford were particularly influential.

Grant was the driving force in our first studies, in which we showed a role for nonreceptor tyrosine kinases in long-term potentiation, and in spatial memory in the hippocampus. Mayford's critical thinking became important somewhat later, as we began to realize the limitations in the first generation of genetically modified mice. The limitations stimulated Mayford to develop regionally restricted promoters that limited the expression of genes to only certain regions of the brain, and methods for controlling the timing of gene expression. Those two technical advances by Mayford proved important in

Kelsey Martin

allowing us, and Susumu Tonegawa (whose laboratory was now also focusing on studying memory in genetically modified mice), to generate mice whose phenotypes were more specific and in whom a genetic defect could be more readily interpreted than in the first generation of genetically modified mice because the defect could be related, somewhat more directly to specific synaptic changes and to behavior. Over the next few years Mayford, Ted Abel, Mark Barad, Isabelle Mansuy, Chris Pittenger, Amy Chen, and Angel Barco created a number of regionally restricted and regulated transgenic animals that allowed us to examine the role of the PKA- CREB-1 and CREB-2 and the protein synthesis-dependent transcriptional switch within the hippocampus, and to find that it was quite similar in principle to what we had encountered in *Aplysia*. Our lab and those of Alcino Silva and Dan Storm found that the cAMP, PKA, and CREB switch were required for long-term forms of synaptic plasticity in the hippocampus, was also required for spatial memory.

A MOLECULAR APPROACH TO THE COGNITIVE MAP OF SPACE IN THE HIPPOCAMPUS: STEPS TOWARD A MOLECULAR BIOLOGY OF ATTENTION

With this background information about genes, LTP, and spatial memory, we now could ask a deeper question: How does an animal learn about extrapersonal space? Why does spatial memory go awry with defects in PKA signaling? What is the function of the transcriptional switch? To address these questions, we turned to studying how space is represented in the hippocampus.

One of the key insights to emerge from the study of higher cognitive functions is that each perceptual or motor act has an internal or neural representation in the brain. These representations can be either simple or complex.

Seth Grant

The simplest internal representations are those evident in the sensory systems where the afferent fibers are arranged as topographic maps of the receptor surface. These are the representations which Wade Marshall, my former mentor at the NIH, had discovered in the 1930s and early 1940s. Marshall showed that this map is most clearly evident in the neural representation of *personal space,* the representation of touch. The neural representation of the space surrounding the body, the *extrapersonal space,* is far more complex. Here the representation is not topographical but encoded in the pattern of firing of cells that do not have any specific topographic relation to one another with respect to the receptor surface. Thus, adjacent cells need not encode adjacent regions of extrapersonal space.

This representation was discovered in 1971, by John O'Keefe at University College London, who made the brilliant observation that the hippocampus has a cognitive map – a complete representation of extrapersonal space. O'Keefe discovered that all the pyramidal cells in the hippocampus, the very same cells that are used to study long-term potentiation have, as a natural function in the intact animal, the ability to encode space. He found that when an animal moves around in a familiar environment, different pyramidal cells in the hippocampus fire as the animal traverses different regions of the environment. This tendency is so marked that O'Keefe referred to the pyramidal cells as *place cells.* Some place cells may fire only when the animal's head enters one position in a given space. Other pyramidal cells will fire when the animal's head enters another position in the same space. Thus, a mouse's brain breaks up the space in which it walks into many small overlapping fields, and each field is assigned to specific cells in the hippocampus, forming a spatial map of the animal's surroundings. When the animal enters a new environment, a new place map is formed within minutes.

Mark Mayford

These observations have given rise to the idea that the hippocampus contains a map-like representation of the animal's current extrapersonal environment, and that the firing of place cells in the hippocampus signals the animal's moment-to-moment location within the environment. This spatial map is the best-understood example of a complex internal representation in the brain, a true cognitive map. It differs in several ways from the classical sensory maps found by Wade Marshall for touch, vision, or hearing. Unlike sensory maps, the map of space is not topographic, that is, neighboring cells in the hippocampus do not represent neighboring regions in the environment. Furthermore, a place cell will fire in the same place regardless of what the animal is looking at. Moreover, the firing of place cells can persist after pertinent sensory cues are removed and even in the dark. Thus, although the activity of a place cell can be modulated by sensory input, it is not determined by sensory input as is the case for the activity of neurons in a sensory system. It appears that the place cells do not map the current sensory input, but the location where the animal *thinks* it is in space.

Place fields are formed in minutes, and once formed the map to which they contribute can remain stable for weeks. It struck me in 1995 that formation of this internal representation – this cognitive map of space – was a learning process and that synaptic plasticity related to LTP might have a role in stabilizing this cognitive representation.

Although place cells have been studied since 1971, nothing was known about the cellular or molecular mechanisms whereby new place fields are formed, and specifically no one had attempted to relate the biology of place cells to the molecular mechanisms of LTP or hippocampal-based memory. To explore this problem, I was fortunate to start a collaboration with Robert Muller at Downstate Medical Center in Brooklyn, who had pioneered the systematic study of place cells. This problem was taken on by Cliff Kentros, a postdoctoral fellow in my lab, by Naveen Agnihotri, a graduate student, and by Alex Rotenberg, a joint student with Muller and myself. Using a combination of pharmacological and genetic approaches, we demonstrated a link between recruitment of PKA and protein synthesis on the one hand, and on the other, the long-term, but *not* short-term stability of the hippocampal representation of space. Thus, PKA and protein synthesis are required for long-term memories of extrapersonal space because that memory is based on a learned internal representation of space whose long-term stability requires PKA and new protein synthesis.

This raised a final question: Explicit memory in humans differs from implicit memory in requiring conscious attention for recall. How does conscious attention come to bear on explicit memory? Indeed, how can one study consciousness in the mouse? In the course of our work on place fields, Kentros, Agnihotri, Hawkins, and I found that the long-term stability of the place field map correlated strongly with the degree to which the animal was required to attend to its environment. This demonstrates that, rather than being an implicit, automatic, process, the long-term recall of a stably formed place cell map requires the mouse to attend to its environment, as would be expected for explicit memory in human beings. The finding that attention, the recruitment of PKA, and new protein synthesis are required to form and recall a stable map in the mouse has opened up a molecular biological approach to an attentional process.

FROM PSYCHOANALYSIS TO *APLYSIA* TO THE ROLE OF ATTENTION IN THE COGNITIVE REPRESENTATION OF EXTRAPERSONAL SPACE

During the past 10 years my career has begun to come full circle. From an initial interest in the complex cognitive problems of psychoanalysis and memory storage, my research on memory led me first to the mammalian hippocampus, which proved too difficult as a first step and forced me to take a more reductionist approach and study initially the simplest forms of memory in *Aplysia,* and then, only much later, the more complex forms of memory in mice. I found that despite important differences in detail, simple implicit and explicit memories have a similar short- and long-term storage form. In each form, short-term storage requires covalent modification of pre-existing proteins leading to the alteration of pre-existing synaptic connections, whereas long-term memory storage requires gene activation, new protein synthesis, and the growth of new synaptic connections.

In the course of this work we began to explore how explicit memory storage for space affects the internal representation of space. We found that on the level of internal representation the storage mechanisms for explicit memory are similar to those in human beings in requiring attention. Attention is a component of conscious response, perhaps the great challenge of all research on mental processes. It thus seems likely that in future decades, the study of memory, perhaps even in mice, is likely to allow molecular insights into even the deepest problems of human behavior.

A PERSONAL PERSPECTIVE

Although doing research on *Aplysia* and the hippocampus and discussing science with colleagues in my lab have given me the greatest intellectual satisfactions, I have loved teaching and have learned a great deal from lecturing to medical and graduate students. It was in the context of the neural science course at Columbia that the idea arose of doing a textbook, *Principles of Neural Science.* In college and medical school I was never a good note-taker. I always preferred sitting back, enjoying the lecture, and just scribbling down a few words here and there. When I came to Columbia to develop the neural science course, I was struck by how much energy students were devoting to writing out every single word of lectures, and I wanted to help them get over that. I therefore encouraged the faculty to provide a syllabus for each lecture, and with time I edited the syllabus, added figures to it and improved it. Then Jimmy Schwartz and I decided that the syllabus was becoming sufficiently useful that we might make a textbook out of it. Our textbook was the first attempt to bridge cell and molecular biology to neural science and neural science to behavior and clinical states. The response to the first edition was so gratifying that we made an effort to make the book better and more complete. With the second edition, not only students but also scientists began to regard our textbook as useful. With the help of Tom Jessell, we further improved the third and fourth editions. The widespread reception of this book, both in the United States and abroad, has been a source of deep satisfaction to me and to the other contributors.

Outside of our work and our family, Denise and I enjoy the visual arts and classical music, especially opera. Our interest in both of these activities is greatly enriched by having within easy reach of our home the great museums and galleries of Manhattan as well as the Metropolitan Opera. We also are inveterate – I am tempted to say *addicted* – collectors of art and antiques. We have lived for 36 years in a now 150-year-old house in the Riverdale section of the Bronx, with wonderful views of the Hudson River and the Palisades. We collect French art nouveau furniture, vases, and lamps, an interest that originated with Denise and her mother, and graphic art of the Austrian and German Expressionists, an interest which originated with me. As I write this, I am beginning to suspect that our collecting may well be an attempt to recapture a part of our hopelessly lost youth.

In the course of my career I have incurred many debts both personal and scientific. First and foremost I owe an enormous personal debt to my parents and my brother Lewis. My parents were able in mid-life to relocate to a foreign country – my father spoke not a word of English when he first arrived in New York – and to create a new life for themselves and their sons. My parents not only succeeded in establishing themselves in their small store in Brooklyn, but were sufficiently successful to support me through college and medical school. They were so occupied with their store that throughout their life in America they did not share in the cultural life of New York, which Lewis and I were beginning to enjoy. Despite their constant labor they were always extremely optimistic and supportive of us, and never tried to dictate decisions about my work or play. Lewis was also an enormous influence on me in my early years, and my interest in classical music and my joy in learning were importantly influenced by him. While a graduate student at Brown University writing his dissertation in linguistics and Middle High German, he was called to service as an intelligence officer in the Korean War. He and his wife, Elise, went first in 1951 to Germany and then in 1953 to Paris, France, where he had a position as a civilian in Air Force Intelligence. He so enjoyed his life in France, that he lost his interest in an academic life and stayed in France for 13 years, where he and Elise raised five children. He eventually returned to the United States and finished his career in a series of administrative positions, in the Health Department of the City of New York. He died in 1979, at age 54 of a recurrence of a cancer of the kidney, which we all thought had been successfully removed when it first presented 10 years earlier.

Second, I have been privileged to enjoy a wonderfully supportive, endlessly interesting, and stable family life with Denise, my partner, best friend, and most honest critic for now 45 years. Throughout our life together she has consistently encouraged my love of research and supported my scientific aspirations. Denise is a professor in the Department of Psychiatry and in the School of Public Health at Columbia University, and has pioneered the study of drug abuse in adolescence. Her work on the epidemiology of drug abuse has become the basis of the current understanding of the developmental sequence whereby adolescents become involved in drugs. I am also greatly indebted to our two children, Paul and Minouche, for the joy they gave Denise and me while growing up and the satisfaction they have given us in seeing what principled and interesting people they have become and how thoughtful they are as parents to their own children. Our son Paul majored in economics at Haverford College and graduated from the Columbia Business School. He now manages a set of investment funds at Dreyfus-Mellon. Paul is married to Emily Kaplan, an interior designer; they live in Scarsdale, N.Y. and have two daughters, Allison (born on January 5, 1992) and Libby (born on October 14, 1995). Our daughter Minouche went to Yale College and Harvard Law School. She practices public interest law in San Francisco specializing in women's rights and family violence. Minouche is married to Rick Sheinfield, also a public interest lawyer, and they have a son, Izak (born on November 10, 1998) and a daughter, Maya (born on March 12, 2001).

In retrospect it seems a very long way for me from Vienna to Stockholm. My timely departure from Vienna made for a remarkably fortunate life in the United States. The freedom that I have experienced in America and in its academic institutions made Stockholm possible for me, as it has for many others.

POSTSCRIPTS: A CURRICULUM VITAE

I began my academic career at the Harvard Medical School, where from 1963 to 1965, I was an instructor in the Department of Psychiatry. In 1965, I moved to New York University as associate professor where, together with Alden

Spencer and James Schwartz, we developed the first group in the country devoted to both cellular neurobiology and behavior. At the time I was recruited to N.Y.U., Denise was recruited to the Columbia University College of Physicians and Surgeons, where she gradually rose to the rank of professor.

In 1974, Harry Grundfest retired and I was recruited to Columbia to replace him. At Columbia I was the founding director of the Center for Neurobiology and Behavior. In 1983, I became a University Professor at Columbia. In 1984, I resigned as director of the Center to become a senior investigator at the newly formed Howard Hughes Medical Research Institute at Columbia.

Since 1974, I have been a member of the National Academy of Sciences USA. Later I became a member of the National Science Academies of Germany and France, the American Academy of Arts and Sciences, the American Philosophical Society, the National Institute of Medicine, and most recently, Germany's Orden Pour Le Mérite für Wissenschaften und Künste. Being invited to join the Orden was for me a particularly great honor. The collection of scholars and scientists in the Orden is extraordinary; as an extra bonus it includes old friends such as the great German historian Fritz Stern, and a sterling group of biologists including Max Perutz, Christiane Nüsslein-Volhard, Bert Sakmann, Erwin Neher, Walter Gehring, Charles Weissman, and Robert Weinberg.

I have been awarded the Lester N. Hofheimer Prize for Research of the American Psychiatric Association (1977), the Karl Spencer Lashley Prize in Neurobiology from the American Philosophical Society (1981), the Dickson Prize in Biology and Medicine from the University of Pittsburgh (1982), the Albert Lasker Award (1983), the Rosenstiel Award of Brandeis University (1984), the Howard Crosby Warren Medal by the Society of Experimental Psychologists (1984), the American Association of Medical Colleges Award for Distinguished Research in the Biomedical Sciences (1985), the Gairdner International Award of Canada for Outstanding Achievement in Medical Science (1987), the National Medal of Science (1988), the J. Murray Luck Award for Scientific Reviewing from the National Academy of Sciences (1988), the American College of Physicians Award in Basic Science (1989), the Robert J. and Claire Pasarow Foundation Award in Neuroscience (1989), the Bristol-Myers Squibb Award for Distinguished Achievement in Neuroscience Research (1991), the Warren Triennial Prize from the Massachusetts General Hospital (1992), the Harvey Prize of the Technion in Haifa (1993), the Stevens Triennial Prize from Columbia University (1995), the Dana Award (1997), the Gerard Prize of the Society of Neuroscience (1997), the Wolf Prize of Israel (1999), and the Dr. A.H. Heineken Prize for Medicine from the Royal Netherlands Academy of Arts and Sciences in Amsterdam (2000).

I have received honorary degrees from nine universities, including three European universities: the University of Vienna, Edinburgh, and Turin. Surprisingly, the first honorary degree I received, in 1983, was from the Jewish Theological Seminary in New York. I was thrilled that they would even know of my work. I suspect they learned of that from my colleague Mortimer

Ostow, one of the psychoanalysts who first stirred my interest in relating psychoanalysis and the brain. My father had already died but my mother came to the graduation ceremony, and in his introductory remarks Gerson D. Cohen, the chancellor of the seminary, referred to my having received a good Hebrew education at the Yeshiva of Flatbush, an acknowledgement which filled my mother's Jewish heart with pride.

As this recitation makes clear, I also owe a profound intellectual debt to my scientific teachers – Harry Grundfest, Dominick Purpura, Wade Marshall, and Ladislav Tauc – who tolerated my naivete and encouraged my brashness. I also benefited greatly from Steve Kuffler's sage insight and advice and from Alden Spencer's generous friendship. I also am indebted to the extraordinary collection of colleagues, fellows, and students that I have had the privilege of interacting and collaborating with and whose individual contributions I describe in more detail in my Nobel Lecture. Finally, I am deeply grateful to Columbia University and the Howard Hughes Medical Research Institute, two great institutions that have created open environments supportive of scholarship and research.

SELECTED REFERENCES

Adey, W.B. (1965) Electrophysiological patterns and cerebral impedance characteristics in orientering and discriminative behavior. Proc. Intern. Congr. Physiol. Soc. 23rd, Tokyo: 324-339.

Berkley, G.E. (1988) Vienna and Its Jews. The Tragedy of Success, 1880-1980, with Forward by Harry Zoh. Abt Boos, Cambridge, Mass.

Bukey, E.B. (2000) Hitler's Austria. Popular Sentiment in the Nazi Era 1938-1945. The University of North Carolina Press, Chapel Hill, N.C.

Burns, B.D. (1958) The Mammalian Celebral Cortex Monographs of the Physiological Society. Edward Arnold (Publishers) Ltd., London.

Ebbinghaus, H. (1885) Memory: A Contribution to Experimental Psychology (1885). Reprinted, New York, Dover, 1963.

Hebb, D.O. (1949) The Organization of Behavior, John Wiley & Sons, Inc., N.Y.

Hyden, H. and Egyhazi, E. (1962) Nuclear RNA changes in nerve cells during learning experiments in rats. Proc. Natl. Acad. Sci. USA 48:1366-1373.

Kershaw, Ian (2000) Hitler 1936-1945 Nemesis, W.W. Norton & Co., New York, N.Y.

Kuffler, S.W. and Fzaguirre, C. (1955) Process of excitation in the dendrites and soma of single isolated sensory nerve cells of the lobsters and crayfish. J. Gen. Physiol. 39:87-119.

Pulzer, Peter (1988) The Rise of Political Anti-Semitism in Germany and Austria. Harvard University Press, Cambridge, Mass.

Ramón y Cajal, S. (1894) The Croonian Lecture: La Fine Structure des Centres Nerveux. Proc. R. Soc. London B., 56, 444-467.

Schorske, Carl E. (1980) Fin de Siecle Vienna. Politics and Culture. Knopf, Inc., Random House, New York, N.Y.

Weiss, D.W. (1999) Reluctant Return. A Survivor's Journey to an Austrian Town. Indiana University Press, Bloomington, Ind.

Zuckmayer, C. (1966) A Part of Myself. Translated from the German by Richard and Clara Winston. Harcourt, Brace, Jovanovich, Inc., New York, N.Y.

THE MOLECULAR BIOLOGY OF MEMORY STORAGE: A DIALOG BETWEEN GENES AND SYNAPSES

Nobel Lecture, December 8, 2000

by

ERIC R. KANDEL

Howard Hughes Medical Institute, Columbia University, College of Physicians and Surgeons, New York, USA.

INTRODUCTION

One of the most remarkable aspects of an animal's behavior is the ability to modify that behavior by learning, an ability that reaches its highest form in human beings. For me, learning and memory have proven to be endlessly fascinating mental processes because they address one of the fundamental features of human activity: our ability to acquire new ideas from experience and to retain these ideas in memory. In fact, most of the ideas we have about the world and our civilization we have learned so that we are who we are in good measure because of what we have learned and what we remember. However, not all learning experiences are positive. Many psychological and emotional problems also result at least in part from our experiences. In addition, specific disorders of learning and memory haunt both the infant and the adult. Down syndrome, fragile X mental retardation, age-related loss of memory, and the devastation of Alzheimer's disease are only the familiar examples of a large number of disorders that affect memory.

Throughout my career I have been interested in the biology of learning. I have been curious to know: What changes occur in the brain when we learn? And, once something is learned, how is that information retained in the brain as memory? I have tried to address these questions by developing a reductionist approach that would allow me to investigate the biological mechanisms underlying elementary forms of learning and memory at a cell and molecular level – as specific molecular activities within specific identified nerve cells.

For a biologist like myself, interested in mental processes, the study of learning has the further appeal that, unlike other mental processes such as thought, language and consciousness, learning is relatively accessible to a cellular and molecular analysis. Elementary forms of learning and memory have been well characterized by classical psychology since the work of Ivan Pavlov and Edgar Thorndike in the first half of the 20th century, and these forms of learning are the most clearly delineated and, for the experimenter, most easily controlled of any mental process.

I first became interested in the study of memory in 1950, as a result of my readings in psychoanalysis while still an undergraduate at Harvard College. Later, during medical training, I found the psychoanalytic approach limiting because it tended to treat the brain, the organ that generates behavior, as a black box. In the mid 1950s, while still in medical school, I began to appreciate that during my generation the black box of the brain would be opened and that the problems of memory storage, once the exclusive domain of psychologists and psychoanalysts, could be investigated with the methods of modern biology. As a result, my interest in memory shifted from a psychoanalytic to a biological approach. As a postdoctoral fellow at the National Institutes of Health (NIH) in Bethesda from 1957 to 1960, I focused on learning more about the biology of the brain, and became interested in knowing how learning produces changes in the neural networks of the brain and how a transient short-term memory is converted to an enduring long-term memory.

From the beginning, my purpose in translating questions about the psychology of learning into the empirical language of biology was not to replace the logic of psychology or psychoanalysis with the logic of cell and molecular biology, but to try to join these two disciplines and to contribute to a new synthesis that would combine the mentalistic psychology of memory storage with the biology of neuronal signaling. As I thought more concretely about the neural mechanisms of memory storage, I hoped further that the biological analysis of memory might carry with it an extra bonus, that the study of memory storage might reveal new aspects of neuronal signaling. Indeed, this has proven true. Time and again, the molecular study of memory has revealed novel aspects of more general biological processes.

DEVISING A RADICAL REDUCTIONIST STRATEGY TO LEARNING AND MEMORY

At first thought, someone interested in learning and memory might be tempted to tackle the problem in its most complex and interesting form. This was the approach that my colleague Alden Spencer and I originally had in 1958 when, at the start of our scientific careers, we joined forces at the NIH to study the cellular properties of the hippocampus, the part of the mammalian brain thought to be most directly involved in aspects of complex memory (1). We were initially interested in a simple question: Are the electrophysiological properties of the pyramidal cells of the hippocampus, which were thought to be the key hippocampal cells involved in memory storage, fundamentally different from other neurons in the brain, such as the well-studied motor neurons in the spinal cord involved in simple movement? In the course of studying the pyramidal cells of the hippocampus, it became clear to us that all nerve cells have similar signaling properties. Therefore, the intrinsic signaling properties of neurons would themselves not give us key insights into memory storage (17).

Thus, the unique functions of the hippocampus had to arise not so much from the intrinsic properties of pyramidal neurons but from the pattern of functional interconnections of these cells, and how those interconnections are affected by learning. To tackle that problem we needed to know how sensory information about a learning task reaches the hippocampus, and how information processed by the hippocampus influences behavioral output. This was a formidable challenge, since the hippocampus has a large number of neurons and an immense number of interconnections. It seemed unlikely that we would be able to work out in any reasonable period of time how the neural networks, in which the hippocampus was embedded, participate in behavior and how those networks are affected by learning.

Thus, to bring the power of modern biology to bear on the study of learning, it seemed necessary to take a very different approach – a radically reductionist approach. Instead of studying the most complex cases, we needed to study the simplest instances of memory storage, and to study them in the elementary reflex behavior of those animals that were most experimentally tractable. To do this we needed to find experimental systems in which a simple behavioral act that was controlled by a small number of large and accessible nerve cells could be modified by learning. Only in this way could we correlate changes in the overt behavior of the animals with molecular events in identifiable neurons and examine how sensory processing in the brain is modified by learning to give rise to memories.

Such a reductionist approach was hardly new in 20th century biology. One need only think of the use of *Drosophila* in genetics, of bacteria and bacteriophages in molecular biology, and of the squid giant axon in the study of the conduction of nerve impulses. Nevertheless, when it came to the study of behavior, many investigators were reluctant to use a reductionist strategy. In the 1950s and 1960s many biologists and psychologists believed that behavior was the one area of biology in which the use of simple animal models, particularly invertebrate ones, was least likely to succeed. They argued that only higher animals exhibit interesting forms of learning and that these forms require neuronal organizations and neuronal mechanisms qualitatively different from those found in simple animals. As a result, an approach to learning based on simple invertebrates was bound to fail because it would lack relevance to mammalian and particularly to human behavior.

It was my belief at the outset, however, that concerns about the use of a simple experimental system to study learning were misplaced. The question was and is not whether there is something special about the human brain; there clearly is. Rather, the question is whether the human brain and human behavior have anything at all in common with the nervous system and behavior of simpler animals. If so, these fundamental, common principles of neuronal organization might well be studied more profitably in simple animals.

The answer to this second question, about commonality, was clear. By 1960, work by students of comparative behavior such as Konrad Lorenz, Niko Tinbergen, and Karl von Frisch had shown that humans share many behavioral patterns and even simple forms of learning with simple animals (reviewed in 2). That the evolution of behavior and learning is conservative should not be surprising, since the evolution of other biological functions is also conservative. There are, for example, no fundamental functional or biochemical differences between the nerve cells and synapses of humans and those of a snail, a worm or a fly. Since behavior and learning is an expression of nerve cell activity, it would be surprising if the learning capability of people did not have some elementary features in common with the learning of snails, worms, or flies. And, if elementary forms of learning are common to all animals with an evolved nervous system, there must be conserved features in the mechanisms of learning at the cell and molecular level, that can be studied effectively even in simple invertebrate animals.

A SIMPLE INVERTEBRATE SYSTEM THAT LENDS ITSELF TO A REDUCTIONIST APPROACH

After an extensive search for a suitable experimental animal, I settled on the giant marine snail *Aplysia* (Figure 1), because it offers three important experimental advantages: its nervous system is made up of a small number of nerve cells; many of these are very large; and (as became evident to me later) many are uniquely identifiable (3,13). Whereas the mammalian brain has a million-million central nerve cells, *Aplysia* has only 20,000. These cells are clustered in 10 anatomical units called ganglia, each of which contains about 2,000 cells (Figure 2). An individual ganglion, such as the abdominal ganglion, mediates not one but a family of behaviors. Thus, the simplest behaviors that can be modified by learning may involve less than 100 cells. This numerical simplification made it possible to identify the specific contribution of individual neurons to the behavior in which they participate (13). In addition to being few in number, these cells are the largest nerve cells in the animal kingdom, reaching up to 1,000 μm in diameter, large enough to be seen with the naked eye (Figure 3). Because of their extraordinary size and their distinctive pigmentation, it is possible to recognize many of the cells as unique individuals. One can record from these large cells for many hours without any difficulty, and the same cell can be returned to and recorded from over a pe-

Figure 1. The giant marine snail *Aplysia californica* belongs to the opisthobranch subclass of the gastropod molluscs. Animals of this species may grow to be 30 cm in length and weigh 1 kilogram.

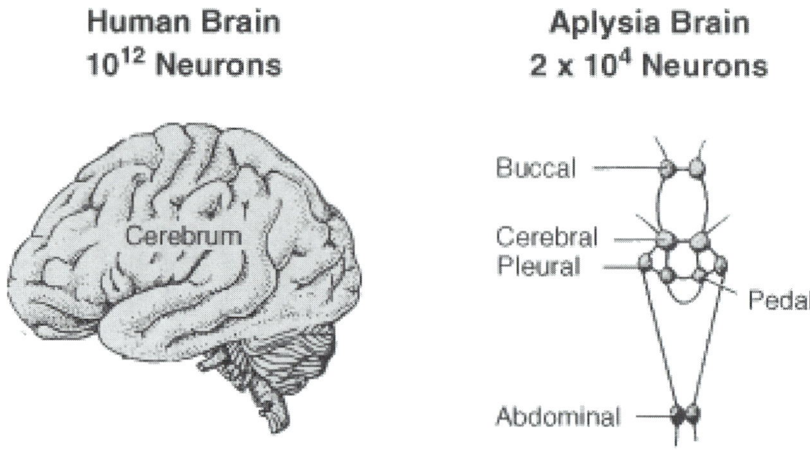

Figure 2. The human brain has a million-million neurons while the brain of *Aplysia* has 20,000 nerve cells. *Aplysia* nerve cells are clustered together in five major bilateral pairs of ganglia, each ganglion containing about 2,000 nerve cells.

riod of days. The cells can easily be dissected out for biochemical studies, so that from a single cell one can obtain sufficient mRNA to make a cDNA library. Finally, these identified cells can readily be injected with labeled compounds, antibodies, or genetic construct procedures which opened up the molecular study of signal transduction within individual nerve cells.

REQUIREMENT FOR A CELL-BIOLOGICAL STUDY OF MEMORY STORAGE

Given a technically advantageous experimental system, how does a cell biologist begin to address the problem of learning? The strategy that my colleagues and I developed involved four sequential steps: (1) We first wanted to define a simple behavior that can be modified by learning and that gives rise to memory storage. (2) We next wanted to identify the cells that make up the neural circuit of that behavior. (3) Within that neural circuit we then wanted to locate the critical neurons and interconnections that had been modified by learning and that store memory. (4) Finally, we wanted to analyze the changes that occur at those sites in response to learning and memory storage, first on

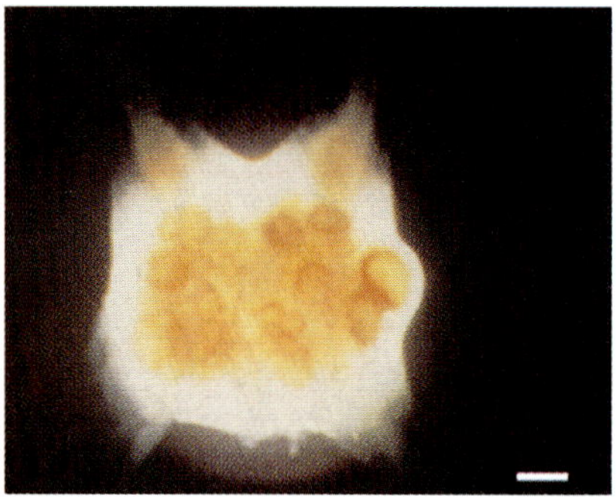

Figure 3. A photomicrograph of an abdominal ganglion of *Aplysia* shows the distinctive pigmentation and positions of its cells. The largest cells are a millimeter in diameter and can be seen with the naked eye. Scale bar = 1 mm. (See also Fig. 6A).

the cellular and then on the molecular level (13). I follow this outline in the discussion below.

DELINEATING A BEHAVIOR IN *APLYSIA* THAT IS CAPABLE OF BEING MODIFIED BY LEARNING

Irving Kupfermann and I first wanted to study the simplest possible behavior of *Aplysia* (4). We examined the animal's behavioral capabilities and delineated a very simple defensive reflex: the withdrawal of the gill upon stimulation of the siphon, an action that is like the quick withdrawal of a hand from a hot object. In *Aplysia*, the gill is a respiratory organ that lies exposed in the mantle cavity. When the animal is in a normal, relaxed state, the gill is partially covered by a sheet of skin (the mantle shelf), which ends in a fleshy spout, the siphon (Figure 1). When a weak tactile stimulus is applied to the siphon, both the siphon and gill are withdrawn into the mantle cavity and for protection under the mantle shelf (Figure 4); ref. 5).

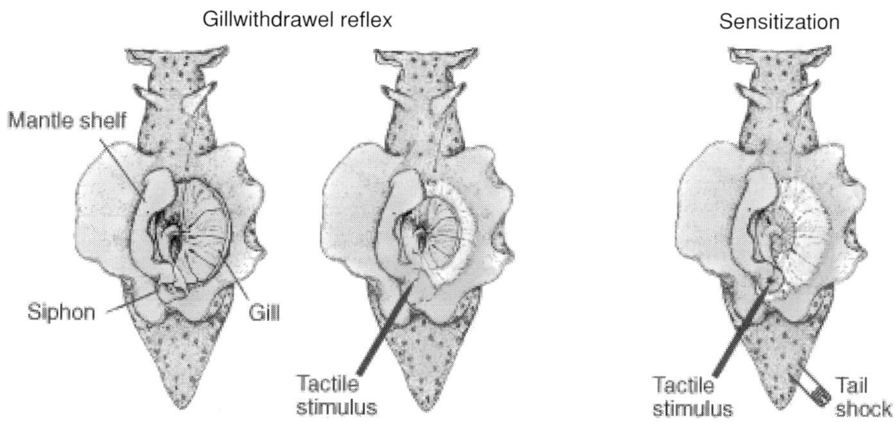

Figure 4. A dorsal view of *Aplysia* showing the gill, the animal's respiratory organ. The gill lies in the mantle cavity and is normally covered partially by the mantle shelf. A light touch to the siphon with a fine probe causes the siphon to contract and the gill to withdraw into the mantle cavity under the protection of the mantle shelf. Here, the mantle shelf is shown to be retracted for a better view of the gill. Sensitization of the gill-withdrawal reflex, produced by applying a noxious stimulus to another part of the body, such as the tail, leads to an enhancement of the withdrawal reflex of both the siphon and the gill.

Kupfermann, Harold Pinsker, and later Tom Carew, Robert Hawkins and I found that this simple reflex could be modified by three different forms of learning: habituation, sensitization, and classical conditioning (5,8,9). As we examined these three forms of learning, we were struck by the resemblance each had to corresponding forms of memory storage in higher vertebrates and humans. As with vertebrate learning, memory storage for each type of learning in *Aplysia* has two phases: a transient memory that lasts minutes and an enduring memory that lasts days. Conversion of short-term to long-term memory storage requires spaced repetition – practice makes perfect even in snails (8,9,10).

We focused initially on one type of learning, *sensitization*, a form of learned fear in which a person or an experimental animal learns to respond strongly to an otherwise neutral stimulus (5,8,10). For example, if somone is suddenly exposed to an aversive stimulus, such as a gunshot going off nearby, that person will be sensitized by the unexpected noise. As a result, that person will be frightened and will now startle to an otherwise innocuous stimulus like a tap on the shoulder. Similarly, on receiving an aversive shock to another part of the body such as the tail (or head), an *Aplysia* recognizes the stimulus as aversive and learns to enhance its defensive reflex responses to a variety of subsequent stimuli applied to the siphon, even innocuous stimuli (Figure 4; 12). The animal now remembers the shock, and the duration of this memory is a function of the number of repetitions of the noxious experience. A single shock gives rise to a memory lasting only minutes; this short-term memory does not require the synthesis of new protein. In contrast, four or five spaced shocks to the tail give rise to a memory lasting several days; this long-term memory does require the synthesis of new protein. Further training gives rise to an even more enduring memory lasting weeks, which also requires new protein synthesis (Figure 5; 10,12).

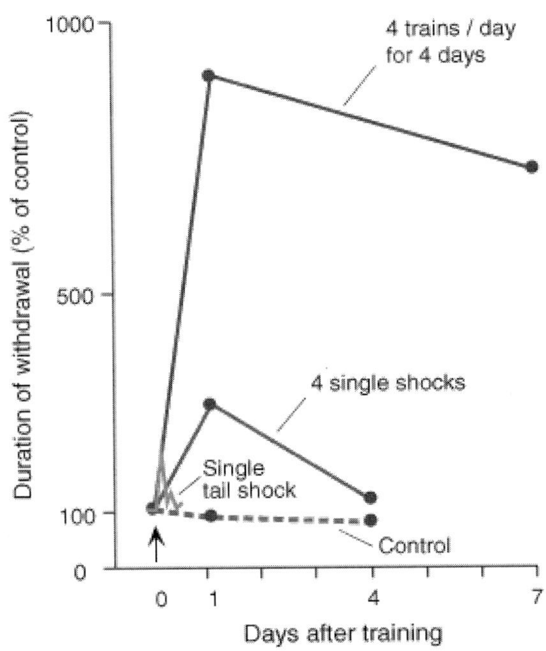

Figure 5. Spaced repetition converts short-term memory into long-term memory in *Aplysia*. In the resting state, before sensitization training, a weak touch to the siphon causes only a weak and brief siphon and gill withdrawal reflex. Following a single noxious, sensitizing, shock to the tail, the siphon and gill withdrawal reflex repsonse to that same weak touch is enhanced and produces a much larger reflex withdrawel response. The memory following this single noxious stimulus lasts about one hour. After four or five single tail shocks, the animal's gill and siphon withdrawel reflex to that same weak touch is enhanced further and now lasts more than a day. If the animal receives further training – four brief trains of single shocks a day over the course of four days – the siphon and gill reflex withdrawal to the same weak siphon touch is enhanced still further and the memory now lasts for several weeks. (Modified from 98.)

Thus, just as in the complex learning in mammals (107,108), long-term sensitization differs from the short-term process in requiring the synthesis of new proteins. This was our first clear evidence for the conservation of biochemical learning mechanisms between *Aplysia* and vertebrates, and it reinforced the hope that a detailed analysis of short-term memory and its transition to long-term memory in *Aplysia* would reveal molecular mechanisms of general importance.

DEFINING THE NEURAL CIRCUIT IN CELLULAR DETAIL

To analyze the cellular mechanisms of sensitization, we needed to identify the neural circuit of the gill withdrawal reflex. Kupfermann and I quickly localized the central neuronal machinery for the reflex behavior in the animal's abdominal ganglion (4,6). Because we soon realized that many cells in *Aplysia* could be identified in every animal of the species (3,4,6,7), we were able to give the cells specific names and, most important, return to the same cell time and again – in both untrained and trained animals. In this way Kupfermann, Castellucci, Carew, Hawkins, John Byrne, and I were able to work out significant components of the neural circuit gill-withdrawal reflex in terms of individual cells and cell clusters. The circuit has 24 mechanoreceptor sensory neurons that innervate the siphon skin and make direct monosynaptic connections with six gill motor cells (7,11,36). The sensory neurons also made indirect connections with the motor cells through small groups of excitatory and inhibitory interneurons (18,19). In addition to being identifiable, individual cells also proved to have surprisingly large effects on behavior (Figure 6C; reviewed in 13,22,36).

As we examined the neural circuit of this reflex in detail, we were struck by its invariance – the cells that make up the circuit and their interconnections are always the same. In every animal we examined, each cell connected only to certain target cells and not to others (Figure 7). Carew, John Koester, Wayne Hening, and I also found this invariance in the neural circuitry of other behaviors in *Aplysia* – inking, control of the circulation, locomotion (reviewed in 13,15) – raising a key question in the cell-biological study of learning: How can learning occur in a neural circuit that is precisely wired?

HOW DOES LEARNING AFFECT THE INVARIANT ELEMENTS OF THE NEURAL CIRCUIT?

In his Croonian Lecture to the Royal Society of 1894, Santiago Ramón y Cajal proposed a theory of memory storage: memory is stored in the growth of new connections (16). This prescient idea was neglected in good part for half a century as students of learning fought over newer competing ideas. First, Karl Lashley, Ross Adey, Wolfgang Köhler, and a number of Gestalt psychologists proposed that learning leads to changes in electric fields or chemical gradients, which they postulated surround neuronal populations and are produced by the aggregate activity of cells recruited by the learning process.

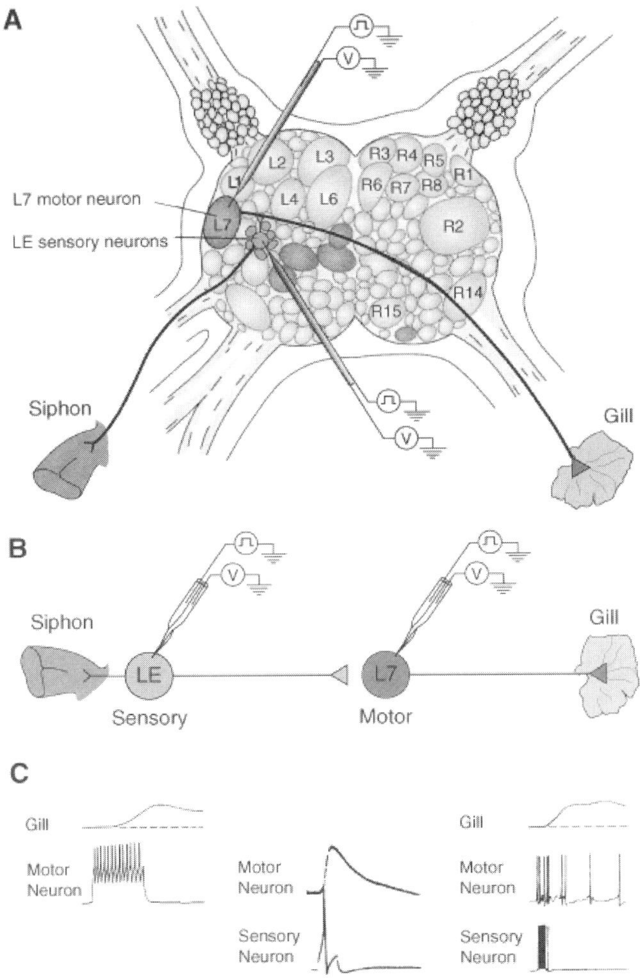

Figure 6. The neural circuit of the gill-withdrawal reflex can be delineated in terms of specific identified cells.

A. This dorsal view of the abdominal ganglion shows some of the identifiable cells (compare with Fig. 3). The six identified motor cells to the gill are shaded brown; seven sensory neurons are shaded blue. In the figure a sensory neuron that synapses on gill motor neuron L7 is being stimulated electrically and a microelectrode in the motor neuron records the synaptic potential produced by the action potential in the sensory neuron (see middle trace in part **C**).

B. The physiological demonstration of the direct connections between the sensory neuron and motor neuron. The sensory neuron receives input from the siphon skin; the motor neuron makes direct connections onto the gill. The fact that the cells are large and identifiable allows for the mapping of connections between specific identified cells. This part shows the experimental arrangement for simultaneously recording in a pre- and postsynaptic cell. The sensory neuron makes a direct connection onto the motor neuron, as is evident in the middle trace in part **C**.

C. Individual cells make significant contributions to the reflex. Stimulating a single motor neuron (traces on the left) produces a detectable change in the gill, and stimulating a single sensory neuron produces a large synaptic potential in the motor neuron (traces in the middle). Repeated stimulation of a single sensory neuron increases the frequency of firing in the motor neuron, leading to a visible reflex contraction of the gill (traces on the right). A single tactile stimulus to the skin normally activates 6 to 8 of the 24 sensory neurons, causing each to fire 1 to 2 action potentials. The repetitive firing of 10 action potentials in a single sensory neuron, designed to simulate the firing of the total population, in fact simulates reasonably well the reflex behavior.

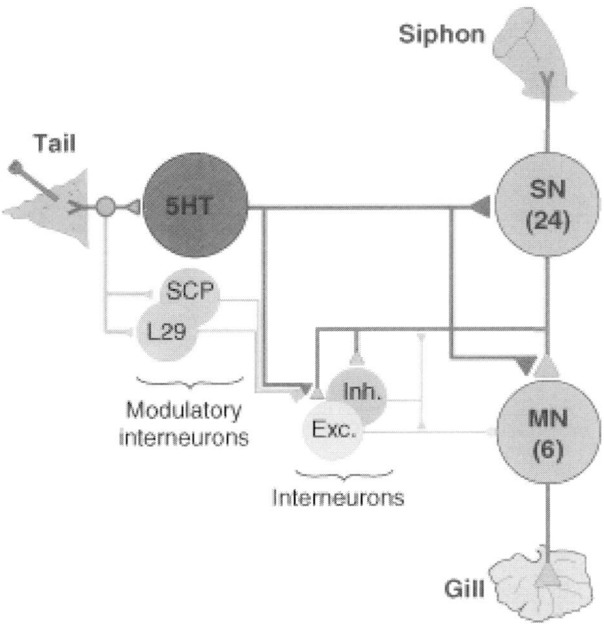

Figure 7. The neural circuit of the gill-withdrawal reflex. The siphon is innervated by 24 sensory neurons that connect directly with the six motor neurons. The sensory neurons also connect to populations of excitatory and inhibitory interneurons that in turn connect with the motor neurons. Stimulating the tail activates modulatory interneurons that act on the terminals of the sensory neurons as well as on those of the excitatory interneurons. There are three classes of modulatory neurons activated by tail stimuli: (1) neurons that release serotonin (5-HT), (2) neurons that release a peptide called the small cardioactive peptides (SCP), and (3) the L29 cells, which release an unidentified modulatory neurotransmitter. The serotonergic modulatory action is the most important. Blocking the action of these serotonergic cells blocks the effects of sensitizing stimuli.

Second, Alexander Forbes and Lorente de Nó proposed that memory is stored dynamically by a self-reexciting chain of neurons. This idea was later championed by Donald Hebb as a mechanism for short-term memory. Finally, Holger Hyden proposed that learning led to changes in the base composition of DNA or RNA. Even though there was much discussion about the merits of each of these ideas, there was no direct evidence to support any of them (reviewed in 17).

We were now in a position to address these alternative ideas by confronting directly the question of how learning cans occur in a circuit with fixed neuronal elements. Kupfermann, Castellucci, Carew, Hawkins, and I examined the neural circuit of the gill-withdrawal reflex while the animal underwent neither sensitization or habituation, a form of learning in which the animal learns to ignore an innocuous stimulus to siphon when given with monotonous repetition. (We later also extended these studies to an examination of classical conditioning (20).) Our studies provided clear evidence for Cajal's idea: learning results from changes in the strength of the synaptic connections between precisely interconnected cells (6,7). Thus, while the organism's developmental program assures that the connections between cells are invariant, it does not specify their precise strength. Rather, the strength and effectiveness of these preexisting chemical connections can be altered by experience. Seen in the perspective of these three forms of learning, synaptic plasticity emerged as a fundamental mechanism for information storage by the nervous system, a mechanism that is built into the very molecular architecture of chemical synapses (95).

We soon appreciated that the synaptic strength of a given chemical synapse could be modified in two ways: homosynaptically and heterosynaptically. *Homosynaptic* changes in synaptic strength occur in a synapse because of activity in the presynaptic and postsynaptic neurons of that very synapse. During habituation, homosynaptic changes occur in the monosynaptic connections between the sensory neurons and the motor neurons of the gill-withdrawal reflex. *Heterosynaptic* changes occur in a synapse where presynaptic and postsynaptic neurons are themselves not active but there is, instead, activity in one or more modulatory interneurons that act on the presynaptic neurons, on the postsynaptic neurons of the synapse, or on both, to modify the strength of their synaptic connections. During sensitization, heterosynaptic changes are induced in the monosynaptic connections between the sensory neurons and motor neurons of the gill-withdrawal reflex. Hawkins, Abrams, and I later found that these two types of regulation are recruited together in classical conditioning (20,21). Classical conditioning therefore illustrated that the elementary forms of homo- and heterosynaptic plasticity form an alphabet of basic mechanisms that can produce combinations of plasticity with novel properties.

In all three forms of learning we found particularly large changes in the synaptic strength of the direct connections between the sensory and motor neurons of the reflex. We therefore focused on this one component of the reflex and found several additional principles that have proven to be quite general. First, we found that the same synaptic connection can be modulated in opposite ways by different forms of learning. For example, habituation leads to a homosynaptic weakening of synaptic connections between the sensory neurons and their target cells, the motor neurons and interneurons, while sensitization leads to heterosynaptic strengthening of these same sets of connections (7,13,33). Second, learning not only leads to changes in synaptic strength, it can also affect the excitability of neurons. In the case of sensitization the excitability of the sensory (presynaptic) neurons is increased (25). Third, the synaptic changes persist, thereby contributing to memory storage (10,33,94). Indeed, the same synaptic connection can store both short- and long-term memory (7,23,33,43,94,98). At a given synapse synaptic plasticity can either be short- or long-lived depending on the number of spaced repetitions of the learning stimulus, and these parallel not only the behavioral changes of short-term memory but also those of long-term memory (7,43,93). Finally, long-term memory storage involves not only a change in synaptic strength, but also anatomical changes, changes in the number of synaptic connections (57,58,59).

The changes at the synapse between the sensory and the motor neuron are only a part of the changes in the neural circuit of the gill-withdrawal reflex. Important changes occur elsewhere in the circuit, but we have studied them less. We have focused on the monosynaptic portion of the circuit in order to probe in depth the molecular mechanisms that contribute to learning and memory.

INITIAL STEPS TOWARD A MOLECULAR CHARACTERIZATION OF SENSITIZATION

What are the molecular mechanisms whereby short-term memory is established, and how is it converted to long-term memory? Initially, we focused on short-term sensitization (Figure 8). In collaboration with James H. Schwartz, we found that the synaptic changes, like the short-term behavior, were expressed even when protein synthesis was inhibited. Since the short-term changes persisted for many minutes, it seemed unlikely that they involved a simple conformational change of one or more proteins. We therefore proposed in 1971, that short-term memory might require a series of sequential reactions similar to that mediated by cAMP-mediated signaling (26). Our attention was drawn to cAMP because Sutherland, Rall and Krebs had found that various neurotransmitters could increase cAMP concentration in the brain and in other tissues, and in the liver cAMP activated the cAMP-dependent protein kinase (26). In 1972, Schwartz, Howard Cedar, and I found that stimulation of the modulatory pathways recruited during heterosynaptic facilitation led to an increase in cAMP in the abdominal ganglion (27). Cedar and Schwartz next applied a number of neurotransmitter candidates and found that serotonin and dopamine could increase levels of cAMP (29). This made us wonder whether the population of modulatory interneurons, which produces the heterosynaptic facilitation that gives rise to sensitization, contains serotonergic or dopaminergic cells.

Later, Hawkins, Castellucci, David Glanzman, and I delineated the modulatory system activated by a sensitizing stimulus to the tail (19,34,35), and confirmed that it contains serotonergic interneurons. The system includes two other classes of heterosynaptic modulators, but the serotonergic neurons are most important for sensitization: blocking their actions alone blocked sensitization (34).

A DISTINCTION BETWEEN MEDIATING AND MODULATORY CIRCUITRY

In 1974, Castellucci and I next attempted to localize the change produced by habituation and sensitization to either the presynaptic or the postsynaptic component of the synapse. We therefore applied a quantal analysis to the synaptic connections between the sensory and motor cells and found that the short-term homosynaptic depression that accompanies habituation is presynaptic, involving a reduction in the amount of transmitter released from the presynaptic sensory neuron (30). This transmitter, Nicholas Dale later

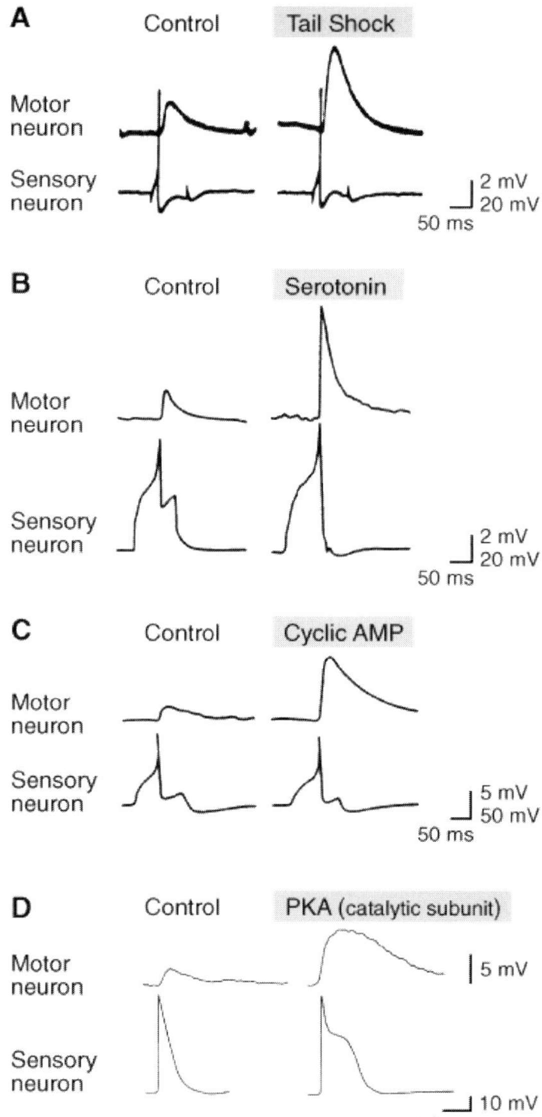

Figure 8. The strength of the synaptic connections between the sensory and the motor neurons is increased response to a single tail shock that produces short-term sensitization, in response to a single pulse of serotonin, and in response to direct injection of either cAMP or PKA into the presynaptic sensory neurons.

A. The increase in synaptic strength produced by a single sensitizing stimulus can be achieved by an electrical stimulus to the tail, to the head, or to the neural pathways from the tail or head.

B. A similar facilitation can also be reproduced by stimulating an individual modulatory cell such as one of the serotonergic cells or by the application of serotonin to the connection between the sensory neurons and motor neurons.

C. Facilitation can also be produced by injecting cyclic AMP into the sensory neurons.

D. Facilitation can also be produced by simply injecting the catalytic subunit of PKA into the sensory neurons. In this experiment the spikes of the sensory neuron have been artificially broadened by the application of tetraethylammonium (TEA) which blocks some of the cell's K^+ currents. This reveals dramatically the further broadening produced by the catalytic subunit of PKA through the reduction of the remaining K^+ currents. A similar broadening is produced by tail stimuli, application of serotonin, or injection of cAMP. (From 13,31,32,37.)

showed, is glutamate (51). Conversely, we found that the short-term heterosynaptic facilitation that accompanies sensitization results from an increased release of this transmitter from the sensory neuron (31). We next found that a single brief application of serotonin could simulate short-term sensitization (Figures 8A and B) – by directly enhancing the release of transmitter from the sensory neurons, much as does the sensitization-induced facilitation – whereas dopamine could not (32). Moreover, pharmacological inhibitors of serotonergic receptors blocked the effect of sensitizing stimuli, providing further evidence for the importance of serotonin as a modulator of synaptic strength (32).

Understanding the role of serotonergic and other modulatory neurons in sensitization allowed us to distinguish two different types of neural circuits important for behavior and learning: mediating and modulating (Figure 7). A mediating circuit controls the gill-withdrawal reflex. It consists of the sensory neurons, interneurons, and motor neurons of the reflex, and it can be modified homosynaptically by habituation. But other, more complex forms of learning, such as sensitization and classical conditioning, involve modulatory circuits, somewhat akin to the arousal or attentional circuitries in higher animals. In *Aplysia* such a modulating circuit consists of serotonergic and other modulatory interneurons that act upon the mediating circuitry to regulate the strength of its connections. For example, modulatory interneurons act on the sensory neurons, including on their presynaptic terminals, to enhance the amount of glutamate released.

SHORT-TERM MEMORY INVOLVES cAMP AND THE ACTIVATION OF PKA

The finding that serotonin acts on specific receptors in the presynaptic terminals of the sensory neuron to enhance transmitter release, coupled with the fact that serotonin is capable of bringing about an increase in cAMP in the abdominal ganglion, caused us to ask whether serotonin facilitates the release of transmitter from the sensory neurons by acting via cAMP. We first explored this idea in 1976, when Brunelli, Castellucci, and I injected cAMP directly into the presynaptic cells and found that it produced presynaptic facilitation (Figure 8C) (32,33). This provided the most direct and compelling evidence then available in the nervous system, that cAMP is involved in a specific physiological function and it gave us our first insight into the molecular mechanisms of short-term memory.

Marc Klein and I next asked: How does cAMP enhance transmitter release? We found that serotonin, or injected cAMP, leads to increased excitability and a broadening of the action potential by reducing a specific K^+ current, allowing greater Ca^{2+} influx into the presynaptic terminal with each action potential (24). The greater Ca^{2+} influx could contribute to the enhanced transmitter release. Following the lead of Paul Greengard, who had proposed that cAMP produces all of its action in the brain through the cAMP-dependent protein kinase (PKA), Klein and I suggested that cAMP may lead to phosphorylation of this K^+ channel by activating PKA (24). In collaborative experiments with Paul Greengard in 1980, Castellucci, Schwartz, and I tested this idea by injecting the active catalytic subunit of PKA and found that the catalytic subunit by itself produced broadening of the action potential and enhancement of glutamate release (Figure 8D) (37). Conversely, the specific peptide inhibitor of PKA (PKI) blocked the actions of serotonin. These findings provided direct evidence for the role of PKA in short-term presynaptic facilitation (38).

In 1981, Steven Siegelbaum arrived at Columbia, having just mastered single-channel recording during a visit to Erwin Neher and Bert Sakmann in Göttingen. In an elegant series of experiments, Siegelbaum and Joseph Camardo identified a novel K^+ channel, the S-type K^+ channel, and showed that it could be modulated by cAMP (39). Then Siegelbaum and Michael Schuster used an isolated inside-out patch of membrane from the sensory neuron to show that PKA could act directly to decrease the number of channels open in that patch (40). Later, Byrne showed that serotonin also modulates a delayed-rectifier K^+ channel whose contribution to spike broadening proved even more important than that of the S-type channel (reviewed in 41). The combined studies indicated that the S-type channel mediated the increase in excitability with a minor contribution to broadening, whereas the delayed-rectifier K^+ channel contributed little to excitability but had a major role in spike broadening. Finally, Hochner, Klein, and I – and independently Jack Byrne and his colleagues – showed that, in addition to spike broadening, serotonin also enhanced release by an as-yet-unspecified action on the release machinery. Thus, serotonin leads to an increase in presynaptic cAMP, which activates PKA and leads to synaptic strengthening through a combination of mechanisms (reviewed in 41).

LONG-TERM MEMORY INVOLVES THE NUCLEAR TRANSLOCATION OF PKA AND MAP KINASE AND THE ACTIVATION OF CREB-1

Even earlier, in 1973, Arnold Kriegstein, an M.D.-Ph.D. student, succeeded in culturing *Aplysia* larvae so that we could raise *Aplysia* in the laboratory (28), providing us with animals of any size or age at any time. This was a major advance and opened up the study of synaptic plasticity in dissociated cell culture, which required very young animals (42,43).

In the intact animal, one tail shock produces short-term sensitization, which does not depend on protein synthesis; four to five tail shocks produce sensitization that lasts several days, and this long-term process requires new protein synthesis. Both are reflected in alterations in strength in the connections between the sensory and motor neurons (Figure 9). By substituting puffs of serotonin for tail shocks, Sam Schacher, Pier Giorgio Montarolo, Philip Goelet, and I could model this behavioral protocol in a culture dish consisting of a single sensory cell making synaptic connections with a single motor cell (43). We were able to induce both short- and long-term facilitation in this culture and found, as we had with the intact animal, that the long-term

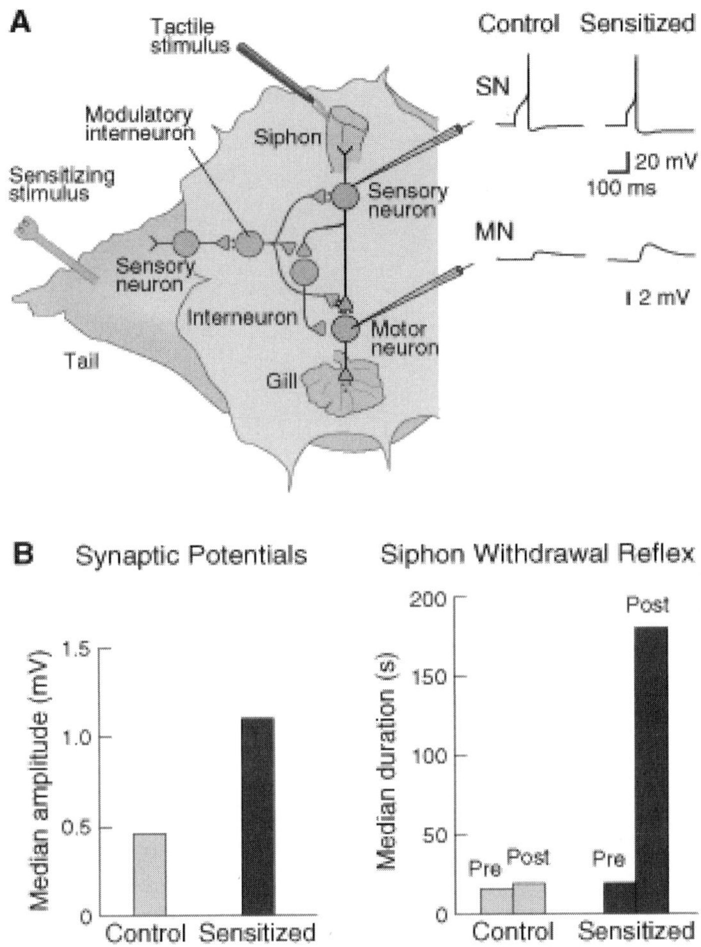

Figure 9. Long-term sensitization of the gill-withdrawal reflex of *Aplysia* involves long-term facilitation of the connections between sensory and motor neurons.

A. Experimental arrangement showing recordings from the sensory and motor neuron. The recordings on the right show representative records of action potentials in a siphon sensory neuron and synaptic potentials in a gill motor neuron in a control animal and in an experimental animal one day after the animal had received long-term sensitization training by stimulating its tail.

B. The median amplitude of the postsynaptic potentials (PSP) in an identified gill motor neuron is greater in sensitized animals than in control animals (histogram on the left). In the same experiment the amplitude of the withdrawal reflex is greatly enhanced in sensitized animals compared to control animals (histogram on the right). The sensitized animals were tested one day after training. The effect of sensitization on the neural circuit of the gill-withdrawal reflex is measured by the median duration of withdrawal of the siphon. (**Pre** = score before training; **post** = score after training.) (Adapted from 98.)

process differed from the short-term process in requiring the synthesis of new proteins (Figure 10). We had now trapped, in a culture consisting of two interconnected cells, the requirement for protein synthesis necessary to establish long-term memory.

We next used this cell culture to ask: What genes are activated to convert the short-term process to the long-term process, and what genes are essential

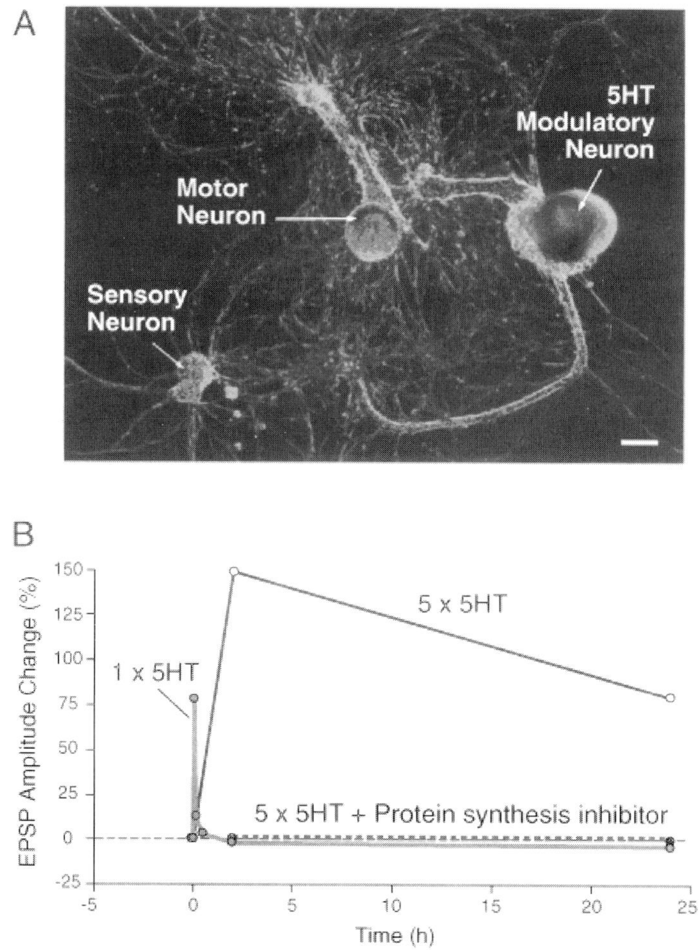

Figure 10. The monosynaptic pathway of the gill-withdrawal reflex reconstituted in cell culture.

A. In dissociated cell culture a single sensory neuron makes direct connections onto a motor neuron, and these connections can be modulated by a single serotonergic neuron. In fact, one does not need the serotonergic cell and can simply puff serotonin into the culture. Scale bar = 50 μm. (From 42.)

B. One puff of serotonin gives a transient facilitation that lasts minutes; this facilitation does not require new protein synthesis. Five puffs of serotonin produce a facilitation that lasts over a day; this does require the synthesis of new protein. (Modified from 53.)

for the maintenance of the long-term process? We found that five spaced puffs of serotonin (simulating five spaced shocks to the tail) activate PKA, which in turn recruits the mitogen-activated protein kinase (MAPK) and both translocate to the nucleus, where they activate a transcriptional cascade. The cascade begins with the transcription factor CREB-1, the cAMP response element binding protein-1), so called because it binds to a cAMP response element (CRE) in the promoters of target genes. CREB-1 leads to the activation of a set of immediate response genes, and in turn leads to the growth of new synaptic connections (44,45,46,47,49,50,56,59).

The first clue to the importance of CREB in long-term memory was provided in 1990 by Pramod Dash and Binyamin Hochner (45). They injected, into the nucleus of a sensory neuron in culture, oligonucleotides carrying the CRE DNA element, thereby titrating out CREB, and found that this selectively blocked long-term but not short-term facilitation (Figure 11). Later, Dusan Bartsch cloned *Aplysia* CREB-1a (ApCREB-1a) and showed that injection of the phosphorylated form of this one transcription factor by itself could ini-

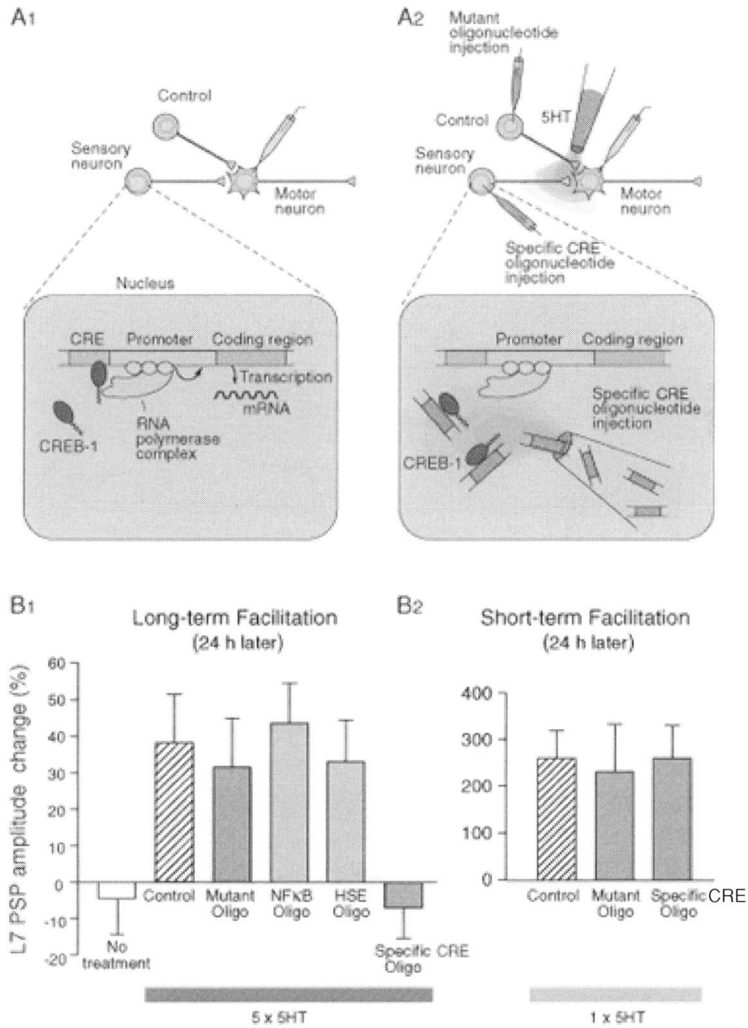

Figure 11. Long-term facilitation requires gene transcription mediated by the transcription factor, the cAMP response element-binding protein (CREB-1). To inhibit the action of CREB, an oligonucleotide encoding the cyclic AMP response element (CRE) was injected into the *Aplysia* sensory neuron in culture. The sensory neuron in these experiments was in synaptic contact with the motor neuron.

A. Experimental arrangement. **A1**. A single motor cell was cultured with two sensory cells. **A2**. One sensory cell was injected with the specific oligonucleotide encoding the CRE, designed to titrate out the CREB transcription factor, and the other was used as a control for injection of mutated or other oligonucleotide.

B. The specific oligonucleotide blocks the serotonin-induced long-term facilitation measured 24 hours after injection (**B1**) but does not affect short-term facilitation (**B2**). A mutated oligonucleotide encoding the CRE or an oligonucleotide encoding the heat shock enhancer and the enhancers of NFKβ do not affect long-term facilitation. (Modified from 45.)

tiate the long-term process. Looking downstream from ApCREB (54), Cristina Alberini and Bartsch also found two additional positive transcription regulators: ApC/EBP and activation factor (Ap/AF) (52,55).

INHIBITORY CONSTRAINTS ON MEMORY: MEMORY SUPPRESSOR GENES

In 1995, Bartsch found that positive regulators are only half the story – there are also inhibitory constraints on memory (53). Long-term synaptic facilitation requires not only activation of memory-enhancer genes, but also inactivation of memory-suppressor genes. One of these, the transcription factor ApCREB-2, can repress ApCREB-1a mediated transcription; relieving this repression lowers the threshold for the long-term process. Indeed, Bartsch found that when antiserum that blocks the action of ApCREB-2 is injected into the sensory neurons, a single application of serotonin, which normally produces facilitation lasting only minutes, now produces facilitation lasting days, accompanied by the growth of new synaptic connections (Figure 12).

Thus, during long-term memory storage, a tightly controlled cascade of gene activation is switched on, with memory-suppressor genes providing a threshold or set of checkpoints for memory storage, presumably to ensure that only salient features are learned. It is clearly an evolutionary advantage for an animal to learn and store in long-term memory only facts important for survival, rather than retaining everything. In addition memory-suppressor genes may well be regulated independently of the activators. Such independent action may be necessary to allow for the modulation of memory storage by emotional stimuli, as occurs in "flashbulb memories," memories of emotionally charged events that are recalled in detail, as if a complete picture had been instantly and powerfully etched in the brain.

Once the repressive action of CREB-2 is removed and CREB-1 is activated, a set of immediate response genes is induced. One of these, the ubiquitin C-terminal hydrolase, is responsible for the proteolytic cleavage of the regulatory subunit of PKA, thereby removing a second inhibitory constraint or memory suppressor gene. Since the regulatory subunit normally inhibits the kinase activity of the catalytic subunit, cleavage extends the activity of the kinase for several hours (Figure 12; refs. 60,61). In 1982, Schwartz and I first proposed this mechanism as the simplest molecular mechanism for long-term memory storage – the long-term process coopts and prolongs the action of the kinase involved in the short-term process but renders it independent of any further signaling by serotonin or PKA (115).

THE STABLE, SELF-MAINTAINED FORM OF LONG-TERM MEMORY IS REFLECTED IN THE GROWTH OF NEW SYNAPTIC CONNECTIONS

However, this persistent kinase is only required for 10 to 12 hours. Craig Bailey and Mary Chen first showed that what makes long-term memory enduring is the growth of new synaptic connections, a structural change that

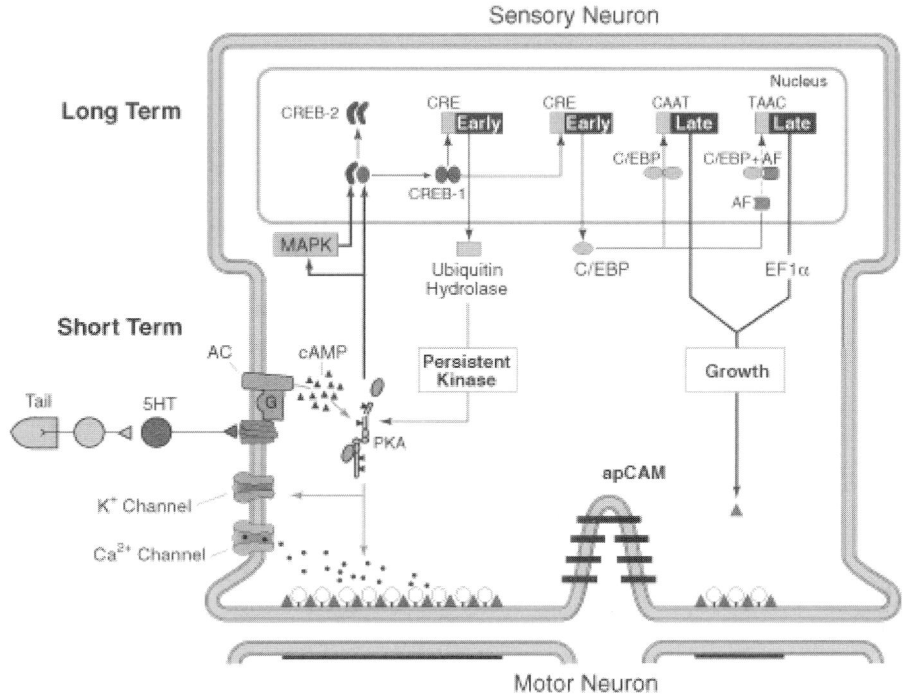

Figure 12. Long-term sensitization of the gill-withdrawal reflex of *Aplysia* leads to two major sets of changes in the sensory neurons of the reflex: (1) persistent activity of protein kinase A and (2) the growth of new synaptic connections.

In short-term sensitization (lasting minutes to hours) a single tail shock causes a transient release of serotonin that leads to covalent modification of preexisting proteins. The serotonin acts on a transmembrane serotonin receptor to activate the enzyme adenylyl cyclase (AC), which converts ATP to the second messenger cyclic AMP. In turn, cAMP recruits the cyclic AMP-dependent protein kinase A (PKA) which has both catalytic (the oval shaped structures) and regulatory subunits (the spindle-shaped structures). The regulatory subunits inhibit the catalytic subunits. When the level of cAMP rises, the cAMP binds to the regulatory subunits, causing them to undergo a conformational change so that they dissociate from and free the catalytic subunits, allowing the feed catalytic subunits to phosphorylate substrate proteins in the presynaptic terminals leading to enhanced transmitter availability and release. These substrate protein include K^+ channels and components of the exocytotic machinery for release. The duration of these modifications parallels the short-term memory. With repeated stimulation of long-term sensitization, the level of cAMP rises and persists for several minutes. This frees the catalytic subunits for a sufficient period of time to allow them to translocate to the nucleus, and in so doing they recruit the mitogen-activated protein kinase (MAPK).

In the nucleus, PKA and MAPK phosphorylate and activate the cyclic AMP response element-binding (CREB-1) protein and remove the repressive action of CREB-2, an inhibitor of CREB-1. Once the inhibitory constraint of CREB-2 is removed and CREB-1 is activated, CREB-1 in turn activates several immediate-response genes. One of these encodes a ubiguitin hydrolase, a component of a specific ubiquitin protease that leads to the regulated proteolysis of the regulatory subunit of PKA. This cleavage of the (inhibitory) regulatory subunit results in persistent activity of PKA, leading to persistent phosphorylation of the substrate proteins of PKA, including proteins involved in the short-term process. A second immediate-response gene activated by CREB-1 is C/EBP, which acts both by itself as a homodimer and as a heterodimer with activating factor (AF) to activate downstream genes including elongation factor 1α (EF1α) that lead to the growth of new synaptic connections.

parallels the duration of the behavioral memory (56,57,58,59). As the connections retract over time, the memory fades. A typical sensory neuron in the intact animal has about 1,200 synaptic varicosities. Following long-term sensitization, the number more than doubles to about 2,600. With time as the memory fades and the reflex response returns to control levels the number returns to about 1,500 synaptic connections (Figure 12).

An important clue to the molecular actions that give rise to these structural changes came from the identification by Mark Mayford of a third class of memory-suppressor genes, the *Aplysia* cell adhesion molecules (ApCAMs) (62,63), which belong to the immunoglobulin family of cell adhesion molecules, which includes mammalian neural cell adhesion molecule (NCAM) and *Drosophila* Fasciclin II. After exposure to five spaced puffs of serotonin, the concentration of ApCAM at the surface membrane of the presynaptic sensory neuron decreases as a result of the internalization of the transmembrane form of ApCAM, a process that requires ongoing protein synthesis. This clathrin-mediated endocytosis is blocked by mutations in the consensus sequence for MAP kinase phosphorylation in the cytoplasmic tail (48). Selective down-regulation of the transmembrane isoform of ApCAM causes defasciculation; it decreases the adhesive interaction of sensory cell neurites with each other, a prerequisite to process outgrowth and the formation of new synaptic connections. Over-expression of the transmembrane form of ApCAM blocks the long-term process.

THE cAMP, PKA, CREB SWITCH IS ALSO IMPORTANT FOR LEARNING AND MEMORY IN *DROSOPHILA*

The pioneering work of Seymour Benzer and the subsequent studies of his students, Chip Quinn, Tim Tully, Jerry Yin, and Ronald Davis, have led to the identification of a number of genes required for memory storage in *Drosophila* (reviewed in 97,102). Many of the genes identified in this way are the same as those implicated in plasticity in *Aplysia.* For example, *Drosophila* genes *dunce, rutabaga,* and *amnesiac* all encode components of the cAMP-PKA cascade. Other genes identified encode participants in cell adhesion molecules similar to ApCAM (97). Moreover, a protein synthesis-dependent phase of learning has been described by Tully and his colleagues, and Yin and Tully have shown that, as in *Aplysia,* CREB activators and repressors have a critical role in induction of long-term memory (97).

IS LONG-TERM FACILITATION SYNAPSE-SPECIFIC?

The finding of a transcriptional cascade illustrates that there is ready communiction between the synapse and the nucleus and why long-term memory requires new protein synthesis immediately after training. But this finding posed a new cell-biological problem. A single neuron makes hundreds of contacts on many different target cells (Figure 13). Short-term synaptic changes are synapse-specific. Since long-lasting synaptic changes require transcription

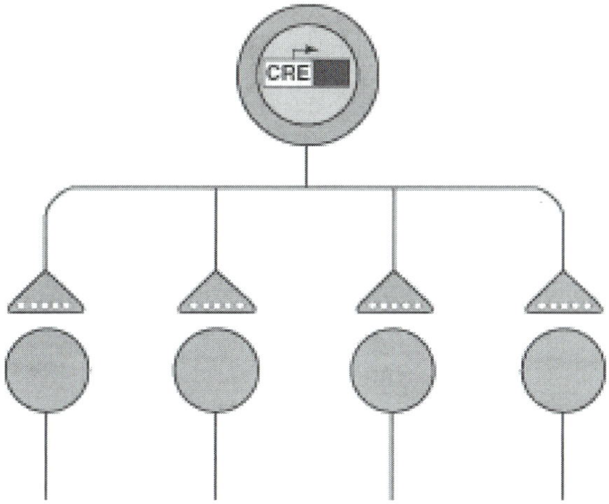

Figure 13. A single sensory neuron connects to many target cells. The requirement of a transcriptional mechanism for long-term memory raises the question: What is the unit of long-term information storage? Is it a single synapse, as with short-term facilitation, or the entire neuron? Is there a mechanism for restricting synaptic facilitation to some synaptic connections?

and thus the nucleus, an organelle shared by all synapses of a neuron, is long-term memory storage a cell-wide process? Or, are there cell-biological mechanisms that maintain the synapse specificity of long-term facilitation?

To examine these questions at the level of individual synapses, Kelsey Martin modified our culture system. She cultured one sensory cell with a bifurcating axon with two motor neurons, forming two widely separated syn-

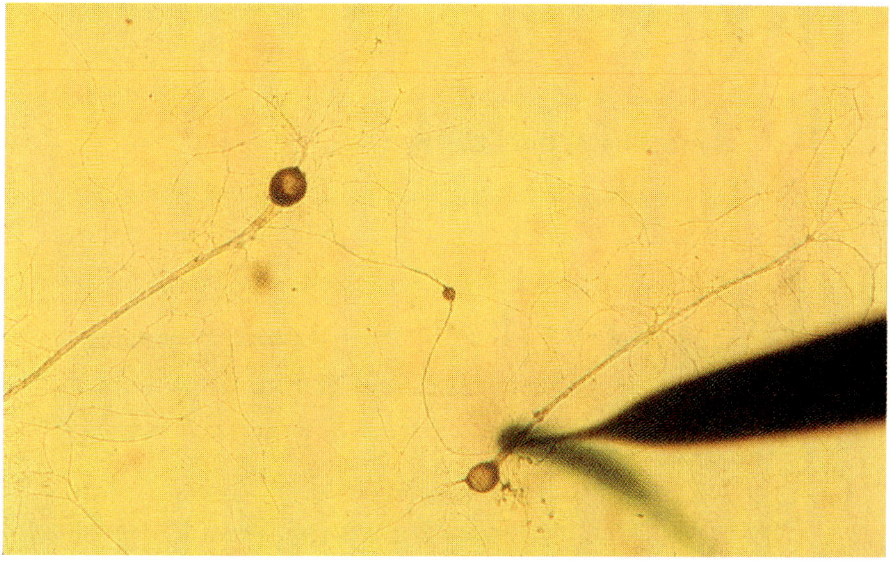

Figure 14. This photomicrograph shows a culture system developed to examine the action of two independent branches of a single in *Aplysia* sensory neuron (the small neuron in the middle) on two different motor neurons (large neurons). Serotonin can be selectively applied to one and not the other of the two branches. The flow of the serotonin can be monitored with the dye, fast green. (From 64.)

apses (Figure 14). In this culture system, a single puff of serotonin applied to one synapse produces transient facilitation at that synapse only, as expected (64,65). Five puffs of serotonin applied to one branch produces long-lasting facilitation (>72 hours) that is also restricted to the stimulated synapse (Figure 15). This long-lasting synapse-specific facilitation requires CREB and also leads to structural changes. Thus, despite recruitment of nuclear processes, long-term changes in synaptic function and structure are confined only to those synapses stimulated by serotonin.

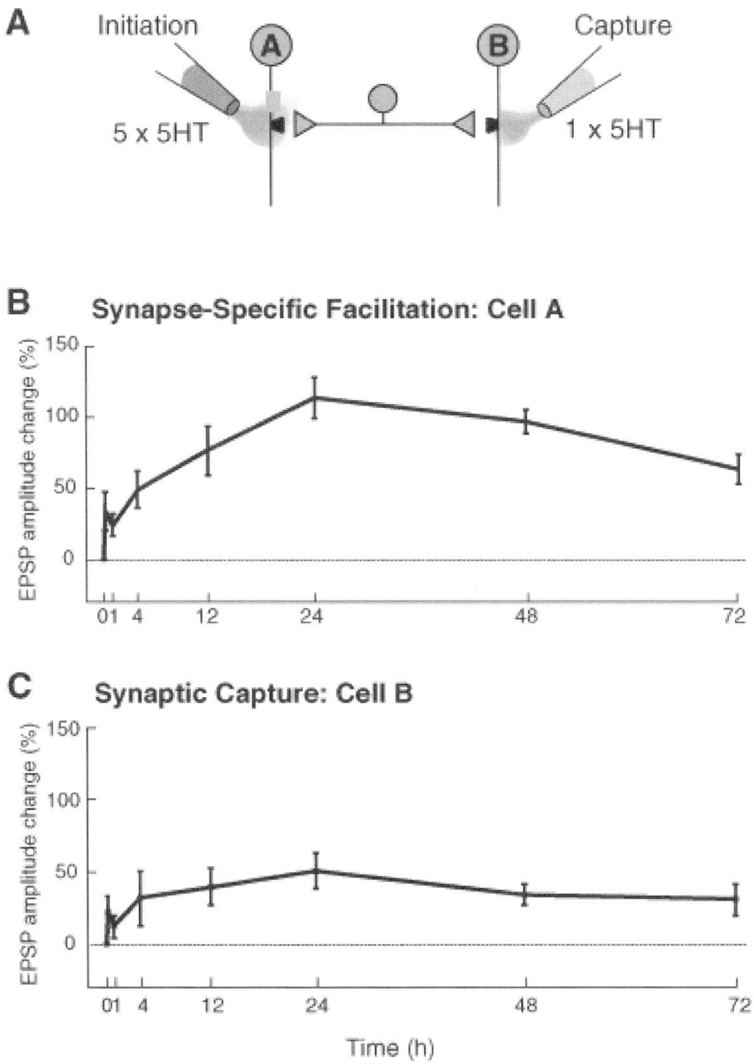

Figure 15. Long-term facilitation is synapse-specific and can be captured at another branch by the stimulus that initiates the short-term process.

A. Five puffs of serotonin are applied at the site of initiation (cell A) and produce a synapse-specific facilitation that is shown in part B. This synapse-specific facilitation is not evident at the synapse of cell B unless that synapse is itself primed with a single puff of serotonin. Under those circumstances the long-term facilitation at synapse A can be captured at synapse B in a reduced form, as illustrated in part C. (From 64.)

How does this come about? Martin, Andrea Casadio, Bailey, and I found that five puffs of serotonin send a signal to the nucleus to activate CREB-1, which then appears to send proteins to all terminals; however, only those terminals that have been marked by serotonin can use these proteins productively for synaptic growth. Indeed, we found that one puff of serotonin to the previously unstimulated synapse is sufficient to mark that synapse so that it can capture a reduced form of the long-term facilitation induced at the other site by five puffs of serotonin (Figure 15).

These results gave us a new and surprising insight into short-term facilitation (Figure 16). The stimulus that produces the short-term process has two functions. When acting alone, it provides a selective, synapse-specific enhancement of synaptic strength, which contributes to short-term memory, lasting minutes. When acting in conjunction with the activation of CREB initiated by a long-term process in either that synapse or in any other synapse on the same neuron, the stimulus that produces the short-term process serves to locally mark those synapses at which it occurs. The marked synapse can then productively utilize the proteins activated by CREB for synaptic growth to produce a persistent change in synaptic strength.

The existence of long range signaling between the synapse and the nucleus and the nucleus and the synapse introduces a new dimension into the integrative action of neurons that alters the rules that govern the short-term process whereby synapses are strengthened or weakened. Although long-term facilitation is *synapse-specific* and restricted, once transcription has been activated by the long-term process, the potential for plastic change of *all* the synapses of the neuron has, in fact, become altered. As a result, following the initiation of transcription in a neuron by long-term activation of one synapse, the action of any other synapse of that neuron is no longer determined simply by the history of that synapse but is also determined by the history of the transcriptional machinery in the nucleus. Thus, the logic of the long-term process is quite different from that of the short-term process.

THE NATURE OF THE LOCAL MARKING SIGNAL

How does one puff of serotonin mark a synapse for long-term change? We found that the synapse is marked for both long-term synaptic facilitation and for growth of new synaptic connections by covalent modifications of preexisting proteins mediated by PKA. However, for the structural change to persist, local protein synthesis is required (65). Oswald Steward's important work in the early 1980s had shown that dendrites contain ribosomes, and that specific mRNAs are transported to the dendrites and translated there locally (101). But the function of these locally translated mRNAs was unknown. Our experiments showed that one function was to stabilize the synapse-specific long-term functional and structural changes.

What proteins might be important locally for the stabilization? To answer this question we needed a preparation in which we could study the local mRNAs and proteins in isolation, without contamination of the cell body or

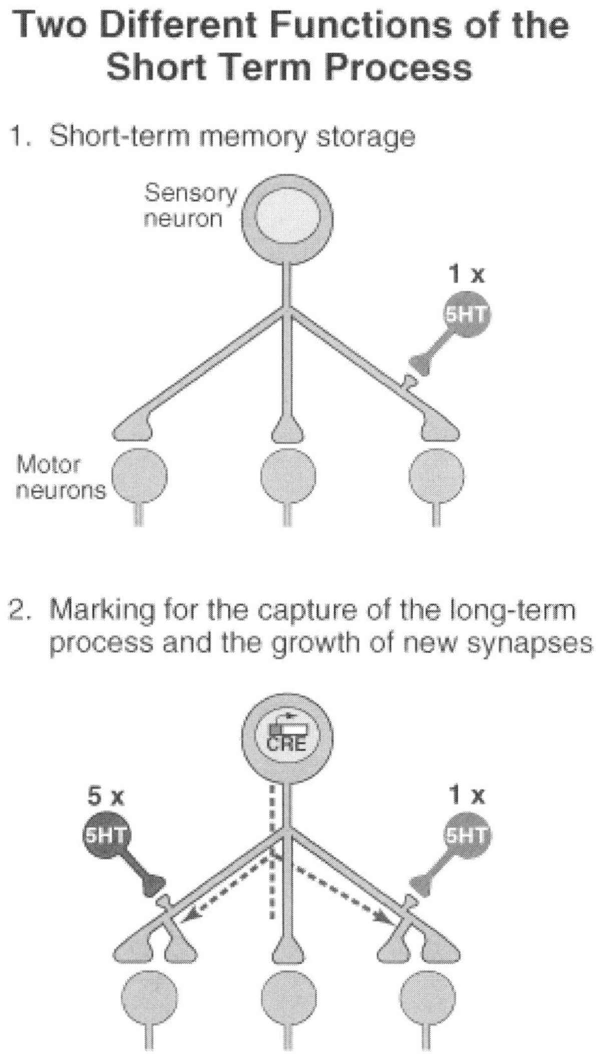

Figure 16. Two different functions of the short-term process by itself and in conjunction with the long-term process occurring at any other part within the neural tree. By itself, short-term facilitation participates in short-term memory storage. In conjunction with long-term facilitation at any other synapse, the short-term process marks the specific synapse to which it is applied and allows it to capture and use productively the proteins necessary to establish the long-term process and to grow new synaptic connections.

surrounding glial cells. Martin was able to culture several hundred sensory neurons in a dish and then cut off their cell bodies, allowing us to study the mRNAs in the processes and how they are regulated (64). Serotonin could stimulate translation in these isolated processes. Moreover, the transcripts in these processes contained regulatory sequences consistent with their using at least two mechanisms for regulating local translation. These two local mechanisms of translation serve different functional roles.

One group of transcripts, such as the elongation factor EF-1α, has an oligopyrimidine tract in the 5′ untranslated region. This tract also is found in a small set of transcripts that are preferentially translated in response to stimulation by growth factors or mitogens. The ability of these growth factors to recruit these transcripts is blocked selectively by the drug rapamycin (reviewed in 99,100,101). Indeed, we found that the stabilization of facilitation and of the growth of new connections, both at the site of capture and the site of initiation, are sensitive to inhibition by rapamycin (Figure 17 and reference 65). Other transcripts are not blocked by rapamycin. This rapamycin-insensitive component of translation is necessary for initiating the long-term process and is blocked by emetine, a general inhibitor of protein synthesis (Figure 17).

A FOURTH CONSEQUENCE OF NEUROTRANSMITTER SIGNALING: REGULATING LOCAL PROTEIN SYNTHESIS

These studies thus revealed a new, fourth type, of synaptic action mediated by neurotransmitter signaling. Three of these four have emerged, at least in part, from the study of learning and memory. First, in 1951, Katz and Fatt opened up the modern study of chemical transmission with their discovery of

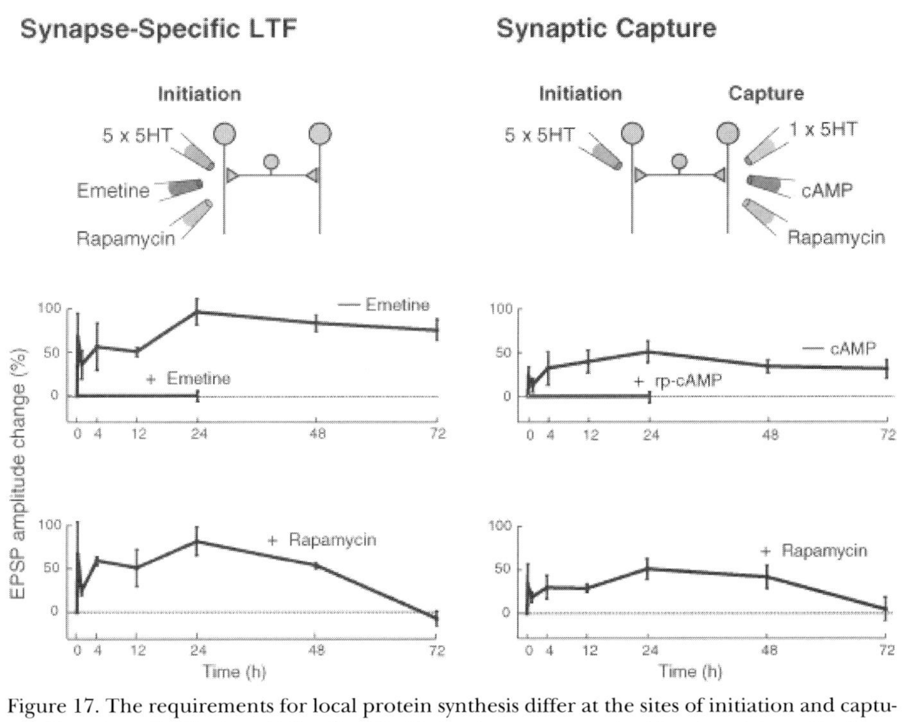

Figure 17. The requirements for local protein synthesis differ at the sites of initiation and capture of long-term facilitation. At the site of initiation two components of local protein synthesis are required: an emetine-sensitive component of local protein synthesis is required to set up the long-term process and a rapamycin-sensitive component of local protein synthesis to maintain facilitation after 48 hours. At the site of capture, inhibitors of cyclic AMP block the capture completely. By contrast, the rapamycin-synthesis-sensitive components of local protein synthesis is only required for maintenance of facilitation. (From 65.)

ionotropic receptors that regulate ion flux through transmitter-gated ion channels to produce fast synaptic actions, lasting milliseconds (110). Second, in the 1970s, metabotropic receptors were found to activate second-messenger pathways, such as the cAMP-PKA pathway, to produce slow synaptic activity lasting minutes (111). As we have seen in *Aplysia*, this slow synaptic action can regulate transmitter release, thereby contributing to short-term memory for sensitization. Third, an even more persistent synaptic action, lasting days, results from repeated action of a modulatory transmitter such as serotonin. With repeated applications of serotonin, second-messenger kinases translo-

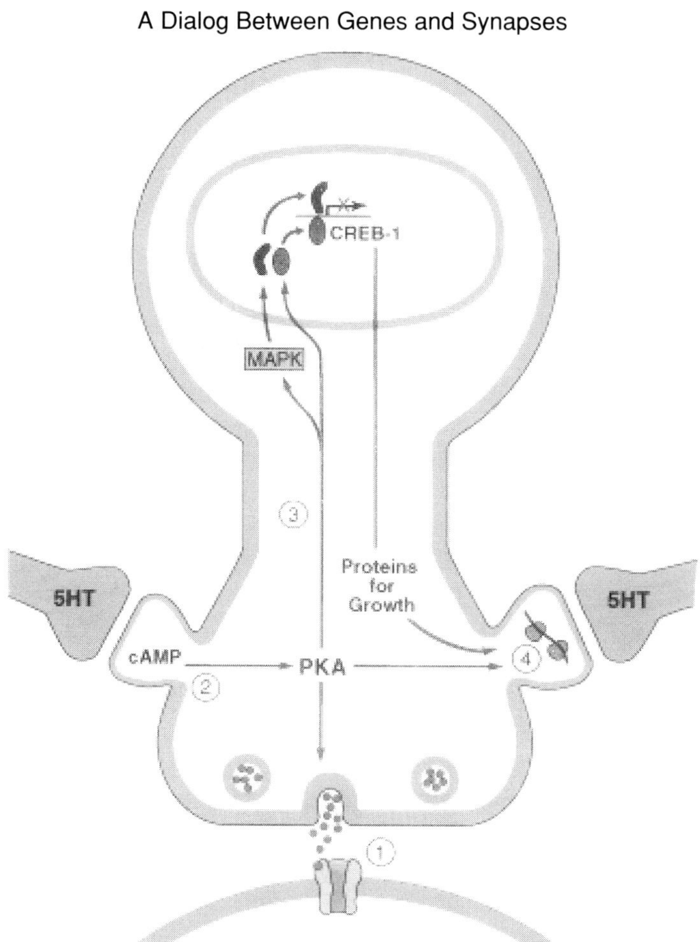

Figure 18. Four different consequences of the action of neurotransmitters. These events show that the synapses and nucleus readily interact. **1.** Transmitter activation of a ligand-gated ion channel leads to a rapid synaptic action lasting milliseconds to seconds. **2.** Transmitter activation of a seven transmembrane receptor and a second messenger kinase leads to a more enduring synaptic action lasting minutes. **3.** Repeated transmitter activation of a seven transmembrane receptor leads to the translocation of the kinase to the nucleus and to activation of transcription, producing a persistent synaptic action lasting days. **4.** Transmitter can also activate local protein synthesis to stabilize the synapse-specific long-term facilitation.

cate to the nucleus, where they activate a cascade of gene induction leading to the growth of new synaptic connections. This of course raises the problem of synapse specificity that we have considered above. Our experiments, in the bifurcated culture system, revealed a novel fourth action of neurotransmitters, the marking of the synapse and the regulation of local protein synthesis which contributes to the solution of that problem and to the establishment of synapse-specific long-term facilitation.

THERE ARE TWO MAJOR TYPES OF MEMORY: THE CASE OF EXPLICIT MEMORY

I have so far considered only the simplest cases of memory storage – those involving reflexes – a form called implicit or procedural memory. Implicit memory is memory for perceptual and motor skills and is expressed through performance, without conscious recall of past episodes. In contrast, the memories we hold near and dear are called explicit (or declarative) memories. These memories require conscious recall and are concerned with memories for people, places, objects, and events. Explicit memory involves a specialized anatomical system in the medial temporal lobe, and a structure deep to it, the hippocampus (Figure 19; reviewed in 66,94,95).

How are explicit memories stored? We had known from the work of Louis Flexner, Bernard Agranoff, Sam Barondes, and Larry Squire that explicit memory, like implicit memory, has a short-term phase that does not require protein synthesis and a long-term phase that does (66). Are these two com-

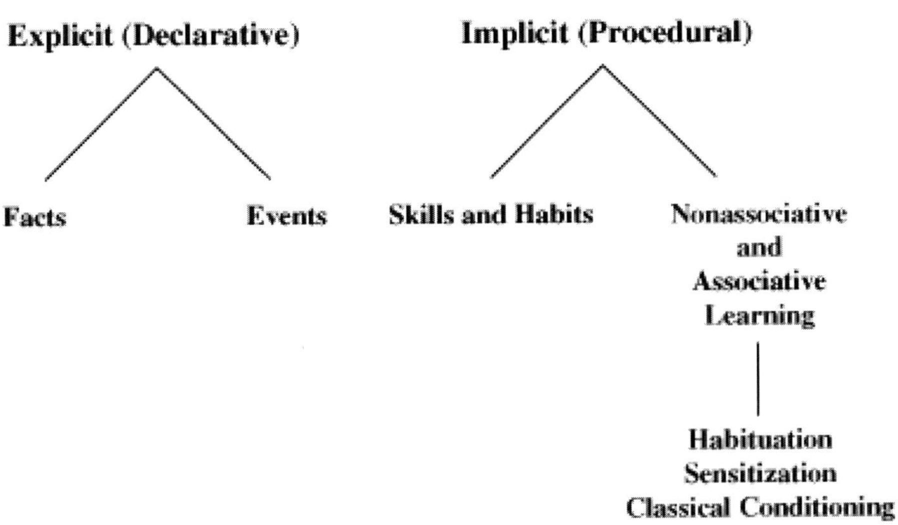

Figure 19. Two major forms of long-term memory: explicit (declarative) and implicit (procedural). Explicit or declarative memory is the memory for persons, places, and objects and requires conscious participation for recall. Implicit or procedural memory is the memory for perceptual and motor skills, which is evident in performance and does not involve conscious participation for recall.

ponents of memory storage also represented at the cellular level? What rules govern explicit memory storage?

A decade ago, when I reached my 60th birthday, I finally gathered up my courage and returned to the hippocampus. A major stimulus for me was the presence in my laboratory of Seth Grant who was eager to work with genetically modified mice (67). When Mario Capecchi and Oliver Smithies succeeded in achieving targeted gene ablation in ES stem cells, it became clear to me that mice would now offer a superb genetic system for relating individual genes to synaptic plasticity, on the one hand, and to complex explicit memory storage on the other. Although mice are relatively simple mammals, they have a medial temporal lobe system, including a hippocampus, that resembles that of humans, and they use their hippocampus much as we do to store memory of places and objects (Figure 20).

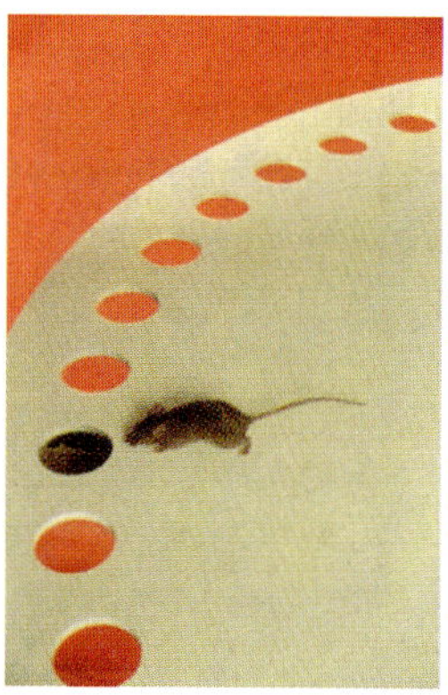

Figure 20. Barnes maze used in studying spatial memory in the mouse. The mouse is put in the center of a wooden platform. Because mice do not like open spaces they try to find some way to move out of it. The only way they can escape from the exposed space is to find the one hole out of 40 that leads to an escape hatch.

In our work with mice we have focused on memory for extrapersonal space as a model of explicit memory, because spatial memory is well represented in rodents and has been particularly well studied. Although we still do not know much about how information is transformed as it gets into and out of the hippocampus, it is well established that the hippocampus contains a cellular representation of extrapersonal space – a cognitive map of space – and that lesions of the hippocampus interfere with spatial tasks (67). Moreover, in 1972, Terje Lømo and Tim Bliss in Per Andersen's laboratory in Oslo, discovered that the perforant path, a major pathway within the hippocampus, exhibits activity-dependent plasticity, a change now called long-term potentiation (LTP) (Figure 21). In the CA1 region of the hippocampus where LTP has been studied most extensively, it had been found to be induced postsynaptically by activation of an NMDA receptor to glutamate. In the late 1980s Richard Morris found that blocking the NMDA receptor pharmacologically not only interfered with LTP but also blocked memory storage (69,70).

We focused primarily on the Schaffer collateral, another pathway in the hippocampus, not only because it had been extensively studied but also because work with a patient R.B. by Larry Squire and his colleagues had shown

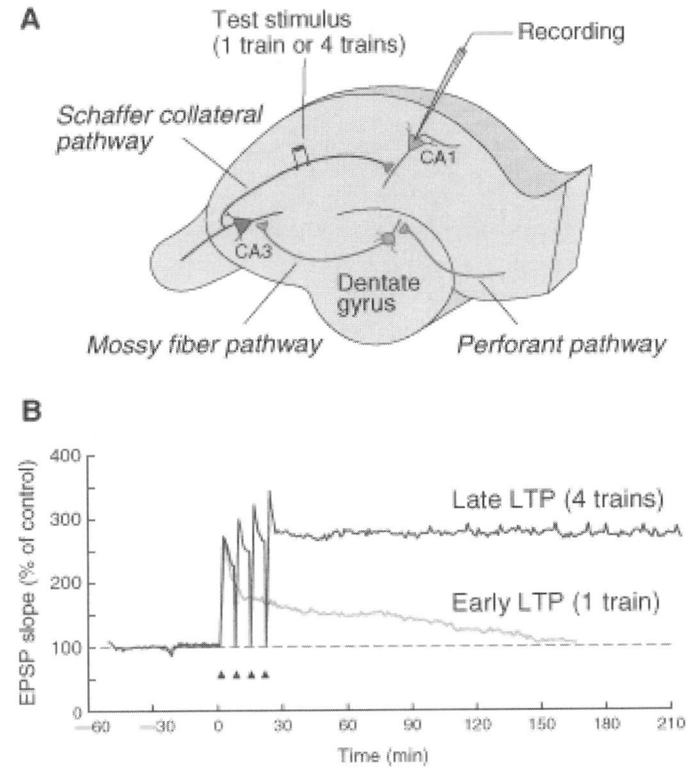

Figure 21. Long-term potentiation (LTP) in the hippocampus.

A. Three major pathways denote the direction of the impulse flow, each of which gives rise to LTP. The *perforant pathway* from the subiculum forms excitatory connections with the granule cells of the dentate gyrus. The *mossy fiber pathway*, formed by the axons of the granule cells of the dentate gyrus, connects the granule cells with the pyramidal cells in area CA3 of the hippocampus. The *Schaffer collateral pathway* connects the pyramidal cells of the CA3 region with the pyramidal cells in the CA1 region of the hippocampus.

B. A schematic time course depiciting the early and late phases of LTP in the Schaffer collateral pathway. A single train of stimuli for one second at 100 Hz elicits an early LTP, and four trains at 10-minute intervals elicit the late phase of LTP. The early LTP lasts about two hours and the late LTP lasts more than 24 hours.

that a lesion restricted to the CA1 region was sufficient to produce a significant loss of explicit memory storage (68).

Early work on LTP in this pathway in hippocampal slices by others had focused on the response to one or two trains of electrical stimuli. But in our work on *Aplysia* we had found long-term memory and the synaptic changes that accompany it emerge most effectively with repeated stimuli. So Uwe Frey, Yan-You Huang, Peter Nguyen, and I examined whether LTP changed with repeated stimulation (71,72,73) and found that in each of the three major hippocampal pathways LTP has phases, much like facilitation in *Aplysia.*

The early phase of LTP, produced by a single train of stimuli, lasts only 1–3 hours and does not require new protein synthesis (reviewed in 73); it involves covalent modifications of preexisting proteins that lead to the strengthening of preexisting connections, similar in principle to short-term facilitation in *Aplysia.* But, as has been well documented by Roger Nicoll and his colleagues, the molecular details of the early phase of LTP in the Schaffer collateral pathway differ from those of short-term facilitation in *Aplysia.* The influx of Ca^{2+} through the NMDA receptor channel leads to the activation of a Ca^{2+} calmodulin-dependent protein kinase and the phosphorylation of pre-existing AMPA receptors, and insertion into the postsynaptic membrane of new AMPA receptors to glutamate (75).

By contrast, repeated trains of electrical stimuli produce a late phase of LTP, which has properties quite different from early LTP and quite similar to long-term facilitation in *Aplysia* (Figure 21). The late phase of LTP persists for at least a day and requires both translation and transcription. Like long-term facilitation in *Aplysia* (71,73), the late phase of LTP is strongly modulated by a heterosynaptic input, in this case mediated by dopamine (79). At the Schaffer collateral synapse the late phase of LTP, like long-term storage of implicit memory, requires PKA, MAPK, and CREB, and appears to lead to the growth of new synaptic connections (Figure 22; see refs. 71,72,73,74,76,77, 78,80,81,82).

Together with Vadim Bolshakov and Hava Golan, Siegelbaum and I examined the late phase of LTP on an elementary level. We stimulated a single presynaptic CA3 neuron and recorded from a single CA1 postsynaptic cell, and found that the late phase requires a coordinated regulation of both pre-

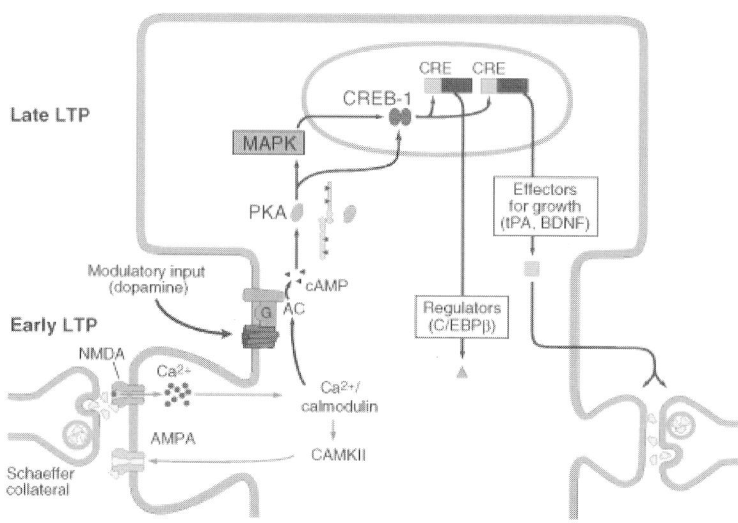

Figure 22. A schematic model of the molecular events contributing to the early and late phase of LTP in the Schaffer collateral pathway. A single train of action potentials initiates early LTP by activating NMDA receptors, leading to Ca^{2+} influx into the postsynaptic cell, and to the recruitment of a set of second messenger kinases including the calcium calmodulin-dependent protein kinase which phosphorylates existing AMPA receptors and also leads to the insertion of new receptors. There is also thought to be a retrograde signal that diffuses from the postsynaptic cell to the presynaptic terminals to enhance transmitter release. With four repeated trains, the Ca^{2+} influx also rectruits adenylyl cyclase, which activates the cAMP-dependent protein kinase (PKA) and MAPK. These kinases translocate to the nucleus where they phosphorylate CREB and other transcription factors. CREB in turn activates target genes that are thougt to lead to structural changes. Mutations in mice that block PKA or CREB reduce the late phase of LTP and compromise long-term memory for space. The adenylyl cyclase can also be modulated by the modulatory transmitter dopamine acting through D1/D5 receptors.

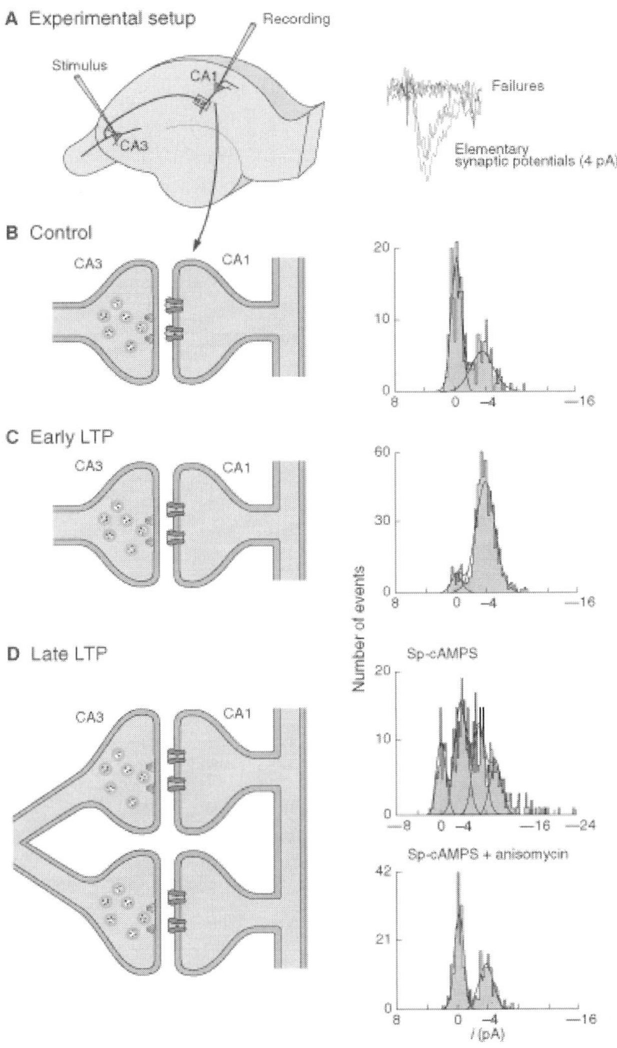

Figure 23. A distinction between the early and late phases of long-term potentiation in the hippocampus is evident at the level of the elementary connection between a single CA3 cell and a single CA1 cell. (Modified from 74.)

A. A single CA3 cell can be stimulated to produce any elementary synaptic potential in a CA1 cell that is comparable in size (abour 4 pA) to a spontaneously released miniatyre synaptic potential.

B. Stimulating a single CA3 cell repeatedly at low frequency produces either an elementary synaptic potential or no response (a failure). The distribution of the amplitudes of many responses can be approximated by two Gaussian curves, one centered on zero (the failures) and the other centered on 4 pA (the successful responses). These histograms are consistent with the synapse between a single CA3 cell and a CA1 cell having a single active zone from which it releases, or fails to release, a single vesicle in an all-or-none manner (successes or failures). In control conditions there are many failures, i.e., the synapse has a low probability of releasing vesicles.

C. In the early phase of LTP, the probability of successful releases rises significantly. But there are still two Gaussian curves in the distribution of responses, consistent with the view that the single release site still releases only a single vesicle but now with a high probability of release.

D. In the late phase of LTP induced by cAMP, the distribution of responses no longer fits two Gaussian curves but instead requires three of four Gaussian curves, consistent with the idea that new presynaptic active zones and postsynaptic receptors have grown. These effects are blocked by anisomycin, an inhibitor of protein synthesis.

and postsynaptic components of the synapse. In the resting state, an action potential in a presynaptic CA3 neuron released either zero or one vesicle onto the CA1 neuron. By contrast, during the late phase of LTP, a single action potential in a CA3 neuron released several vesicles of transmitter onto the CA1 neuron (Figure 23). This increase in the number of vesicles released would seem to entail a coordinated growth of new presynaptic release sites as well as the insertion of new clusters of postsynaptic receptors (105,106). Consistent with this idea, and with the properties of the late phase of LTP, these long-term changes require new protein synthesis.

THE LATE PHASE OF LTP REQUIRES PKA AND CONTRIBUTES TO LONG-TERM EXPLICIT MEMORY FOR EXTRAPERSONAL SPACE

To explore further the specific role of PKA in late LTP and to determine its role in memory, Ted Abel, Mark Barad, Rusiko Bourtchouladze, Nguyen, and I generated transgenic mice that express R(AB), a mutant form of the regulatory subunit of PKA that inhibits enzyme activity (84). To restrict expression of R(AB) to the postnatal hippocampus and other forebrain regions, we used the promoter from the Ca^{2+}/calmodulin protein kinase IIα (CaMKIIα) gene which Mayford had isolated and characterized. In these R(AB) transgenic mice the reduction in hippocampal PKA activity was paralleled by a significant decrease in late LTP, while basal synaptic transmission and early LTP remained unchanged (Figure 24). Most interestingly, this deficit in the late phase of LTP was paralleled by behavioral deficits in hippocampus-dependent long-term memory for context, for extrapersonal space, whereas learning,

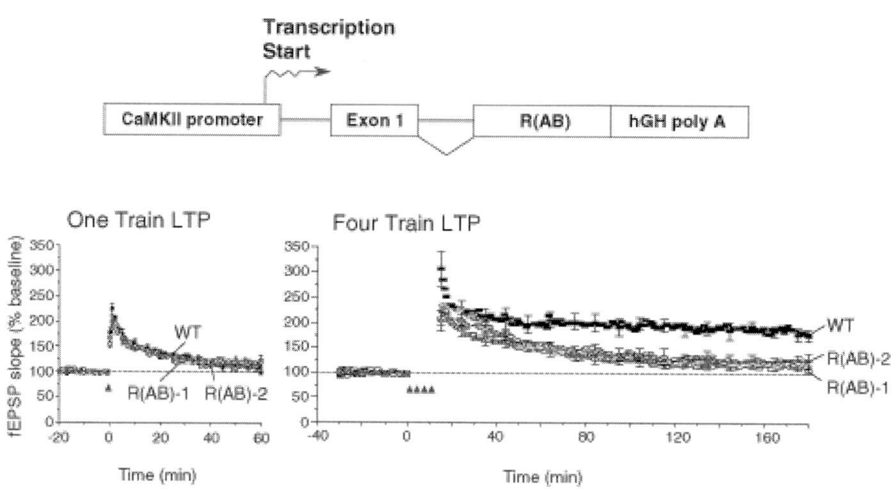

Figure 24. A comparison of LTP in wild-type mice and mice in which PKA has been compromised by the expression in the hippocampus of a transgene, a dominant negative inhibitor of PKA. This inhibitor R(AB) is a mutated form of the regulatory subunit of PKA that inhibits the catalytic subunit but does not recognize cAMP. Two lines of mutant mice (R(AB)-1 and R(AB)-2) are compared to the wild-type mice (**WT**). The mutant mice have perfectly good early LTP comparable to that of wild-type mice. Mutations in mice that block PKA reduce or eliminate late LTP. (From 84.)

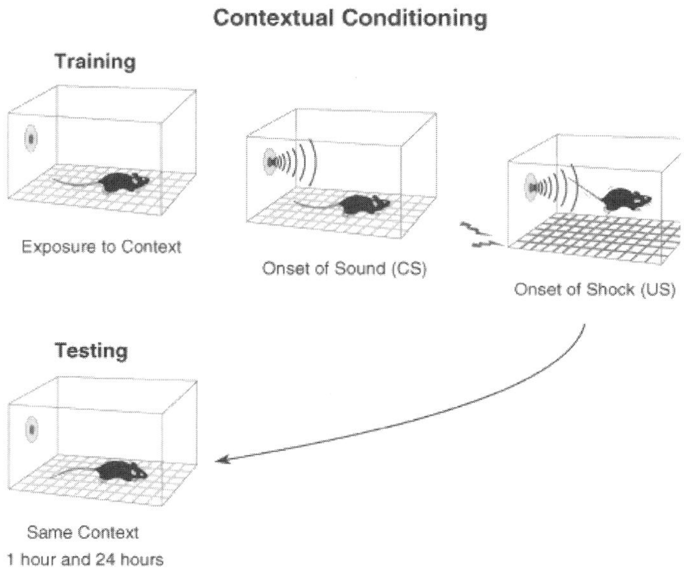

Figure 25. The protocol for context conditioning. The conditioning consists of exposure to the context followed by a tone and then a shock. The animals are then tested 1 hour and 24 hours after training. (From 84.)

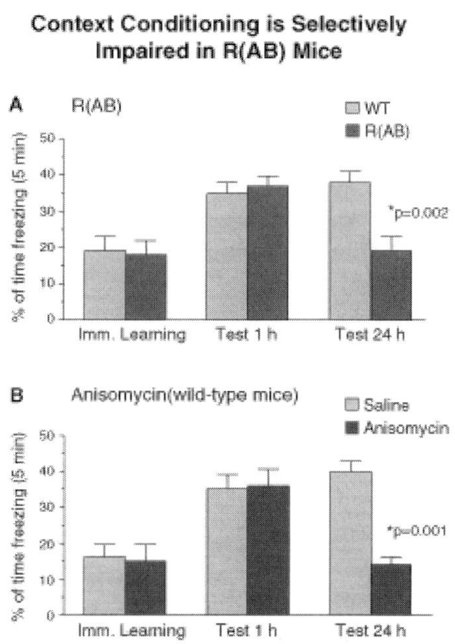

Figure 26. Mutant mice that express the R(AB) gene in the hippocampus, blocking the action of PKA, have a selective defect for long-term contextual memory (**A**). Mice that express R(AB) were conditioned to freeze to the context in the form of a box illustrated in Figure 25. The mice first walked around for a brief period of time and became familiar with the context in which they walked. They then heard a sound and received a shock delivered through the electrified grid in the floor. As a result the animals learned to associate the context of the space with shock and to freeze when placed in the box at a future time. These mice learned well and had good short-term memory at one hour for freezing to context. However, they no longer froze to context at 24 hours after conditioning, indicating a defect in a form of long-term explicit (declarative) memory that requires the hippocampus. Wild-type mice exposed to anisomycin, an inhibitor of protein synthesis, during training show a similar defect for long-term memory when tested 24 hours after conditioning (**B**), indicating that the same phenotype is obtained with R(AB) mutation as with inhibition of protein synthesis. (From 84.)

and short-term memory, are unimpaired (Figures 25 and 26). Thus, in the storage of explicit memory of extrapersonal space in the mammalian hippocampus, PKA plays a critical role in the transformation of short-term memory into long-term memory, much as it does in the storage of implicit memory in *Aplysia* and *Drosophila.*

TOWARD A MOLECULAR BIOLOGY OF COGNITION: PKA IS REQUIRED FOR AN INTERNAL REPRESENTATION OF EXTRAPERSONAL SPACE

Using the R(AB) mice we could now ask: What are the specific functions in spatial memory of PKA and the late phase of LTP? Why do animals with compromised PKA signaling have difficulty with space (84)? In addressing these questions we were influenced by the classic studies of John O'Keefe and John Dostrovsky, who in 1971 discovered that the pyramidal cells of the hippocampus – the cells one examines artificially using electrical stimuli to the Schaffer collateral pathway while studying LTP – are "place cells;" they actually encode extrapersonal space in real life (85). A given pyramidal cell will fire only when the head of the mouse is in a certain part of an enclosed space: the cell's place field. Thus, when an animal walks within an enclosed space, a particular subset of pyramidal cells in the hippocampus becomes active (103). When the animal is in different regions, different sets of pyramidal cells become active (Figure 27).

These findings led O'Keefe and Nadel to develop the idea that the pyramidal cells of the hippocampus form an internal neural representation, or "cognitive map," of the space surrounding the animal (85). This holistic neural representation was thought to permit the animal to solve spatial problems efficiently. When placed in a new environment, an animal develops an internal representation of the space (the coordinated firing of a population of place cells) within minutes, and once this representation is formed it is normally stable for days. The same cell will have the same firing field each time the animal is reintroduced to that environment. When now placed in a second environment, a new map is formed – again in minutes – in part from some of the cells that made up the map of the first environment and in part from pyramidal cells that had been silent previously (reviewed in 85).

It struck me that the formation of a new map resembled a learning process. The map develops with time as the animal walks around for several minutes to familiarize itself with the space, and once learned, the map of space is retained for days and weeks. I therefore wondered whether protein synthesis and the molecular pathways underlying the late phase of LTP were important for the long-term stabilization of this map.

Even though both LTP and place cells had been discovered in the early 1970s, there had been no earlier attempt to link neural plasticity to a place cell map pharmacologically or genetically by exploring the role of molecules important for LTP in the formation and stabilization of place fields (86,87,88,89,103). In one of our initial experiments, Cliff Kentros, Robert Muller, Hawkins, and I simply blocked LTP pharmacologically using an

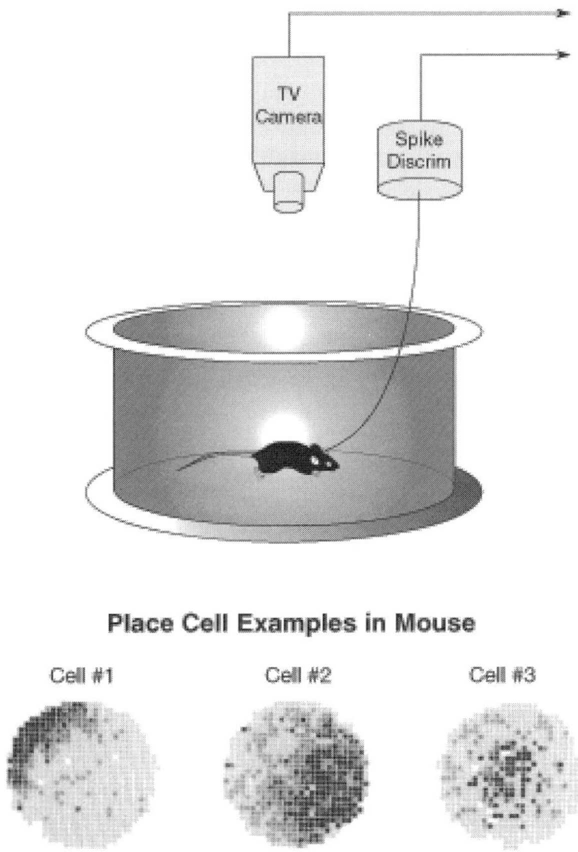

Figure 27. Spatial memory can be studied in the mouse by recording from individual pyramidal cells in the hippocampus. The firing pattern of these "place" cells create an internal representation of the animal's location within its surrounding. A mouse is attached to a recording cable and placed inside a cylinder (49 cm in diameter by 34 cm high). The other end of the cable goes to a 235-channel commutator attached to a computer-based spike-discrimination system. The cable is also used to supply power to a light-emitting diode mounted on the headstage the mouse carries. The entire apparatus is viewed with an overhead TV camera whose output goes to a tracking device that detects the position of the mouse. The output of the tracker is sent to the same computer used to detect spikes, so that parallel time series of positions and spikes are recorded. The occurrence of spikes as a function of position is extracted from the basic data and is used to form two-dimensional firing-rate patterns that can be numerically analyzed or visualized as color-coded firing-rate maps. Dark areas indicate regions in the circular enclosure in which the cell fires at high rates. (Based on Kandel and Squire, 1998.)

NMDA receptor antagonist (87). We found that when placed in a new environment, animals with blocked NMDA receptors formed a good spatial map that was still stable one hour later. However, by 24 hours, the map had become completely unstable. Most pyramidal cells no longer retained the representation of the field they had initially. This suggested that activation of NMDA receptors – perhaps a step in modifying the strength of the synapse – is required for the long-term stabilization of a place cell map, a result consistent with the role for the late phase of LTP in the stabilization of a place cell map.

We next asked: Does a selective deficit that affects only the late phase of

LTP and leaves the early phase completely intact, cause a selective abnormality in the long-term stability of place cells? Since only the late phase of LTP requires PKA, Alex Rotenberg, Muller, Abel, Hawkins, and I returned to the R(AB) transgenic mice with diminished PKA activity and a diminished form of late LTP (88). If reduced activity of PKA affected the stability of place cells, R(AB) mice should be able to form a stable map of space in a novel environment, as in normal animals, and to be stable for at least 1 hour. However, the cell field should be unstable if recorded 24 hours later. This is precisely what we found (Figure 28).

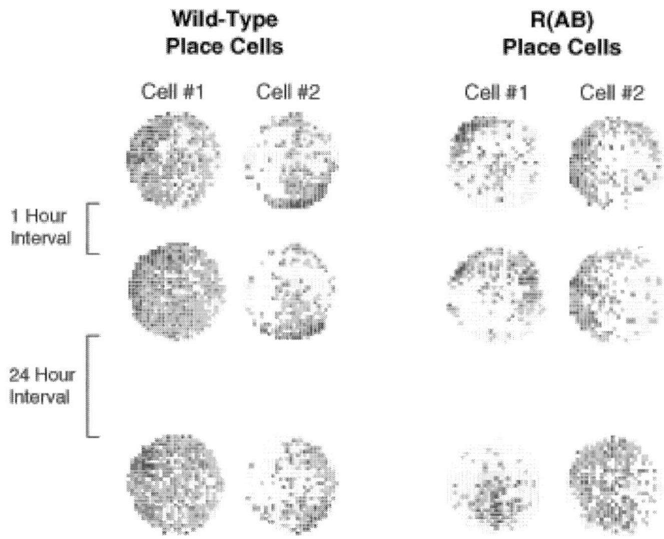

Figure 28. The long-term stability of the place field of a pyramidal cell in a wild-type mouse and in a mutant mouse. Two simultaneous recordings from pyramidal cells in the hippocampus are illustrated in a wild-type mouse and a mutant mouse. In the wild-type mouse the spatial fields form and are stable when the mouse is taken out of the test environment and put back one hour later. When the wild-type mouse is taken out and put back 24 hours later, the fields of the pyramidal cells are still quite stable. By contrast, in the R(AB) mice in which PKA has been compromised, the pyramidal cells form good fields and these are stable at one hour, but the fields are not stably maintained and are altered at 24 hours.

The fact that long-term instability in the spatial map and the deficit in long-term memory paralleled the deficit in the late phase of LTP suggested that PKA-mediated gene activation and the synthesis of new protein might be essential for the stabilization of the spatial map. Naveen Agnihotri, Kentros, Hawkins, and I tested this idea, and found that inhibiting protein synthesis indeed destabilized the place fields in the long-term much as does inhibiting PKA (Figure 29) (88,89).

In the course of this work, we found, remarkably, that, as is the case with explicit memories in humans, a key feature in the stabilization of PKA and protein synthesis-dependent phase of memory is attention (90). When a mouse does not attend to the space it walks through, the map forms but is unstable after three to six hours. When the mouse is forced to attend to the space, however, the map invariably is stable for days. How does this attentional

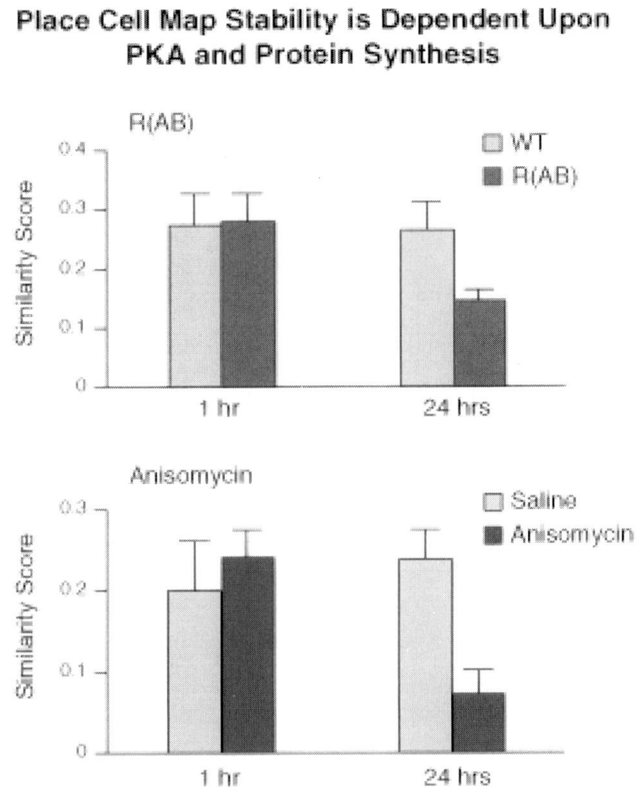

Figure 29. The group data for mice expressing R (AB) in which PKA expression is compromised, and those in which protein synthesis is inhibited. The top graph shows the group data for R(AB) and wild-type mice and the bottom shows that these phenotypes are also obtained by inhibiting protein synthesis with anisomycin. (From 88, 89.)

mechanism work? Our recent work suggests that one component of the attentional system necessary for stability of the spatial maps is mediated by the dopaminergic modulatory input acting through a D1/D5 receptor that activates adenylyl cyclase, cAMP, and PKA. The actions of dopamine and other modulatory systems might, among other things, trigger the protein synthesis-dependent steps that stabilize the map.

THERE ARE ALSO INHIBITORY CONSTRAINTS ON LTP AND EXPLICIT MEMORY STORAGE

Recently our laboratory (92) and Emmanuel Landau (116) and his colleagues have found that the threshold for hippocampal synaptic plasticity and memory storage is determined by the balance between protein phosphorylation and dephosphorylation. This balance importantly involves PKA and the Ca^{2+}-sensitive phosphatase, calcineurin, the initial step in a phosphatase cascade (91,92). To determine whether endogenous calcineurin acts as a constraint on this balance, we inhibited calcineurin and examined the effects on synaptic plasticity and memory storage. Using the CaM kinase promoter and the doxycycline-dependent rtTA system that Mayford had first successfully applied to the brain, Isabelle Mansuy, Gael Malleret, Danny Winder, Tim Bliss, and I found that a transient reduction of calcineurin activity resulted in facilitation of LTP both *in vitro* and *in vivo* (Figure 30) (92). This facilitation persisted over several days in the intact animal, and was accompanied by enhanced learning and strengthening of short- and long-term memory on several spatial and non-spatial tasks requiring the hippocampus. The LTP and memory improvements were reversed fully by suppressing expression of the transgene through withdrawal of doxycycline. These results, together with previous findings by Winder and Mansuy showing that overexpression of calcineurin impairs PKA-dependent components of LTP and memory (83,91), demonstrate that endogenous calcineurin can act as a negative regulator of synaptic plasticity, learning, and memory (Figure 30).

AN OVERALL VIEW

The problem of memory storage is conveniently divided into two components: (1) the *storage component* of memory, the molecular mechanism whereby information is stored, and (2) the *systems component* of memory, the mechanisms whereby the storage sites at each point in the explicit or implicit neural system that mediates memory interact to encode, store, and recall memory. I here have addressed primarily the storage component of memory. Our studies of this component have led to two general conclusions.

First, the research I have reviewed suggests that aspects of the storage mechanisms – the cellular and molecular strategies used in *Aplysia* for storing

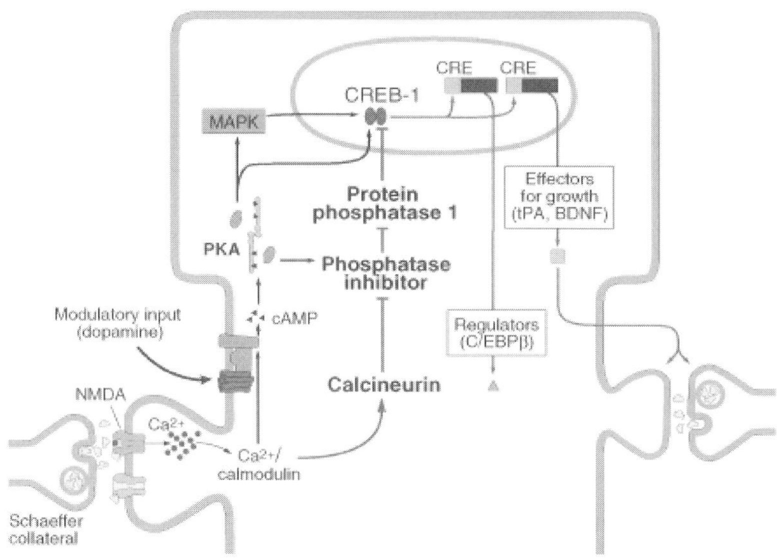

Figure 30. Long-term potentiation requires regulation not only of kinases but also of phosphatases. The phosphatase cascade initiated by calcineurin shuts off a phosphatase inhibitor and thereby disinhibits the protein phosphatase, which can now inhibit the kinase cascade. (Based on 92.)

short- and long-term memory – are conserved from mollusk to mammals, and that the same molecular strategies are employed in both implicit and explicit memory storage. With both implicit and explicit memory there are stages in memory that are encoded as changes in synaptic strength and that correlate with the behavioral phases of short- and long-term memory. The short-term synaptic changes involve covalent modification of preexisting proteins, leading to modification of pre-existing synaptic connections, whereas the long-term synaptic changes involve activation of gene expression, new protein synthesis, and the formation of new connections. Whereas short-term memory storage uses different signaling kinases for implicit and explicit memory, long-term storage of both implicit and explicit memory uses a conserved signaling sequence consisting of PKA, MAPK, and CREB-1. We presume that this represents only the core signaling pathway, and that at least in the mouse, additional components are recruited.

Finally, in both implicit and explicit memory the switch from short-term to long-term memory is regulated by a number of inhibitory constraints. One such constraint, the balance between protein phosphorylation and dephosphorylation, determines the threshold for hippocampal synaptic plasticity and hippocampal-dependent explicit memory storage. Removing this constraint lowers the threshold of late LTP and enhances storage of explicit memory.

Second, in addition to providing new insights into the molecular mechanisms of learning and memory storage, the molecular-biological study of learning has revealed new features of synaptic transmission and new cell-biological functions of synaptic signaling. For example, we have learned that modulatory transmitters of the brain serve as reinforcing stimuli important for synaptic plasticity related to learning and memory storage. Different forms of learning recruit these modulatory transmitters which can then act in one of three ways: (1) they activate second-messenger kinases that are transported to the nucleus where they initiate processes required for neuronal growth and long-term memory; (2) they mark the specific synapses for capture of the long-term process and regulate local protein synthesis for stabilization; and (3) they appear to mediate, in ways we are just beginning to understand, attentional processes required for explicit memory formation and recall.

Most important, the study of long-term memory has made us aware of the extensive dialog between the synapse and the nucleus and the nucleus and the synapse (Figure 24). The long range interactions between the nucleus and synapse introduce a new level of non-Sherringtonian integration into neuronal functioning for long-term synaptic plasticity that is different from that used by the short-term process. In the long-term process the response of a synapse is not simply determined by its own history of activity. It is also significantly determined by the history of transcriptional activation in the nucleus.

I started this essay by pointing out that 40 years ago, at the beginning of my career, I thought that a reductionist approach based on the use of a simple experimental system such as *Aplysia* might allow us to address fundamental questions in the study of learning and memory. That was a leap of faith for which I have been rewarded beyond my fondest hopes. Still, the complexity of explicit memory is formidable, and we have only begun to explore it. We as yet know little about the molecular mechanisms that initiate or stabilize the synaptic growth associated with long-term memory. What are the signaling molecules that lead to the cytoskeletal rearrangements during synaptic remodeling? How do they relate to the molecules involved with synapse formation during development? In addition, we have here only considered the molecular mechanisms of *memory storage* – mechanisms that appeared to be shared, at least in part, by both explicit (declarative) and implicit (procedural) memory. But the storage mechanism is only one part of the memory problem. The more difficult part of memory – especially explicit memory – is the *systems problem* of memory. We still need to seek answers to a family of important questions. How do different regions of the hippocampus and the medial temporal lobe – the subiculum, the entorhinal, parahippocampal and perirhinal cortices – interact in the storage of explicit memory? How is information in any of these regions is transferred for ultimate consolidation in the neocortex? We do not, for example, understand why the initial storage of long-term memory requires the hippocampus, whereas the hippocampus is not required once a memory has been stored for weeks or months (reviewed in 95,96). What critical information does the hippocampus convey to the neocortex?

We also know very little about the nature of recall of explicit (declarative) memory, a recall that requires attention and in humans, conscious effort. These systems problems of the brain will require more than the bottoms-up approach of molecular biology. They will also require the top-down approaches of cognitive psychology, neurology, and psychiatry. Finally, we will require a set of syntheses that bridge the two approaches.

Despite these complexities, we have every reason to believe that these and other questions in the biology of learning will be vigorously addressed in the near future. For the biology of the mind has now captured the imagination of the scientific community of the 21st century, much as the biology of the gene fascinated the scientific community of the 20th century. As the biological study of the mind comes to assume the central position within biology and medicine that it deserves, we have every reason to expect that a succession of brain scientists will be called to Stockholm and honored for their own leaps of faith.

ACKNOWLEDGEMENTS

I have had the privilege to work with and to learn from many gifted students, fellows, and collaborators, and I have tried throughout this lecture to acknowledge their contributions. But beyond specific collaborators, my science has benefited enormously from the interactive environment created by the Center for Neurobiology and Behavior at the College of Physicians and Surgeons of Columbia University. It would be hard to find a more ideal environment in which to mature as a scientist. Specifically, I have benefited greatly from my long-standing friendship with Richard Axel, Craig Bailey, Jane Dodd, Robert Hawkins, John Koester, Thomas Jessell, James H. Schwartz, Steven Siegelbaum, and Gerald Fischbach, the current Dean of the College of Physicians and Surgeons. I am further grateful to John Koester for his excellent leadership of the Center for Neurobiology and Behavior, and to David Hirsh, Sam Silverstein, and John Oldham, chairs of the three departments to which I belong. Finally, I am indebted to Herbert Pardes who, until recently, served as Dean of the College of Physicians and Surgeons.

My research has been generously supported by the Howard Hughes Medical Institute, the NIH, the Mathers Foundation, FRAXA, and The Lieber Trust. I am particularly indebted to the Howard Hughes Medical Institute and its leadership, Donald Fredrickson, George Cahill, Purnell Chopin, Max Cowan, Donald Harter, and more recently Tom Cech and Gerry Rubin, whose farsighted vision has encouraged Hughes investigators to take a long-term perspective so as to be able to tackle challenging problems. Research on learning and memory certainly meets both of these criteria!

REFERENCES

1. Kandel, E.R. and Spencer, W.A. (1961) Electrophysiological properties of an archicortical neuron. In Problems in Electrobiology, Ann. N.Y. Acad. Sci. 94:570–603.
2. Thorpe, W.H. (1963) *Learning and Instinct in Animals.* Revised Edition. Cambridge, Mass: Harvard University Press.
3. Frazier, W.T., Kandel, E.R., Kupfermann, I., Waziri, R., and Coggeshall, R.E. (1967) Morphological and functional properties of identified neurons in the abdominal ganglion of *Aplysia californica.* J. Neurophysiol. 30:1288–1351.
4. Kupfermann, I. and Kandel, E.R. (1969) Neuronal controls of a behavioral response mediated by the abdominal ganglion of *Aplysia.* Science 164:847–850.
5. Pinsker, H., Kupfermann, I., Castellucci, V., and Kandel, E.R. (1970) Habituation and dishabituation of the gill-withdrawal reflex in *Aplysia.* Science 167:1740–1742.
6. Kupfermann, I., Castellucci, V., Pinsker, H., and Kandel, E.R. (1970) Neuronal correlates of habituation and dishabituation of the gill-withdrawal reflex in *Aplysia.* Science 167:1743–1745.
7. Castellucci, V., Pinsker, H., Kupfermann, I., and Kandel, E.R. (1970) Neuronal mechanisms of habituation and dishabituation of the gill-withdrawal reflex in *Aplysia.* Science 167:1745–1748.
8. Pinsker, H.M., Hening, W.A., Carew, T.J., and Kandel, E.R. (1973) Long-term sensitization of a defensive withdrawal reflex in *Aplysia.* Science 182:1039–1042.
9. Carew, T.J., Pinsker, H.M., and Kandel, E.R. (1972) Long-term habituation of a defensive withdrawal reflex in *Aplysia.* Science 175:451–454.
10. Frost, W.N., Castellucci, V.F., Hawkins, R.D., and Kandel, E.R. (1985) Monosynaptic connections made by the sensory neurons of the gill- and siphon-withdrawal reflex in *Aplysia* participate in the storage of long-term memory for sensitization. Proc. Natl. Acad. Sci. USA 82:8266–8269.
11. Byrne, J., Castellucci, V., and Kandel, E.R. (1974) Receptive fields and response properties of mechanoreceptor neurons innervating skin and mantle shelf of *Aplysia.* J. Neurophysiol. 37:1041–1064.
12. Castellucci, V.F., Blumenfeld, H., Goelet, P., and Kandel, E.R. (1989) Inhibitor of protein synthesis blocks long-term behavioral sensitization in the isolated gill-withdrawal reflex of *Aplysia.* J. Neurobiol. 20:1–9.

13. Kandel, E.R. (1976) Cellular Basis of Behavior: An Introduction to Behavioral Biology. San Francisco: Freeman.
14. Kandel, E.R. (1978) A Cell-Biological Approach to Learning. Grass Lecture Monograph I. Bethesda, Md.: Society of Neuroscience.
15. Kandel, E.R. (1979) The Behavioral Biology of *Aplysia*: A Contribution to the Comparative Study of Opisthobranch Molluscs. San Francisco: Freeman.
16. Cajal, S.R. (1894) La fine structure des centres nerveux, The Croonian Lecture. Proc. R. Soc. Lond. 55:444–468.
17. Kandel, E.R. and Spencer, W.A. (1968) Cellular neurophysiological approaches in the study of learning. Physiol. Rev. 48:65–134.
18. Hawkins, R.D., Castellucci, V.F., and Kandel, E.R. (1981) Interneurons involved in mediation and modulation of the gill-withdrawal reflex in *Aplysia*. I. Identification and characterization. J. Neurophysiol. 45:304–314.
19. Hawkins, R.D., Castellucci, V.F., and Kandel, E.R. (1981) Interneurons involved in mediation and modulation of the gill-withdrawal reflex in *Aplysia*. II Identified neurons produce hetero-synaptic facilitation contributing to behavioral sensitization. J. Neurophysiol. 45:315–326.
20. Carew, T.J., Hawkins, R.D., and Kandel, E.R. (1983) Differential classical conditioning of a defensive withdrawal reflex in *Aplysia californica*. Science 219:397–400.
21. Hawkins, R.D., Abrams, T.W., Carew, T.J., and Kandel, E.R. (1983) A cellular mechanism of classical conditioning in *Aplysia*: Activity-dependent amplification of presynaptic facilitation. Science 219:400–405.
22. Byrne, J.H.,Castellucci, V.F., and Kandel, E.R. (1978) Contribution of individual mechanoreceptor sensory to defensive gill-withdrawal reflex in *Aplysia*. J. Neurophysiol. 41:418–431.
23. Carew, T.J., Castellucci, V.F., and Kandel, E.R. (1979) Sensitization in *Aplysia*: Restoration of transmission in synapses inactivated by long-term habituation. Science 205:417–419.
24. Klein, M. and Kandel, E.R. (1980) Mechanism of calcium current modulation underlying presynaptic facilitation and behavioral sensitization in *Aplysia*. Proc. Natl. Acad. Sci. USA 77:6912–6916.
25. Klein, M., Hochner, B., and Kandel, E.R. (1986) Facilitatory transmitters and cAMP can modulate accommodation as well as transmitter release in *Aplysia* sensory neurons. Evidence for parallel processing in a single cell. Proc. Natl. Acad. Sci. USA 83:7994–7998.
26. Schwartz, J.H., Castellucci, V.F., and Kandel, E.R. (1971) Functioning of identified neurons and synapses in abdominal ganglion of *Aplysia* in absence of protein synthesis. J. Neurophysiol. 34:939–953.
27. Cedar, H., Kandel, E.R., and Schwartz, J.H. (1972) Cyclic adenosine monophosphate in the nervous system of *Aplysia californica*. I. Increased synthesis in response to synaptic stimulation. J. Gen. Physiol. 60:558–569.
28. Kriegstein, A.R., Castellucci, V., and Kandel, E.R. (1974) Metamorphosis of *Aplysia californica* in laboratory culture. Proc. Natl. Acad. Sci. USA 71:3654–3658.
29. Cedar, H. and Schwartz, J.H. (1972) Cyclic adenosine monophosphate in the nervous system of *Aplysia californica*. II. Effect of serotonin and dopamine. J. Gen. Physiol. 60:570–587.
30. Castellucci, V. and Kandel, E.R. (1974) A quantal analysis of the synaptic depression underlying habituation of the gill-withdrawal reflex in *Aplysia*. Proc. Natl. Acad. Sci. USA 71:5004–5008.
31. Castellucci, V. and Kandel, E.R. (1976) Presynaptic facilitation as a mechanism for behavioral sensitization in *Aplysia*. Science 194:1176–1178.
32. Brunelli, M., Castellucci, V., and Kandel, E.R. (1976) Synaptic facilitation and behavioral sensitization in *Aplysia*: Possible role of serotonin and cyclic AMP. Science 194:1178–1181.
33. Kandel, E.R., Brunelli, M., Byrne, J., and Castellucci, V. (1976) A common presynap-

tic locus for the synaptic changes underlying short-term habituation and sensitization of the gill-withdrawal reflex in *Aplysia*. Cold Spring Harbor Symp. Quant. Biol. 40:465–482.

34. Glanzman, D.L., Mackey, S.L., Hawkins, R.D., Dyke, A.M., Lloyd, P.E., and Kandel, E.R. (1989) Depletion of serotonin in the nervous system of *Aplysia* reduces the behavioral enhancement of gill withdrawal as well as the heterosynaptic facilitation produced by tail shock. J. Neurosci. 9:4200–4213.
35. Mackey, S.L., Kandel, E.R., and Hawkins, R.D. (1989) Identified serotonergic neurons LCB1 and RCB1 in the cerebral ganglia of *Aplysia* produce presynaptic facilitation of siphon sensory neurons. J. Neurosci. 9:4227–4235.
36. Byrne, J.H., Castellucci, V.F., Carew, T.J., and Kandel, E.R. (1978) Stimulus-response relations and stability of mechanoreceptor and motor neurons mediating defensive gill-withdrawal reflex in *Aplysia*. J. Neurophysiol. 41:402–417.
37. Castellucci, V.F., Kandel, E.R., Schwartz, J.H., Wilson, F.D., Nairn, A.C., and Greengard, P. (1980) Intracellular injection of the catalytic subunit of cyclic AMP-dependent protein kinase simulates facilitation of transmitter release underlying behavioral sensitization in *Aplysia*. Proc. Natl. Acad. Sci. USA 77:7492–7496.
38. Castellucci, V.F., Nairn, A., Greengard, P., Schwartz, J.H., and Kandel, E.R. (1982) Inhibitor of adenosine 3':5'-monophosphate-dependent protein kinase blocks presynaptic facilitation in *Aplysia*. J. Neurosci. 2:1673–1681.
39. Siegelbaum, S., Camardo, J.S., and Kandel, E.R. (1982) Serotonin and cAMP close single K^+ channels in *Aplysia* sensory neurons. Nature 299:413–417.
40. Shuster, M.J., Camardo, J.S., Siegelbaum, S.A., and Kandel, E.R. (1985) Cyclic AMP-dependent protein kinase closes the serotonin-sensitive K^+ channels of *Aplysia* sensory neurones in cell-free membrane patches. Nature 313:392–395.
41. Byrne, J H. and Kandel, E.R. (1995) Presynaptic facilitation revisited: State and time dependence. J. Neurosci. 16:425–435.
42. Rayport, S.G. and Schacher, S. (1986) Synaptic plasticity in vitro: Cell culture of identified *Aplysia* neurons mediating short-term habituation and sensitization. J. Neurosci. 6:759–763.
43. Montarolo, P.G., Goelet, P., Castellucci, V.F., Morgan, J., Kandel, E.R., and Schacher, S. (1986) A critical period for macromolecular synthesis in long-term heterosynaptic facilitation in *Aplysia*. Science 234:1249–1254.
44. Schacher, S., Castellucci, V.F., and Kandel, E.R. (1988) cAMP evokes long-term facilitation in *Aplysia* sensory neurons that requires new protein synthesis. Science 240:1667–1669.
45. Dash, P.K., Hochner, B., and Kandel, E.R. (1990) Injection of cAMP-responsive element into the nucleus of *Aplysia* sensory neurons blocks long-term facilitation. Nature 345:718–721.
46. Bacskai, B.J., Hochner, B., Mahaut-Smith, M., Adams, S.R., Kaang, B.-K., Kandel, E.R., and Tsien, R.Y. (1993) Spatially resolved dynamics of cAMP and protein kinase A subunits in *Aplysia* sensory neurons. Science 260:222–226.
47. Martin, K.C., Michael, D., Rose, J.C., Barad, M., Casadio, A., Zhu, H., and Kandel, E.R. (1997) MAP kinase translocates into the nucleus of the presynaptic cell and is required for long-term facilitation in *Aplysia*. Neuron 18:899–912.
48. Bailey, C.H., Kaang, B.-K., Chen, M., Martin, K.C., Lim, C.-S., Casadio, A., and Kandel, E.R. (1997) Mutation in the phosphorylation sites of MAP kinase blocks learning-related internalization of apCAM in *Aplysia* sensory neurons. Neuron 18:913–924.
49. Kaang, B.-K., Kandel, E.R., and Grant, S.G.N. (1993) Activation of cAMP-responsive genes by stimuli that produce long-term facilitation in *Aplysia* sensory neurons. Neuron 10:427–435.
50. Glanzman, D.L., Kandel, E.R., and Schacher, S. (1990) Target-dependent structural changes accompanying long-term synaptic facilitation in *Aplysia* neurons. Science 249:799–802.

51. Dale, N. and Kandel, E.R. (1993) L-glutamate may be the fast excitatory transmitter of *Aplysia* sensory neurons. Proc. Natl. Acad. Sci. USA 90:7163–7167.
52. Alberini, C., Ghirardi, M., Metz, R., and Kandel, E.R. (1994) C/EBP is an immediate-early gene required for the consolidation of long-term facilitation in *Aplysia*. Cell 76:1099–1114.
53. Bartsch, D., Ghirardi, M., Skehel, P.A., Karl, K.A., Herder, S.P., Chen, M., Bailey, C.H., and Kandel, E.R. (1995) *Aplysia* CREB2 represses long-term facilitation: Relief of repression converts transient facilitation into long-term functional and structural change. Cell 83:979–992.
54. Bartsch, D., Casadio, A., Karl, K.A., Serodio, P., and Kandel, E.R. (1998) CREB1 encodes a nuclear activator, a repressor, and a cytoplasmic modulator that form a regulatory unit critical for long-term facilitation. Cell 95:211–223.
55. Bartsch, D., Ghirardi, M., Casadio, A., Giustetto, M., Karl, K.A., Zhu, H., and Kandel, E.R. (2000) Enhancement of memory-related long-term facilitation by ApAF, a novel transcription factor that acts downstream from both CREB1 and CREB2. Cell 103:595–608.
56. Bailey, C.H., Montarolo, P., Chen, M., Kandel, E.R., and Schacher, S. (1992) Inhibitors of protein and RNA synthesis block structural changes that accompany long-term heterosynaptic plasticity in *Aplysia*. Neuron 9:749–758.
57. Bailey, C.H. and Chen, M. (1988) Long-term memory in *Aplysia* modulates the total number of varicosities of single identified sensory neurons. Proc. Natl. Acad. Sci. USA 85:2373–2377.
58. Bailey, C.H. and Chen, M. (1989) Time course of structural changes at identified sensory neuron synapses during long-term sensitization in *Aplysia*. J. Neurosci. 9:1774–1780.
59. Bailey, C.H. and Kandel, E.R. (1993) Structural changes accompanying memory storage. Annu. Rev. Physiol. 55:397–426.
60. Hegde, A.N., Inokuchi, K., Pei, W., Casadio, A., Ghirardi, M., Chain, D.G., Martin, K.C., Kandel, E.R., and Schwartz, J.H. (1997) Ubiquitin C-terminal hydrolase is an immediate-early gene essential for long-term facilitation in *Aplysia*. Cell 89:115–126.
61. Chain, D.G., Casadio, A., Schacher, S., Hegde, A.N., Valbrun, M., Yamamoto, N., Goldberg, A.L., Bartsch, D., Kandel, E.R., and Schwartz, J.H. (1999) Mechanisms for generating the autonomous cAMP-dependent protein kinase required for long-term facilitation in *Aplysia*. Neuron 22:147–156.
62. Mayford, M., Barzilai, A., Keller, F., Schacher, S., and Kandel, E.R. (1992) Modulation of an NCAM-related adhesion molecule with long-term synaptic plasticity in *Aplysia*. Science 256:638–644.
63. Bailey, C.H., Chen, M., Keller, F., and Kandel, E.R. (1992) Serotonin-mediated endocytosis of apCAM: An early step of learning-related synaptic growth in *Aplysia*. Science 256:645–649.
64. Martin, K.C., Casadio, A., Zhu, H., Yaping, E., Rose, J., Bailey, C.H., Chen, M., and Kandel, E.R. (1998) Synapse-specific transcription-dependent long-term facilitation of the sensory to motor neuron connection in *Aplysia*: A function for local protein synthesis in memory storage. Cell 91:927–938.
65. Casadio, A., Martin, K.C., Giustetto, M., Zhu, H., Chen, M., Bartsch, D., Bailey, C.H., and Kandel, E.R. (1999) A transient neuron-wide form of CREB-mediated long-term facilitation can be stabilized at specific synapses by local protein synthesis. Cell 99:221–237.
66. Bacskai, B.J., Hochner, B., Mahaut-Smith, M., Adams, S.R., Kaang, B.-K., Kandel, E.R., and Tsien, R.Y. (1993) Spatially resolved dynamics of cAMP and protein kinase A subunits in *Aplysia* sensory neurons. Science 260:222–226.
67. Grant, S.G.N., O'Dell, T.J., Karl, K.A., Stein, P.L., Soriano, P., and Kandel, E.R. (1992) Impaired long-term potentiation, spatial learning, and hippocampal development in fyn mutant mice. Science 258:1903–1910.

68. Squire, L.R. and Kandel, E.R. (1999) *Memory: Mind to Molecules.* New York: Scientific American Books.
69. Bliss, T.V. and Lømo, T. (1973) Long-lasting potentiation of synaptic transmission in the dentate area of the anaesthetized rabbit following stimulation of the perforant path. J. Physiol. 232:331–356.
70. Morris, R.G.M., Anderson, E., Lynch, G.S., and Baudry, M. (1986) Selective impairment of learning and blockade of long-term potentiation by an N-methyl-D-aspartate receptor antagonist, AP5. Nature 319:774–776.
71. Frey, U., Huang, Y.-Y., and Kandel, E.R. (1993) Effects of cAMP simulate a late stage of LTP in hippocampal CA1 neurons. Science 260:1661–1664.
72. Huang, Y.-Y. and Kandel, E.R. (1994) Recruitment of long-lasting and protein kinase A-dependent long-term potentiation in the CA1 region of hippocampus requires repeated tetanization. Learning & Memory 1:74–82.
73. Nguyen, P.V., Abel, T., and Kandel, E.R. (1994) Requirement of a critical period of transcription for induction of a late phase of LTP. Science 265:1104–1107.
74. Bolshakov, V.Y., Golan, H., Kandel, E.R., and Siegelbaum, S.A. (1997) Recruitment of new sites of synaptic transmission during the cAMP-dependent late phase of LTP at CA3-CA1 synapses in the hippocampus. Neuron 19:635–651.
75. Malenka, R.C. and Nicoll, R.A. (1997) Silent synapses speak up. Neuron 19:473–476.
76. Ma, L., Zablow, L., Kandel, E.R., and Siegelbaum, S.A. (1999) Cyclic AMP induces functional presynaptic boutons in hippocampal CA3-CA1 neuronal cultures. Nature Neurosci. 2:24–3.
77. Bourtchouladze, R., Franguelli, B., Blendy, J., Cioffi, D., Schutz, G., and Silva, A.J. (1994) Deficient long-term memory in mice with a targeted mutation of the cAMP-responsive element-binding protein. Cell 79:59–68.
78. Impey, S., Obrietan, K., Wong, S.T., Poser, S., Yano, S., Wayman, G., Deloulme, J.C., Chan, G., and Storm, D.R. (1998) Cross talk between ERK and PKA is required for Ca^{2+} stimulation of CEB-dependent transcription and ERK nuclear translocation. Neuron 21:869–883.
79. Huang, Y.-Y. and Kandel, E.R. (1995) D1/D5 receptor agonists induce a protein synthesis-dependent late potentiation in the CA1 region of the hippocampus. Proc. Natl. Acad. Sci. USA 92:2446–2450.
80. Muller, D. (1997) Ultrastructural plasticity of excitatory synapses. Rev. Neurosci. 8:77–93.
81. English, J.D. and Sweat, J.D. (1996) Activation of p42 mitogen-activated protein kinase in hippocampal long term potentiation. J. Biol. Chem. 271:24329–24332.
82. Engert, F. and Bonhoeffer, T. (1999) Dendritic spine changes associated with hippocampal long-term synaptic plasticity. Nature 399:66–70.
83. Winder, D.G., Mansuy, I.M., Osman, M., Moallem, T.M., and Kandel, E.R. (1998) Genetic and pharmacological evidence for a novel, intermediate phase of long-term potentiation (I-LTP) suppressed by calcineurin. Cell 92:25–37.
84. Abel, T., Nguyen, P.V., Barad, M., Deuel, T.A.S., Kandel, E.R., and Bourtchouladze, R. (1997) Genetic demonstration of a role for PKA in the late phase of LTP and in hippocampus-based long-term memory. Cell 88:615–626.
85. O'Keefe, J. and Nadel, L. (1978) *The Hippocampus as a Cognitive Map.* Oxford: Clarendon Press.
86. Rotenberg, A., Mayford, M., Hawkins, R.D., Kandel, E.R., and Muller, R.U. (1996) Mice expressing activated CaMKII lack low frequency LTP and do not form stable place cells in the CA1 region of the hippocampus. Cell 87:1351–1361.
87. Kentros, C., Hargreaves, E., Hawkins, R.D., Kandel, E.R., Shapiro, M., and Muller, R.U. (1998) Abolition of long-term stability of new hippocampal place cell maps by NMDA receptor blockade. Science 280:2121–2126.
88. Rotenberg, A, Abel, T., Hawkins, R.D., Kandel, E.R., and Muller, R.U. (2000) Parallel instabilities of long-term potentiation, place cells, and learning causes by decreased protein kinase A activity. J. Neurosci. 20:8096–8102.

89. Agnihotri, N., Hawkins, R.D., Kandel, E.R., and Kentros, C. (2001) Protein synthesis inhibition selectively abolishes long-term stability of hippocampal place cell maps. Soc. Neurosci. Abstr., in press.
90. Kentros, C., Agnihotri, N., Hawkins, R., Muller, R., and Kandel, E. (2001) Stabilization of the hippocampal representation of space in mice requires attention. Soc. Neurosci. Abstr., in press.
91. Mansuy, I.M., Mayford, M., Jacob, B., Kandel, E.R., and Bach, M.E. (1998) Restricted and regulated overexpression reveals calcineurin as a key component in the transition from short-term to long-term memory. Cell 92:39–49.
92. Mallaret, G., Haditsch, U., Genoux, D., Jones, M.W., Bliss, T.V.P., Vanhoose, A.M.,Weitlauf, C., Kandel, E.R., Winder, D.G., and Mansuy, I.M. (2001) Inducible and reversible enhancement of learning, memory, and long-term potentiation by genetic inhibition of calcineurin. Cell 104:675–686.
93. Frey, U. and Morris, R.G. (1997) Synaptic tagging and long-term potentiation. Nature 385:533–536.
94. Castellucci, V.F., Carew, T.J., and Kandel, E.R. (1978) Cellular analysis of long-term habituation of the gill-withdrawal reflex in *Aplysia*. Science 202:1306–1308.
95. Milner, B., Squire, L.R. and Kandel, E.R. (1998) Cognitive neuroscience and the study of memory. Neuron 20:445–468.
96. Squire, L.R. and Zola-Morgan, S. (1991) The medial temporal lobe memory system. Science 253:1380–1386.
97. Dubnau, J. and Tully, T. (1998) Gene discovery in *Drosophila*: New insights for learning and memory. Annu. Rev. Neurosci. 21:407–444.
98. Frost, W.N., Castellucci, V.F., Hawkins, R.D., and Kandel, E.R. (1985) Monosynaptic connections made by the sensory neurons of the gill- and siphon-withdrawal reflex in *Aplysia* participate in the storage of long-term memory for sensitization. Proc. Natl. Acad. Sci. USA 82:8266–8269.
99. Martin, K.C., Barad, M., and Kandel, E.R. (2000) Local protein synthesis and its role in synapse-specific plasticity. Curr. Opin. Neurobiol. 10:587–592.
100. Schuman, E.M. (1997) Synapse specificity and long-term information storage. Neuron 18:339–342.
101. Steward, O. (1997) mRNA localization in neurons: A multipurpose mechanism? Neuron 18:9–12.
102. Weiner, J. (1999) *Time, Love, Memory: A Great Biologist and His Quest for the Origins of Behavior.* New York: Knopf.
103. Muller, R. (1996) A quarter of a century of place cells. Neuron 17:813–822.
104. McHugh, T.J., Blum, K.I., Tsien, J.Z., Tonegawa, S., and Wilson, M.A. (1996) Impaired hippocampal representation of space in CA1 specific NMDAR1 knockout mice. Cell 87:1339–1349.
105. Bolshakov, V.Y. and Siegelbaum, S.A. (1995) Regulation of hippocampal transmitter release during development and long-term potentiation. Science 269:1730–1734.
106. Stevens, C.F. and Wang, Y. (1994) Changes in reliability of synaptic function as a mechanism for plasticity. Nature 371:704–707.
107. Ebbinghaus, H. (1885) Memory: A Contribution to Experimental Psychology (1885). Reprinted, New York, Dover, 1963.
108. Flexner, L.B., and Flexner, J.B. (1966) Effect of acetoxycycloheximide and of an acetoxycycloheximide-puromycin mixture on cerebral protein synthesis and memory in mice. Proc. Natl. Acad. Sci. USA 55:369–374.
109. Brink, J.J., Davis, R.E., and Agranoff, B.W. (1966) Effects of puromycin, acetoxycycloheximide and actinomycin D on protein synthesis in goldfish brain. J. Neurochem. 13:889–896.
110. Fatt, P. and Katz, B. (1951) An analysis of the end-plate potential recorded with an intracellular electrode. J. Physiol. (Lond.) 115:320–370.
111. Greengard, P. (1976) Possible role for cyclic nucleotides and phosphorylated membrane proteins in postsynaptic actions of neurotransmitters. Nature 260:101–108.

112. Monsour, S.L., Thomas, K.R., and Capecchi, M.R. (1988) Disruption of the proto-oncogene int-2 in mouse embryonic-derived stem cells: A general strategy for targeting mutations to non-selectible genes. Nature 336:348–352.
113. Reid, L.H., Shesely, E.G., Kim, H.S., and Smithies, O. (1991) Cotransformation and gene targeting in mouse embryonic stem cells. Mol. Cell Biol. 11:2769–2777.
114. Koller, B.H. and Smithies, O. (1989) Inactivating the beta 2-microglobulin locus in mouse embryonic stem cells by homologous recombination. Proc. Natl. Acad. Sci. USA 86:8932–8935.
115. Kandel, E.R. and Schwartz, J.H. (1982) Molecular biology of learning: Modulation of transmitter release. Science 218:433–443.
116. Blitzer, R.D., Conner, J.H., Brown, G.P., Wong, T., Shenolikar, S., Iyengar, R., and Landau, E.M. (1998) Gating of CaMKII by cAMP-regulated protein phosphatase activity during LTP. Science 280:1940–1942.